# RUMINANT PHYSIOLOGY
# Digestion, Metabolism, Growth and Reproduction

## Dedication

This volume is dedicated to the memory of the late Dr F.M.C. Gilchrist.

# Ruminant Physiology
## Digestion, Metabolism, Growth and Reproduction

Edited by

**P.B. Cronjé**

*Department of Animal and Wildlife Sciences*
*University of Pretoria*
*Pretoria*
*South Africa*

Associate Editors

**E.A. Boomker**
**P.H. Henning**
**W. Schultheiss**
**J.G. van der Walt**

CABI *Publishing*

**CABI *Publishing* is a division of CAB *International***

CABI Publishing
CAB International
Wallingford
Oxon OX10 8DE
UK

Tel: +44 (0)1491 832111
Fax: +44 (0)1491 833508
Email: cabi@cabi.org
Web site: http://www.cabi.org

CABI Publishing
10 E 40th Street
Suite 3203
New York, NY 10016
USA

Tel: +1 (212) 481 7018
Fax: +1 (212) 686 7993
Email: cabi-nao@cabi.org



A catalogue record for this book is available from the British Library, London, UK

**Library of Congress Cataloging-in-Publication Data**
Ruminant physiology : digestion, metabolism, growth, and reproduction / edited by P. Cronje ; assoc. editors, E.A. Boomker ... [et al.].
p. cm.
Includes bibliographical references and index.
ISBN 0-85199-463-6 (alk. paper)
1. Ruminants--Physiology--Congresses. I. Cronjé, P. (Pierre) II. Boomker, E. A.

QL737.U5 R868 2000
571.1′963--dc21 00–023661

ISBN 0 85199 463 6

Typeset in 10/12pt Garamond by Columns Design Ltd, Reading
Printed and bound in the UK by Biddles Ltd, Guildford and King's Lynn

# Contents

# Contributors

**N.R. Adams**, *CSIRO Division of Animal Production and CRC for Premium Quality Wool, Wembley, 6014 Western Australia*

**R.I. Aminov**, *Department of Animal Sciences and Division of Nutritional Sciences, University of Illinois at Urbana-Champaign, Urbana, IL 61801, USA*

**J.J. Bass**, *AgResearch, Ruakura Agricultural Research Centre, Private Bag 3123, Hamilton, New Zealand*

**D.E. Bauman**, *Department of Animal Science, Cornell University, Ithaca, NY 14853, USA*

**C.R. Baumrucker**, *Department of Dairy and Animal Science, Penn State University, 302 Henning Building, University Park, PA 16802, USA*

**A.W. Bell**, *Department of Animal Science, Cornell University, Ithaca, NY 14853, USA*

**F. Bocquier**, *Adipose Tissue and Milk Lipids Team, Herbivore Research Unit, INRA-Theix, 63122 St Genès Champanelle, France*

**J.M. Brameld**, *Division of Nutritional Biochemistry, School of Biological Sciences, University of Nottingham, Sutton Bonington Campus, Loughborough, Leicestershire LEI2 5RD, UK*

**B.H. Breier**, *Research Centre for Developmental Medicine and Biology, Faculty of Medicine and Health Science, University of Auckland, Private Bag 92019, Auckland, New Zealand*

**P.J. Buttery**, *Division of Nutritional Biochemistry, School of Biological Sciences, University of Nottingham, Sutton Bonington Campus, Loughborough, Leicestershire LE12 5RD, UK*

**A. Chesson**, *Rowett Research Institute, Greenburn Road, Bucksburn, Aberdeen AB21 9SB, UK*

**Y. Chilliard**, *Adipose Tissue and Milk Lipids Team, Herbivore Research Unit, INRA-Theix, 63122 St Genès Champanelle, France*

**I.G. Colditz**, *CSIRO Animal Production, Pastoral Research Laboratory, Armidale, NSW 2350, Australia*

**P.B. Cronjé**, *Department of Animal and Wildlife Sciences, University of Pretoria, Pretoria 0002, South Africa*

**J.M. Dawson**, *Division of Nutritional Biochemistry, School of Biological Sciences, University of Nottingham, Sutton Bonington Campus, Loughborough, Leicestershire LE12 5RD, UK*

**R.A. Ehrhardt**, *Department of Animal Science, Cornell University, Ithaca, NY 14853, USA*

**D.L. Emery**, *CSIRO Animal Production, McMaster Laboratory, Prospect, NSW 2148, Australia*

**A. Faurie**, *Department of Physiology and Brain Function Research Unit, University of the Witwatersrand, 7 York Road, Parktown, Johannesburg 2193, South Africa*

**H.J. Flint**, *Rowett Research Institute, Greenburn Road, Bucksburn, Aberdeen AB21 9SB, UK*

**E. Forano**, *Laboratoire de Microbiologie, INRA CR de Clermont-Ferrand – Theix, 63122 St Genès Champanelle, France*

**J.M. Forbes**, *Centre for Animal Sciences, Leeds Institute of Biotechnology and Agriculture, University of Leeds, Leeds LS2 9JT, UK*

**C.W. Forsberg**, *Department of Microbiology, University of Guelph, Guelph, Ontario N1G 2W1, Canada*

**B.W. Gallaher**, *Research Centre for Developmental Medicine and Biology, Faculty of Medicine and Health Science, University of Auckland, Private Bag 92019, Auckland, New Zealand*

**M. Gill**, *NR International, Central Avenue, Chatham Maritime, Kent ME4 4TB, UK*

**J.S. Gonzalez**, *Department of Animal Production, University of Leon, 24071 Leon, Spain*

**W.L. Grovum**, *Department of Biomedical Sciences, Ontario Veterinary College, University of Guelph, Guelph, Ontario, N1G 2W1, Canada*

**K.L. Houseknecht**, *Animal Health Drug Discovery, Pfizer Inc., Groton, CT 06340-8002, USA*

**A.W. Illius**, *Division of Biological Sciences, University of Edinburgh, West Mains Road, Edinburgh EH9 3JT, UK*

**N.S. Jessop**, *Division of Biological Sciences, University of Edinburgh, West Mains Road, Edinburgh EH9 3JT, UK*

**R. Kambadur**, *AgResearch, Ruakura Agricultural Research Centre, Private Bag 3123, Hamilton, New Zealand*

**H. Laburn**, *Department of Physiology and Brain Function Research Unit, University of the Witwatersrand, 7 York Road, Parktown, Johannesburg 2193, South Africa*

**S. Liu**, *CSIRO Division of Animal Production and CRC for Premium Quality Wool, Wembley, 6014 Western Australia*

**G.E. Lobley**, *Rowett Research Institute, Greenburn Road, Bucksburn, Aberdeen AB21 9SB, UK*

**R.I. Mackie**, *Department of Animal Sciences and Division of Nutritional Sciences, University of Illinois at Urbana-Champaign, Urbana, IL 61801, USA*

**D.G. Masters**, *CSIRO Division of Animal Production and CRC for Premium Quality Wool, Wembley, 6014 Western Australia*

**S.J. McClure**, *CSIRO Animal Production, McMaster Laboratory, Prospect, NSW 2148, Australia*

**J.P. McNamara**, *Department of Animal Sciences, 233 Clark Hall, Washington State University, PO Box 646351, Pullman, WA 99164-6351, USA*

**C.S. McSweeney**, *CSIRO Tropical Agriculture, Long Pocket Laboratory, Indooroopilly, Queensland 4068, Australia*

**G.D. Milano**, *Departamento de Fisiopatologia, Facultad de Ciencias Veterinarias UNCP-BA, Campus Iniversitario, Paraje Arroyo Seco (7000) Tandil, Argentina*

**D. Mitchell**, *Department of Physiology and Brain Function Research Unit, University of the Witwatersrand, 7 York Road, Parktown, Johannesburg 2193, South Africa*

**M. Morrison**, *Department of Animal Science, School of Biological Sciences and Center for Biotechnology, University of Nebraska, Lincoln, NE 68583-0908, USA*

**J. Oldham**, *AgResearch, Ruakura Agricultural Research Centre, Private Bag 3123, Hamilton, New Zealand*

**M.H. Oliver**, *Research Centre for Developmental Medicine and Biology, Faculty of Medicine and Health Science, University of Auckland, Private Bag 92019, Auckland, New Zealand*

**D.S. Parker**, *Department of Biological and Nutritional Sciences, Faculty of Agriculture and Biological Science, University of Newcastle upon Tyne, Newcastle upon Tyne NE1 7RU, UK: present address; Novius Europe s.a./n.v., Rue Gulledellestraat 94, B-1200 Brussels, Belgium*

**K. Persson Waller,** *Swedish University of Agricultural Sciences, Faculty of Veterinary Medicine, Department of Obstetrics and Gynaecology, P0 Box 7039 Uppsala, S-750 07 Sweden*

**F.D. Provenza**, *Department of Rangeland Resources, Utah State University, Logan, Utah 84322–5230, USA*

**K.P. Scott**, *Rowett Research Institute, Greenburn Road, Bucksburn, Aberdeen AB21 9SB, UK*

**C.J. Seal**, *Department of Biological and Nutritional Sciences, Faculty of Agriculture and Biological Science, University of Newcastle upon Tyne, Newcastle upon Tyne NE1 7RU, UK*

**M. Sharma**, *AgResearch, Ruakura Agricultural Research Centre, Private Bag 3123, Hamilton, New Zealand*

**J.W. Steel**, *CSIRO Animal Production, McMaster Laboratory, Prospect, NSW 2148, Australia*

**R.L. Tellam**, *CSIRO Tropical Agriculture, Longpocket Laboratory, Indooroopilly, Queensland 4068, Australia*

**R.G. Vernon**, *Hannah Research Institute, Ayr KA6 5HL, UK*

**J.G. van der Walt**, *Department of Veterinary Physiology, Faculty of Veterinary Science, Private Bag X04, Onderstepoort 0110, South Africa*

**K.A. Ward**, *CSIRO Animal Production, LB1, Delivery Centre, Blacktown, NSW 2148 Australia*

**B.A. White**, *Department of Animal Sciences and Division of Nutritional Sciences, University of Illinois at Urbana-Champaign, 1207 W. Gregory Drive, Urbana, IL 61801, USA*

**K.E. Webb, Jr**, *Department of Animal and Poultry Sciences, Virginia Polytechnic Institute and State University, Blacksburg, VA 24061-0306, USA*

# Foreword

The IX International Symposium on Ruminant Physiology was held in Pretoria, South Africa, during October 1999, and followed the traditions set by the previous meetings held in Nottingham (1960), Ames (1965), Cambridge (1969), Sydney (1974), Clermont-Ferrand (1979), Banff (1984), Sendai (1989) and Willingen (1994). The event was attended by 250 delegates from 28 countries. The plenary papers are published in this volume, and the 242 poster communications were published in the *South African Journal of Animal Science.*

The central issue that emerged from this symposium was that new technologies, notably molecular biology and modelling, have become important research tools for the physiologist. It was, however, apparent that more research is needed to relate advances in these technologies and in our understanding of fundamental physiological mechanisms to the solution of practical problems. The papers reiterated that animals, as free-living organisms, have an inherent ability to select for different nutrients and to control nutrient partitioning between different tissues, but that this can be modified and manipulated by human intervention. The important question of whether tissues are in competition for nutrients or whether the partitioning of nutrients is an integral part of coordinating the optimal use of nutrients will, no doubt, elicit much new research. There has been a clear shift from the traditional nutritional input–fermentation approach to rumen microbiology towards a molecular ecology approach, and a new horizon has appeared with regard to our quest to understand host–parasite relationships.

Thanks are expressed to the members of the organizing committee, sub-committees and the many willing helpers. The sponsors deserve a special mention: their contributions made it possible to give financial support to deserving delegates, to publish the poster abstracts and to produce this volume.

An international guiding committee was constituted to consider the future of the ISRP and the offers to host the X ISRP. The members of the committee are: Dr John Bass, Prof. Alan Bell, Prof. Giuseppi Bertoni, Prof. Peter Buttery, Prof. Norman Casey (convenor), Dr Yves Chilliard, Prof. Pierre Cronjé, Prof. Jong Ha, Dr Jan Hofmeyr

(WAAP Vice-president), Dr Heinz Meissner, Prof. Y Obara and Prof. Wolfgang von Engelhardt. The committee formulated a number of guiding principles for future ISRP meetings. These are: that the meeting should retain the character of previous symposia; that the focus should remain on the physiology of livestock, reviewing advances over the previous 5 years and setting directions for the next period; that comparative physiology and the impact of advances in physiology on products and sensitive consumer issues are important; that the venue for meetings should be situated where there is a core of established ruminant physiologists who could organize the symposium and, in particular, attend to the scientific programme and publish the proceedings; that the symposium should be easily accessible to young scientists and scientifically developing communities. After considering several invitations and taking the principles agreed upon into consideration, the hosting of the X ISRP was awarded to Denmark.

Norman H. Casey
(Chairman: Organizing Committee of the IX ISRP)

# I Regulation of Feed Intake

# 1 Integration of Learning and Metabolic Signals into a Theory of Dietary Choice and Food Intake

J.M. FORBES[1] AND F.D. PROVENZA[2]

[1]Centre for Animal Sciences, Leeds Institute of Biotechnology and Agriculture, University of Leeds, Leeds, UK; [2]Department of Rangeland Resources, Utah State University, Logan, Utah, USA

## Introduction

The challenge of understanding how diet selection and food intake are controlled is one that occupies an important place in the fields of nutrition, physiology and psychology. In the case of ruminant animals there are two special reasons for our interest in the subject: the complexities of the digestive system and consequent metabolic peculiarities; and the agricultural and ecological importance of the sub-order. Despite several decades of intensive study there is still no consensus on how intake is controlled (Fisher, 1996), nor is there agreement about the way in which animals determine which food(s) to eat when a choice is available. The past few years have seen the publication of sufficient new evidence to allow us to advance our hypotheses about the control of food intake and diet selection.

Firstly we review advances in our understanding of the role of learning in determining preferences and aversions for foods by ruminants; we then summarize the ways in which the central nervous system (CNS) is informed about digestive and metabolic processes; discuss the day-to-day variation in intake as an enabling factor in the linking of learning with the physiological consequences of eating; and finally propose how learning and metabolic information are brought together to provide testable hypotheses of the control of diet selection and voluntary food intake. We take it as axiomatic that long- and short-term regulation of intake are interwoven and do not attempt to differentiate between the two.

## Learned associations between the sensory properties of a food and the metabolic consequences of eating that food

This section presents recent evidence to reinforce the concept that ruminant animals learn to associate the post-ingestive consequences of eating a food with the sensory properties of that food and that they use such conditioned preferences and aversions to direct their selection between foods.

 

### Adaptation of choices of foods in order to avoid excessive intakes of toxins and to ensure adequate intakes of essential nutrients

In establishing that ruminants can learn to choose between foods to avoid toxicity it is logical to start with an overtly toxic substance, i.e. LiCl, which has been widely used in conditioned aversion studies. Sheep find LiCl, injected or in the food, to be unpleasant (Dutoit *et al.*, 1991) as it induces a conditioned taste aversion, the strength of which is proportional to the dose administered. Feeding neophobia also increases as a function of the LiCl dose associated with the last novel food encountered. When sheep and goats were offered food containing 2% LiCl, their daily intake after the third day fluctuated about a level that resulted in a LiCl dose of 39 mg $kg^{-1}$ for sheep and 27 mg $kg^{-1}$ for goats, i.e. similar to the doses causing mild aversion in rats and human beings.

The word 'toxin' is usually reserved for a substance that causes obvious signs of discomfort or distress. However, all dietary components are capable of acting as toxins, if present in great excess over requirements. Even a mild excess can generate aversion as toxins do not have to be consciously sensed in order for their effects to be relayed to the CNS and to have the potential to influence learned aversion. Equally, a deficiency of an essential nutrient can form the unconditioned stimulus for the development of food aversions. An example is provided by Hills *et al.* (1998) in which sheep either replete or depleted in sulphur were offered foods with different contents of sulphur. Replete sheep given high- and low-sulphur foods initially ate at random but within 2 days reduced the proportion of the high-sulphur food to achieve a sulphur concentration in the total diet very close to that thought to be optimal. Conversely, depleted sheep initially ate a high proportion of the high-sulphur food but later reduced the sulphur content chosen until it stabilized at the optimum level.

Other examples of non-random diet selection in order to control the intake of a 'nutrient' are: protein (Kyriazakis and Oldham, 1993), sodium (Denton, 1982), energy (Burritt and Provenza, 1992) and oxalic acid (Kyriazakis *et al.*, 1998).

### Ruminants learn preferences for a food flavour associated with infusions that correct deficiency; the same nutrient given to excess leads to avoidance of the associated flavour

In order to demonstrate unequivocally that such appetites are dependent on learned associations between the sensory properties of the foods and their nutritive value it is necessary to divorce the flavour of the food from its yield of nutrients. This can be done by offering animals a distinctive food and at the same time giving a nutrient by a route that bypasses the mouth, usually intraruminal infusion. In one such experiment with lambs (Villalba and Provenza, 1997a) one flavour was paired with rumen infusion of starch (2.5–9.4% of daily digestible energy (DE) intake) and another flavour with control. Subsequent preference was strongly for the starch-paired flavour, even 8 weeks after infusions had stopped. Starch is rapidly fermented to volatile fatty acids, predominantly propionic, in the rumen. Propionate absorption is likely to be insufficient for glucose synthesis in straw-fed animals so the hypothesis was tested that the supply of this limiting nutrient would induce a preference for the flavour of food eaten during supplementation (Villalba and Provenza, 1996). Even though the propionate supplied was equivalent to no more than 1.4% of the daily metabolizable energy (ME) intake, after 8 days of conditioning the sheep had developed a strong preference for food flavoured with that flavour given during supplementation. It was shown that the preference was induced by the propionate rather than the sodium or osmolality of the infusions (Villalba and Provenza, 1996; Villalba and Provenza, 1997b).

We can conclude that a single nutrient can induce a preference or an aversion to the flavour it was paired with during training, depending on the rate of administration in relation to the animal's requirements.

## Continuum from deficiency, through sufficiency, to excess for each nutrient

Several recent experiments have addressed the question: Do ruminants prefer a flavour associated with an intermediate, optimum content of a nutrient over flavours associated with the same nutrient present in excess or deficiency?

Arsenos and Kyriazakis (1999) have demonstrated a continuum between conditioned preferences and aversions in sheep to flavoured foods associated with doses of casein from 9 to 53 g given by gavage. The lower two doses led to conditioned flavour preferences, presumably because they alleviated a N deficiency, while the higher two doses led to conditioned flavour aversions, presumably being sensed as toxic overdoses. The authors observe that the existence of a continuum of flavour preferences and aversions created by different amounts of the same nutrient source could be the basis of how ruminants select a diet which meets their nutrient requirement at a particular point in time.

Sheep preferred a flavour paired with intraruminal administration of acetate at several doses (4, 8 or 12% of daily DE intake) or propionate (4% of daily DE intake), but became averse to a flavour paired with higher doses of propionate (12% of daily DE intake). This again suggests a role for learning about different concentrations of metabolites in the control of diet selection (Villalba and Provenza, 1997b).

There is thus evidence that a food that the animal believes alleviates a deficiency becomes preferred over other foods, while one thought to be excessive in the same nutrient becomes aversive. Under natural conditions such responses would lead to 'nutritional wisdom', i.e. eating a mixture of foods which most closely meets the animal's nutrient requirements.

## Ratio in which nutrients are supplied by different foods affects dietary choice

There are some situations in which the ratio of nutrients being absorbed from the digestive tract is such as to induce metabolic imbalance. When acetate and propionate were infused together into the rumen of sheep, conditioned preferences were demonstrated for the associated flavoured wheat straw but the preference was greater when the ratio of acetate:propionate in the infusate was 55:45 than when it was 75:25 (Villalba and Provenza, 1997b). It is likely that straw-fed lambs, with a high ratio of acetate:propionate produced by normal ruminal fermentation, would be deficient in glucose and this would be better alleviated by the mixture with the higher proportion of propionate.

It has been proposed that certain types of diet provide imbalances between energy and protein supply at different times of day, even though they may be balanced overall. Kyriazakis and Oldham (1997) set out to test whether such asynchrony of nutrient supply would influence diet selection, relative to a food designed to provide the same nutrient supply in a synchronous manner. Foods were formulated to provide rapidly- or slowly-fermentable energy with high or low rumen degradable protein (RDP) – all foods had the same calculated contents of ME and metabolizable protein (MP). When choices were offered, the proportion of the low-RDP food in the selected diet was lower when the carbohydrate source was rapidly – rather than slowly – fermentable, which is consistent with the hypothesis that ruminants learn to select a mixture of foods that minimizes metabolic imbalance.

Lambs discriminate between the post-ingestive effects of energy and protein and associate those effects with a food's flavour to modify food choices (J.J. Villalba and F.D. Provenza, personal communication). Lambs acquire a preference for a poorly nutritious flavoured food eaten during intraruminal infusions of energy (starch) or protein (casein), and shortly after an intraruminal infusion of energy or protein (preload), lambs decrease their preferences for the flavour previously conditioned with starch or casein, respectively. Preloads of casein decreased preferences for flavours previously paired with casein, and increased preferences for flavours paired with starch. Preloads of energy had the opposite effect. These results show that lambs discriminate between the post-ingestive effects of energy (starch) and protein (casein) and associate the effects with specific external cues (i.e. added flavours) to regulate macronutrient ingestion.

There is thus accumulating evidence that ruminants prefer to avoid a food, with an adequate nutrient balance overall, that delivers different nutrients at different rates and results in temporary imbalances.

## *Temporal relationships between ingestion and toxicosis*

The above raises the question about the time-scale of conditioning. Conditioned aversions to foods are unusual in that it is not necessary for the conditioning and the unconditioned stimuli to be very close in time. In contrast to classical conditioning, ingestion of a flavoured food can become associated with abdominal or metabolic discomfort several hours later (Provenza *et al.*, 1993). In acquiring strong preferences for flavoured straw paired with starch, the delay between straw ingestion and starch infusion affects preferences by lambs (Villalba *et al.*, 1999). Lambs that received starch 1 h after beginning to eat straw had lower preferences than lambs that received starch immediately after beginning to eat straw.

The shorter the delay between food ingestion and post-ingestive feedback, the stronger the preference, which is why non-structural carbohydrates that are rapidly digested in the rumen condition strong food preference (Dove, 1996). Sheep actively seek plant parts high in soluble carbohydrates (Simpson and Dove, 1994), which leads to improved efficiency of microbial protein synthesis and increased organic matter intake (Dove and Milne, 1994). The stronger the aversion the longer it persists (Kyriazakis *et al.*, 1998).

It is necessary, on the one hand, for conditioned taste aversions to persist (otherwise they have no function), but also to be flexible (otherwise an animal might be saddled with an unnecessary aversion for the rest of its life).

### Maintenance of stable rumen function

The ruminant must maintain a stable rumen environment if the incumbent microbes are not to be threatened. Of particular importance is the regular intake of fibrous material and it is relevant to consider to what extent ruminants exhibit an 'appetite' for fibre. Given free choice between forage and concentrates, cattle and goats take about 20% of their DM intake as forage. Campion and Leek (1997) found that sheep offered a diet free of long fibre ate significant amounts of fibre when available separately, and that a polyethylene fibre pompom introduced via the rumen cannula and placed so as to brush the reflexogenic areas of the reticulorumen, reduced voluntary polyethylene fibre intake.

Indeed, sheep reduced their intake of a high-energy food, but maintained their intake of a low-energy food when infused with acid or alkali to increase rumen osmolality. It was concluded that the sheep were responding in order to maintain the stability of conditions in the rumen (Engku Azahan and Forbes, 1992; Cooper *et al.*, 1995). Further, Cooper *et al.* (1996) found that inclusion of sodium bicarbonate in the diet of sheep given a choice between foods with low and high energy density increased the proportion of high-energy feed selected – the bicarbonate ameliorated the low pH induced by the rapid fermentation of the high-energy feed. Sheep self-regulate intake of sodium bicarbonate when fed diets varying in grain content (Phy and Provenza, 1998). These observations are all consistent with a desire by ruminants to maintain a certain degree of fibrousness in their diet in order to stabilize conditions in the rumen.

### Trade-offs

There are many situations in which the animal must balance its intake or choice of foods in order to trade off the intake of a toxin against the need to obtain nutrients (bearing in mind that, in general, the only difference between a toxin and a nutrient is the concentration in the diet). For example, lambs preferred barley to lucerne pellets in the absence of added toxin (LiCl), but this preference was reversed the higher the content of LiCl in the barley (Wang and Provenza, 1997). Intake of foods containing a constant concentration of LiCl increased as the proportion of barley in the food increased, i.e. animals were better able to tolerate a higher concentration of toxin the higher the yield of energy per kg. Likewise, sheep fed diets high in sagebrush, which contains various terpenes, eat more sagebrush as the macronutrient content of their diet increases (R.E. Banner, J. Rogosic, E.A. Burritt and F.D. Provenza, personal communication). When this finding is translated to the natural situation, in which both toxin and nutrient contents of plants vary with season and location, it emphasizes how nutrients interact to influence food intake. It also illustrates the ability of the body to integrate signals from nutrients and those from toxins.

Many forages contain toxic phenolics but these are sometimes the available plant species with the highest yield of digestible nutrients and grazing animals must trade nutrients off against toxins. When mule deer were offered pairs of foods with different phenolic:digestible energy ratios they ate high-energy foods when these were low in phenolics, but low-energy foods when the high-energy ones were high in phenolics (McArthur *et al.*, 1993). Similarly, goats chose to eat more low-digestibility blackbrush twigs when the high-digestibility twigs were high in toxin (either naturally-occurring condensed tannins or added LiCl) (Provenza *et al.*, 1994a). Also, lambs preferred foods higher in readily-available carbohydrate but when a high-energy food had tannin added to it then preference switched to the lower-energy food (C.H. Titus, F.D. Provenza, E.A. Burritt, A. Perevolotsky and N. Silanikove, personal communication). When polyethylene glycol (PEG) was given each morning lambs also shifted their preferences to lower-energy foods as tannin levels increased, but to a significantly lesser extent than their counterparts which did not receive the PEG supplement. Those supplemented with PEG ate more than unsupplemented animals of the tannin-containing foods, particularly as the abundance of nutritious alternatives diminished.

Hutchings *et al.* (1999) have shown how lambs trade off high nutritive value of herbage against the risk of being parasitized with *Ostertagia*; animals which had been feed-restricted took greater risks of parasitism in order to obtain more nutritious grass than unrestricted lambs.

It can be concluded, therefore, that diet selection is directed towards stabilizing conditions in the rumen and the rest of the body, i.e. avoiding disease, metabolic imbalance and upset. It is also presumably directed against over-stimulation of the rumen by physical means, although an attempt to condition a flavour aversion by inflation of a balloon in the rumen of cows did not succeed (Klaiss and Forbes, 1999).

## Afferent pathways for conditioning stimuli

This section presents evidence that signals from metabolic receptors converge with those from other classes of receptor, such as those monitoring visceral distension, osmolality and adiposity, to provide the CNS with a composite picture of comfort/discomfort and thus to play a vital part in influencing preferences for foods. This will be done mainly by reference to reviews rather than a detailed description.

### *Visceral organs*

From the mouth to the anus, food and digesta provide a continuously changing set of stimuli to stretch and tactile receptors. The complexity of the information provided to the CNS is formidable even though there is considerable convergence in the afferent branches of the autonomic nervous system which relays the impulses (Forbes, 1996).

The seminal work of Leek and colleagues, summarized by Forbes and Barrio (1992), showed how physical and chemical stimulation of receptors in the digestive tract activates vagal afferent pathways to the gastric centre of the medulla oblongata. In addition, the liver is sensitive to propionate, once again relaying its information to the CNS via autonomic afferents (Anil and Forbes, 1987) and providing a comprehensive assessment of the energy supply via the liver.

If a simple experimental procedure such as inflating a balloon in the rumen, or infusing the salt of a volatile fatty acid at a constant rate, has complex effects then how much more complex are the effects of the passage of a meal through the digestive tract, with ever-changing physical and chemical properties? It seems unlikely that the CNS could interpret individually the nature and degree of stimulation of each receptor as there is both polymodal and polytopic integration of visceral signals with convergence in hind, mid and fore brains (Forbes, 1996). Thus, severe stimulation of receptors in a specific part of the digestive tract is felt as a dull, mid-line sensation by humans so presumably milder stimulation by physiological stimuli are also not ascribed by the CNS to specific sites in the abdomen.

It is not necessary to be consciously aware of abdominal discomfort in order for the CNS to be able to respond to events in gut and liver, i.e. to participate appropriately in the control of gut motility and secretion, and to remember the sensory properties of the food eaten shortly before the discomfort occurred (Provenza *et al.*, 1994b; Forbes, 1998). Abdominal pain is likely to be an extension of normal sensitivity to stimuli such as distension of a viscus or irritation of a mucosal surface by chemicals so that metabolic imbalance is relayed as a mild form of toxicity.

### *Body tissues*

Some parts of the CNS, such as the nucleus of the solitary tract (NTS) and hypothalamus, are sensitive to a shortage of available energy (Grill, 1986). Starving cells of glucose by treatment with 2-deoxy-D-glucose, which blocks glucose uptake by cells, has shown this. An excess of glucose in the blood does not influence the NTS activity, only a shortage. For the rest of the body to fail to protect the CNS from starvation must be seen as an unusual and serious situation and one that would be likely to be a potent conditioning stimulus to eat more food in future.

At a broader level, adipose tissue has recently been shown to produce a hormone, leptin or ob-protein, increasingly as adipocyte size increases. Leptin is taken up in the CNS where it inhibits feeding via the neuropeptide Y system (Houseknecht *et al.*, 1998). It is reasonable to assume, for the time being, that any negative feedback due to leptin is integrated with stimuli emanating from the viscera so that the normal meal-to-meal regulation is modulated to a slight extent, but one which, being persistent, results in significant long-term reduction in intake in obese animals. We see no need to consider short-term and long-term controls of intake as being different in nature, just in the rate of change of the signals involved.

No muscle hormone equivalent to leptin has so far been found and it seems likely that protein deposition exerts its effect on feeding by taking amino acids out of the circulation and thus influencing liver metabolism.

### *'Metabolic discomfort' as a concept to describe the sensations relayed to the CNS from receptors sensitive to supply and balance of nutrients*

We propose, therefore, that there is a complex set of signals coming from many parts of the body, the strength of which is dependent on previous nutrition (i.e. feeding behaviour), both short- and long-term. Short-term excesses of nutrient intake result in over-stimulation of gut and liver receptors, and long-term excesses result in over-production of leptin, which generate a state of 'metabolic discomfort', otherwise known as 'satiety'. Conversely, under-stimulation of liver and hind-brain receptors due to under-eating (relative to demands) generates a different type of metabolic discomfort, especially when coupled with low levels of leptin, usually known as 'hunger'. We view the range from extreme hunger to excessive satiety as a continuum. The animal directs its behaviour to achieving the most 'comfortable' situation, in which comfort is not only induced by the appropriate supply of nutrients, but also by social pressures to eat or not to eat, and by learned associations (see above).

The anatomical and physiological mechanisms underlying affective and cognitive systems have been fairly well established (Provenza *et al.*, 1998). Taste afferents converge with visceral afferents in the solitary nucleus of the brain stem and proceed to the limbic system, where the hypothalamus and related structures maintain homeostasis in the internal environment through the endocrine system, the autonomic nervous system, and the neural system concerned with motivation and drive (i.e. incentive modification). Higher cortical centres interact with the hypothalamus through the limbic system, and regulate the internal environment primarily by indirect action on the external environment (i.e. behaviour modification). These alternative means of regulating the internal environment generally function in parallel. For example, the taste of food is adjusted according to that food's effect on the internal environment; on this basis, animals use thalamic and cortical mechanisms to select foods that are nutritious and avoid those that are toxic.

Given a choice of foods, and the ability to learn to eat that ratio between the foods which balances nutrient supply with demand, we interpret the evidence presented above to suggest that animals adjust their food choice to minimize metabolic discomfort. Thus, in meeting its 'requirements' it is not fulfilling some long-term goal, but rather reacting to hard- and soft-wired programmes, i.e. signals from body organs and tissues and the associations previously established between these signals and the sensory properties of the food (Provenza *et al.*, 1998).

## *If nutrient requirements change then level of comfort changes and choice is adjusted to regain a state of maximum comfort*

If it is true that the animal eats to minimize discomfort then any change in the nutrient uptake by tissues should lead to a change in the selection made between a choice of foods.

Protein requirements are higher for pregnant than for non-pregnant ewes and when Cooper *et al.* (1994) offered non-pregnant and late-pregnant ewes the choice between high (HP) and low-protein (LP) foods, both of high energy content, they found that the pregnant animals selected a significantly greater proportion of HP than those not pregnant.

Lactating cows offered free access to grass silage and a restricted amount of concentrates, offered as a choice one with greater (HP) and another with less (LP) than the estimated requirement for MP, ate a greater proportion of HP the higher their output of milk protein (Lawson *et al.*, 1999), i.e. they appeared to be directing their selection between foods to supply their 'requirements' for protein.

Growing lambs chose a diet well-matched to their requirements for growth (Hou *et al.*, 1991a) and worked to maintain this balanced diet when one of the foods required up to 30 responses to obtain a reinforcement (Hou *et al.*, 1991b).

Thus, diet selection is not only driven by the composition of the foods on offer, but also by the requirements of the animal, which change in a systematic manner with reproductive cycles.

## What is true for choice between foods is, we propose, also true for a single food

### *An animal 'experiments' with different levels of intake of a single food until its comfort levels are optimized*

In much of the previous discussion we have concentrated on the control of dietary choice, but we now propose that a logical extension of the proposal that dietary choice is directed to minimizing metabolic discomfort, is that the intake of a single food is also directed to achieving the same end. If our hypothesis is true then we should be able to find evidence that animals given a single food 'experiment' by increasing and decreasing their intake in a cyclical manner in order to continually ascertain whether their comfort is maximized.

#### *Cyclical intake of foods containing toxins*

Such cyclicity might be most easily seen with a food containing an overt toxin such as tall larkspur. Pfister *et al.* (1997) showed that cattle given pelleted food containing 27% larkspur *ad libitum* showed distinct cyclic patterns of intake, where increased larkspur consumption on 1 or 2 days was followed by reduced consumption on the following day. The amount of larkspur (mean 2.0 kg day$^{-1}$; 17.8 mg toxic alkaloid kg$^{-1}$ body weight) consumed was just below a level that would produce overt signs of toxicity. Further experiments with LiCl and larkspur given in a maize supplement confirmed this cyclical pattern of intake.

#### *Evidence of cyclicity for 'normal' foods*

Feeding behaviour is, by its very nature, cyclical and examination of daily feed intakes of individual animals shows considerable day-to-day variation, sometimes by as much as twofold, too large to be due simply to any changes in the environment or the physical and chemical properties of the forage.

An animal that eats exactly the same amount of food each day has no opportunity to find out whether its overall comfort would be better served by eating a little more, or a little less, than on the previous days. External factors affect food intake, including the composition of the food, climatic variables and changes in management procedures. These include variations in the time of day at which the day's refusals are measured and the time spent without access to food at milking time or when fresh food is being delivered. Such variations in daily intake, if they can be monitored and associated with changes in feedback from visceral and other receptors, could allow the animal to assess whether it should eat more or less than before. However, the effects of such variables on daily intake should be similar for all animals. The extent to which there is endogenous variation can be investigated by examining the residual variation from regression of daily intake of one animal's intake against the mean intake of the rest of the group.

A difficulty in describing day-to-day variation in intakes is that few papers, if any, publish daily intakes of individuals. We therefore obtained unpublished daily intakes of a maize silage/concentrate mixed feed by 12 lactating cows over a 31-day period (EMBRAPA, Juiz de Fora, Brazil, 1999, unpublished results). There was an approximately 5-day cycle in the mean intake of the group but the intakes of individuals show some considerable additional variation. The mean standard deviation of the residuals for the linear regressions of the intakes of each individual against the group intake for this data set was 3.63 with a range 2.00 to 5.13 kg day$^{-1}$, i.e. there is considerable individual variation in addition to the variation in daily intake shown by the whole group. Anyone who works with animals knows that voluntary intake by individuals varies day by day.

### *An hypothesis to account for cyclicity*

When given a chance, animals eat a variety of foods and they forage in many different locations. We contend that animals satiate (or get bored) from eating the same foods in the same places (Provenza, 1996). The more adequate the food(s) or location(s) nutritionally, the less animals satiate, but they satiate none the less. For example, cattle fed either plain or ammoniated straw in one flavour or another (either maple or coconut) for 1 day strongly preferred the alternate flavour when offered a choice for 2 h day$^{-1}$ for the next 5 days. The change in preference was stronger for plain than for ammoniated straw (S.B. Atwood, F.D. Provenza, R.D. Wiedmeier and R.E. Banner, personal communication). For similar reasons, sheep prefer foods in a variety of flavours and nutrient contents (Early and Provenza, 1998; J.J. Villalba and F.D. Provenza, personal communication), and they forage in locations with a variety of foods (Scott and Provenza, 1998). The same is true for toxins: the higher the toxin content of a food, the more readily animals satiate on the food (E.A. Burritt and F.D. Provenza, personal communication). Thus, satiety causes animals to eat different foods and to forage in different locations. Access to a single food induces boredom for that food and a reduction in intake, followed by hunger and increased intake; hence the cyclical pattern of intake observed. Variety is the spice of life because bodies satiate on familiarity and thrive on variety. In a changing world, it is the body's way to encourage old (sheep)dogs to learn new tricks.

## *A change in the composition of the food, or in the requirements for one or more nutrients, results in a gradual re-learning of the amount to be eaten to re-establish optimum comfort*

Changes in food composition lead to changes in food intake. If the change is in the nutritional value then the animals may well find that a higher or lower daily dry matter (DM) intake gives it more comfort than simply continuing to eat the same daily weight as before the change. On the other hand, if the sensory properties of the food change without significant change in nutrient yield any change in intake level is usually short-lived. Conversely, a change in flavour causing a temporary drop in acceptability can be masked by the inclusion of a familiar flavour (Frederick *et al.*, 1988).

## *Change of diet*

A particular situation in which it is necessary for animals to learn about food is when the food to which they have been accustomed is suddenly changed to another with different sensory and nutritional properties. In most practical and experimental situations sudden changes are avoided by a slow change of diet; in changeover experiments it is usual to wait for animals to stabilize on a new diet before recommencing recording of feeding behaviour. Mean daily intakes of 32 individual sheep accustomed to oaten chaff and then suddenly changed to barley straw are shown in Fig. 1.1, together with intakes of two individuals, selected at random (J. Hills, 1998, unpublished results). On the first day after the change the sheep ate very little which suggests that they were not familiar with the new food and were showing neophobia – a wariness of new food in which only small amounts are taken in order to assess whether there are unpleasant consequences to eating it. A combination of increasing hunger and the realization that the new food does not cause illness encouraged increased acceptance of the food and a steadily increasing intake. On the 4th and 5th days there is a decline in intake, perhaps due to rumen disturbance if the microflora has not adapted to the new food, followed by a gradual climb to a plateau, stabilizing some 10 days after introduction of the barley straw. The two individuals shown on Fig. 1.1 conform quite well to the mean for all 32 animals, but with greater day-to-day fluctuations than the mean.

When the more stable data from the last 15 days were analysed, as for the cow data above, there were again large day-to-day fluctuations in intake by individuals, in comparison with fluctuations in the mean for all animals. The mean standard deviation of the residuals was 82, with a range from 31 to 162 g $day^{-1}$. Only in 15 out of the 32 individual sheep was daily intake in the last 15 days significantly ($P < 0.05$) correlated with the daily mean intake of all 32 sheep. For the two individuals in Fig. 1.1, the one represented by the dashed line was significantly correlated, but not the other. This

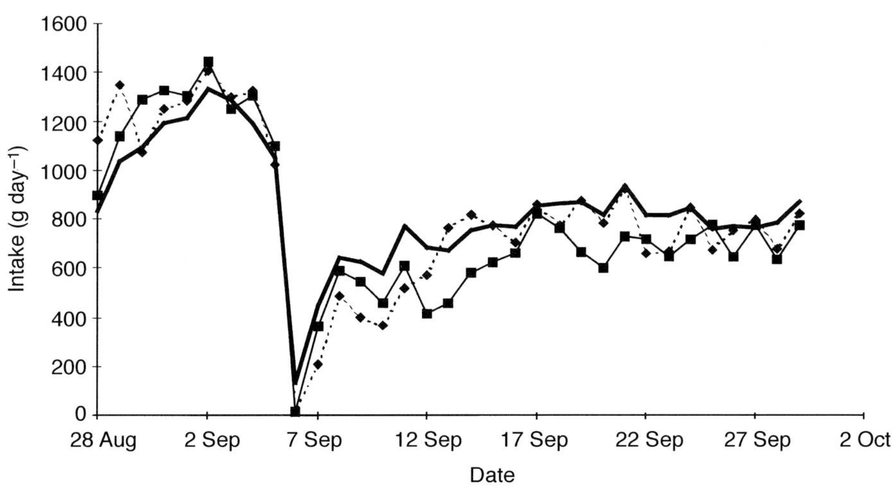


**Fig. 1.1.** Mean daily intakes for 32 sheep (solid line) and two individual sheep (lines with symbols) selected at random; on 6 September the oaten chaff was replaced with barley straw (J. Hills, unpublished results).

again is clear evidence for large variations in daily intake that are unrelated to external factors such as food quality or climatic variables. Even if this day-to-day variation is not purposeful, it still serves the animals' intake control system by providing information about whether a small increase or decrease in intake improves metabolic comfort.

*Changes in the animal*

After weaning, the intake of DM usually increases in proportion to energy requirements for growth. As maturity approaches and the amount of fat in the body increases, food intake reaches a plateau, although in many ruminants there is an annual cycle of intake and body fat content (Forbes, 1995). These changes occur slowly and there is plenty of time for animals to learn whether small increases or decreases in intake are necessary to conserve minimal discomfort.

On the other hand, as a cow progresses from late pregnancy into early lactation there are much more rapid physical and metabolic changes including a reduction in the competition for abdominal space, changes in amounts and ratios of nutrients required, oestrogens increasing in late pregnancy and then falling at parturition, and the distraction of parturition itself. Thus, there is a whole new balance of factors required in order to achieve optimal metabolic comfort which takes the animal a long time to get right, hence the slow increase in early lactation.

## Minimal total discomfort – a modelling approach

There are numerous models of voluntary food intake by ruminants (see Chapter 2), but integration of metabolic, physical, social and other factors, where attempted, has been dealt with somewhat superficially. Although we have advocated the concept of additivity of different feedback factors, we are not aware of any models that have properly taken this into account (but see Fisher, 1996), although some have acknowledged the need to do so (Poppi *et al.*, 1994).

Our thesis is that signals from the various families of visceral receptors (including signals generated by numerous metabolites interacting at the level of the liver and other organs that are sensitive to metabolic fluxes), together with signals from adipose tissue, social stimuli and environmental factors, are integrated by the CNS in an approximately additive manner to generate a total signal of 'discomfort'. The animal then adjusts its intake and/or choice, continuing in the direction that results in a reduction in discomfort and learning as it goes.

In a complex version of such a model there would be representation of numerous nutrients and metabolites as well as a detailed description of physical effects in the digestive tract and external factors such as social and environmental effects. The relationship between the strength of each of these signals and the frequency of impulses in afferent nerves would need to be known and such information is not available in most cases. For purposes of the simple model presented here we have taken ME and crude protein (CP) to represent important classes of nutrients. We assumed that the effect of a deviation in nutrient supply from the food from the animal's 'requirement' has an effect on comfort which is proportional to the square of the deviation (this has the advantage of getting rid of negatives and also emphasizes large deviations from optimal). Therefore a deficiency of a nutrient has the same unbalancing effect as an excess. A third factor affecting intake in this model is neutral detergent fibre (NDF), as an index of the bulk of the food and thus rumen fill. Fibre is included because an excess inhibits food intake while the animal's digestion suffers if there is a deficiency, in a manner parallel to the way that energy and protein supply are proposed to generate discomfort when either greater or less than the rate at which they are being utilized by the animal. Fig. 1.2 gives an example in which discomfort is plotted against intake of a typical forage. This illustrates how the sum of the discomforts due to several properties of the food, in relation to the 'requirements' of the animal, generates a total discomfort signal which the animal attempts to minimize, according to the hypotheses presented here. Iteration proceeds until the intake that produces the minimum metabolic discomfort is reached. Forbes (1999) describes the model in further detail.

When the ME, CP and NDF contents of the foods used by Kyriazakis and Oldham (1993) are used as parameters in the model, together with the requirements of their lambs, good agreement of daily intake was obtained for the high- and medium-protein foods. Predicted intakes of the low-protein foods were much greater than observed, however, suggesting that greater weighting be put on protein deficiency as a contributor to metabolic discomfort and no doubt the weighting of the signals from deviations in ME, CP and NDF will need refinement.

When two foods with different composition are available the model 'experiments' by changing the rates of intake of each until minimum discomfort is reached, thereby predicting both daily food intake and the proportion of each of the two foods eaten. The choices made by sheep when offered two foods with different protein contents in the same experiment as that discussed in the previous paragraph (Kyriazakis and Oldham, 1993) were modelled. Where possible the lambs chose foods in a ratio that

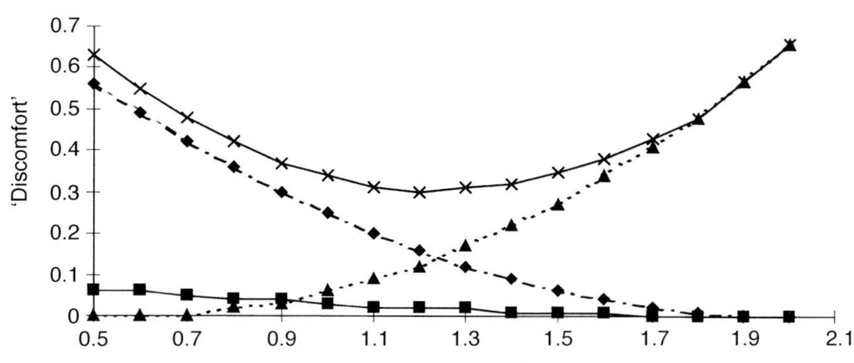


**Fig. 1.2.** Discomfort due to: ◆, metabolizable energy (ME); ■, crude protein (CP); ▲, neutral detergent fibre (NDF); ×, total; plotted against food intake. Discomfort is the square of the weighted proportional deviation of the supply of each 'nutrient' from the animal's 'requirement'. In this example the requirements were those of a growing lamb for 20 Mj ME, 250 g CP and 350 g NDF day$^{-1}$ and the forage food provided 10 Mj ME, 120 g CP and 600 g NDF kg$^{-1}$. Given the hypothesis that the animal eats that amount of food at which total discomfort is minimized, the predicted intake in this example is 1.2 kg day$^{-1}$.

gave approximately an optimum protein content in the diet while the model provides a 'perfect' dietary protein content and minimal 'metabolic discomfort' with choice feeding. Choice-fed lambs were observed to eat more in total than similar sheep offered single foods while the model predicts daily intakes of single- and choice-fed animals to be similar. Thus, the model behaves in a similar way to animals in some respects but not in others.

A number of improvements are immediately apparent. The model could be refined mathematically to avoid iteration in arriving at a solution; however, this iteration does, according to our hypothesis, mimic what is going on in the animal. Many more qualities of the food (nutrients) should be included once there is a better understanding of how they are involved in pathways that contribute to metabolic comfort/discomfort. Feedbacks from body reserves (insulin, leptin) should be included as signals added to those emanating from receptors in the digestive tract and liver to incorporate the long-term aspects of intake control. We believe, however, that this novel approach to simulating and predicting intake and choice by ruminants and other animals allows a flexibility not previously possible in incorporating whatever level of detail in metabolism is required by the user – the form of the model presented here is the simplest possible.

## Conclusions and future research

A very important task of the ruminant nutritionist is to minimize the cost of a diet (by including cheap, fibrous materials) while ensuring adequate intake of digestible nutrients. We need to know more about how ruminants trade off the various factors tending to stimulate and inhibit feeding, especially the balance between nutrients and fibre.

We do not envisage any demarcation between 'short-term' and 'long-term' controls of intake or diet selection. It is important, therefore, to establish experimentally whether there is additivity between effects of rapidly-changing visceral signals, such as distension and volatile fatty acids, and slow-changing indicators of adiposity such as leptin.

Experimental approaches need to be devised to discover whether animals can remember how much they have eaten in a day. Can they integrate feedback signals over a whole day? 24 h seems to be a sensible time over which to integrate knowledge about food and its effects as there are diurnal rhythms of feeding due to the dark–light cycle, and the regular times of feeding and milking in housed animals.

Different individuals meet their similar requirements by very different patterns of feeding behaviour within the day (Tolkamp and Kyriazakis, 1997). There is also considerable variation between individuals in their choice between foods and, as outlined above, in the intake in an individual on consecutive days. Such variations tend to have been overlooked with the advent of statistics during the 20th century which has advanced our ability to design and analyse experiments, and thus to better understand biological processes, but has obscured the vital importance of variation among individuals in biological processes (Provenza *et al.*, 1999). In studies of behaviour and nutrition we typically determine nutritional needs and formulate rations for the average member of the herd and not for specific individuals. Nevertheless, variation among individuals could affect productivity of the group if the ration diverges too much from what individuals at the extremes – which can be as much as half of the group (Provenza *et al.*, 1996) – prefer and can tolerate. Diets that enable animals to select among foods may better enable each individual to best meet its nutritional needs, and under some circumstances, may lower daily feed costs.

Above all, we emphasize the need to acknowledge that metabolic factors, physical factors and learning all have important roles to play in the complexities of the control of food intake in ruminant animals. We need to recognize the true multifactorial nature of the control of voluntary intake and diet selection if we are to advance understanding and predictive ability.

## Acknowledgements

The authors are grateful to EMBRAPA, Juiz de Fora, Brazil, and to James Hills, University of New England, Australia, for access to their unpublished data of daily food intakes of individual animals; also to Ilias Kyriazakis for critically reviewing the manuscript.

## References

Anil, M.H. and Forbes, J.M. (1987) Neural control and sensory functions of the liver. *Proceedings of the Nutrition Society* 46, 125–133.

Arsenos, G. and Kyriazakis, I. (1999) The continuum between preferences and aversions for flavoured foods in sheep conditioned by administration of casein doses. *Animal Science* 68, 605–616.

Burritt, E.A. and Provenza, F.D. (1992) Lambs form preferences for nonnutritive flavors paired with glucose. *Journal of Animal Science* 70, 1133–1136.

Campion, D.P. and Leek, B.F. (1997) Investigation of a 'fibre appetite' in sheep fed a 'long fibre-free' diet. *Applied Animal Behaviour Science* 52, 79–86.

Cooper, S.D.B., Kyriazakis, I. and Oldham, J.D. (1994) The effect of late pregnancy on the diet selections made by ewes. *Livestock Production Science* 40, 263–275.

Cooper, S.D.B., Kyriazakis, I. and Nolan, J.V. (1995) Diet selection in sheep – the role of the rumen environment in the selection of a diet from 2 feeds that differ in their energy density. *British Journal of Nutrition* 74, 39–54.

Cooper, S.D.B., Kyriazakis, I. and Oldham, J.D. (1996) The effects of physical form of feed, carbohydrate source, and inclusion of sodium-bicarbonate on the diet selections of sheep. *Journal of Animal Science* 74, 1240–1251.

Denton, D.A. (1982) *The Hunger for Salt.* Springer Verlag, Berlin.

Dove, H. (1996) Constraints to the modeling of diet selection and intake in the grazing ruminant. *Australian Journal of Agricultural Research* 47, 257–275.

Dove, H. and Milne, J.A. (1994) Digesta flow and rumen microbial protein production in ewes grazing perennial ryegrass. *Australian Journal of Agricultural Research* 45, 1229–1245.

Dutoit, J.T., Provenza, F.D. and Nastis, A. (1991) Conditioned taste-aversions – how sick must a ruminant get before it learns about toxicity in foods. *Applied Animal Behaviour Science* 30, 35–46.

Early, D. and Provenza, F.D. (1998) Food flavor and nutritional characteristics alter dynamics of food preference in lambs. *Journal of Animal Science* 76, 728–734.

Engku Azahan, E.A. and Forbes, J.M. (1992) The effects of intraruminal infusions of sodium salts on the selection of hay and concentrate foods by sheep. *Appetite* 18, 143–154.

Fisher, D.S. (1996) Mathematical integration of distension and chemostatic feedbacks for regulation of ruminant intake in diets with varied dietary protein. *Journal of Animal Science* 74, 3076–3081.

Forbes, J.M. (1995) *Voluntary Food Intake and Diet Selection in Farm Animals.* CAB International, Wallingford, 520 pp.

Forbes, J.M. (1996) Integration of regulatory signals controlling forage intake in ruminants. *Journal of Animal Science* 74, 3029–3053.

Forbes, J.M. (1998) Dietary awareness. *Applied Animal Behaviour Science* 57, 287–297.

Forbes, J.M. (1999) Minimal total discomfort as a concept for the control of food intake and selection. *Appetite* 33, 371.

Forbes, J.M. and Barrio, J.P. (1992) Abdominal chemo- and mechanosensitivity in ruminants and its role in the control of food intake. *Experimental Physiology* 77, 27–50.

Frederick, G., Forbes, J.M. and Johnson, C.L. (1988) Masking the taste of rapeseed meal in dairy compound food. *Animal Production* 46, 518.

Grill, H.J. (1986) Caudal brainstem contributions to the integrated neural control of energy homeostasis. In: Ritter, R.C., Ritter, S. and Barnes C.D. (eds) *Feeding Behavior – Neural and Humoral Controls.* Academic Press, New York, pp. 103–129.

Hills, J., Nolan, J.V., Hinch, G.N. and Lynch, J.J. (1998) Selection of diets differing in sulphur content by sheep in sulphur adequate or depleted states. *Proceedings of the British Society of Animal Science* 1998, 28.

Hou, X.Z., Emmans, G.C., Anderson, D., Illius, A.W. and Oldham, J.D. (1991a) The effect of different pairs of feeds offered as a choice on food selection by sheep. *Proceedings of the Nutrition Society* 50, 94A.

Hou, X.Z., Lawrence, A.B., Illius, A.W., Anderson, D. and Oldham, J D. (1991b) Operant studies on feed selection in sheep. *Proceedings of the Nutrition Society* 50, 95A.

Houseknecht, K.L., Baile, C.A., Matteri, R.L and Spurlock, M.E. (1998) The biology of leptin: A review. *Journal of Animal Science* 76, 1405–1420.

Hutchings, M.R., Kyriazakis, I., Gordon, I.J. and Jackson, F. (1999) Trade-offs between nutrient intake and faecal avoidance in herbivore foraging decisions: the effect of animal parasitic status, level of feeding motivation and sward nitrogen content. *Journal of Animal Ecology* 68, 310–323.

Klaiss, M. and Forbes, J.M. (1999) Effect of conditioning feed flavour to rumen distension in cows. *Proceedings of the British Society of Animal Science*, p. 85.

Kyriazakis, I. and Oldham, J.D. (1993) Diet selection in sheep: the ability of growing lambs to select a diet that meets their crude protein (nitrogen × 6.25) requirements. *British Journal of Nutrition* 69, 617–629.

Kyriazakis, I. and Oldham, J.D. (1997) Food intake and diet selection in sheep: The effect of manipulating the rates of digestion of carbohydrates and protein of the foods offered as a choice. *British Journal of Nutrition* 77, 243–254.

Kyriazakis, I., Anderson, D.H. and Duncan, A.J. (1998) Conditioned flavour aversions in sheep: the relationship between the dose rate of a secondary plant compound and the acquisition and persistence of aversion. *British Journal of Nutrition* 79, 55–62.

Lawson, R.E., Redfern, E.J. and Forbes, J.M. (2000) Choices by lactating cows between concentrates high or low in digestible undegraded protein. *Animal Science* (in press).

McArthur, C., Robbins, C.T., Hagerman, A E. and Hanley, T.A. (1993) Diet selection by a ruminant generalist browser in relation to plant chemistry. *Canadian Journal of Zoology* 71, 2236–2243.

Pfister, J.A., Provenza, F.D., Manners, G.D., Gardner, D.R. and Ralphs, M.H. (1997) Tall larkspur ingestion – Can cattle regulate intake below toxic levels? *Journal of Chemical Ecology* 23, 759–777.

Phy, T.S. and Provenza, F.D. (1998) Eating barley too frequently or in excess decreases lambs' preference for barley but sodium bicarbonate and lasalocid attenuate the response. *Journal of Animal Science* 76, 1578–1583.

Poppi, D.P., Gill, M. and France, J. (1994) Integration of theories of intake regulation in growing ruminants. *Journal of Theoretical Biology* 167, 129–145.

Provenza, F.D. (1996) Acquired aversions as the basis for varied diets of ruminants foraging on rangelands. *Journal of Animal Science* 74, 2010–2020.

Provenza, F.D., Nolan, J.V. and Lynch, J.J. (1993) Temporal contiguity between food ingestion and toxicosis affects the acquisition of food aversions in sheep. *Applied Animal Behaviour Science* 38, 269–281.

Provenza, F.D., Lynch, J.J., Burritt, E.A. and Scott, C.B. (1994a) How goats learn to distinguish between novel foods that differ in postingestive consequences. *Journal of Chemical Ecology* 20, 609–624.

Provenza, F.D., Lynch, J.J. and Nolan, J.V. (1994b) Food aversion conditioned in anesthetized sheep. *Physiology and Behavior* 55, 429–432.

Provenza, F.D., Scott, C.B., Phy, T.S. and Lynch, J.J. (1996) Preference of sheep for foods varying in flavors and nutrients. *Journal of Animal Science* 74, 2355–2361.

Provenza, F.D., Villalba, J.J., Cheney, C.D. and Werner, S.J. (1998) Self-organization of foraging behavior: From simplicity to complexity without goals. *Nutrition Research Reviews* 11, 199–222.

Provenza, F.D., Villalba, J.J. and Augner, M. (1999) The physics of foraging. In: Buchanan-Smith, J.G., Bailey, L.D. and McCoughey, P. (eds) *Proceedings of the XVIII International Grassland Congress.* Extension Service, Saskatchewan Agriculture and Food, Saskatoon, Saskatchewan, pp. 99–107.

Scott, L.L. and Provenza, F.D. (1998) Variety of foods and flavors affects selection of foraging locations by sheep. *Applied Animal Behaviour Science* 61, 113–122.

Simpson, R.J. and Dove, H. (1994) Plant non-structural carbohydrates, diet selection and intake. *Proceedings of the Australian Society of Animal Production* 20, 59–61.

Tolkamp, B.J. and Kyriazakis, I. (1997) Measuring diet selection in dairy cows: effect of training on choice of dietary protein level. *Animal Science* 64, 197–207.

Villalba, J.J. and Provenza, F.D. (1996) Preference for flavored wheat-straw by lambs conditioned with intraruminal administrations of sodium propionate. *Journal of Animal Science* 74, 2362–2368.

Villalba, J.J. and Provenza, F.D. (1997a) Preference for flavored wheat straw by lambs conditioned with intraruminal infusions of acetate and propionate. *Journal of Animal Science* 75, 2905–2914.

Villalba, J.J. and Provenza, F.D. (1997b) Preference for wheat straw by lambs conditioned with intraruminal infusions of starch. *British Journal of Nutrition* 77, 287–297.

Villalba, J.J., Provenza, F.D. and Rogosic, J. (1999) Preference for flavored wheat straw by lambs conditioned with intraruminal infusions of starch administered at different times after straw ingestion. *Journal of Animal Science* 77, 3185–3190.

Wang, J. and Provenza, F.D. (1997) Dynamics of preference by sheep offered foods varying in flavors, nutrients, and a toxin. *Journal of Chemical Ecology* 23, 275–288.

# 2 Mathematical Models of Food Intake and Metabolism in Ruminants

A.W. Illius[1], N.S. Jessop[1] and M. Gill[2]

[1]*Division of Biological Sciences, University of Edinburgh, Edinburgh, UK;*
[2]*NR International, Central Avenue, Chatham Maritime, Kent, UK*

## Introduction

The daily rate of food intake is the single most important factor affecting animal performance and productive efficiency. Knowledge of food intake is necessary for diet formulation, for the prediction of animal performance, for the design and control of production systems and for the assessment of animal–resource interactions in grazing ecosystems. Modelling the control of food intake and nutrient supply is a way of furthering our understanding of mechanisms, of testing the consequences of our assumptions, and of developing a mechanistic framework capable of accurate prediction.

There are, broadly, two types of models: 'digesta kinetics' and 'metabolic' models. The first type is concerned with the prediction of intake and digestion, on the assumption that rate of intake is limited by the rate of decrease in volume of rumen digesta by digestion and passage. The second type is concerned with the production, absorption and utilization of nutrients via microbial and animal metabolism. Whilst the former type often relies on empirical estimates of parameters, and is mostly concerned with prediction of intake and digestion, the latter type of model is rarely used to predict intake, and indeed generally requires intake as a model input. Instead, it is most often used in the pursuit of knowledge. We will exclude from this review statistical models used for data analysis.

The distinction between these types of model goes back to the early 1970s, when two models were published that had a profound influence on subsequent modelling: Baldwin *et al.* (1970) and Waldo *et al.* (1972). These models are very different in complexity and represent divergent schools of thought. Baldwin *et al.* (1970) modelled the rumen with chemically defined substrates and emphasized the stoichiometry of fermentation and the prediction of fermentation end-products. Digestion was a second-order process affected by microbial mass. This model influenced many subsequent metabolic models which describe the process of digestion as the appearance of fermentation end-products, or are purely concerned with intermediary metabolism (Gill *et al.*, 1984).

The simple model of Waldo *et al.* (1972) was concerned with cellulose disappearance rather than appearance of fermentation end-products, and partitioned cellulose into a fraction that is potentially digestible by rumen microbes and an indigestible fraction. The substrate was thus defined biologically rather than chemically, and the rates of digestion and passage were assumed to be first order.

At present, the major uncertainty in the prediction of intake lies less in the cases where intake is constrained by low digesta clearance rate than where voluntary intake is limited by feedback from metabolic factors. Thus, variation in daily intake can more clearly be related to the kinetics of digestion and passage, assessed on daily timescales, than to shorter-term metabolic events. Voluntary intake is reduced by low diet digestibility and passage rate, and it is therefore assumed that low rates of ruminal digestion and passage lead to physical limitations on daily intake (Laredo and Minson, 1973). More highly digestible feeds can, potentially, be eaten in greater quantities before the presumed physical constraints of gut clearance apply, and voluntary intake is then more likely to be determined by metabolic constraints which are related to the animal's ability to utilize absorbed nutrients. The role of metabolic constraints in controlling intake of low-quality forages has been demonstrated (Egan, 1977), resulting in reappraisal of whether physical constraints are the major determinant of intake of such forages (Tolkamp and Ketelaars, 1992; Weston, 1996; Pitroff and Kothman, 1999). Although the interaction between physical and metabolic factors is of evident interest and importance for our understanding of the control food intake, there have been few attempts to address it using modelling, perhaps because of the different time-scales and data derivation methods adopted between the two modelling approaches outlined above.

This chapter reviews the main themes and issues in modelling the processes governing food intake in ruminants. It addresses the two broad objectives of modelling: making predictions and pursuing knowledge. An important application is to make accurate predictions of intake and digestion for specified combinations of food and animal characteristics. Ruminant research is, in part, conducted in order that livestock industries and society will benefit, and models are an effective means of consolidating and applying diverse information from ruminant research programmes. However, modelling is also an important investigative procedure in its own right. It allows us to investigate mechanisms, test the consequences of our assumptions and hypotheses, show where knowledge or data are lacking, and may be used to define hypotheses and treatments in an experimental programme generating new data.

## Review of the task facing modellers

Modellers aim to represent the phenomenon of interest in mathematical form. The simplest phenomenological description is a regression model, in which variance in intake is accounted for by independent variables without their necessarily having a causal relationship with intake. These empirical relationships between feed factors and voluntary intake have the merit that they are the best-fit explanations of variation in intake. Depending on how much of the variance they explain, they may be a sound basis for prediction provided that extrapolation beyond the range of the independent variables is avoided. Phenomenological and statistical models should aim to describe the system's behaviour with the smallest number of parameters required to reach the prescribed level of accuracy. Far from attempting a thorough description of the underlying processes, the aim of this type of modelling is to achieve clarity by eliminating unnecessary detail. On the other hand, mechanistic models attempt to represent the underlying causal relationships, and are, in principle, capable of depicting the complexities of the system more faithfully. However, the depth of knowledge across all factors is very uneven, and it should be remembered that increasing complexity is likely to reduce accuracy and tractability. Certainly, any attempt at a complete representation of every causal link is futile and has to be abandoned at some point. At that point, a mechanistic model becomes phenomenological, and relies on empirical relationships. For example, even the most detailed models of rumen function examined by Bannink *et al.* (1997a) inevitably employ empirical descriptions at some point, for example to describe particle dynamics and feed degradation characteristics. The point is often made (Forbes and France, 1993) that a mechanistic model of some phenomenon occurring at level $i$ is one that may consist of empirical relationships at level $i - 1$. Despite what might be regarded as a weakness (Faichney, 1993), the use of empirical descriptions in otherwise mechanistic models should be recognized as a strength: a representation of mechanisms is supported by a robust parameter description. The usual proviso applies about only using empirical estimates within the range of the independent variables used to derive the estimate.

Rumen and intake modelling have been pursued by gradually increasing the amount of detailed description of mechanisms, as they become elucidated. But the important question of how much detail is necessary and justified in a model depends on the model's purpose. Accurate prediction requires robustness, and that requires simplicity and a sound basis for parameter estimation. Exploratory modelling justifies the inclusion of more detail, and the use of parameters whose value cannot readily be estimated. Thus, the parameter values become the experimental variable, and the model's performance is then related to systematic variation in those values (Bannink *et al.*, 1997a).

Inadequate detail in models can lead to erroneous predictions. From consideration only of the degradation rates of feed constituents it was predicted by Sinclair *et al.* (1993) that asynchrony in the supply of fermentable nitrogen and carbohydrate would cause rumen microbes to be limited by the supply of fermentable nitrogen for much of the day, with negative consequences for the production of fermentation end-products. Empirical results have failed to support this. Sinclair *et al.* (1993) did not, apparently, consider endogenous contributions (urea recycling, sloughing of cells from rumen epithelium and recycling of microbial matter in the rumen) to fermentable nitrogen. Inclusion of the effect of these endogenous N-sources in a model suggested that ammonia and microbial recycling, and the contribution of hind gut fermentation, would reduce the asynchrony in the balance of substrates available to microbes and of nutrients absorbed into the bloodstream (Illius and Jessop, 1996). The model predicted that microbial production is mostly limited by the rate of supply of fermentable carbohydrate, the converse of that predicted by ignoring supply of endogenous N.

In principle, metabolic constraints on food intake result when there is a build-up in concentration of one or more nutrients. This occurs when the rate of metabolism is limited, either by an imbalance of the other metabolites required to synthesize protein or fat, or because the rate is approaching the genetic potential of the animal. This will vary with the animal's stage of growth, reproductive status, physical and climatic environment, and ability to store or otherwise dispose of any surpluses. The balance of nutrients required will depend on the mix of purposes to which they are put, such as for protein or fat deposition, lactation, thermoregulation or locomotion. Some deviation from this ratio can be accepted, provided that body stores can be added to or depleted to balance intake with requirements at maximal production. Surplus amino acids or minerals may also be de-aminated or excreted. Further deviations from the optimal diet, such as would cause a deficit of an essential amino acid relative to the other nutrients absorbed, may cause the animal to compensate by eating more, provided it can dispose of the nutrients in surplus. More extreme dietary imbalances may result in reduced intake. Based on this framework, Illius and Jessop (1996) developed a conceptual model of metabolic constraints on intake of diets varying in energy and protein yield of absorbed nutrients.

It is not at all clear how the principles of metabolic control over intake are actually expressed by the animal, in terms of meal size and frequency, and daily intake, although it is clear that cessation of eating a single meal is not controlled by post-absorptive signals. Throughout the animal kingdom, the satiation process appears to be under tight, pre-absorptive sensory control (Smith and Gibbs, 1979). Most animals eat discrete meals which end before absorption of all the ingested nutrients can take place, suggesting that the origin of satiety signals is the gastrointestinal tract (Houpt, 1982). Meals are terminated in expectation of the post-absorptive consequences rather than being solely the result of them, and thus satiety is a state partly specified by the stimulus conditions (Booth, 1985). For example, the sense of taste plays a role in the termination of feeding (Swithers and Hall, 1994).

The complexity of these responses emphasizes why voluntary intake is so difficult to predict from first principles: it is, ultimately, a psychological phenomenon. It involves the neural integration of many signals, and is subject not only to the interplay of positive and negative physical and metabolic signals, but also to psychological phenomena such as perceptual constraints and learning (Provenza, 1995). The elucidation of how the integration of signals is affected by the animal's physiological and mental state remains an important challenge.

Attempting a truly mechanistic model of the regulation of food intake (combining taste, pre-absorptive sensation, hormonal responses, conditioning) would present problems of overwhelming complexity, even if all the relevant parameters could be evaluated. A much more parsimonious approach would be to model the functional aspects of the system by addressing the functions that short-term regulation of nutrient intake has evolved to perform.

## Prediction of intake and digestion using digesta kinetics models

The basic structure of models of intake and digestion is an elaboration of that first proposed by Waldo *et al.* (1972) and Mertens and Ely (1979), and is illustrated in Fig. 2.1. Clearance of a food constituent from the site of fermentation is the outcome of the competing processes of digestion and passage (which are usually assumed to be first-order), with respective rates $k_d$ and $k_p$, and the fraction digested, $D$, is therefore a simple function of the relative rates: $D = k_d/(k_d + k_p)$. Allen and Mertens (1988) and

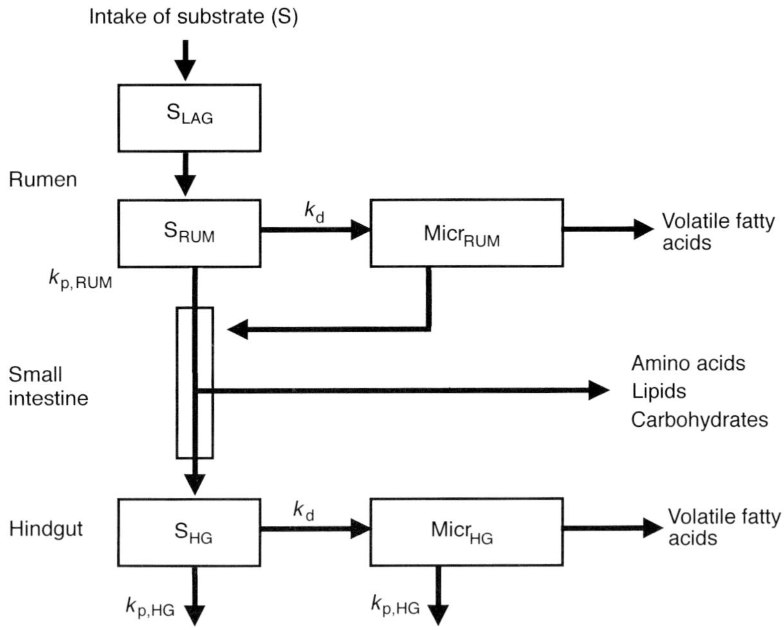


**Fig. 2.1.** Flow diagram of a compartment model sub-unit describing the kinetics of passage and digestion of a substrate (S) as it progresses through a lag phase, undergoes passage and fermentation in the rumen and hindgut. Fermentation causes the substrate to be converted to microbial biomass (Micr) and fermentation products, absorbed from the rumen, small intestine and hindgut. $k_p$ and $k_d$ are first-order rate constants for passage and digestion.

Mertens (1993) provided comprehensive reviews of elaborations of this basic model. The central assumption of digesta kinetics models is that intake is determined by the physical capacity of the rumen and that the clearance of digesta from the rumen allows further intake to occur. Illius and Allen (1994) made a detailed comparison of the structure and assumptions of published models, which differ largely in their fractionation of food and descriptions of digestion and passage kinetics.

### *Substrate fractionation*

Most recent models divide foods into compartments for cell contents, digestible cell wall and indigestible cell wall, in recognition of their different rates of digestion and passage, and assign first-order rate constants to these processes. Compartments should be defined as subsets of the whole that have homogeneous kinetic properties, and these may not necessarily correspond with any physically or chemically definable compartment in the real system. Further disaggregation of cell wall by particle size is usually thought necessary to account for selective retention in the rumen, and possibly different digestion rate, of large particles. The heterogeneity of potentially digestible cell wall as a fermentation substrate was established by Van Milgen *et al.* (1993). They showed differences in the potentially digestible fraction, discrete lag time and digestion rates between cellulose and hemicellulose in both lucerne and wheat straw, with some additional differences in these parameters between particles of different size. Representation of the carbohydrate, protein and lipid components of cell contents may be required to interface with microbial sub-models specified in biochemical terms (Gordon and Illius, 1996). In essence, foods need to be described down to the level of detail that is consistent both with differences in digestion or passage rate or effects on metabolism of sufficient magnitude to affect the system's dynamics. However, the degree to which information on food composition and component dynamics is available must also be considered. Models that demand a high degree of specification of substrates are limited in their application, due to a shortage of information about food composition and the expense of routinely analysing foods for a wide range of components. Evaluation of highly specified models is also hampered by the limited number of empirical observations conducted at a commensurate level of detail, i.e. of the supply of nutrients to the animal.

## Digestion kinetics

From a methodological standpoint, ruminal digestion may be regarded as two related processes: the disappearance of feed constituents (as contributors to the dry matter) due to solubilization and microbial fermentation, and the utilization of feed substrates by the microbial population. Studies of digestion kinetics *in situ* and *in vitro* typically measure substrate disappearance, which is the natural starting point for the digestion component in digesta kinetics models.

In principle, a detailed assessment of digestion kinetic parameters for multiple substrates from *in situ* studies could be applied to the prediction of digestibility in an appropriate model. In practice, methodological issues and the costs of estimating separate parameters for fractions of each feed present problems for this approach. The many sources of error in estimating lag duration and the rate and potential extent of digestion by *in situ* techniques have been summarized by Nocek (1988) and others. Artefacts arise due to effects of, for example, pH in the bags, efflux of finely ground material and unduly long lag times of unmasticated forage samples. Not accounting for and estimating the lag time and degradation constant simultaneously can reduce estimated degradation rates markedly, especially if the first bags are removed from the rumen during the lag phase (Dhanoa, 1988). Methodological limitations such as these led Firkins *et al.* (1998) to stress that more work needs to be done to improve accuracy of estimation of kinetic parameters if models are to predict digestion properly. Until then, uncertainties remain about translating digestion kinetics studied *in situ* into events occurring *in vivo*.

## Animal factors

Most digesta kinetics models have addressed a particular animal type and have not attempted to describe animal effects due to body size and physiological state. The size-related effects of species differences can, broadly, be described by allometric relationships, due to the striking association between physiological processes and animal mass (Taylor and Murray, 1987). The duration of physiological events is longer in large animals, and is expected to scale as $M^{0.27}$. Examples given in Table 2.1 show that mean particulate retention time and the time to comminute large particles scale with exponents not significantly different from 0.27, in the manner of other temporal variables such as the time between successive heart beats or intestinal contractions (Clark, 1927). It will be noted that, despite the high proportion of variance explained ($r^2$), there is still appreciable prediction error, as indicated by residual coefficient of variation (cv) in the range 13–25% (Table 2.1).

Illius and Gordon (1991) showed that the scaling approach works well for interspecific differences. It has a less clear theoretical basis and is less likely to be successful with intraspecific variation (within a group of dairy cows). Animal effects due to physiological state, age and environment are much harder to account for than effects due to size. Chilibroste *et al.* (1997) adjusted animal effects for physiological state (pregnancy and lactation) according to the Agricultural Research Council (1980). Physiological effects are potentially a major source of error in model predictions, and are of considerable economic importance. Attention needs to be directed to isolating these sources of between-animal variation and modelling the causal variables.

## Passage kinetics

The values for passage rate that are used in models may be: derived from observation of the food in question; generalized from a range of empirical observations (Sniffen *et al.*, 1992); derived from an allometric function of body mass (Illius and Gordon, 1991); or be a set value for all animals and forages (Mertens and Ely, 1979). Empirical estimates of passage rate are the norm in digesta kinetics models, because mechanistic modelling of passage rate has seldom been attempted. Sauvant *et al.* (1996) linked particulate outflow to a number of functions such as chewing during rumination and reticular activity, but the approach is still at an investigative stage and the model has not yet

**Table 2.1.** Allometric relationships between physiological variables and body mass.

| Variable | Allometric expression | Range in M (kg) | $n$ | $r^2$ | Residual cv (%) | Reference |
|---|---|---|---|---|---|---|
| Rumen digesta load (kg DM) | $Y = 0.01W^{1.15(\pm 0.039)}$ | 5.5–725 | 21 | 0.98 | 25 | Gordon and Illius (1994) |
| Whole gut mean retention time (h) | $Y = 16.7M^{0.23(\pm 0.026)}$ | 1.4–907 | 48 | 0.89 | 13.3 | Update of Illius and Gordon (1991) |
| Retention time of large particles (h) | $Y = 7.2N^{0.69(\pm 0.10)}M^{0.22(\pm 0.058)}$ | | 40 | 0.71[a] | 25 | Illius and Gordon (1991) |

W is body mass less digesta fresh weight; M is body mass; N is indigestible neutral detergent fibre (NDF) concentration; DM is dry matter.
[a]After accounting for variation between experiments.

been tested on many forages. Better information on the pattern of outflow and its relation to chewing and gut motility is needed.

The assumption of first-order particle kinetics, when fluxes are a constant proportion of a homogeneous compartment, is increasingly under question. A high flow rate of all nutrients would be expected in the duodenum immediately after feeding if homogeneous pools of specific nutrients are assumed to exist. However, food components arriving in the rumen are not immediately mixed with their respective ruminal compartments and available for digestion and passage. This, together with the rumen's sacculated structure, the stratification of particulate matter and the effects of particle buoyancy suggest that any food component, such as small particulate matter, may reside in a number of sub-compartments. Therefore, considerable sub-compartmentation may be required to satisfy the requirement for homogeneity. Various mathematical approaches have been used to differentiate between components which are immediately available for passage and those which require hydration, rumination, digestion, etc. Even if a homogeneous pool can be identified close to the reticulo-omasal orifice, constant fractional outflow is only likely to occur if rumen contractions (strength and frequency) are relatively constant throughout the day. Thiago *et al.* (1992) identified increased numbers of contractions in the first 5 h after consumption of large meals by steers, but total myoelectric activity peaked later in the feeding cycle. The consequence of this later peak was apparent when fractional outflow rates of neutral detergent fibre (NDF) were calculated from data obtained by emptying the rumen at different times after feeding (Gill, 1990). Fractional outflow rate varied between 0.0140 $h^{-1}$ (2–6 h) and 0.0410 $h^{-1}$ (20–23 h after feeding). P.H. Robinson and M. Gill (unpublished) went a step further and calculated the fractional outflow rates of a number of nutrients during different time periods. The weighted mean fractional rate for NDF was the lowest (0.0242 $h^{-1}$) and that for crude protein the highest (0.0825 $h^{-1}$). Little diurnal variation was observed for NDF fractional rates, while that for crude protein declined sharply from over 0.115 $h^{-1}$ to less than 0.055 $h^{-1}$ in immediate response to feeding a protein meal.

The passage of heterogeneous digesta was modelled by Matis *et al.* (1989) as a probabilistic process, using gamma-distributed residence times. They developed a model of passage of heterogeneous digesta with, essentially, time-delay or mixing compartments followed by a final homogeneous compartment from which first-order outflow occurs. Models of this form agree closely with faecal marker flow patterns, and suggest the existence of two particulate pools: a mixing pool that represents a time delay of about 10 h in dairy cows followed by a homogeneous first-order outflow pool. This view is supported by the much closer agreement between observed cell wall digestibility and that predicted by using a two-compartment model with time-dependence in the first compartment than by using a model with a single compartment (Huhtanen and Vanhatalo, 1997).

Phenomena explaining such age dependency in particulate passage are comminution and buoyancy. Poppi *et al.* (1981) demonstrated retention times of large particles of tropical forages of *c.* 11 h and 18 h in sheep and cattle, respectively. Studies with plastic particles indicate that buoyancy also affects passage in both sheep and cattle. Particles with high buoyancy have a lower fractional rate of passage than those with lower buoyancy (Campling and Freer, 1962; Kaske and Engelhardt, 1990). Changes in buoyancy over the time-course of particle digestion and comminution would result in passage of particles deviating from first-order kinetics. Such deviation could possibly explain why the model of Illius and Gordon (1991) predicted intake but underestimated digestibility of low-quality forages (Allen, 1996). To examine the possible effects of particle buoyancy, Jessop and Illius (1999) modified the model of Illius and Gordon (1991) to include separate sets of compartments for components of each meal, thereby allowing passage rate to vary with particle age. The rate of passage of small particles, $k_p'$, was a function of the proportion of indigestible neutral detergent fibre (INDF), $Q$, remaining in each particle pool (Allen, 1996). The baseline passage rate, $k_p$, was varied over range $R$, with threshold INDF concentration $Q_d$ and rate $s$:

$$k_p' = k_p\left(1 + R\left(\frac{1}{1+e^{-s(Q-Q_d)}} - \frac{1}{2}\right)\right)$$

The parameters $R$ and $s$ were given values of 1.5 and 10 respectively, based on the data of Campling and Freer (1962).

Predicted intakes and digestibilities were found to be only slightly sensitive to $Q_d$ and rather insensitive to $R$ and $s$. The optimal value of $Q_d$, determined as the value which gave the best agreement between observed and predicted intake and digestibility, was found not to be constant across foods but to vary consistently with both forage INDF and plant part (leaf or stem). This suggests that chemical and physical factors determine the time-course of buoyancy in a more complex manner than suggested by Allen (1996). The buoyancy model increased predicted digestibility and reduced intake on a forage with low digestion rate (0.02 $h^{-1}$) but increased both digestibility and intake rate at a higher digestion rate (0.08 $h^{-1}$). The predicted marker excretion pattern was more realistic than from a model omitting the time-delay pool.

## Model performance and sensitivity

Illius and Allen (1994) reviewed the performance of digesta kinetics models. Few models have been used to predict intake over a wide range of circumstances. The model of Mertens and Ely (1979) under-predicted low dry matter (DM) intakes, over-predicted high DM intakes, and explained only 26% of the variation in the intake observed in the 166 forages tested. The poor relationship between predicted and observed intakes independent of bias is probably due to the use of the same constants for fractional rates of passage and particle size reduction for all forages and animals. The amount of variation in intake explained by the model is less than that observed by using dietary concentrations of fibre (NDF, acid detergent fibre) alone (Mertens and Ely, 1982). Hyer *et al.* (1991), using a reference data set of 42 forages from the literature, found that intake was systematically under-predicted for low digestibility forages and over-predicted for high digestibility forages. Both Mertens and Ely (1982) and Hyer *et al.* (1991) suggested that energy demand and not ruminal capacity limits intake of highly digestible forages. That being so, good agreement of prediction with observation from a digesta kinetics model would hardly be expected for high digestibility forages. Illius and Gordon (1991) used data from low digestibility forages and found predicted intake agreed well with observed values, without bias. The variation in intake explained by the model was greater than that predicted by NDF alone ($r^2$ = 0.10, not significant) or

INDF ($r^2$ = 0.18). Chilibroste *et al.* (1997) used data on digesta and passage kinetics from a variety of sources in the literature and reported close agreement between predicted and observed intake ($r^2$ = 0.92). Their model attempts to set digesta load as a function of physiological state, to adjust passage for rumen fill, and to model associative effects on feed degradation.

Forage digestibility predicted by digesta kinetics models is generally within 15% of observed values (Illius and Allen, 1994), with $r^2$ in the range 0.5–0.7. The bias in digestibility estimates that frequently occur with these models is more likely to result from poor parameter estimation of digestion kinetics of forages than from inaccurate passage rate or from fundamental flaws in model structure (Illius and Gordon, 1991).

The sensitivity of model performance to variation in parameter values is seldom reported, but is of interest as to how accurate models need to be. If model output is sensitive to a particular parameter, then more attention needs to be paid to estimating it accurately than to parameters making little difference to output. The model of Illius and Gordon (1991) showed appreciable sensitivity of intake prediction only to digesta load, particulate passage rate and the proportion of the cell wall that is digestible. Parameters such as digestion rate, lag time, microbial growth efficiency and particle comminution rate caused output to vary by only about 10% of a variation in parameter value. Digestibility predictions were sensitive to the proportion of the cell wall that is digestible, but to little else.

### Causal role of digesta load and passage rate in digesta kinetics models

In modelling, as in any other form of scientific investigation, progress depends on correctly identifying causal factors. Digesta kinetics models of intake commonly assume that intake is limited by the physical capacity of the rumen and determined by the clearance rate of ruminal digesta, which is dependent on the processes of digestion, particle breakdown and passage rate. It is not necessary to assume that the rumen is ever completely full, merely that some set point of digesta load, normally expressed as mass of DM, serves to regulate intake. Provided that this set-point can be specified, and that digesta turnover is accurately modelled, the amount of intake required to return to the set-point can be predicted. Thus, Illius and Gordon (1991) used a model with a digesta load set-point to achieve a 1:1 relationship ($r^2$ = 0.61, $n$ = 25, residual cv = 14.5%) between predicted and observed intake (expressed on a metabolic weight basis) in cattle and sheep fed temperate and tropical forages ranging in NDF concentration from 622 to 875 g $kg^{-1}$. They derived the digesta load set-point from an allometric analysis (Table 2.1).

Despite the apparent success of digesta kinetics models at predicting intake, there are doubts about the validity of the underlying assumptions. Mathison *et al.* (1995) argued that digesta passage is not merely a property of foodstuffs but is also a function of the propulsive activities of the forestomach, with the implication, at least, that the animal may exert some control over passage. Doubts about the role of digesta load in regulating intake of poor-quality forages have been raised by Weston (1996) and Pitroff and Kothman (1999), arguing that digesta load is not normally at a fixed upper limit, that it varies with the animal's physiological state and environment, and increases with energy deficit. Over a range of forage quality, Weston (1996) observed a negative correlation between net energy (NE) intake and digesta load. However, the forages affording the lowest NE intake still allowed the lambs to meet their maintenance requirements for energy at a digesta load of 26% of live mass, suggesting that forages were insufficiently poor to really test for a constant upper limit to digesta load. The best forages were associated with the digesta loads of 11% of live mass, and provided 2.25 times the maintenance energy intake – when energy-based regulation would be expected. Weston (1996) accepted that digesta load must have some upper limit that constrains intake, but argued that metabolic factors such as energy status modify the digesta load the animal will tolerate. Digesta kinetics models would need to incorporate such flexibility in the digesta load constraint to account for variations in intake due to the animal's physiological status.

It is harder to address the question of whether passage rate is the driving variable in digesta residence in the fermentation compartment, or whether the animal's demand for nutrients drives both intake rate and passage rate. Illius and Gordon (1991) showed that retention time scales approximately as expected, with $M^{0.27}$. But since digesta load scales as about $M^1$ and energy and protein requirements scale as $M^{0.73}$, it could equally be argued that retention time must scale as $M^1/M^{0.73}$, or $M^{0.27}$ when animals are eating to requirement. The argument that other temporal variables (see above) scale as $M^{0.27}$ does not really resolve this debate about which is the fundamental causal variable because the reason why metabolic rates and time scale allometrically as they do is unclear. Given the frequently observed negative correlation between NDF concentration and intake (Jung and Allen, 1995), which implies that something other than energy demand constrains intake, we are still inclined to the view that retention time is not largely under the control of the animal, and that constraint to passage rate induced by feed characteristics is a causal factor in the intake of low-quality forages.

In summary, the current generation of digesta kinetics models can apparently predict intake of low-to-medium quality forages by animals with modest nutrient requirements, but deeper knowledge of the mechanisms underlying control of digesta load and passage rate are needed to allow such models to predict how intake varies with physiological state and nutrient demand.

## Metabolic models

### *Microbial metabolism*

The objective of metabolic models that focus on substrate metabolism by microbes is the prediction of nutrient supply to the animal (Black *et al.*, 1980–1981; Baldwin *et al.*, 1987b; Dijkstra *et al.*, 1992, 1998). In the earliest model, substrate supply to a single microbial pool followed first-order kinetics (Black *et al.*, 1980–1981). The energy (ATP) released from fermentation of substrate is used to meet the microbial maintenance requirement and in excess of this is used to drive microbial growth. Growth can be limited by either ATP yield or fermentable nitrogen supply. In the models of Baldwin *et al.* (1987b) and Dijkstra *et al.* (1992, 1998), supply of substrate is a second-order process, being influenced by both substrate concentration and microbial mass following Michaelis–Menten kinetics, but for which the derivation of appropriate parameter values is problematic. A single pool of microbes is divided between large particles, small particles and water in the model of Baldwin *et al.* (1987b), and therefore microbes flow out of the rumen at a composite rate. The models of Dijkstra *et al.* (1992, 1998) describe three microbial pools representing amylolytic and fibrolytic microorganisms and protozoa. Microbial maintenance requirements vary between these pools, as do substrate specificities. Amylolytic and protozoal pools had variable composition such that starch levels could increase at high nutrient availability and provide a reserve of carbohydrate at times of nutrient shortage. Protozoa engulf amylolytic and fibrolytic microorganisms in relation to their relative pool sizes although engulfment rate was reduced as protozoal starch content increased. The outflow rates of each pool differ, amylolytic microorganisms being washed out of the rumen at the liquid passage rate, fibrolytic ones at the solid passage rate and protozoa at a lower rate. None of the models represent effects of low ruminal pH on cellulolytic activity although such effects are being studied (Sauvant, 1998).

These models therefore place less emphasis on digesta kinetics. However, the shift in focus from digesta disappearance to the metabolism of feed constituents does not liberate such models from the exigencies of properly defining the digestion kinetics of those constituents. These constituents are the substrates that are inputs to microbial models, which are usually modelled at a given steady-state rate of intake. Bannink *et al.* (1997a) compared three sophisticated models of rumen function (Baldwin *et al.*, 1987b; Danfær, 1990; Dijkstra *et al.*, 1992) using inputs from seven diets with very complete observations available of rumen dynamics, and found that the models gave markedly different results. Much of the reason for this stems from the use of some concepts and parameter inputs, for instance concerning particle dynamics and the partitioning of carbohydrate fermentation, that can not be estimated from rumen observations. Although the model of Dijkstra *et al.* (1992) uses digestion kinetics parameters derived from *in situ* studies, the other two do not, relying instead on the microbial utilization of substrate to drive digestion. Bannink *et al.* (1997a) concluded that, without any input from *in situ* studies, microbial metabolism models are unlikely to be able to replicate observed differences in feed degradation. Explanations of the divergence between model predictions of volatile fatty acids (VFA) production and values measured *in vivo*, may relate both to inadequacies in modelling fermentation stoichiometry and absorption kinetics (Bannink *et al.*, 1997b) and to artefacts in estimation of VFA production (Beever, 1993). In general, rumen models are highly sensitive to changes in passage rates, because of their effects on the time for which substrates are accessible by the microbes in the rumen. However, the study of passage rates has received much less attention than that of digestion rates and thus it is difficult to assess the degree of error introduced. For example, over a meal cycle it has been shown that there is marked variation in the liquid outflow rate from the rumen (Warner and Stacy, 1968). Depending on the characteristics of the feed this can have quite different effects on the pattern of nutrient absorption. There are few data describing the patterns of liquid and particulate outflow from the rumen during a meal cycle, i.e. in non-steady-state conditions. Nearly all experimental work has attempted to maintain steady-state by constant, low-level feeding over the 24 h period where there is unlikely to be large fluctuations in rumen dynamics.

Considering the large number of parameter values required as inputs by sophisticated models of rumen function, they have limited application for predictive purposes at present, but are valuable research tools which need to be developed further (Bannink *et al.*, 1997a).

## Models of animal response to nutrients

Such models use, explicitly or implicitly, the nutrients produced from fermentation and digestion of ingested food as inputs. They aim to predict the consequences in terms of change in state of the animal (gain or loss of body mass) and synthesis of products such as milk. The models differ widely in the level of detail used, ranging from those that operate using metabolizable energy (ME) and protein (MP, e.g. AFRC, 1993) to those that require a detailed description of the profile of absorbed nutrients (Baldwin *et al.*, 1987a).

The ME and MP system provides an example of an aggregated, phenomenological approach in which ME and MP are used following a set of rules determined from empirical observations. Although distinction is drawn between protein and energy supply, ME includes the energy content of absorbed amino acids regardless of whether the amino acids will be catabolized or the efficiency with which energy-yielding metabolites will be used. The empirical relationships on which the model is based vary depending on mature size and sex of the animal to account for such differences in maintenance requirements, nutrient partitioning and composition of gain.

A simple and elegant model of nutrient use has recently been published by Emmans (1997) that has general applicability to a wide range of animal species. This approach has been used as the basis of predicting requirements and intake (Oldham, 1999) and the description of nutrient supply and requirement can be at a relatively aggregated (protein, and non-protein energy) or more detailed (individual amino acids, carbohydrate and lipid) level depending on the information available.

The most detailed models produced are based on a description of the major metabolic pathways (e.g. Baldwin *et al.*, 1980, 1987a; Gill *et al.*, 1984; Illius and Jessop, 1995). These models consist of a set of metabolic steps that vary from describing a single enzymic reaction to a complete metabolic pathway. Metabolites are state variables and their oxidation or conversion into products (such as tissue protein or milk protein) is determined by the rates of the metabolic pathways involved. Michaelis–Menten kinetics are used such that the overall flux through a pathway depends on the potential activity of that pathway, $V_{max}$, the Michaelis–Menten constant, $k_m$, and the concentration of the substrate. Additional terms representing end product inhibition or sigmoidal responses to substrate or hormonal effects on $V_{max}$ and $k_m$ may be added. Such models are parameterized (i.e. the values of $V_{max}$ and $k_m$ derived) from experiments *in vitro* performed on tissue slices or isolated cells. The level of aggregation varies between models. Gill *et al.* (1984) treated all transactions as occurring in a single pool. Baldwin *et al.* (1987a) separated metabolism between viscera, adipose tissue and the remaining lean body and allow perfect communication between them. Illius and Jessop (1995) divided metabolism between viscera and the rest of the body with transfer of nutrients determined by rates of blood flow and diffusion through extracellular water. The number of parameters required for such models is large and it is unclear as to how these might vary between species or even within species across different physiological states or degrees of maturity.

The main strengths of metabolic modelling are in testing where knowledge is inadequate, and in the coordination of experimental work. One example of synergistic links between modelling and an experimental programme was research aimed at improving understanding of voluntary intake by growing ruminants. Initially, a simplified model of growth (Gill *et al.*, 1984) represented all the protein in the body as one pool, although it is known that rates differ between organs. A separate model was then developed (Gill *et al.*, 1989) using experimental data on protein turnover rates generated by work at the tissue level. This second model identified the importance of the contribution of protein metabolism in the liver and gut to overall energy maintenance, which identified the need for research on the metabolism of the liver and gut. These initial studies increased understanding of the differences in metabolism of these tissues in steers fed forage and concentrate, and thus indirectly to an increased understanding of factors controlling intake.

## Modelling studies of metabolic constraints on intake

There have been few attempts to model the interaction of digestive and metabolic constraints on feed intake, despite the probable importance of this interaction. One of the reasons is the lack of data to parameterize metabolism models at the level of the whole animal. The most detailed models, resulting from research programmes integrating *in vitro* studies of tissues with a modelling exercise at the whole animal level, still require that intake be specified.

The first attempt at integrating concepts of metabolic control and physical limitation of eating behaviour and feed intake in ruminants was made by Forbes (1980). The model assumed that animals would satisfy their energy requirements, subject to a physical constraint imposed by rumen capacity. Literature estimates of the relation between food digestion and passage rates were used to derive estimates of the rates of energy absorption and the quantity of food residues. Meal patterning arose from the comparison of the rate of energy supplied by digestion with the rate of energy utilization in maintenance, lactation and growth; thresholds being assigned to initiate and terminate eating, subject to physical constraints. The model was not particularly sensitive to the values used for these thresholds. An alternative mechanism to that of dual thresholds is that of positive feedback, in which positive reinforcement of eating increases the motivation to eat and temporarily overcomes negative stimuli arising from nutrient absorption (Houston and Sumida, 1985). Since the upper limit on energy utilization and the lower limit imposed by gut capacity and passage rate are inputs to the Forbes model, thereby placing bounds on food intake, intake is predicted well by the model. The pattern of meals was also realistic, suggesting that the model gave a good representation of the surface phenomena, and the model clearly had heuristic value in showing that ruminant feeding behaviour could be simulated numerically.

The model of Fisher *et al.* (1987) was also used to study the integration of chemostatic and physical constraints on ruminant feed intake, one of the objectives being to redress the omission of chemostatic feedback from the model of Mertens and Ely (1979). The method of incorporating feedback controls was to strike a balance between negative and positive stimuli, rather than there being separate set-points or thresholds for each. They achieved this by a mathematical formulation expressing intake as a function of rumen distension and nutrient flow. The function contains a double exponential term intended to relate the strength of each stimulus in relation to the other (i.e. chemostasis being generally weaker than distension signals, especially at high distension, but stronger at low distension). Investigation of the effect of varying the parameter values demonstrated how the relative strength of the stimuli could be altered, and values were found which gave good approximations between observed and predicted intakes. This model requires the use of some imaginary parameters, and thus can be said to be exploratory. The same applies to the model of Sauvant *et al.* (1996) insofar as it represents feeding motivation in a 'feeding decision submodel' which includes energy status and food palatability. The model determines eating and rumination behaviour to predict intake and passage in an innovative manner, based around more conventional rumen and microbial submodels. The model incorporates a function to relate feeding motivation to fill, rather than using an upper limit to fill as a fixed constraint, but requires careful parameterization, particularly for assumed effects of palatability.

Poppi *et al.* (1994) used the model of animal metabolism developed by Gill *et al.* (1984) to examine the integration of intake regulation, the approach being to identify pathways that could limit intake and to calculate the first limiting pathway or factor. Both physical and metabolic pathways were examined: namely, instantaneous intake rate, faecal output, rumen fill, genetic potential for protein deposition, heat dissipation and ATP degradation. The 'ATP degrader' required by their model to avoid a build-up of ATP was used as an indicator of excessive energy intake or nutrient imbalance. Modifications of the model by Illius and Jessop (1995) removed the requirement for an ATP degrader, instead of which a build-up of acetate indicated excessive energy intake or inadequate glycerol precursors for acetate clearance. Poppi *et al.* (1994) found energy excess to be limiting, sometimes simultaneously with other pathways, on diets ranging from poor-quality forages to cereals, and the authors concluded that this is indicative of the wide range of dietary conditions under which nutrient balance is implicated in intake regulation.

## Future challenges to modelling intake and metabolism

Optimizing the quantity, ratio and diurnal pattern of nutrients absorbed by ruminants remains an important goal for nutritional and modelling research. One clear requirement is further integration of digesta kinetics and metabolic models for the purpose of incorporating both animal and diet effects on intake, nutrient supply and utilization. This will further require a suitable framework for modelling the effects of metabolic signals on eating and digesta kinetics. It is likely that functional approaches will be more successful than wholly mechanistic ones, given the complexity of integrating psychological, biochemical and physical signals.

Combining empirical estimates of feed degradation rates with models of microbial metabolism acknowledges the effects of physical feed characteristics, but such models still require improvements in the understanding and parameterization of fermentation pathways if VFA production is to be predicted more accurately. Models that describe microbial or animal metabolism in terms of known biochemical pathways generally use parameter values estimated to yield sensible results in particular circumstances. The clear problem facing these very detailed models is the difficulty of reaching satisfactory estimates of critical parameter values, especially when experimental work and modelling are not carried out in harmony (Baldwin and Sainz, 1995). Even when they are (Gill and Beever, 1991), it is not clear that single, static values for parameters is a satisfactory representation of dynamic and adaptive metabolic systems. Metabolic modelling also needs to get away from steady-state conditions, in order to accord with the reality of variable inputs and the many time-delaying components of the system (Sauvant and Van Milgen, 1995).

Further work is required on the mechanistic modelling of passage, and on the significance of variable fractional outflow rates. Where daily nutrient supply is the output of interest, the effect of variable outflow will only be important for nutrients with very short mean retention times, which are also digestible, e.g. highly degradable proteins. For models which include representation of hormonal effects dependent on nutrient absorption, hour-by-hour changes in the flow of nutrient may be of considerable significance, but flows alone are inappropriate, since further delays or changes in profiles may occur during absorption through the intestinal wall. The consequences of variable flow rates on these different processes need to be examined.

## References

AFRC (1993) *Energy and Protein Requirements of Ruminants.* CAB International, Wallingford, UK.

Agricultural Research Council (1980) *The Nutritional Requirements of Ruminant Livestock.* Commonwealth Agricultural Bureaux, Slough.

Allen, M.S. (1996) Physical constraints on voluntary intake of forages by ruminants. *Journal of Animal Science* 74, 3063–3075.

Allen, M.S. and Mertens, D.R. (1988) Evaluating constraints on fiber digestion by rumen microbes. *Journal of Nutrition* 118, 261–270.

Baldwin, R.L. and Sainz, R.D. (1995) Bottlenecks for improvement of future models in animal nutrition science. In: Danfaer, A. and Lescoat, P. (eds) *Proceedings of IV International Workshop on Modelling Nutrient Utilization in Farm Animals.* National Institute for Animal Science, Jele, Denmark, pp. 163–171.

Baldwin, R.L., Lucas, H.L. and Cabrera, R. (1970) Energetic relationships in the formation and utilization of fermentation end-products. In: Phillipson, A.T. (ed.) *Physiology of Digestion and Metabolism in the Ruminant.* Oriel Press Ltd, Newcastle upon Tyne, UK, pp. 319–334.

Baldwin, R.L., Smith, N.E., Taylor, J. and Sharp, M. (1980) Manipulating metabolic parameters to improve growth rate and milk secretion. *Journal of Animal Science* 51, 1416.

Baldwin, R.L., France, J. and Gill, M. (1987a) Metabolism of the lactating cow. I. Animal elements of a mechanistic model. *Journal of Dairy Research* 54, 77–105.

Baldwin, R.L., Thornley, J.H.M. and Beever, D.E. (1987b) Metabolism of the lactating cow. II. Digestive elements of a mechanistic model. *Journal of Dairy Research* 54, 107–131.

Bannink, A., DeVisser, H., Dijkstra, J. and France, J. (1997a) Impact of diet-specific input parameters on simulated rumen function. *Journal of Theoretical Biology* 184, 371–384.

Bannink, A., DeVisser, H., Klop, A., Dijkstra, J. and France, J. (1997b) Causes of inaccurate prediction of volatile fatty acids by simulation models of rumen function in lactating cows. *Journal of Theoretical Biology* 189, 353–366.

Beever, D.E. (1993) Rumen function. In: Forbes, J.M. and France, J. (eds) *Quantitative Aspects of Ruminant Digestion and Metabolism.* CAB International, Wallingford, pp. 187–215.

Black, J.L., Beever, D.E., Faichney, G.J., Howarth, B.R. and Graham, N.McC. (1980–81)

Simulation of the effects of rumen function on the flow of nutrients from the stomach of sheep: part 1 – Description of a computer program. *Agricultural Systems* 6, 195–219.

Booth, D.A. (1985) Food-conditioned eating preferences and aversions with interoceptive elements: Conditioned appetites and satieties. *Annals of the New York Academy of Science* 443, 22.

Campling, R.C. and Freer, M. (1962) The effect of specific gravity and size on the mean time of retention of inert particles in the alimentary tract of the cow. *British Journal of Nutrition* 16, 507–518.

Chilibroste, P., Aguilar, C. and García, F. (1997) Nutritional evaluation of diets. Simulation model of digestion and passage of nutrients through the rumen-reticulum. *Animal Food Science and Technology* 68, 259–275.

Clark, A.J. (1927) *Comparative Physiology of the Heart.* Cambridge University Press, Cambridge, UK.

Danfær, A. (1990) A dynamic model of nutrient digestion and metabolism in lactating dairy cows. PhD Thesis, Report 671, National Institute of Animal Science, Denmark.

Dhanoa, M.S. (1988) On the analysis of dacron bag data for low degradability feeds (Research Note). *Grass and Forage Science* 43, 441–444.

Dijkstra, J., Neal, H.D.St.C., Beever, D.E. and France, J. (1992) Simulation of nutrient digestion, absorption and outflow in the rumen: model description. *Journal of Nutrition* 122, 2239–2256.

Dijkstra, J., France, J. and Tamminga, S. (1998) Quantification of the recycling of microbial nitrogen in the rumen using a mechanistic model of rumen fermentation processes. *Journal of Agricultural Science, Cambridge*, 130, 81–94.

Egan, A.R. (1977) Nutritional status and intake regulation in sheep. VIII Relationships between the voluntary intake of herbage by sheep and the protein/energy ratio in the digestion products. *Australian Journal of Agricultural Research* 28, 907.

Emmans, G.C. (1997) A method to predict the food intake of domestic animals from birth to maturity as a function of time. *Journal of Theoretical Biology* 186, 189–199.

Faichney, G. (1993) Digesta flow. In: Forbes, J.M. and France, J. (eds) *Quantitative Aspects of Ruminant Digestion and Metabolism.* CAB International, Wallingford, pp. 53–86.

Firkins, J.L., Allen, M.S., Oldick, B.S. and St-Pierre, N.R. (1998) Modeling ruminal digestibility of carbohydrates and microbial protein flow to the duodenum. *Journal of Dairy Science* 81, 3350–3369.

Fisher, D.S., Burns, J.C. and Pond, K.R. (1987) Modeling *ad libitum* DM intake by ruminants as regulated by distension and chemostatic feedbacks. *Journal of Theoretical Biology* 126, 407.

Forbes, J.M. (1980) A model of the short-term control of feeding in the ruminant: effects of changing animal or feed characteristics. *Appetite* 1, 21.

Forbes, J.M. and France, J. (1993) *Quantitative Aspects of Ruminant Digestion and Metabolism.* CAB International, Wallingford, pp. 1–10.

Gill, M. (1990) Future for whole-animal metabolism models. In: Robson, A.B. and Poppi, D.P. (eds) *Proceedings of Third International Workshop on Modelling Digestion and Metabolism in Farm Animals.* Lincoln University, New Zealand, pp. 333–345.

Gill, M. and Beever, D.E. (1991) Modelling as an aid to nutrition research. In: Isotope and Related Techniques in Animal Production and Health. *Proceedings of an International Symposium on Nuclear and Related Techniques in Animal Production and Health* jointly organised by IAEA and FAO, Vienna, April 1991. Vienna: IAEA, pp. 171–182.

Gill, M., Thornley, J.H.M., Black, J.M., Oldham, J.D. and Beever, D.E. (1984) Simulation of the metabolism of absorbed energy-yielding nutrients by the young sheep. *British Journal of Nutrition* 52, 621.

Gill, M., France, J., Summers, M., McBride, B.W. and Milligan, L.P. (1989) Mathematical integration of protein metabolism in growing lambs. *Journal of Nutrition* 119, 1269–1286.

Gordon, I.J. and Illius, A.W. (1994) The functional significance of the browser-grazer dichotomy in African ruminants. *Oecologia* 98, 167–175.

Gordon, I.J. and Illius, A.W. (1996) The nutritional ecology of ruminants: a reinterpretation. *Journal of Animal Ecology* 65, 18–28.

Houpt, K.A. (1982) Gastrointestinal factors in hunger and satiety. *Neuroscience and Biobehavioral Reviews* 6, 145.

Houston, A. and Sumida, B. (1985) A positive feedback model for switching between two activities. *Animal Behaviour* 33, 315.

Huhtanen, P. and Vanhatalo, A. (1997) Ruminal and total cell-wall digestibility estimated by a combined *in situ* method utilizing mathematical models. *British Journal of Nutrition* 78, 583–598.

Hyer, J.C., Oltjen, J.W. and Galyean, M.L. (1991) Evaluation of a feed intake model for the grazing beef steer. *Journal of Animal Science* 69, 836–842.

Illius, A.W. and Allen, M.S. (1994) Assessing forage quality using integrated models of intake and digestion by ruminants. In: Fahey, G.C. (ed.) *Forage Quality, Evaluation, and Utilization*. University of Nebraska, Lincoln, Nebraska, pp. 869–890.

Illius, A.W. and Gordon, I.J. (1991) Prediction of intake and digestion in ruminants by a model of rumen kinetics integrating animal size and plant characteristics. *Journal of Agricultural Science, Cambridge* 116, 145–157.

Illius, A.W. and Jessop, N.S. (1995) Modelling metabolic costs of allelochemical ingestion by foraging herbivores. *Journal of Chemical Ecology* 21, 693–719.

Illius, A.W. and Jessop, N.S. (1996) Metabolic constraints on voluntary intake in ruminants. *Journal of Animal Science* 74, 3052–3062.

Jessop, N.S. and Illius, A.W. (1999) Modeling the influence of buoyancy on particle dynamics in the foregut of ruminants. In: Stone, G., Forbes, T.D.A., Stuth, J.W. and Byers, F.M. (eds) *Veterinary International Symposium on the Nutrition of Herbivores* 10–16 April, San Antonio, Texas (CD-ROM).

Jung, H.G. and Allen, M.S. (1995) Characteristics of cell walls affecting intake and digestibility of forages by ruminants. *Journal of Animal Science* 73, 2774–2790.

Kaske, M. and Engelhardt, W. von (1990) The effect of size and density on mean retention time of particles in the gastrointestinal tract of sheep. *British Journal of Nutrition* 63, 457–465.

Laredo, M.A. and Minson, D.J. (1973) The voluntary intake, digestibility, and retention time by sheep of leaf and stem fractions of five grasses. *Australian Journal of Agricultural Research* 24, 875–888.

Mathison, G.W., Okine, E.K., Vaage, A.S., Kaske, M. and Milligan, L.P. (1995) Current understanding of the contribution of the propulsive activities in the forestomach to the flow of digesta. In: Englehardt, W.V., Leonhard-Marek, S., Breves, G. and Gieseck, D. (eds) *Ruminant Physiology: Digestion, Metabolism, Growth and Reproduction*. Ferdinand Enke Verlag, Stuttgart, p. 23.

Matis, J.H., Wehrly, T.E. and Ellis, W.C. (1989) Some generalized stochastic compartment models for digesta flow. *Biometrics* 45, 703–720.

Mertens, D.R. (1993) Rate and extent of digestion. In: Forbes, J.M. and France, J. (eds) *Quantitative Aspects of Ruminant Digestion and Metabolism.* CAB International, Wallingford, pp. 13–52.

Mertens, D.R. and Ely, L.O. (1979) A dynamic model of fiber digestion and passage in the ruminant for evaluating forage quality. *Journal of Animal Science* 49,1085–1095.

Mertens, D.R. and Ely, L.O. (1982) Relationship of rate and extent of digestion to forage utilization – A dynamic model evaluation. *Journal of Animal Science* 54, 895–906.

Nocek, J.E. (1988) *In situ* and other methods to estimate ruminal protein and energy digestibilty: A review. *Journal of Dairy Science* 71, 2051–2069.

Oldham, J.D. (1999) Genotype × nutrition interactions in herbivores. In: Jung, H.-J.G. and

Fahey, G.C., Jr (eds) *Vth International Symposium on the Nutrition of Herbivores.* American Society of Animal Science, Savoy, Illinois, pp. 482–504.

Pitroff, W. and Kothman, M.M. (1999) Regulation of intake and diet selection by herbivores. In: Jung, H.-J.G. and Fahey, G.C., Jr (eds) *Vth International Symposium on the Nutrition of Herbivores.* American Society of Animal Science, Savoy, Illinois, pp. 366–422.

Poppi, D.P., Minson, D. and Ternouth, J.H. (1981) Studies of cattle and sheep eating leaf and stem fractions of grasses. III. The retention time in the rumen of large feed particles. *Australian Journal of Agricultural Research* 32,123–137.

Poppi, D.P., Gill, M. and France, J. (1994) Integration of theories of intake regulation in growing ruminants. *Journal of Theoretical Biology* 167, 129.

Provenza, F.D. (1995) Postingestional feedback as an elementary determinant of food preference and intake in ruminants. *Journal of Range Management* 48, 2.

Sauvant, D. (1998) In vitro *Techniques for measuring Nutrient Supply to Ruminants.* British Society of Animal Science Occasional Publication No. 22.

Sauvant, D. and Van Milgen, J. (1995) Dynamic aspects of carbohydrate and protein breakdown and the associated microbial matter synthesis. In: Englehardt, W. von, Leonhard-Marek, S., Breves, G. and Gieseck, D. (eds) *Ruminant Physiology: Digestion, Metabolism, Growth and Reproduction.* Ferdinand Enke Verlag, Stuttgart, pp. 71–91.

Sauvant, D., Baumont, R. and Faverdin, P. (1996) Development of a mechanistic model of intake and chewing activities of sheep. *Journal of Animal Science* 74, 2785–2802.

Sinclair, L.A., Garnsworthy, P.C., Newbold, J.R. and Buttery, P.J. (1993) Effect of synchronizing the rate of dietary energy and nitrogen release on rumen fermentation and microbial protein synthesis in sheep. *Journal of Agricultural Science, Cambridge* 120, 251.

Sniffen, C.J., O'Connor, J.D., Van Soest, P.J., Fox, D.G. and Russell, J.B. (1992) A net carbohydrate and protein system for evaluating cattle diets: II. Carbohydrate and protein availability. *Journal of Animal Science* 70, 3562–3577.

Smith, G.P. and Gibbs, J. (1979) Postprandial satiety. *Progress in Psychobiology and Physiological Psychology* 8, 179.

Swithers, S.E. and Hall, W.G. (1994) Does oral experience terminate ingestion? *Appetite* 23, 113.

Taylor, St.C.S. and Murray, J.I. (1987) Genetic aspects of mammalian survival and growth in relation to body size. In: Hacker, J.B. and Ternouth, J.H. (eds) *The Nutrition of Herbivores.* Academic Press, Sydney, pp. 487–533.

Tolkamp, B.J. and Ketelaars, J.J.M.K. (1992) Toward a new theory of feed intake regulation in ruminants 2. Costs and benefits of feed consumption: an optimization approach. *Livestock Production Science* 30, 297.

Thiago, L.R.L., Gill, M. and Sissons, J.W. (1992) Studies of method of conserving grass herbage and frequency of feeding in cattle. 2. Eating behaviour, rumen motility and rumen fill. *British Journal of Nutrition* 67, 319–336.

Van Milgen, J., Berger, L.L. and Murphy, M.R. (1993) Digestion kinetics of alfafa and wheat straw assuming heterogeneity of the potentially digestible fraction. *Journal of Animal Science* 71, 1917–1923.

Waldo, D.R., Smith, L.W. and Cox, E.L. (1972) Model of cellulose disappearance from the rumen. *Journal of Dairy Science* 55, 125–129.

Warner, A.C.I. and Stacy, B.D. (1968) The fate of water in the rumen 2. Water balances throughout the feeding cycle in sheep. *British Journal of Nutrition* 22, 389–410.

Weston, R.H. (1996) Some aspects of constraints to forage consumption by ruminants. *Australian Journal of Agricultural Research* 47, 175–197.

# 3 Control of Salivation and Motility of the Reticulorumen by the Brain in Sheep

W.L. GROVUM[1] AND J.S. GONZALEZ[2]

[1]Department of Biomedical Sciences, Ontario Veterinary College, University of Guelph, Guelph, Ontario, Canada; [2]Department of Animal Production, University of Leon, Leon, Spain

## Introduction

Brain–gut interactions are critical in salivation, gastric motility and ingestive behaviour in ruminants but the sites in the brain controlling these functions are largely unknown. Better understanding in these areas may help to resolve some of the numerous problems associated with feeding high concentrate diets to grazing ruminants in intensive production systems. Examples of the problems include acidosis, inappetence, vagal indigestion and stasis. The basic mechanisms limiting roughage consumption may also be addressed to enhance the productivity of ruminants when they are at pasture or being fed preserved forages.

No papers exist on the sites in the ruminant brain which control salivation, except for Grovum and Gonzalez (1994), whereas the central controls over motility of the reticulorumen are limited to the brain stimulation work of Bell and Lawn (1955), Dussardier (1960) and Andersson *et al.* (1958) and to the single unit studies of Beghelli *et al.* (1963), Howard (1970) and Harding and Leek (1971). The bases for sequencing and pacing reticuloruminal contractions are largely unknown. Although it is known that the reticulo-omasal orifice and the reticular groove are under vagal control (Titchen and Newhook, 1975), the origins of orifice constrictor and dilatory neurons and of the groove closure neurons in the brainstem have never been reported. Most of the brain stimulation work reported here has been published recently as short communications (Grovum and Gonzalez, 1994, 1998a,b,c; Grovum, 1998).

## Methodology used to identify central sites controlling salivation and gastric motility

The simple premise for using electrical stimulation was that the preganglionic cell bodies of facial, glossopharyngeal and vagal neurons were located in discrete regions of the brainstem and that systematic stimulation would eventually excite them, induce

 *Ruminant Physiology: Digestion, Metabolism, Growth and Reproduction* (ed. P.B. Cronjé)

saliva secretion and/or contractions of the stomach and hence disclose their location in a three-dimensional map relative to obex.

All experiments were acute, the sheep being anaesthetized with sodium pentobarbital. Fluids, consisting of normal saline with 5% glucose and 20 mEq KCl $l^{-1}$ added, were given at 1 drop $s^{-1}$ (about 3 l per sheep). Normal body temperature was maintained by circulating warm water through a radiator under the operating table. Saliva, dripping from cannulae inserted into the parotid and submandibular ducts via the mouth, was measured with infrared drop counters and their outputs were recorded as vertical deflections on a grass polygraph. Pressures, measured in anaesthetic breathing bags inserted surgically into the reticulum (400 ml inflation) and the ventral sac of the rumen (1400 ml) and in a catheter flushed with heparinized saline and inserted into the saphenous artery, were also recorded. The sheep's head was secured in a stereotaxic instrument and, after opening the skull and cutting the dura, the brainstem was exposed by removing the cerebellum with suction and ligating the tentorium on the right. The palate-bar was adjusted vertically to make the floor of the IV ventricle horizontal. This was not done when the forebrain was studied because the skull and dura were opened only over the cerebrum. Current spread from the tip of the concentric electrode was found to be less than 1 mm downward in studies on salivation.

## Investigations into the central control of saliva secretion

### *Brainstem sites from which parotid and mandibular secretions were evoked*

The three-dimensional characteristics of the parotid and submandibular centres found in two of six sheep in the work of Grovum and Gonzalez (1994) are illustrated in Fig. 3.1. The parotid centres were located an average of 2 mm caudal to the submandibular centres and otherwise overlapped them in the rostral–caudal plane (see lower portions of each graph extended below the grid representing the base of the brain). In three-dimensional coordinates, the combined centres were located on average between 7 and 16 mm rostral to obex and from 1 to 11 mm lateral from the midline (see lower portions again). The centres were 2–5 mm deep (see upper graphs), the dorsal edges being on average 1 mm below the floor of the fourth ventricle near midline and 6 mm below it at the lateral extremes. Here, the excitable parotid tissue extended in most cases down

**Fig. 3.1.** *(opposite)* The three-dimensional topography of sites (centres) in the brainstem of two sheep where electrical stimulation evoked saliva secretion by the parotid (○) and the mandibular (▲) glands. The grid in the middle of each graph delimits the base of brain. The points below the grid are simply extensions from those above to highlight the coordinates of the centres in the rostral–caudal and lateral planes. These results also demonstrate the variability that exists in the locations of the sites in different sheep.
Vertical scale – 0 mm is the floor of the IV ventricle.
Rostral–caudal or anterior–posterior (A–P) stereotaxic plane – 0 is obex and positive numbers are rostral to it.
Lateral plane – 0 is midline.
VII and IX – Facial and glossopharyngeal nerves respectively.

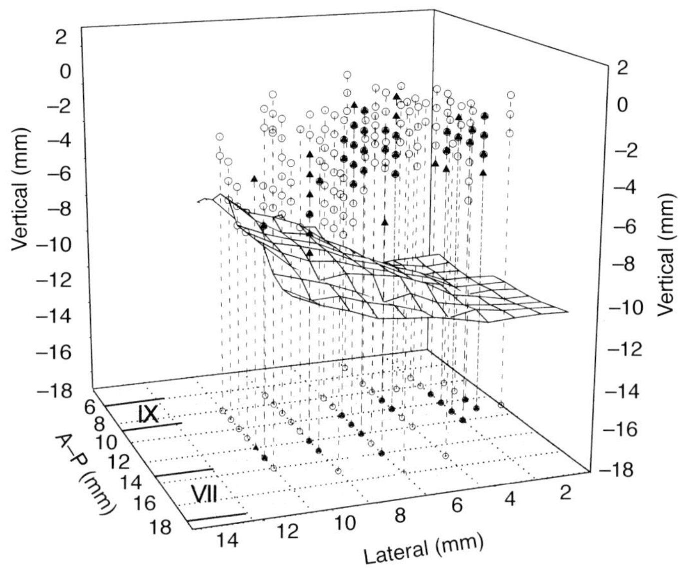
Vertical (mm)
A–P (mm)
Lateral (mm)
IX
VII

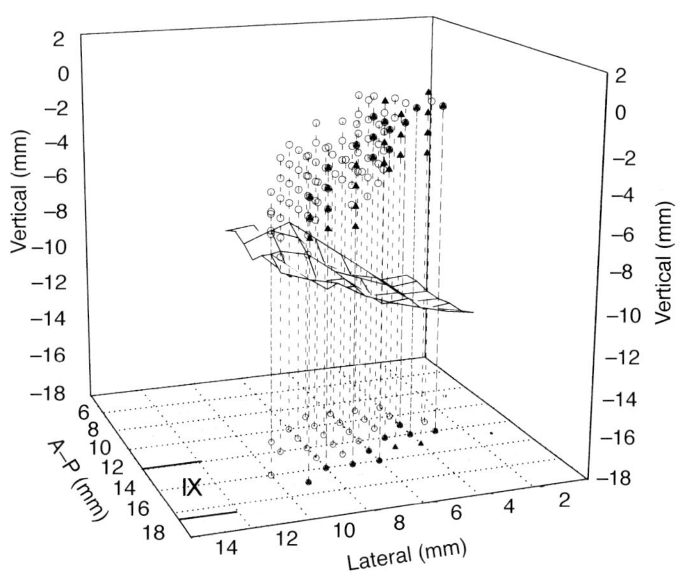
Vertical (mm)
A–P (mm)
Lateral (mm)
IX

to the base of the brain. From a dorsal perspective (see lower portions), the orientation of most of the centres was essentially in the lateral plane except that the submandibular centres angled slightly cranially from midline. The combined centres in four sheep were located between cranial nerves VII and IX in two, adjacent to VII in one and adjacent or slightly caudal to IX in the other. In the two sheep in which only the submandibular centres were found (not illustrated), one was located between VII and IX and the other was more adjacent to IX than between VII and IX. This variability with respect to the cranial nerves was in addition to that for the stereotaxic coordinates of landmarks in the sheep brain of up to 5 mm in the vertical and anterior–posterior (A–P) planes observed by McKenzie and Smith (1973). The practical significance of such variability for future work is that the secretion sites will have to be localized with stimulation before single-unit recordings are made.

The finding of functionally separate sites controlling parotid and submandibular secretion is in agreement with the fact that the parotid gland secretes during resting, eating and rumination, whereas the submandibular gland produces saliva only during eating (Denton, 1957; Carr, 1984). A functional separation of these two central systems existed, even though the neurons controlling these glands were anatomically largely intermingled.

### Frontal lobe site where parotid secretion was evoked

Stimulating the apex of the frontal cortex with bipolar side-by-side electrodes 15–20 mm from midline and 0–15 mm on the underside of the curvature of the brain elicited secretions from the parotid glands in ten sheep (Grovum and Gonzalez, 1998b). The effect was increased when the strength of stimulation was increased from 8–20 V (Fig. 3.2). Chewing movements, swallowing, and an inhibition of respiration, were frequently but not always associated with the secretory responses described. Forestomach motility was not affected. The normal inputs, which excite or inhibit this 'parotid' site, are unknown except for the fact that stimulating the olfactory bulbs (8–20 V with bipolar side-by-side electrodes) where they attached to the frontal lobe had no effect on salivation.

### A putative osmosensitive site in the amygdala inhibiting parotid secretion

When 2 ml of a 2 Osm solution of glucose, polyethylene glycol (PEG)-200, NaCl, sucrose or urea were injected separately into the jugular vein, or into the circulations of the parotid gland (isolated surgically except for one artery, one vein and nerve supply) and the brain (retrograde injections through the mandibular artery exposed in jaw and cannulated), parotid secretion was decreased on average by 1, 20 and 34% respectively. All site means differed from one another ($P < 0.001$; Grovum, 1998). The effects in the gland were substrate specific and difficult to understand, and furthermore, were achieved at unphysiological tonicity levels. However, when the solutes were introduced directly into the circulation to the brain, via retrograde injections into the mandibular artery, the depressions in parotid secretion appeared to be mediated by an osmosensitive mechanism, because the inhibition by NaCl (% decrease from control values =

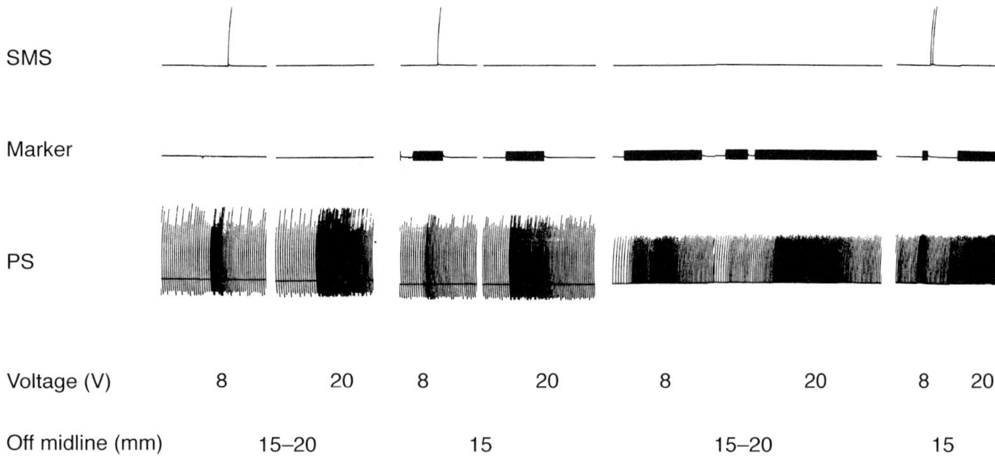


**Fig. 3.2.** Parotid secretion (PS) evoked in four sheep by stimulating the frontal cortex in the sheep brain on average 15–20 mm from midline and either on its apex or down to 15 mm on its underside using 8 and 20 V. The vertical deflections in the traces indicate drops. Secretions by the submandibular gland (SMS) were not affected. The marker line (middle trace) showed when stimuli were applied (marker signal was not activated for sheep 15).

$36^{bc}$) did not differ from that of glucose ($32^{ab}$), PEG-200 ($35^{ab}$), or sucrose ($43^{c}$) (the means with similar superscripts were not different ($P > 0.05$)). This effect of NaCl could therefore be accounted for completely by its osmotic characteristics. The fact that the inhibition by urea ($29^{a}$) was 83% of that for the other solutes was also noteworthy. In fact, the strong responses to both glucose and urea indicated that the inhibition of parotid secretion was mediated by an *osmosensitive system* different from the *osmoreceptors* in the organum vasculosum of the lateral terminalis (OVLT) near the III ventricle which evoke thirst and water conservation through antidiuretic hormone (ADH) release. The osmoreceptor terminology is justified by a direct effect of solute on the sensors due to the fact that the OVLT region is devoid of a blood–brain barrier. The reason for claiming that the system inhibiting salivation is different, is that thirst and ADH release are affected either little or not at all by injections of glucose and urea according to Fitzsimons (1989), because they can apparently diffuse into the *osmoreceptors* and fail to excite them as intracellular water is not withdrawn. On the other hand, NaCl and sucrose remain outside such cells, withdraw water and evoke neuronal responses, which lead to thirst and ADH release. An action of the solutes on salivation, via the ventricular system, was also ruled out. There was no differential effect between intracarotid NaCl, sucrose and urea on the one hand, which have been shown to withdraw water from the ventricles (and concentrate its $Na^{+}$) and glucose on the other hand, which did not withdraw water (McKinley *et al.*, 1978).

Continuous infusions of NaCl into the circulations of the brain and the parotid glands indicated that the inhibition of parotid secretion mediated centrally was achieved well within physiological limits of hypertonicity (threshold was 5 mOsm $kg^{-1}$) whereas that mediated by the gland was not (threshold was > 42.4 mOsm $kg^{-1}$). Since the central osmosensitive neurons could be expected to be functional in day-to-day life, it was important to locate them for further study.

The central osmosensitive site inhibiting parotid secretion was sought by injecting hypertonic saline into the circulation of the left half of the sheep brain (via the left mandibular artery), noting the inhibition, and then removing 5 mm slices of left brain from the front end until the inhibition was lost (Grovum, 1998). The inhibition was reduced largely when the ninth slice was removed which exposed the cranial 2 mm of the pituitary gland. It was assumed that the effect was due to the removal of many of the osmosensitive neurons. To pinpoint them, the brains in other sheep were stimulated with a concentric electrode in the region of the ninth slice. The rationale was that the osmosensitive neurons should inhibit parotid secretion whether they were excited by hypertonic solutes or electrical stimulation. Figure 3.3 illustrates such a response in the right gland to electrical stimulation on the right. On average, stimulating sites from 5 mm above the base to the base of the right brain, 13.7 mm lateral to midline and 1.9 mm rostral to the pituitary gland (29.5 mm rostral to earbar zero) inhibited right parotid secretion. Left parotid secretion was affected variably due to a variety of connections between the right and left cerebrum. This putative osmosensitive site (in the ventro-medial amygdala) inhibiting parotid secretion was found bilaterally and is new. It may help to protect body fluid levels along with the osmoreceptors in the OVLT involved in thirst and ADH release.

Efferent activity in the parotid nerve was observed to decrease in multi-unit preparations after hypertonic NaCl injections into the carotid artery, but not after identical injections into the jugular vein. It appeared that the increase in plasma osmotic pressure from the carotid injections was detected in the brain and that, as a result, the parotid 'centre' in the medulla was inhibited. The decreased parasympathetic activity to the parotid gland then reduced or inhibited parotid secretion.

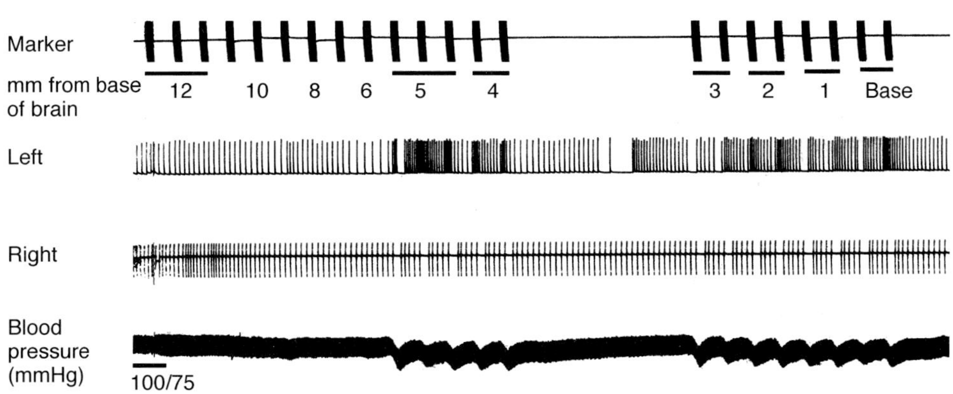


**Fig. 3.3.** Inhibition of right parotid secretion (right – ||| represents drops) by electrical stimulation of the right ventro-medial amygdala (from 5 mm above the base of the brain at earbars +26.7 mm, 13 mm to the right of midline) in sheep 272. Parotid secretion on the left and blood pressure are also shown. The top trace shows when stimuli (15 V, 40 Hz, 1 ms, 8 s every 30 s) were delivered through the concentric electrode and the numbers indicate the distance of the electrode tip from the base of the brain.

The specialized osmosensitive system described above, which apparently inhibited saliva secretion to help maintain fluid balance, has manifested itself in ruminants, rats and humans. It is particularly important for ruminants because they secrete voluminous quantities of saliva compared with other species, probably to facilitate the processing of roughage particulate matter in the reticulorumen and the emptying of undigested residues into the lower gut. Dehydration reduces their saliva production (Silanikove and Tadmor, 1989) and their food intakes (Silanikove and Tadmor, 1989; Langhans *et al.*, 1995) possibly because the associated systemic hypertonicity inhibits parotid as well as non-parotid salivation (Warner and Stacy, 1977). High-producing dairy cattle in hot arid environments have a threefold problem with water balance because they lose large quantities of water in the circulation to milk and saliva (temporary but none the less a loss) and, in addition, they must use water to keep cool. The problems are compounded when elevated body temperatures also reduce motility of the reticulorumen, slow emptying and increase rumen fill (Silanikove, 1992). Making the blood of rats hypertonic essentially abolished saliva secretion and diminished their heat tolerance because they cool themselves by spreading saliva on their body surfaces (Stricker and Hainsworth, 1970). A number of small fur-bearing animals do likewise. A similar mechanism may also exist in people as injections of 5% NaCl intravenously evoked thirst and reductions in salivary flow sensed as 'dry mouth' (Holmes and Gregersen, 1947).

The phenomenon of postprandial hypertonicity is well established in ruminants. For example, osmotic pressures of plasma increased by 15–22 mOsm $kg^{-1}$ after meals in the work of Stacy and Warner (1966). This is likely to be exacerbated by feeding more concentrated diets. The dilemma is that, when high-producing ruminants need the buffering capacity of saliva the most to deal with an acidotic rumen (after meals), saliva secretion may in fact be inhibited by hypertonicity in blood. Feeding buffers to alleviate acidosis is common but questionable because the cations from the buffers, when absorbed, may elevate plasma osmotic pressure further, thus exacerbating the postprandial hypertonicity and causing further reductions in salivation. In addition, the ions from the buffer will elevate osmotic pressures in the reticulorumen fluid and hence, may reduce food intake (Grovum and Bignell, 1989). An alternative solution to acidosis in ruminants is to discover how to increase their intrinsic buffer supply by stimulating parotid secretion.

## Investigations into brainstem control of gastric motility

### *Review of background*

#### *Involvement of the brain*

The role of the brain in motility of the reticulorumen was evident when stimulation of the peripheral cut end of the vagus nerve made it contract (Marschall, 1910 as cited by Andersson *et al.*, 1959). When cyclic motility was eliminated by vagotomy (Mangold and Klein, 1927 as cited by Clark, 1953) but preserved in decerebrate sheep (Iggo, 1951), it was clear that the brainstem caudal to the intercollicular plane was important in programming motility sequences. The remaining brainstem was said to contain a 'reticuloruminal motor centre' which could be activated by distending the reticulum or by stimulating the central cut end of one vagus (Iggo, 1951). Dussardier and Albe-Fessard (1954) were even able to maintain cyclic motility of the reticulorumen with reticular distension and to stimulate it reflexly by vagal stimulation when the decerebration was made 5 mm below the junction of the pons and the medulla.

### *Brain stimulation*

Contractions of the rumen were evoked by stimulating the interior of the brainstem electrically from midline to the lateral edge and from 2 mm caudal to 6 mm rostral to obex (Bell and Lawn, 1955). This site traversed many nuclei and fibre tracts and involved much of the reticular formation. Later, Andersson *et al.* (1958) elicited contractions of the reticulorumen in conscious goats by stimulating next to midline considerably below the dorsal motor nucleus of the vagus from obex −1 mm to obex +3 mm. They argued that this nucleus might contain the 'pace-maker' for reticuloruminal motility even though stimulations close to it inhibited contractions. Curiously too, most of the points stimulated by Bell and Lawn (1955) which evoked contractions were outside this nucleus, and the two points within it were associated with oesophageal contractions, not with contractions of the rumen. Dussardier (1960) made the reticulum, but not the rumen, contract when he stimulated below the surface of the medulla in medial planes from obex −2 mm to obex +1mm (the data can be seen in his Table No. 5). He postulated that circuitry existed in the reticular formation which initiated contractions of the reticulum. The stimulation of efferent fibres did not appear to account for many of these contractions as they were evoked from sites well below the fibre tracts from the dorsal vagal motor nucleus to the vagus (see Figure 61 in Dussardier, 1960). Dussardier (1960) further differentiated evoked contractions from the ability of brainstem stimulations to excite the rate circuit and affect the frequency of cyclic contractions. The loci which increased and decreased rate were distributed from 5 mm caudal to 5 mm rostral to obex over about two-thirds of the brainstem mass. The only clear groupings that may have constituted sites for decreasing rate were located first near the surface surrounding the dorsal vagal motor nucleus 3–4 mm caudal to obex and secondly, perhaps dorsolateral to the nucleus between 2 and 3 mm rostral to obex. This agreed with Andersson *et al.* (1959) who had arrested cyclic motility in conscious goats by stimulating the brainstem dorsolateral to this nucleus from 1 mm caudal to 2 mm rostral to obex. Furthermore, Howard (1970) found an inhibitory area in the interior of the medulla 2 mm from midline extending from the midpoint of the dorsal vagal motor nucleus cranially and ventrally. The abundance of inhibitory effects elicited by stimulation in and around the nucleus may explain why stimulating it directly failed to evoke contractions of the reticulorumen in the work of Howard (1970). The inhibition in some cases was short-lived as it was followed immediately by a reticular contraction (Howard, 1970).

The dorsal vagal motor nucleus appears to be the origin of final efferent pathways to the different structures in the reticulorumen. The gastric motor fibres appear to originate in this nucleus (Dussardier, 1960: Figure 61; Howard, 1970) and bilateral lesions in these nuclei rostral to obex have abolished all motility in the reticulorumen (Beghelli *et al.*, 1964 as cited by Howard, 1970).

### *Single-unit recording*

Beghelli *et al.* (1963) used rather large electrodes (30 μm at the tip) to record bursts of electrical activity from the dorsal nucleus of the vagus before and during reticular contractions in young lambs (12–18 kg). However, they concluded that the centres initiating contractions may be located in the reticular formation as postulated by Dussardier (1960). This was because there was a delay of up to 10 s between the start of faradic stimulation of the central cut end of the vagus and records of bursts of electrical activity from the vagal nucleus, which preceded the evoked reticular contractions. Leek and Harding (1975) localized the 'gastric centres' in the dorsal vagal nucleus and up to 1 mm dorsal and lateral to it in the medulla from 2 mm caudal to 6 mm rostral to obex and 1.5–2.5 mm lateral from midline. The gastric motor neurons were found 2–2.5 mm below the surface at points 1 and 2 mm rostral to obex. However, this site was tiny compared with the sites where stimulations evoked contractions of the reticulorumen.

## *Surface sites from which reticular and ruminal contractions were evoked separately*

Stimulating the surface of the brainstem evoked contractions of the reticulum at one site (from 2 mm caudal to 2 mm rostral to obex and 1.8–5.6 mm lateral to midline) and the rumen at another (6.0–8.0 mm rostral to obex and 1.5–3.5 mm lateral to midline). These effects are illustrated in Fig. 3.4 and were first reported by Grovum and Gonzalez (1998c). Howard (1970) applied unipolar stimulation superficially to the

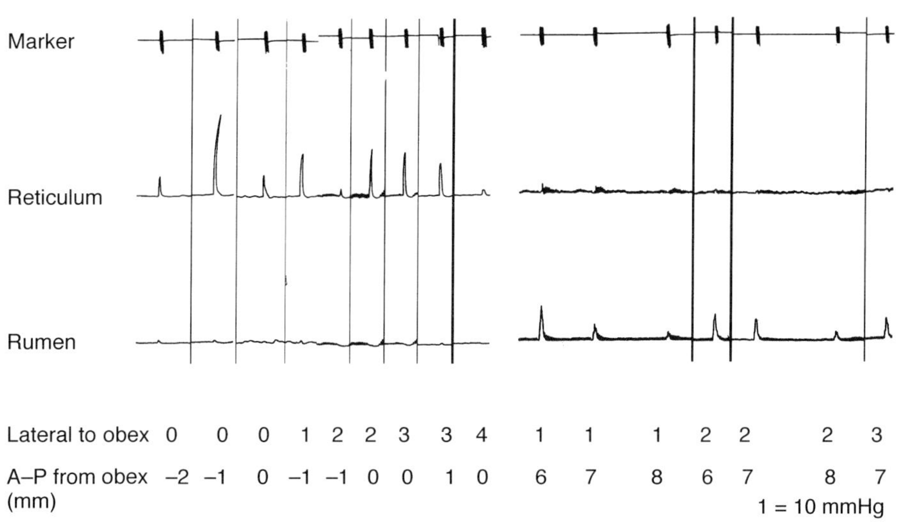


**Fig. 3.4.** Examples of stimulation on the surface of the brainstem, which evoked contractions of the reticulum only or of the rumen only, in sheep. The coordinates of the electrode tip are given relative to obex at 0 in the rostral–caudal or anterior–posterior stereotaxic plane (A–P – positive numbers go in the rostral direction) and 0 in the lateral plane. A concentric electrode was used (8 V, 1 ms, 40 Hz, 4 s). These results show a 'reticular' site lying 0–3 mm lateral and −2 to +1 mm rostral to obex and a separate 'ruminal' site at 1–3 mm lateral and +6 to +8 mm rostral to obex.

medulla but did not find such responses. The tentative interpretation is that the gastric centres may receive inputs from these surface satellite sites to determine the amplitude and form of one of three reticular and of one of two ruminal contraction types seen during mixing, rumination or eructation.

### *Sites in the interior of the brainstem from which only ruminal contractions were evoked*

The data in Fig. 3.5 illustrate sites at considerable distances below the surface of the brainstem where stimulation evoked only ruminal contractions in 11 sheep. Most sites were between 4 and 8 mm rostral to obex (Grovum and Gonzalez, 1998c). There were once again large differences between sheep. The data are evidence for a second satellite site integrated with the 'gastric centres' to determine ruminal contractions. However, the extent of this ruminal site needs clarification. Stimulations nearby in the interior of the medulla usually evoked ruminal and reticular contractions together. The stimulations evoking the responses in Fig. 3.5 may therefore have identified only the edges of those ruminal sites which protruded beyond the overlapping reticular sites. The true ruminal sites may then be much larger than shown in Fig. 3.5. According to this

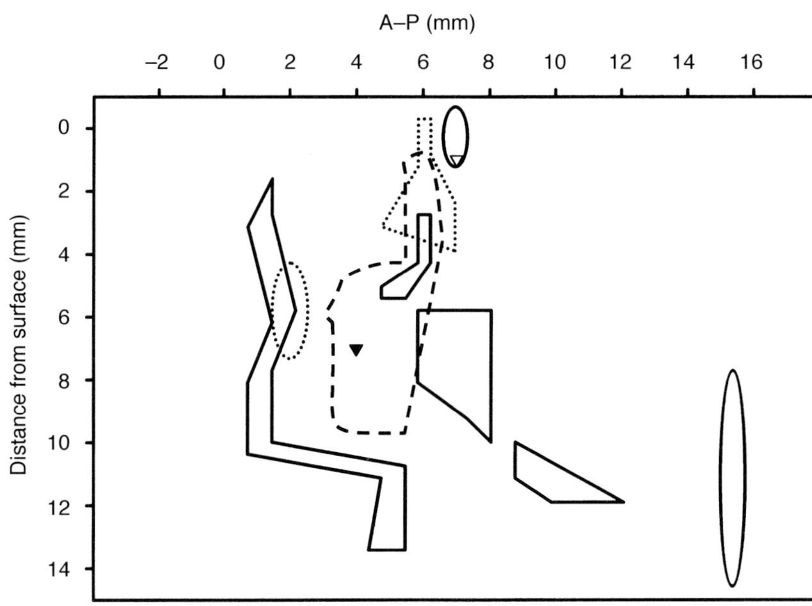


**Fig. 3.5.** A lateral view of the sites in the brainstem of individual sheep where stimulation with a concentric electrode evoked contractions of the rumen but not of the reticulum. The coordinates of the electrode tip are given relative to obex at 0 in the rostral–caudal or anterior–posterior stereotaxic plane (A–P; positive numbers are rostral to obex) and to the surface of the brain at 0 in the vertical plane. All points in the lateral plane in each sheep have been collapsed into the sites depicted.

explanation, the neuronal circuits for the reticulum and the rumen were largely intermixed but functionally separate as seemed to be the case for the neighbouring salivary centres.

## *Two interior brainstem sites from which only reticular contractions were evoked*

One of these sites was usually oriented laterally close to obex (0.3–5.1 mm rostral to obex and 2.7–7.5 mm lateral to midline) and was 1.3–8.0 mm below the surface (Grovum and Gonzalez, 1998c). Stimulations alternating between this site and adjacent regions evoked either just reticular or reticular and ruminal contractions together. This is illustrated in one sheep in Fig. 3.6 even though the site was atypical in that it was oriented more in the cranial caudal direction than laterally as in most other sheep. The points marked with an asterisk indicate loci where just the reticulum contracted. Further, if one looks at the loci with rostral–caudal or anterior–posterior (A–P) stereotaxic readings between −1 and +4, where the reticulum and the rumen contracted together, it is clear that the reticular contractions were often at least five times stronger than the ruminal contractions (e.g. 7 mm right, obex −1 and depths 5–7 mm). Whereas, at obex +5 and +6, the difference was much smaller or even reversed (in second last

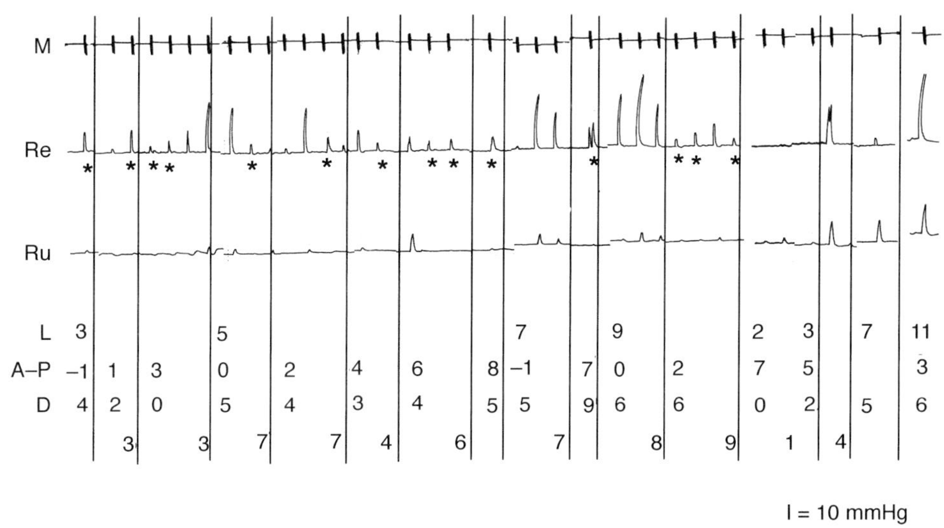


**Fig. 3.6.** Stimulation of the brainstem in sheep in the interior of the medulla (M) near obex evoked contractions of the reticulum (Re) alone (*). The rumen (Ru) may have contracted by itself also, albeit weakly (see 4th and 3rd last columns). A concentric electrode was used (8 V, 1 ms, 40 Hz, 4 s). The coordinates of the electrode tip are given relative to obex at 0 mm in the rostral–caudal or anterior–posterior (A–P) and lateral (L) stereotaxic planes and to the surface of the brain at 0 mm in depth (D). The top line indicates when stimuli were applied. The last column on the right shows the effects of stimulating the vagal rootlets inside the cranium with the electrode tip.

column on the right, the reticular strength was approximately one-third that of the ruminal strength). This may indicate a transition from a reticular site near obex to a ruminal site between 4 and 8 mm rostral to obex in accordance with Fig. 3.5. A careful inspection of the third and fourth last columns of Fig. 3.6 shows that some stimulations between 5 and 7 mm rostral to obex evoked weak ruminal, but no reticular contractions. The independence of the reticular and ruminal contractions is also evident in the middle of Fig. 3.6 at 5 mm lateral, A–P +6 and 4–6 mm depth. At a depth of 4 mm, the reticular and ruminal contractions were of equal size whereas at 5 and 6 mm depths, only reticular contractions were evoked. The contractions at 4 mm depth may therefore have resulted from stimulating functionally separate but intermixed circuits for the reticulum and the rumen. The last column on the right shows the effect of stimulating a point on the vagal rootlets within the cranium.

Reticular contractions were also elicited on their own in five sheep at another location cranially distant from obex (8.5–15.4 mm; Grovum and Gonzalez, 1998c), beyond the most cranial sites from which contractions were evoked by Bell and Lawn (1955) (6 mm); and Dussardier (1960) (5 mm). This site was 4.5–9.3 mm lateral from midline and was 2.0–8.0 mm below the surface. Figure 3.7 illustrates representative data from one sheep in which most of the reticular contractions were evoked 8–14 mm rostral to obex. At the end of the experiment on this sheep both rumen and reticular contractions of normal strength were produced by stimulating the medulla just rostral to obex. The rumen was therefore capable of contracting but it did not respond at the loci shown in Fig. 3.7.

### *Sites controlling the lips of the reticular/oesophageal groove and the reticulo-omasal orifice*

A long air-filled balloon sutured loosely between the lips of the groove and a small pear-shaped balloon dilated in the reticulo-omasal orifice of decerebrate sheep were monitored during brainstem stimulation to identify sites both on the surface and in the interior of the medulla which made these structures contract (Grovum, 1999). In one sheep out of four, three different combinations of groove and orifice responses existed to stimulation of the surface of the brainstem. First, both the groove and the orifice contracted in an oval area oriented caudo-laterally from 1 mm right (R) of midline at obex +6 mm at one end of the oval to 5R, obex +3 at the other. Laterally, in an adjacent oval area (response 2; from 3R, obex +7 to 9R, obex +3), the groove contracted and the orifice relaxed as would be expected during suckling. Finally, stimulations evoked only orifice dilations (response 3; balloon pressure decreased) in a large irregular area centred around a spot 8 mm lateral to obex. From the literature, sites would be expected which make the lips of the groove contract and the orifice either constrict or dilate because efferent vagal stimulation produced such responses (Titchen and Newhook, 1975). It may be that one groove site spanning responses 1 and 2 gave rise to the groove contractions and that one other site spanning responses 2 and 3 governed the orifice dilations. Stimulation where the sites overlapped caused both effects (response 2). However, the normal interactions between these sites are unknown. A third surface site in this sheep may have involved orifice constriction.

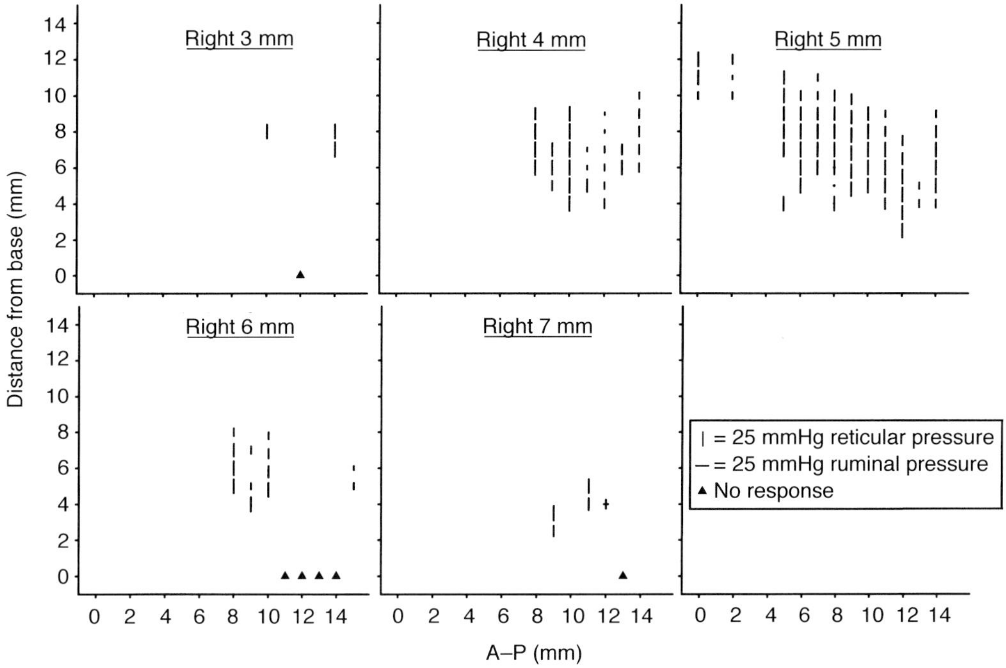


**Fig. 3.7.** Stimulating the interior of the brainstem 8–14 mm rostral to obex evoked contractions of the reticulum but not of the rumen (representative data from one of five sheep). Obex was at 0 mm in the rostral–caudal or anterior–posterior (A–P) stereotaxic plane and at 0 mm lateral (midline). The vertical plane was illustrated with 0 mm being on the base of the brain because the surface was arbitrary due to the cerebellum having been removed by suction. Each of the five sagittal sections shows the magnitude of the reticular contractions as vertical bars with length being proportional to strength. Where the rumen contracted (7 mm right), the strength is indicated by a horizontal bar.

### Brainstem site from which rumination and hypermotility has been evoked

Rumination was evoked by stimulating the medulla in conscious sheep a few mm below the surface just dorsolateral to the dorsal nucleus of the vagus (from obex +2 mm to obex −1 mm in the rostral–caudal plane; Andersson *et al.*, 1959). This may have arisen from stimulation of epithelial receptor input pathways to the brainstem. Motility of the reticulorumen was inhibited at the same time. A high frequency of cyclic motility (5 cycles $min^{-1}$) followed this inhibition. A similar result, except for the rumination, was obtained by Howard (1970) in halothane-anaesthetized sheep. Whether the marked stimulation of frequency had anything to do with the rate circuit of Leek and Harding (1975) is not known. Dussardier (1960) found that excitatory loci for cyclic motility were scattered throughout the brainstem ventro-laterally from the vagal nucleus.

Electrical stimulation of the medulla near obex (1.8–4.1 mm lateral and from 0.6 mm caudal to 2.0 mm cranial) in anaesthetized sheep caused gas to be eliminated from the reticulorumen (Grovum and Gonzalez, 1998a). This site was thought to be an eructation centre and was linked in later work (W.L. Grovum, unpublished data) to relaxation of the upper oesophageal sphincter. Nevertheless, the loss of gas could also have been simply associated with relaxation of this sphincter because the stimulation activated circuits involved with rumination or emesis. An emetic area was located by Andersson *et al.* (1959) near the surface of the medulla at obex +1 and obex +2 mm, dorsal to their 'rumination' area. Clearly, more work is required to understand the function of the site reported by Grovum and Gonzalez (1998a).

A forebrain involvement in rumination is certain since it was evoked in conscious sheep by stimulating the ventrolateral part of the anterior hypothalamus electrically (Andersson, 1951; Larsson, 1954). Rumination could be stimulated with pentagastrin injections into the ventricular fluid (Honde and Bueno, 1984); it occurred as a conditioned response to milking stimuli in the goat (Andersson *et al.*, 1958); and lobotomy and lesions in the ventral forebrain rostral to the optic chiasma increased activity to 24 h $day^{-1}$ in one sheep (Clark, 1953). The latter may be the result of removing a brake on the rumination centre, wherever it is, or of altering afferent input to that centre by altering descending inhibition (Urabe *et al.*, 1968). The present authors are not aware of rumination being reported in decerebrate preparations but eructation can be induced (Titchen and Reid, 1965).

## *A putative mechanism for brainstem control of reticuloruminal motility*

The framework in Fig. 3.8 is highly speculative but it accommodates a diverse body of experimental findings. The main concept advanced is that separate satellite sites in the brainstem may initiate the separate contractions of the reticulum and the rumen during mixing, eructation and rumination by channelling their outputs in a coordinated sequence through the gastric centres. The following situations are addressed:

**1.** To initiate a normal motility cycle, A-type interneurons of the form and amplitude circuit in the gastric centres excite their respective motor neurons for the reticulum and the rumen at intervals of about 1 min (Harding and Leek, 1971; Leek and Harding, 1975). This appears to occur through disinhibition when the C-type interneurons (inhibitory and tonically active) are in turn inhibited by the B-type interneurons from the rate circuit. The A interneurons and the motor neurons are then released to discharge impulses in proportion to the balance of excitatory and inhibitory inputs being received from the rest of the brain and from the periphery. Further, it is not known if the inhibitory effects of stimulation on motility reported by Andersson *et al.* (1959), Dussardier (1960) and Howard (1970) are related to the inhibitory C interneurons of Harding and Leek (1971).

**2.** The three reticular contractions during rumination may originate from the three different satellite sites in the brainstem (one on the surface and two in the interior apart from the gastric centres) where stimulation evoked contractions of the reticulum only (see left half of Fig. 3.8). The first contraction associated with rumination would have to involve the rumination centre. There is no basis presently for linking specific contractions with specific sites.

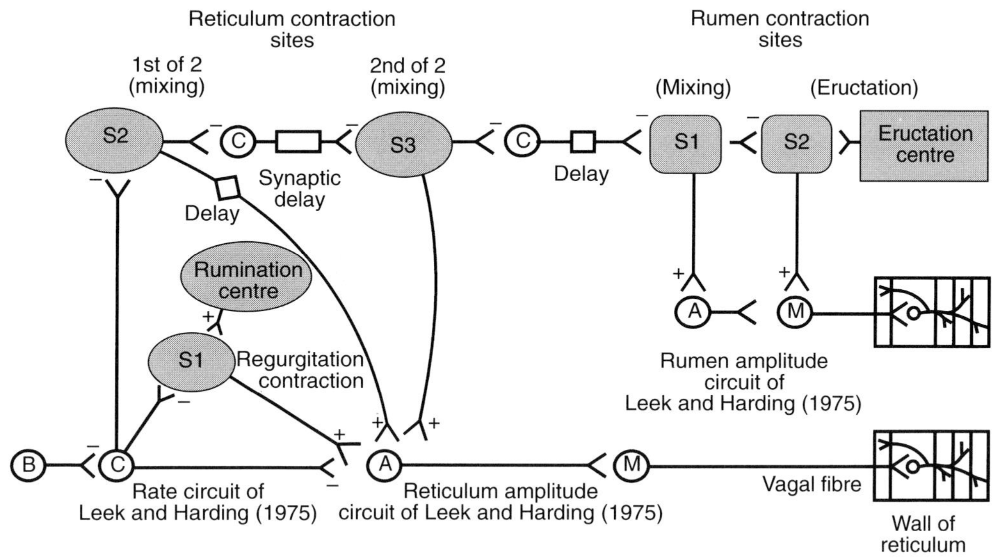


**Fig. 3.8.** A putative system in the brainstem of ruminants for controlling contractions of the reticulorumen during mixing, rumination and eructation (see the text for the details). There are no experimental data linking the three 'reticular' sites to any of the reticular contractions during mixing or rumination. Similarly, there are no data linking the two 'ruminal' sites to ruminal contractions during mixing and eructation.

**3.** The two ruminal brainstem sites (on surface and in the interior) may act independently to trigger ruminal contractions during either mixing or eructation (see top right of Fig. 3.8) but once again, there are no data to link a specific contraction to a specific site. The contraction during eructation would have to be activated by an eructation centre located elsewhere in the brainstem. Separate sites have not been identified for the dorsal and ventral sacs of the rumen. Furthermore, there is no basis for the backward and forward moving contractions in the dorsal sac during mixing and eructation respectively.

**4.** The 'early' motor neurons of Harding and Leek (1971) may act like a final common pathway to the reticulum as their discharges occurred in two phases, one of low frequency associated with the weaker first contraction in mixing and another of greater frequency associated with the stronger second reticular contraction. Accordingly, the scheme in Fig. 3.8 has the outputs from all reticular satellite sites routed through one set of motor neurons for the reticulum.

**5.** The 'late' motor neurons of Harding and Leek (1971) and Leek and Harding (1975) for the rumen may also act as a final common pathway as they were active during both mixing and eructation. This is accommodated in Fig. 3.8 by having the outputs from two satellite sites for the rumen routed through the A interneurons and the motor neurons for the rumen.

**6.** The amplitude of the second reticular contraction during mixing was inhibited more than the first as a result of stimulations starting from the midpoint of the dorsal nucleus of the vagus (2 mm from midline in the region of obex) in a line going in a cranial–ventral direction toward the nucleus ambiguous (Howard, 1970). This could result if the satellite site for the second reticular contraction was inhibited rather than the A type interneurons or the motor neurons in the gastric centres. The satellite site for the first reticular contraction could then continue to operate normally.

**7.** Cyclic motility, which is generated in anaesthetized sheep, often starts with a single rather than a bi-phasic reticular contraction. This could be accounted for if the anaesthetic selectively inhibited one of the two satellite sites on the top left in Fig. 3.8 associated with mixing in the reticulum.

**8.** The strength of secondary dorsal sac contractions and their associated Type A activity exceeded that of primary contractions considerably (Iggo and Leek, 1967; Harding and Leek, 1971). Secondary contractions are more resistant than primary to drugs like xylazine (Grovum, 1986). The possibility of separate satellite controls for ruminal contractions in mixing and eructation could account for such observations.

**9.** The fact that ruminal motility is more depressed by anaesthetics (Iggo and Leek, 1967; Harding and Leek, 1971) and various abdominal stimuli (Titchen, 1960) than reticular motility, could be explained by a differential susceptibility of satellite sites.

**10.** The neural circuits in the brain which give rise to the synaptic delays, shown in Fig. 3.8, and produce a sequence of contractions in the reticulorumen have not been studied. The delays, however they are organized, must be variable as the mixing cycle is completed much faster when the stomach is empty than when it is full.

**11.** Research is needed to link the rate circuit of Leek and Harding (1975) to stimulation sites that markedly increased the frequency of cyclic motility.

## Conclusions regarding salivation and reticuloruminal motility

To improve our understanding of brain–gut interactions in ruminants, recordings of single-unit neural activity in the salivary and satellite motility sites will be a necessary but daunting task. Ultimately, this should help to improve rumen function and hence deal with the digestive problems associated with intensive production systems. When recordings are made from areas of the brain involved in controlling food intake, new ways may be found to increase roughage intakes and hence the productivity of ruminants at pasture.

## References

Andersson, B. (1951) The effect and localisation of electrical stimulation of certain parts of the brain stem in sheep and goats. *Acta Physiologica Scandinavica* 23, 8–23.

Andersson, B., Kitchell, R. and Persson, N. (1958) A study of rumination induced by milking in the goat. *Acta Physiologica Scandinavica* 44, 92–102.

Andersson, B., Kitchell, R.L. and Persson, N. (1959) A study of central regulation of rumination and reticulo-ruminal motility. *Acta Physiologica Scandinavica* 46, 319–338.

Beghelli, V., Borgatti, G. and Parmeggiani, P.L. (1963) On the role of the dorsal nucleus of the vagus in the reflex activity of the reticulum. *Archives Italiennes de Biologie* 101, 365–384.

Bell, F.R. and Lawn, A.M. (1955) Localization of regions in the medulla oblongata of sheep associated with rumination. *Journal of Physiology London* 128, 577–592.

Carr, D.H. (1984) The regulation of parotid and submandibular salivary secretion in sheep. *Quarterly Journal of Experimental Physiology* 69, 589–597.

Clark, C.H. (1953) The nerve control of rumination and reticulo-ruminal motility. *American Journal of Veterinary Research* 14, 376–384.

Denton, D.A. (1957) A gregarious factor in the natural conditioned salivary reflexes of sheep. *Nature* 179, 341–344.

Dussardier, M. (1960) Recherches sur le contrôle bulbaire de la motricité gastrique chez les ruminants. Doctoral thesis, Faculty of Science, University of Paris, France.

Dussardier, M. and Albe-Fessard, D. (1954) Quelques propriétés du centre vagal contrôlant l'activité réflexe de l'estomac des ruminants. *Journal de Physiologie* 46, 354–357.

Fitzsimons, J.T. (1989) Bengt Andersson's pioneering demonstration of the hypothalamic "drinking area" and the subsequent osmoreceptor/sodium receptor controversy. *Acta Physiologica Scandinavica* 583, 215–225.

Grovum, W.L. (1986) The control of motility of the ruminoreticulum. In: Milligan, L.P., Grovum, W.L. and Dobson, A. (eds) *Control of Digestion and Metabolism in Ruminants.* Reston book, Prentice Hall, Englewood Cliffs, New Jersey, pp. 18–40.

Grovum, W.L. (1998) A putative osmosensitive site in the forebrain inhibiting parotid secretion in sheep. *Journal of Physiology* London 509, 8P.

Grovum, W.L. (1999) Stimulation of the brainstem evoked opening and closing of the reticulo-omasal orifice and closure of the reticular/esophageal groove in sheep. *South African Journal of Animal Science* 29(ISRP).

Grovum, W.L. and Bignell, W.W. (1989) Results refuting volatile fatty acids per se as signals of satiety in ruminants. *Proceedings of the Nutrition Society* 48, 3A.

Grovum, W.L. and Gonzalez, J.S. (1994) Centres in the brain stem controlling the secretion of parotid and mandibular saliva in sheep. *Proceedings Society Nutritional Physiology* 3, 76.

Grovum, W.L. and Gonzalez, J.S. (1998a) A putative eructation centre in the sheep brain. *European Journal of Neuroscience* 10 (Suppl. 10), 250 (Abstract 97.01).

Grovum, W.L. and Gonzalez, J.S. (1998b) Stimulation of the brain's frontal lobe elicited body movements and parotid secretion in sheep. *Journal of Animal Science* 76 (Suppl. 1), 357.

Grovum, W.L. and Gonzalez, J.S. (1998c) The reticulum and the rumen contracted separately to brainstem stimulation in sheep. *Journal of Physiology* London, 511, 54P.

Harding, R. and Leek, B.F. (1971) The locations and activities of medullary neurones associated with ruminant forestomach motility. *Journal of Physiology London* 219, 587–610.

Holmes, J.H. and Gregersen, M.I. (1947) Relation of the salivary flow to the thirst produced in man by intravenous injection of hypertonic salt solution. *American Journal of Physiology* 151, 252–257.

Honde, C. and Bueno, L. (1984) Evidence for central neuropeptidergic control of rumination in sheep. *Peptides* 5, 81–83.

Howard, B.R. (1970) The dorsal nucleus of the vagus as a centre controlling gastric motility in sheep. *Journal of Physiology London* 206, 167–180.

Iggo, A. (1951) Spontaneous and reflexly elicited contractions of the reticulum and rumen in decerebrate sheep. *Journal of Physiology London* 115, 74–75P.

Iggo, A. and Leek, B.F. (1967) An electrophysiological study of single vagal efferent units associated with gastric movements in sheep. *Journal of Physiology London* 191, 177–204.

Langhans, W., Rossi, R. and Scharrer, E. (1995) Relationships between feed and water intake in ruminants. In: Engelhardt, W. von, Leonhardt-Marek, S., Breves, G. and Giesecke, D. (eds) *Ruminant Physiology: Digestion, Metabolism, Growth and Reproduction.* Ferdinand Enke Verlag, Stuttgart, pp. 199–216.

Larsson, S. (1954) Hyperphagia from stimulation of the hypothalamus and medulla in sheep and goats. *Acta Physiologica Scandinavica* 32 (Suppl. 115), 1–40.

Leek, B.F. and Harding, R.H. (1975) Sensory nervous receptors in the ruminant stomach and the reflex control of reticulo-ruminal motility. In: McDonald, I.W. and Warner, A.C.I. (eds) *Digestion and Metabolism in the Ruminant.* The University of New England Publishing Unit, Armidale, pp. 60–76.

McKenzie, J.S. and Smith, M.H. (1973) Stereotaxic method and variability data for the brain of the merino sheep. *Journal fur Hirnforschung* 14, 355–366.

McKinley, M.J., Denton, D.A. and Weisinger, R.S. (1978) Sensors for antidiuresis and thirst-osmoreceptors or CSF sodium detectors. *Brain Research* 141, 89–103.

Silanikove, N. (1992) Effects of water scarcity and hot environment on appetite and digestion in ruminants. *Livestock Production Science* 30, 175–194.

Silanikove, N. and Tadmor, A. (1989) Rumen volume, saliva flow rate, and systemic fluid homeostasis in dehydrated cattle. *American Journal of Physiology* 256, R809-R815.

Stacy, B.D. and Warner, A.C.I. (1966) Balances of water and sodium in the rumen during feeding: osmotic stimulation of sodium absorption in the sheep. *Quarterly Journal of Experimental Physiology* 51, 79–93.

Stricker, E.M. and Hainsworth, F.R. (1970) Evaporative cooling in the rat: Effects of dehydration. *Canadian Journal of Physiology and Pharmacology* 48, 18–27.

Titchen, D.A. (1960) The production of rumen and reticulum contractions in decerebrate preparations of sheep and goats. *Journal of Physiology London* 151, 139–153.

Titchen, D.A. and Newhook, J.C. (1975) Physiological aspects of suckling and the passage of milk through the ruminant stomach. In: McDonald, I.W. and Warner, A.C.I. (eds) *Digestion and Metabolism in the Ruminant.* The University of New England Publishing Unit, Armidale, pp. 15–29.

Titchen, D.A. and Reid, C.S.W. (1965) The reflex control of the motility of the ruminant stomach. In: Dougherty, R.W. (ed.) *Physiology of Digestion in the Ruminant.* Butterworth, Washington, pp. 68–77.

Urabe, M., Tsubokawa, T. and Hamabe, N. (1968) Studies of the activities of vagal afferents in the medulla oblongata and thalamus with special reference to a central regulatory mechanism. *Physiology and Behavior* 3, 17–28.

Warner, A.C.I. and Stacy, B.D. (1977) Influence of ruminal and plasma osmotic pressure on salivary secretion in sheep. *Quarterly Journal of Experimental Physiology* 62, 133–142.

# II Rumen Microbiology and Fermentation

# 4 Molecular Ecology and Diversity in Gut Microbial Ecosystems

R.I. MACKIE[1], R.I. AMINOV[1], B.A. WHITE[1] AND C.S. MCSWEENEY[2]

[1]Department of Animal Sciences and Division of Nutritional Sciences, University of Illinois at Urbana-Champaign, Urbana, Illinois, USA; [2]CSIRO Tropical Agriculture, Long Pocket Laboratory, Indooroopilly, Queensland, Australia

## Introduction

The earliest convincing evidence for life derives from 3.8 billion-year-old rock formations which have yielded microfossils including cellular filamentous forms and sheathed colonies resembling blue–green bacteria (Schopf, 1993). Also, all basic evolution in terms of biochemistry and types of energy metabolism must have occurred earlier than 3.5 billion years ago during periods for which we have no direct evidence. Early evolution took place on earth in which most habitats (the atmosphere and oceans) were anaerobic. Thus, bacteria are nearly as old as the planet earth and events such as the oxygen forming atmosphere and even the age of dinosaurs are comparatively recent (Woese, 1994). Also, anaerobic habitats have existed continuously throughout the history of the earth, the gastrointestinal tract being a contemporary microniche (Fenchel and Finlay, 1995). The most obvious impact of fermentation in the modern world is on human and animal nutrition. The annual cellular production of prokaryotes based on population size ($4.3 \times 10^{24}$ cells) and turnover (once daily) in the gut of domestic animals is $2 \times 10^{27}$, which is 2% of that in soil and only 0.1% of that in the oceans (Whitman *et al.*, 1998).

All animals, including humans, are adapted to life in a microbial world. The complexity of animal–microbe relationships varies tremendously, ranging from competition to cooperation (Hungate, 1976, 1984). The animal alimentary tract has evolved as an adaptation enabling the animal to secure food and limit consumption by other animals. This allows the retention and digestion of ingested food, followed by absorption and metabolism of digestion products, whilst feeding and other activities continue. Since microorganisms grow rapidly under favourable conditions in the gut they could become serious competitors for the animal's food. This microbial challenge has modified the course of evolution in animals, resulting in selection for varied animal–microbe relationships. The evolutionary strategy in the first case has been to compete with the resident microbes and in the second to cooperate with them. The third case incorporates a combination of the first two avoiding some of the disadvantages of the cooperation while still obtaining some benefit from fermentative digestion. These models of existing animal–microbe relationship are useful when trying to determine the evolution of mutualistic fermentative digestion in the gastrointestinal tract.

## The first mammalian herbivores

During the late Cretaceous and early Paleocene periods plant-eating mammals were frugivores, presumably because fruit can be more easily processed than foliage. Mammals did not invade the herbivorous niche until the middle Paleocene (Collinson and Hooker, 1991). Evolution of large size was a prerequisite for the exploitation of leaves because of the need for a longer residence time in the gut for bacterial fermentation to obtain sufficient nutrients from foliage and herbage. In the late Cretaceous, dinosaurs occupied the herbivorous niche although grazers were still absent (Farlow, 1987). The appearance of grazers in the Miocene is coincident with a similar radiation of grassland-forming grasses (Thomasson and Voorhies, 1990). Thus herbivore browsers first appear in the middle Paleocene but they did not become significant until the late Eocene (Collinson and Hooker, 1991). Frugivory declined first with the appearance of herbivore browsing followed by an increase in grazers in post-Miocene at the expense of herbivore browsers. The earliest herbivores were ground dwelling (LGMs, large ground mammals) and achieved their dietary specializations largely through evolution from already large, ground-dwelling frugivores or, in the Paleocene, by a size increase from small insectivorous ancestors (Collinson and Hooker, 1991). Large size limited them to the ground. Most browsing herbivores in other locomotor niches (SGMs, small ground mammals) changed their diet from frugivory without changing their locomotor adaptation. A period of nearly 30 million years existed in the vertebrate exploitation of leaves after dinosaur extinction and before the first few mammalian herbivores in the middle Paleocene. This was followed by expansion of herbivores in the late Eocene when climates cooled and more open vegetation became established.

Hume and Warner (1980) published an excellent discussion on the evolution of microbial digestion in mammals. Since the fossil record provides no information on the morphology, physiology, biochemistry or microbiology of the gut, much of the knowledge must be deduced from what is known about present-day animals coupled with the fossil record of animals and their probable feedstuffs and is therefore highly speculative. Microbial digestion surely arose long before mammals evolved. Large luminal populations of microbes develop in regions of the gut with relative stasis where retention time of digesta allows adequate microbial growth. In most deliberations only nutritional contributions to the host animal are considered. These are based on digestion of the plant cell wall by cellulases and hemicellulases provided by the microbial partner, the synthesis of microbial protein from poor-quality dietary proteins and non-protein nitrogen mainly via ammonia as precursor, and the synthesis of B-vitamins and vitamin K. However, little consideration has been given to the protection provided by foregut fermentation resulting in transformation or modification of phytotoxins and mycotoxins in the diet (Guthrie, 1984; Mackie, 1987; McSweeney and Mackie, 1997). Other contributions not normally considered in these deliberations are immunological, physiological and protective (Gaskins, 1997; Rolfe, 1997; Tannock, 1997). These arguments also support the theory that the development of foregut fermentation must have come after an initial development of the hindgut and that all foregut fermenters should have some fermentation in the hindgut (Hume and Warner, 1980; Langer, 1991).

The above mentioned adaptations, modifications and specializations in anatomy and physiology of animals for herbivory would be ineffective in the absence of endosymbiotic microbes to digest the plant cell wall. This raises the question of the actual mode of acquisition of these endosymbionts by their hosts. Since it is unlikely that the fossil record will ever provide direct clues for resolution of this issue, Hotton *et al.* (1996) have hypothesized that microbes were picked up by detritivorous animals foraging in plant litter and those that could survive in the gut environment have assumed a role in the digestive processes of the host. However, it is equally plausible that ingested insects, especially herbivores that harbour such bacteria in their own guts, provided the original source for fermentative endosymbionts (Sues and Reisz, 1998). This concept is compatible with contemporary studies on acquisition of gut microbiota by neonates which document the importance of diet and environment in the early development of microbial populations (Mackie *et al.*, 1999). It is also consistent with development of hindgut fermentation and omnivory that included consumption of plant material prior to development of foregut fermentation. This consumption of insect herbivores by mammalian hosts may be involved in coevolution, the continual process of evolutionary change in the synthesis of secondary compounds by plants followed by comparable detoxification mechanisms in animals that consume them (McSweeney and Mackie, 1997).

The opportunity to examine the intestinal contents of an extinct mammal is rare. Recent excavations at two separate sites in the Great Lakes region of North America revealed assemblages of plant material preserved in late-Pleistocene pond sediments associated with skeletal remains of American mastodons (*Mammut americanum*).The plant material from these assemblages varied in degree of comminution but differed in texture and colour from surrounding sediment and resembled gut content from modern herbivores. The shape and size of the mass of plant material was consistent with the intestinal dimensions of extant elephants (Lepper *et al.*, 1991). The plant material was dated to 11,500 years before present and the masses are thought to be remnants of the small and large intestines. General and selective media were used to cultivate and identify bacteria from the intestinal contents, bone-associated sediments and sediments located some distance from the remains (Rhodes *et al.*, 1998). In all, 295 isolates were cultivated and 38 individual taxa identified. Sequencing of 16S rDNA was used to confirm the taxonomic positions of selected enteric and obligately anaerobic bacteria. Members of the family Enterobacteriaceae represented 41% of all isolates (*Enterobacter cloacae* was the most commonly identified isolate) in the intestinal masses. However, no *Bacteroides* spp. or expected intestinal anaerobes were recovered. In fact, the only obligate anaerobes recovered were clostridia and these were not from the intestinal masses. It was concluded that microbiological evidence supported other macrobotanical data indicating the intestinal origin of these plant masses but it was not possible to establish whether these organisms are direct descendents of the original intestinal microbiota (Rhodes *et al.*, 1998). Although interesting, the results of this study are ambiguous and a more direct method of studying ancient bacteria is required. It is apparent, using rigorous DNA technology protocols, that a picture of the composition of the original microbiota can be obtained at least under special circumstances which allow differentiation of ancient and modern DNA and that mammoths and similarly well-preserved bodies are attractive candidates for future palaeomicrobiological investigations (Rollo and Marota, 1999).

## Molecular ecology and diversity

The microbial community inhabiting the gastrointestinal tract is characterized by its high population density, wide diversity and complexity of interactions. All major groups of microbes are represented in the gut and it contains representatives of the three domains, *Bacteria*, *Archaea* and *Eucarya*, articulated by Woese *et al.* (1990). The rumen, the most extensively studied gut ecosystem, contains large numbers of bacteria (up to $10^{11}$ viable cells $ml^{-1}$ comprising 200 species), ciliate protozoa ($10^4$–$10^6$ $ml^{-1}$ spread over 25 genera), anaerobic rumen fungi (zoospore population densities of $10^3$–$10^5$ $ml^{-1}$ divided into five genera) and bacteriophages ($10^7$–$10^9$ particles $ml^{-1}$) (Klieve and Swain, 1993; Hespell *et al.*, 1997; Orpin and Joblin, 1997; Stewart *et al.*, 1997; Williams and Coleman, 1997). Despite this vast amount of knowledge, the basic prerequisites for ecological studies, namely enumeration and identification of community members, have tremendous limitations. The two major problems faced by microbial ecologists studying the gastrointestinal ecosystem are the inevitable bias introduced by culture-based enumeration and characterization techniques and the lack of a phylogenetically-based classification scheme (Ward, 1989; Amann *et al.*, 1990, 1994; Stahl and Amann, 1991; Stahl,1997).

Modern molecular ecology techniques based on sequence comparisons of nucleic acids (DNA or RNA) can be used to provide molecular characterization while at the same time providing a classification scheme which predicts natural evolutionary relationships. In principle, nucleic acid probes can be designed to hybridize with a complementary target sequence and thus provide a complete description independent of the growth conditions and media used (Ward *et al.*, 1992; Amann, 1995; Amann *et al.*, 1995; Raskin *et al.*, 1997). An example of the power of these modern molecular ecology techniques is provided by the analysis of 16S rRNA sequences (average length 1500 nucleotides). The highly conserved regions of the rRNA molecule can serve as primer binding sites for *in vitro* amplification by PCR (Ludwig *et al.*, 1994). The more conserved regions are also useful, serving as targets for universal probes that react with all living organisms or for discriminating between broad phylogenetic groups such as the domains *Archaea*, *Bacteria* and *Eucarya*. The more variable sequence regions are more appropriate for genus, species and sometimes even strain-specific hybridization probes (Stahl and Amann, 1991; Odenyo *et al.*, 1994a; Lin and Stahl, 1995). The use of molecular ecology techniques based on nucleic acid probes is likely to revolutionize our approach to microbial ecology in the gastrointestinal tract and will provide, not simply a refinement or increased understanding but, a complete description of the gastrointestinal ecosystem for the first time (Mackie and Doré, 1997; Raskin *et al.*, 1997).

In molecular ecology it is important to distinguish between identification, quantification and monitoring of activity or function. This information is highly dependent on the type of target nucleic acid employed and the conceptual and technical basis used for nucleic acid probe design. Generally these can be divided into DNA-based methods employing empirically characterized probes and rRNA-based methods based on comparative sequence analysis for design and interpretation of 'rational' probes (Pace *et al.*, 1986; Stahl and Amann, 1991; Ward *et al.*, 1992; Stahl, 1993b; Raskin *et al.*, 1997).

### DNA-based studies of gut ecology

The earliest nucleic-acid-based analyses of ruminal ecology were published more than a decade ago in order to study survival of bacteria reintroduced into the rumen. Attwood *et al.* (1988) reported on the use of a 1.9 kb randomly cloned, labelled fragment of genomic DNA from the laboratory strain *Prevotella* (*Bacteroides*) *bryantii* to determine survival time of this organism when introduced into the rumen. Results showed that strain $B_14$ had a half-life of 9 h for *in vitro* mixed cultures but only 30 min following inoculation into the rumen. In fact, the organism dropped below the detection limit ($2 \times 10^7$ cells $ml^{-1}$ rumen fluid) within 3 h, most likely due to bacteriocin-like activity in fresh ruminal fluid. A number of studies have used randomly cloned DNA fragments targeting either plasmid or chromosomal DNA as hybridization probes. *Streptococcus bovis*, a normal inhabitant of the rumen, has been implicated in human colonic carcinoma. Whitehead and Cotta (1993) tested a cloned amylase gene as a DNA probe for rapid and accurate identification of *S. bovis* derived from human and bovine sources.

In general, the DNA-based methods for studying gastrointestinal molecular ecology can have high specificity and sensitivity. However, it is necessary to identify those methods that require pure culture isolates (restriction fragment length polymorphism and ribotyping) and therefore have the disadvantage of cultural bias. In addition, DNA probes have also been used to investigate species diversity in the gastrointestinal tract including studies on *Fibrobacter* (Flint *et al.*, 1990), *Selenomonas ruminantium* (Ning *et al.*, 1991), and *Butyrivibrio fibrisolvens* (Mannarelli *et al.*, 1990). Furthermore, these techniques are less suitable for detailed characterization of community structure of gastrointestinal ecosystems compared to rRNA techniques which employ rational probes designed within the framework of comparative sequence analysis and are therefore highly specific and span a greater taxonomic diversity.

### RNA-based studies of gut ecology

The principles and practice involved in 16S rRNA-based methods have been extensively reviewed (Pace *et al.*, 1986; Sayler and Layton, 1990; Ward *et al.*, 1992; Stahl, 1993a,b; Raskin *et al.*, 1997). The use of small subunit rRNA methods, referred to as 16S rRNA-based methods, in gastrointestinal microecology has become well established over the past 10 years. These methods are also applicable to 23S rRNA which will probably be used more extensively in the future since it contains more information (3000 bp compared with 1500 bp for 16S rRNA). At present their use is limited because of the less extensive sequence collection available. Extensive comparative sequence analysis of 16S rRNA molecules representing a wide diversity of organisms shows that different regions of the molecule vary in sequence conservation.

Oligonucleotides complementing regions of universally conversed 16S rRNA sequence are used as universal probes while those complementing more variable regions of sequence are useful as selective probes targeting species, genus, or phylogenetic groups.

The use of 16S rRNA-based methods for ecological studies in the gastrointestinal tract was first demonstrated by Stahl *et al.* (1988). Species-specific 16S rRNA-targeted oligonucleotide probes were developed to enumerate strains of *Fibrobacter* (*Bacteroides*) *succinogenes* and *Lachnospira multiparus* in the bovine rumen to monitor shifts in population abundance or changes in population activity in complex microbial communities. Subsequently, *F. succinogenes* was shown to be a genetically diverse yet phylogenetically coherent genus (Amann *et al.*, 1992; Ogata *et al.*, 1997). Culture-based enumeration of *F. succinogenes* was largely unsuccessful in the same study. These techniques formed the basis for further studies of bacterial populations in gut environments. Group-specific 16S rRNA-targeted oligonucleotide probes specific for bacterial, eukaryotic, and archaeal rRNA were used to quantitate these three groups in the gastrointestinal tract of various domestic animals (bovine, ovine, caprine and porcine). Bacterial, eukaryotic and archaeal populations varied from approximately 60 to 90%, 3 to 30% and 0.5 to 3%, respectively, in the gastrointestinal tract of most of the animals examined (Lin *et al.*, 1997). By using different probes for classes of methanogenic bacteria, it was also shown that the predominant methanogens differed in different animal hosts (Lin *et al.*, 1997). Moreover, these studies revealed a dominant archaeal group that was unaccounted for by the current methanogen-specific probes, suggesting another assemblage of rumen methanogens that have yet to be characterized (Lin *et al.*, 1997). Krause and Russell (1996) developed probes for obligate amino-acid-fermenting ruminal bacteria in order to examine the effect of monesin addition on the levels of these bacteria and their role in amino acid deamination. Probes were developed for the other major ruminal cellulolytic bacteria *Ruminococcus albus* and *Ruminococcus flavefaciens* and used to study the dynamics of bacterial interactions during fermentation of cellobiose, cellulose and alkaline hydrogen peroxide treated wheat straw (Odenyo *et al.*, 1994a,b). The results showed that 16S rRNA-targeted oligonucleotide hybridization probes were effective in quantifying specific bacteria in defined mixed cultures and provided useful information on bacterial competition during growth on insoluble substrates. Furthermore, this study revealed for the first time the production of a bacteriocin-like substance by a ruminal bacterium, a mechanism which may be used to compete for nutrients. These probes have also been used to study the competition for cellobiose or cellulose by *R. albus*, *R. flavefaciens* and *F. succinogenes* in chemostats under substrate-excess and substrate-limited conditions (Shi and Weimer, 1997; Shi *et al.*, 1997). The ruminal bacterium *Synergistes jonesii* has attracted considerable interest due to its limited geographical distribution and its ability to degrade 3-hydroxy-4-1(H)-pyridone, the toxic principle of the tropical leguminous shrub *Leucaena leucocephala* (Allison *et al.*, 1990, 1992). The 16S rRNA sequence of *S. jonesii* was not closely related to any among those bacteria so far characterized and was an ideal candidate to evaluate the use of oligonucleotide probes for tracking bacteria in the ruminal ecosystem. Radiolabelled and fluorescent-dye-conjugated probes were developed for quantitation of *S. jonesii* in a mixed-culture chemostat (McSweeney *et al.*, 1993). Further ecological studies on transmission, colonization, persistence and population studies *in vivo* are still required.

Surprisingly, a limited number of studies have used rRNA-based methods to study ruminal protozoal and fungal populations. Finlay *et al.* (1994) used a fluorescently labelled oligonucleotide probe targeting the 16S rRNA of the Archaea to demonstrate that *Entodinium* species and *Dasytricha ruminantium* contained methanogenic endosymbionts outside digestive vacuoles. Exosymbiotic methanogens had been well documented previously, based on characteristic autofluorescence of these bacteria (Vogels *et al.*, 1980; Stumm *et al.*, 1982; Krumholz *et al.*, 1983). Recent research using a small subunit (SSU) rRNA probe approach targeting ruminal methanogens revealed a taxon-specific association between protozoal and methanogen populations both in the rumen and a continuous culture fermentor system (Sharp *et al.*, 1998). Methanobacteriaceae were the most abundant population in the rumen comprising 89.3% of total Archaea and 99.2% in the protozoal fraction. This value decreased to 54% of Archaeal signal after 48 h of fermentor operation and was correlated with the loss of protozoa from the system. In contrast, the Methanomicrobiales, the most abundant Archaeal population in this study, accounted for 12.1% of Archaeal signal in rumen fluid and was not detected in the protozoal fraction suggesting a free-living lifestyle. This group increased to 26.3% of Archaeal signal in fermentor content without protozoa. These studies suggest the importance, and perhaps specificity, of Methanobacteriaceae as symbionts of rumen protozoa.

Initially, sequence analysis of the 18S rRNA genes from *Polyplastron multivesiculatum* (originally deposited as *Entodinium simplex*) and *D. ruminantium* was used to phylogenetically position these protozoa within the hydrogenosome containing protozoa (Embley *et al.*, 1995). More recently, seven other rumen ciliate protozoa, as well as two additional isolates of *P. multivesiculatum* and *D. ruminantium*, have been analysed and these studies indicate that the rumen ciliates are monophyletic and fall into three distinct groups within the Class Litostomatea (Wright and Lynn, 1997a,b; Wright *et al.*, 1997). Signature probes for *Entodinium caudatum*, *Epidinium caudatum* and *P. multivesiculatum* have been developed and are currently being analysed for use in the study of rumen microbial ecology (Wright *et al.*, 1997). Also, comparative sequence analysis of the 18S rRNA genes was used to phylogenetically position *Neocallimastix* in the Chytridiomycete class of fungi (Doré and Stahl, 1991). We have recently determined the 18S rRNA sequences for four additional rumen fungi and these data support the previous comparative analysis and suggest that the anaerobic rumen fungi are monophyletic (Thill *et al.*, 1997).

## Recent advances and new approaches

### *Sequence-based analysis of rumen microbial diversity*

Development of robust and simple DNA cloning techniques and the PCR have allowed higher resolution analyses of complex communities using SSU rRNA comparative sequence analysis (Hugenholtz and Pace, 1996). This approach has been strongly advocated by Pace (1997). This technique has not been widely applied to the gut ecosystem but studies of the termite gut (Okhuma and Kudo, 1996), the human colon (Wilson and Blitchington, 1996; Doré *et al.*, 1998) and the rumen (Whitford *et al.*, 1998; Tajima *et al.*, 1999) have been published. Results of these studies demonstrate the high degree of genetic diversity in gut ecosystems. Whitford *et al.* (1998) prepared two libraries from strained rumen fluid obtained from dairy cows fed a total mixed ration containing 26% lucerne, 30% maize silage and 35% concentrate. Sequences which clustered with *Prevotella ruminicola* represented the majority of clones (101 of 133 total sequences) isolated in each of the PCR sets. However, many members of this cluster represent phylogenetically distinct groups and are at least different species. Relatively few 16S rRNA sequences similar to the commonly isolated *B. fibrisolvens* were found in this study. Importantly, the majority of rDNA sequences analysed in this study represented novel bacterial diversity which has not yet been cultivated or isolated. A recent study (Tajima *et al.*, 1999) described bacterial diversity by direct retrieval of 16S rDNA sequences in a culture-independent manner. Three SSU rDNA libraries were constructed; the first from total DNA extracted from strained rumen fluid of representative samples obtained before feeding from cows fed a mixed ration (lucerne–timothy hay and maize+barley based concentrate in a 4:1 ratio), the second from the remaining feed particles from the first pooled sample, and the third from strained rumen fluid of cows fed a high roughage (lucerne–timothy hay) diet. The three libraries containing almost full length SSU rDNA sequences (about 1.5 kb long) were completely sequenced and analysed (a total of 161 clones). Only 10 sequences (6.2%) could be identified, six as *B. fibrisolvens,* two as *P. ruminicola,* one as *S. ruminantium* and one as *Succiniclasticum ruminis.* For 34.6% of the sequences the similarity with database sequences ranged from 90 to 98% while for the remaining 59.2% the similarity was less than 90%. Because of the large bacterial diversity, few operational taxonomic units (OTUs) represented a large percentage of the clones. Phylogenetic placement of sequences from the mixed ration/rumen fluid library showed the following affiliations: low G+C Gram-positive bacteria (52.4%), *Cytophaga–Flexibacter–Bacteroides* (38.1%), *Proteobacteria* (4.7%) and *Spirochaetes* (2.4%). Values for the analysis of sequences from the solids remaining (library two) were low G+C Gram-positive bacteria (71.4%), *Cytophaga–Flexibacter–Bacteroides* (26.2%) and *Spirochaetes* (2.4%). Corresponding sequences from the third library (high roughage rumen fluid) were low G+C Gram-positive bacteria (44.2%), *Cytophaga–Flexibacter–Bacteroides* (50.6%) and *Spirochaetes* (3.9%). In addition, 418 randomly isolated colonies from a number of non-selective media were characterized and SSU rDNA sequences obtained. Results showed that 59.6% of strains were identified as *P. ruminicola*, 10.8% as *Prevotella* spp. and 0.8% as *Bacteroides*. The remaining 27.0 and 1.8% of isolates were affiliated with low G+C and high G+C Gram-positive bacteria, respectively. In contrast to the library-based analysis, the cultivation-based phylogenetic approach revealed close clustering with strains that have already been isolated, characterized and sequenced. Similar results for the human colonic ecosystem in terms of great diversity, large proportion of OTUs represented by single clones, and a large proportion of sequences distantly related to deposited sequences and so far uncultivated (Wilson and Blitchington, 1996; Doré *et al.*, 1998).

These approaches all provide useful and novel information but also have limitations which need to be recognized and resolved. The limitations relate to the extraction of nucleic acids from environmental samples, biases, artifacts associated with enzymatic amplification of nucleic acids and cloning of PCR products, and sensitivity and target site accessibility in whole-cell hybridization techniques. These biases and limitations have been well documented and reviewed (Wintzingerode *et al.*, 1997; Head *et al.*, 1998; Muyzer and Smalla, 1998). However, of importance is interpretation of information derived from molecular ecology studies (Stackebrandt, 1997). For example the 16S rRNA sequence variations due to interspecific and intraspecific rRNA operon heterogeneity make analysis of clone libraries or gel electrophoresis patterns derived from environmental samples difficult to interpret (Hunter-Cevera, 1998). In addition, it is hard to draw conclusions about physiological and biochemical properties and the ecological role of unknown (and known) microbes. However, considerable progress in addressing and resolving these limitations is being made.

### Application of denaturing gradient gel electrophoresis (DGGE) and temperature gradient gel electrophoresis (TGGE) in gut microbial ecology

In order to study population structure and dynamics, genetic fingerprinting techniques that provide a pattern or profile of genetic diversity are needed. Recently DGGE of PCR-amplified rDNA fragments has attracted the attention of microbiologists and has been applied in a variety of environmental studies for the analysis of microbial communities (Muyzer *et al.*, 1993). In DGGE, as well as in TGGE, DNA fragments of the same length but with different sequences can be separated. Separation is based on the decreased electrophoretic mobility of a partially melted double-stranded DNA molecule in polyacrylamide gels containing a linear gradient of denaturant (a mixture of urea and formamide) or a linear temperature gradient. Sequence variation within such domains causes the melting temperatures to differ and molecules with different sequences will stop migrating at different positions in the gel. DNA bands in DGGE and TGGE profiles can be visualized using ethidium bromide, SYBR Green I or silver staining which is more sensitive but also stains single-stranded DNA. This procedure allows direct identification of the presence and relative abundance of different species and provides a semi-quantitative estimation of the genetic diversity of microbial populations. DGGE and TGGE of PCR-amplified DNA fragments has been used widely in environmental microbiology to study community complexity, monitor population shifts, analyse enrichment cultures and isolation of bacteria, detect sequence heterogeneity of 16S rRNA genes, compare DNA extraction methods, screen clone libraries and determine PCR and cloning biases (reviewed by Muyzer *et al.*, 1998; Muyzer and Smalla, 1998). This technique has had surprisingly limited application to the gut ecosystem most probably because of the complexity of banding patterns which makes it difficult to analyse.

We have applied this procedure to pure cultures of ruminal fibrolytic bacteria *F. succinogenes*, *R. albus* and *R. flavefaciens* as well as other members of the genus *Ruminococcus* which may not be true members of this genus (*R. bromii*, *R. gnavus*, *R. lactaris*, *R. obeum* and *R. torques*). DGGE profiling of the nine *R. flavefaciens* strains showed that the PCR-amplified V3 region of the 16S rRNA gene from these strains all migrated to the same location in the DGGE gel. These results suggest a lack of sequence diversity in this region of the 16S rRNA gene from *R. flavefaciens*. In contrast, DGGE profiling of the nine *R. albus* strains suggested a high degree of sequence diversity in the PCR-amplified V3 region of the 16S rRNA gene from these strains. These results suggest that there is significant sequence diversity and phylogenetic heterogeneity within the V3 region of the 16S rRNA gene from *R. albus*, and that the V3 region of the 16S rRNA gene from *R. albus* SY3 has a much lower G+C content than other representatives of this species (Cann *et al.*, 1996). All of the *Ruminococcus* DGGE profiles differed from the profile for *F. succinogenes*, and mixtures of PCR-amplified V3 regions from the strains of *F. succinogenes*, *R. albus*, and *R. flavefaciens* were easily resolved in DGGE gel. Further studies with different dilutions of input DNAs from these strains showed that the DGGE technique is quantitative. These results show that DGGE can be used to differentiate between closely related bacterial strains. Thus this sensitive technique is highly suitable to the analysis of microbial diversity and population dynamics for the major fibrolytic bacteria from the rumen.

We have also applied this procedure to rumen samples from steers fed different diets in a preliminary study to determine the utility of these techniques for the analysis of a complex microbial community (Kocherginskaya *et al.*, 1997). Rumen samples were collected from four steers fed a medium-quality grass–legume hay at maintenance intake, and four steers fed a diet of 20% hay, 52% maize, 5% corn steep liquor, 3% minerals and 20% of maize byproducts. Rumen samples were harvested approximately one hour prior to feeding, passed through cheesecloth, and centrifuged. Total genomic DNA was isolated from cell pellets and used for amplification of either the V3 or V9 region of the 16S rDNA gene. When the different PCR profiles obtained from amplification of the V3 region of the 16S rDNA were compared for samples from those animals fed the medium-quality grass–legume hay diet, the patterns were remarkably similar. None the less, the DGGE profiles demonstrated at least 16 distinguishable bands, with five of them being more predominant than the others. Banding profiles obtained from rumen samples of animals fed the maize-based diet were different from those obtained from the medium-quality grass–legume hay diet. Profiles from each of the four maize-based-diet-fed animals also differed from each other. These results demonstrate the utility of this technique in describing the genetic diversity and population structure of the rumen community, both *in vitro* and *in vivo*.

Bacterial diversity in human faeces was analysed using PCR-amplification of the V6–V8 regions of 16S rDNA by TGGE (Zoetendahl *et al.*, 1998). Faecal samples from two individuals showed remarkably stable profiles over a period of at least 6 months and were unique for each individual. TGGE profiles derived from 16S rRNA (by reverse transcriptase PCR) and rDNA amplicons showed similar banding patterns although the intensities of bands with similar mobilities differed in some cases indicating a different contribution to the total active fraction of the predominant faecal bacteria. These results confirm that TGGE analysis of 16S rDNA amplicons, combined with cloning and sequencing of these amplicons, is a reliable approach to relative levels and abundance in complex microbial communities such as faecal bacteria. Bacterial genetic diversity in pig faecal samples was determined using DGGE (Simpson *et al.*, 1999). Optimization of the protocol resulted in a doubling of product bands visualized in the gels. Unique and stable banding patterns were generated from faecal samples of pigs on different diets and of different ages as well as from lumenal and mucosal samples obtained from each gut segment between the stomach and colon. Analysis of the DGGE banding profiles using PHYLIP showed that the patterns grouped according to gut location and that lumenal and mucosal samples from each compartment had the highest similarity to each other (Simpson *et al.*, 1999). DGGE can be applied effectively to monitor changes in bacterial populations and for evaluation of bacterial diversity. However, many of the specific gut populations are minor constituents and there are detection limits for populations comprising less than 1% of total template DNA and PCR amplification with semi-conserved primer sets (Muyzer and Smalla, 1998).

### Whole-cell hybridization

Whole-cell *in situ* hybridization with fluorescently-labelled oligonucleotide probes for studies in microbial ecology was first developed 10 years ago (DeLong *et al.*, 1989; Amann *et al.*, 1990). This technique was used successfully to analyse a wide variety of ecosystems (Amann *et al.*, 1995, 1996). However, applications to analysis of the rumen ecosystem are limited. In short, the procedure involves cell fixation to permeabilize cells while maintaining their morphological integrity. Hybridization with a fluorescent probe to bind with complementary rRNA sequence is carried out on bacterial suspensions or after attachment to coated microscope slides. Following hybridization, the sample is washed to remove unbound probe and the sample viewed by epifluorescence microscopy (Amann *et al.*, 1995). Recent developments that have improved our ability to address structure and function of microbial communities *in situ* have been reviewed (Amann and Kuhl, 1998).

Localization of microorganisms on mucosal or cell surfaces is usually performed by classical histological and immunohistological techniques, which restrict identity of the bacteria involved to morphological features. Even when specific antibodies are available for *in situ* studies, the thick mucous layer above epithelial cells can block penetration of antibodies and extensive washing can remove the mucus layer displacing organisms. Poulsen *et al.* (1994) applied *in situ* 16S rRNA hybridization to investigate the microbiota of the large intestine of streptomycin-treated mice and were able to determine the spatial distribution of *Escherichia coli* in thin sections of intestinal tissue. This allows rapid detection of bacteria which may be difficult to cultivate and their relationship to other cells either host or bacterial. *In situ* growth rates determined by single-cell analysis of intracellular concentrations of DNA and RNA revealed that adherent and mucosal bacteria were growing with generation times of 30–80 min while those in the lumen were static (Poulsen *et al.*, 1995).

An advantage of the fluorescent *in situ* hybridization approach is the ability to both identify and enumerate single cells within a complex ecosystem with specific 16S rRNA-based oligonucleotide probes. Thus, six 16S rRNA-targeted oligonucleotide probes were designed, validated and used to quantify predominant groups of anaerobic bacteria in human faecal samples (Franks *et al.*, 1998). The combination of the two *Bacteroides* probes for the *B. fragilis* and *B. distasonis* groups detected a mean of $5.4 \times 10^{10}$ cells $g^{-1}$ (dry weight) of faeces. The *Clostridium coccoides–Eubacterium rectale* group-specific probe detected a mean of $7.2 \times 10^{10}$ cells $g^{-1}$ (dry weight) of faeces. *Streptococcus–Lactobacillus* group-specific probes detected cells ranging in number from $1.7 \times 10^7$ to $7 \times 10^8$. The future aim of this work is to have a set of about ten probes that can detect more than 90% of the colonic biota in large phylogenetic groups. This approach which combines the power of molecular techniques with modern image analysis at the single cell level will provide much insight into structure–function relationships within gastrointestinal microbial ecosystems.

## Conclusions

The use of molecular ecology techniques based on nucleic acid probes is likely to revolutionize our approach to microbial ecology in the gastrointestinal tract and will provide, not simply a refinement or increased understanding but a complete description of gastrointestinal community for the first time. Modern molecular ecology techniques based on sequence comparisons of nucleic acids (DNA or RNA) can be used to provide molecular characterization while at the same time providing a classification scheme which predicts natural evolutionary relationships. In principle, nucleic acid probes can be designed to hybridize with a complementary target sequence and thus provide a complete description independent of the growth conditions and media used. An example of the power of these modern molecular ecology techniques is provided by the analysis of SSU rRNA sequences. The highly conserved regions of the SSU rRNA molecule can serve as primer binding sites for *in vitro* amplification by PCR. The more conserved regions are also useful, serving as targets for universal probes that react with all living organisms or for discriminating between broad phylogenetic groups such as the domains *Archaea*, *Bacteria* and *Eucarya*. The more variable sequence regions are more appropriate for genus, species and sometimes even strain-specific hybridization probes. Thus nucleic acid probes serve to evaluate the presence of specific sequences in the environment and provide a link between knowledge obtained from pure cultures and the microbial populations they represent in the gastrointestinal tract. Furthermore, whole-cell hybridization using *in situ* PCR is a powerful technique which can be used to describe an organism's expression of key enzymes. Thus development of these procedures and techniques will result in greater insights into community structure and activity of gut microbial communities in relation to functional interactions between different bacteria, spatial and temporal relationships between different microorganisms and between microorganisms and feed particles. The successful development and application of these methods promises to provide the first opportunity to link distribution and identity of gastrointestinal microbes in their natural environment with their genetic potential and *in situ* activities.

## References


Allison, M.J., Hammond, A.C. and Jones, R.J. (1990) Detection of ruminal bacteria that degrade toxic dihydroxypyridine compounds produced from mimosine. *Applied and Environmental Microbiology* 56, 590–595.

Allison, M.J., Mayberry, W.R., McSweeney, C.S. and Stahl, D.A. (1992) *Synergistes jonesii*, gen. nov., sp. nov.: a rumen bacterium that degrades toxic pyridinediols. *Systematic and Applied Microbiology* 15, 522–529.

Amann, R.I. (1995) Fluorescently labelled, rRNA-targeted probes in the study of molecular ecology. *Molecular Ecology* 4, 543–554.

Amann, R.I. and Kuhl, M. (1998) *In situ* methods for the assessment of microorganisms and their activities. *Current Opinion in Microbiology* 1, 352–358.

Amann, R.I., Krumholz, L. and Stahl, D.A. (1990) Fluorescent-oligonucleotide probing of whole cells for determinative, phylogenetic, and environmental studies in microbiology. *Journal of Bacteriology* 172, 762–770.

Amann, R.I., Lin, C., Key, R., Montgomery, L. and Stahl, D.A. (1992) Diversity among *Fibrobacter* isolates: towards a phylogenetic classification. *Systematic and Applied Microbiology* 15, 23–31.

Amann, R.I., Ludwig, W. and Schleifer, K.H. (1994) Identification of uncultured bacteria: a challenging task for molecular taxonomists. *ASM News* 60, 360–365.

Amann, R.I., Ludwig, W. and Schleifer, K.H. (1995) Phylogenetic identification and *in situ* detection of individual microbial cells without cultivation. *Microbiological Reviews* 59, 143–169.

Amann, R.I., Snaidr, J., Wagner, M., Ludwig, W. and Schleifer, K.H. (1996) *In situ* visualization of high genetic diversity in a natural microbial community. *Journal of Bacteriology* 178, 3496–3500.

Attwood, G., Lockington, R.A., Xue, G.P. and Brooker, J.D. (1988) Use of a unique gene sequence as a probe to enumerate a strain of *Bacteroides ruminicola* introduced into the rumen. *Applied and Environmental Microbiology* 54, 534–539.

Cann, I.K.O., Kocherginskaya, S.A. and White, B.A. (1996) Denaturing gradient gel electrophoresis analysis of polymerase chain-reaction amplified genes coding for 16S rRNAs from ruminal fibrolytic bacteria. *Proceedings of Japanese Society of Rumen Metabolism and Physiology* 7, 10.

Collinson, M.E. and Hooker, J.J. (1991) Fossil evidence of interactions between plants and plant-eating mammals. *Philosophical Transactions of the Royal Society*, London, Series B 333, 197–200.

DeLong, E.F., Wickham, G.S. and Pace, N.R. (1989) Phylogenetic stains: ribosomal RNA-based probes for the identification of microbial cells. *Science* 243, 1360–1363.

Doré, J. and Stahl, D.A. (1991) Phylogeny of anaerobic rumen Chytridiomycete inferred from small subunit ribosomal RNA sequence comparisons. *Canadian Journal of Botany* 69, 1964–1971.

Doré, J., Gramet, G., Goderel, I. and Pochart, P. (1998) Culture independent characterization of human faecal flora using rRNA-targeted hybridization probes. *Genetic Selection and Evolution* 30(suppl.1), 287–296.

Embley, T.M., Finlay, B.J., Dyal, P.L., Hirt, R.P., Wilkinson, W. and Williams, A.G. (1995) Multiple origins of anaerobic ciliates with hydrogenosomes within the radiation of aerobic ciliates. *Proceedings of the Royal Society of London* 262, 87–93.

Farlow, J.O. (1987) Speculations about the diet and digestive physiology of herbivorous dinosaurs. *Paleobiology* 13, 60–72.

Fenchel, T. and Finlay, B.J. (1995) *Ecology and Evolution in Anoxic Worlds*. Oxford University Press, New York.

Flint, H.J., McPherson, C.A., Argustin, G. and Stewart, C.S. (1990) Use of a cellulose-encoding gene probe to reveal restriction fragment length polymorphisms among ruminal strains of *Bacteroides succinogenes*. *Current Microbiology* 20, 63–67.

Finlay, B.J., Esteban, G., Clarke, K.J., Williams, A.G., Embly, T.M. and Hirt, R.P. (1994) Some rumen ciliates have endosymbiotic methanogens. *FEMS Microbiology Letters* 117,157–162.

Franks, A.H., Harmsen, H.J.M., Raangs, G.C., Jansen, G.J., Schut, F. and Welling, G.W. (1998) Variations of bacterial populations in human feces measured by fluorescent *in situ* hybridization with group specific 16S rRNA-targeted oligonucleotide probes. *Applied and Environmental Microbiology* 64, 3336–3345.

Gaskins, H.R. (1997) Immunological aspects of host/microbiota interactions at the intestinal epithelium. In: Mackie, R.I., White, B.A. and Isaacson, R.E. (eds) *Gastrointestinal Microbiology*, Volume 2. Chapman and Hall, New York, pp. 537–587.

Guthrie, R.D. (1984) Mosaics, allelochemics, and nutrients: an ecological theory of late Pleistocene megafaunal extinctions. In: Martin, D.S. and Klein, R.G. (eds) *Quartenary Extinctions: a Prehistoric Revolution*. University Arizona Press, Tucson, pp. 259–298.

Head, I.M., Saunders, J.R. and Pickup, R.W. (1998) Microbial evolution, diversity, and ecology: a decade of ribosomal RNA analysis of uncultivated microorganisms. *Microbial Ecology* 35, 1–21.

Hespell, R.B., Akin, D.E. and Dehority, B.A. (1997) Bacteria, fungi and protozoa of the rumen. In: Mackie, R.I., White, B.A. and Isaacson, R. (eds) *Gastrointestinal Microbiology*, Volume 2. Chapman and Hall, New York, pp. 59–186.

Hotton, N., Olson, E.C. and Beerbower, R. (1996) Amniote origins and the discovery of herbivory. In: Sumida, S.S. and Martin, K.L.M. (eds) *Amniote Origins.* Academic Press, New York, pp. 207–264.

Hugenholtz, P. and Pace, N.R. (1996) Identifying microbial diversity in the natural environment: a molecular phylogenetic approach. *Trends in Biotechnology* 14, 190–197.

Hume, I.D. and Warner, A.C.I. (1980) Evolution of microbial digestion in mammals. In: Ruckebusch, Y. and Thievend, P. (eds) *Digestive Physiology and Metabolism in Ruminants.* MTD Press, Lancaster, pp. 665–684.

Hungate, R.E. (1976) Microbial activities related to mammalian digestion and absorption of food. In: Spiller, G.A. and Amen, R.J. (eds) *Fiber in Human Nutrition.* Plenum Press, New York, pp. 131–149.

Hungate, R.E. (1984) Microbes of nutritional importance in the alimentary tract. *Proceedings of the Nutrition Society* 43, 1–11.

Hunter-Cevera, J. (1998) The value of microbial diversity. *Current Opinion in Microbiology* 1, 278–285.

Klieve, A. and Swain, R.A. (1993) Estimation of ruminal bacteriophage numbers by pulsed-field gel electrophoresis and laser densitometry. *Applied and Environmental Microbiology* 59, 2299–2303.

Kocherginskaya, S.A., Simpson, J.M. and White, B.A. (1997) Microbial community structure of the rumen as assessed by denaturing gradient gel electrophoresis of polymerase chain-reaction amplified 16S rDNA genes. *Evolution of the Rumen Microbial Ecosystem, RRI-INRA Rumen Microbiology Symposium*, p. 37.

Krause, D.O. and Russell, J.B. (1996) An rRNA approach for assessing the role of obligate amino acid-fermenting bacteria in ruminal amino acid deamination. *Applied and Environmental Microbiology* 62, 815–821.

Krumholz, L.R., Forsberg, C.W. and Veira, D.M. (1983) Association of methanogenic bacteria with ruminal protozoa. *Canadian Journal of Microbiology* 29, 676–680.

Langer, P. (1991) Evolution of the digestive tract in mammals. *Verhandlungen der Deutschen Zoologischen Gesellschaft* 84, 169–193.

Lepper, B.T., Frolking, T.A., Fisher, D.C., Goldstein, G., Sanger, J.E., Wymer, D.A., Ogden, J.G. and Hooge, P.E. (1991) Intestinal contents of a late Pleistocene mastodon from midcontinental North America. *Quaternary Research* 36, 1–6.

Lin, C. and Stahl, D.A. (1995) Taxon-specific probes for the cellulolytic genus *Fibrobacter* reveal abundant and novel equine-associated populations. *Applied and Environmental Microbiology* 61, 1348–1351.

Lin, C.Z., Raskin, L. and Stahl. D.A. (1997) Microbial community structure in gastrointestinal tracts of domestic animals: Comparative analyses using rRNA-targeted oligonucleotide probes. *FEMS Microbiology Ecology* 22, 281–294.

Ludwig, W., Dorn, S., Springer, N., Kirchof, G. and Schleifer, K.H. (1994) PCR-based preparation of 23S rRNA-targeted group-specific polynucleotide probes. *Applied and Environmental Microbiology* 60, 3236–3244.

Mackie, R.I. (1987) Microbial digestion of forages in herbivores. In: Hacker, J.B. and Ternouth, J.H. (eds) *The Nutrition of Herbivores.* Academic Press, Sydney, pp. 233–265.

Mackie, R.I. and Doré, J. (1997) Molecular ecology of gastrointestinal ecosystems. In: Martins, M.T., Sato, M.I.Z., Tiedje, J.M., Hagler, L.C.N., Dobereiner, J. and Sanchez, P.S. (eds) *Progress in Microbial Ecology.* SBM/ICOME, Sao Paulo, Brazil, pp. 317–328.

Mackie, R.I., Sghir, A. and Gaskins, H.R. (1999) Developmental microbial ecology of the neonatal gastrointestinal tract. *American Journal of Clinical Nutrition* 69 (suppl), 1035S-1045S.

McSweeney, C.S. and Mackie, R.I. (1997) Gastrointestinal detoxification and digestive disorders in ruminant animals. In: Mackie, R.I. and White, B.A. (eds) *Gastrointestinal Microbiology,* Vol. 1. Chapman and Hall, New York, pp. 583–634.

McSweeney, C.S., Mackie, R.I., Odenyo, A.A. and Stahl, D.A. (1993) Development of an oligonucleotide probe and its application for detection and quantitation of the ruminal bacterium *Synergistes jonesii* in a mixed-population chemostat. *Applied and Environmental Microbiology* 59, 1607–1612.

Mannarelli, B., Stack, R.J., Lee, D. and Ericsson, L. (1990) Taxonomic relatedness of *Butyrivibrio, Lachnospira, Roseburia*, and *Eubacterium* species as determined by DNA hybridization and extracellular-polysaccharide analysis. *International Journal of Systematic Bacteriology* 40, 370–378.

Muyzer, G. and Smalla, K. (1998) Application of denaturing gradient gel electrophoresis (DGGE) and temperature gradient gel electrophoresis (TGGE) in microbial ecology. *Antonie van Leeuwenhoek Journal of Microbiology* 73, 27–141.

Muyzer, G., Brinkoff, T., Nubel, U., Santegoeds, C., Schafer, H. and Wawer, C. (1998) Denaturing gradient gel electrophoresis (DGGE) in microbial ecology. *Molecular Microbial Ecology Manual* 3.4.4:1–27. Kluwer Academic Publishers, The Netherlands.

Muyzer, G., de Waal, E.C. and Uitterlinden, A.G. (1993) Profiling of complex microbial populations by denaturing gradient gel electrophoresis analysis of polymerase chain reaction-amplified genes coding for 16S rRNA. *Applied Environmental Microbiology* 59, 695–700.

Ning, Z., Attwood, G.T., Lockington, R.A. and Brooker, J.D. (1991) Genetic diversity in ruminal isolates of *Selenomonas ruminantium*. *Current Microbiology* 22, 279–284.

Odenyo, A.A., Mackie, R.I., Stahl, D.A. and White, B.A. (1994a) The use of 16S rRNA-targeted oligonucleotide probes to study competition between ruminal fibrolytic bacteria: development of probes for *Ruminococcus* species and evidence for bacteriocin production. *Applied and Environmental Microbiology* 60, 3688–3696.

Odenyo, A.A., Mackie, R.I., Stahl, D.A. and White, B.A. (1994b) The use of 16S rRNA-targeted oligonucleotide probes to study competition between ruminal fibrolytic bacteria: pure-culture studies with cellulose and alkaline peroxide-treated wheat straw. *Applied and Environmental Microbiology* 60, 3697–3703.

Ogata, K., Aminov, R.I., Nagamine, T., Sugiura, M., Tajima, K., Mitsumori, M., Sekizaki, T., Kudo, H., Minato, H. and Benno, Y. (1997) Construction of a *Fibrobacter succinogenes* genomic map and demonstration of diversity at the genomic level. *Current Microbiology* 35, 22–27.

Okhuma, M. and Kudo, T. (1996) Phylogenetic diversity of the intestinal bacterial community in the termite *Reticulotermes speratus*. *Applied and Environmental Microbiology* 62, 461–468.

Orpin, C.G. and Joblin, K.N. (1997) The rumen anaerobic fungi. In: Hobson, P.N. and Stewart, C.S. (eds) *The Rumen Microbial Ecosystem*, 2nd edn. Chapman and Hall, New York, pp. 140–195.

Pace, N.R. (1997) A molecular view of microbial diversity and the biosphere. *Science* 276, 734–740.

Pace, N.R., Stahl, D.A., Lane, D.J. and Olsen, G.J. (1986) The analysis of natural microbial populations by ribosomal RNA sequences. *Advances in Microbial Ecology* 9, 1–55.

Poulsen, L.K., Lon, F., Kristensen, C.S., Hobolth, P., Molin, S. and Krogfelt, K.A. (1994) Spatial distribution of *Escherichia coli* in the mouse large intestine inferred from rRNA *in situ* hybridization. *Infection and Immunity* 62, 5191–5194.

Poulsen, L.K., Licht, T.R., Rang, C., Krogfelt, K.A. and Molin, S. (1995) The physiological state of *E. coli* growing in the large intestine of streptomycin-treated mice. *Journal of Bacteriology* 177, 5840–5845.

Raskin, L., Capman, W.C., Sharp, R., Poulson, L.K. and Stahl, D.A. (1997) Molecular ecology of gastrointestinal systems. In: Mackie, R.I., White, B.A. and Isaacson, R. (eds) *Gastrointestinal Microbiology*, Vol. 2. Chapman and Hall, New York, pp. 243–298.

Rhodes, A.N., Urbance, J.W., Youga, H., Corlew-Newman, H., Reddy, C.A., Klug, M.J., Tiedje,

J.M. and Fisher, D.C. (1998) Identification of bacterial isolates obtained from intestinal contents associated with 12,000-year-old mastodon remains. *Applied and Environmental Microbiology* 64, 651–658.

Rolfe, R.D. (1997) Colonization resistance. In: Mackie, R.I., White, B.A. and Isaacson, R.E. (eds) *Gastrointestinal Microbiology*, Vol. 2. Chapman and Hall, New York, pp. 501–536.

Rollo, F. and Marota, I. (1999) How microbial ancient DNA, found in association with human remains, can be interpreted. *Philosophical Transactions of the Royal Society*, London Series B 354, 111–119.

Sayler, G.S. and Layton, A.C. (1990) Environmental application of nucleic acid hybridization. *Annual Review of Microbiology* 44, 625–648.

Schopf, J.W. (1993) Microfossils of the early archaean apex chert: new evidence of the antiquity of life. *Science* 260, 640–646.

Sharp, R., Zeimer, C.J., Stern, M.D. and Stahl, D.A. (1998) Taxon-specific associations between protozoal and methanogen populations in the rumen and a model rumen system. *FEMS Microbiology Ecology* 26, 71–78.

Shi, Y. and Weimer, P.J. (1997) Competition for cellobiose among three predominant ruminal cellulolytic bacteria under substrate-excess and substrate-limited conditions. *Applied and Environmental Microbiology* 63, 743–748.

Shi, Y., Odt, C.L. and Weimer, P.J. (1997) Competition for cellulose among three predominant ruminal cellulolytic bacteria under substrate-excess and substrate-limited conditions. *Applied and Environmental Microbiology* 63, 734–742.

Simpson, J.M., McCracken, V.J., White, B.A., Gaskins, H.R. and Mackie, R.I. (1999) Application of denaturant gradient gel electrophoresis for the analysis of the porcine gastrointestinal microbiota. *Journal of Microbiological Methods* 36, 167–179.

Stackebrandt, E. (1997) The biodiversity convention and its consequences for the inventory of prokaryotes. In: Martins, M.T., Sato, M.I.Z., Tiedje, J.M., Hagler, L.C.N., Dobereiner, J. and Sanchez, P.S. (eds) *Progress in Microbial Ecology*. SBM/ICOME, Sao Paulo, Brazil.

Stahl, D.A. (1993a) The natural history of microorganisms. *ASM News* 59, 609–613.

Stahl, D.A. (1993b) Comparison of nucleic acids from microorganisms: sequencing approaches. Molecular evolution: producing the biochemical data. *Methods in Enzymology* 224, 373–391.

Stahl, D.A. (1997) Molecular approaches for the measurement of density, diversity and phylogeny. In: McInerney, M.J., Walter, M.V., Stetzenback, L.D. and Walter, M.V. (eds) *Manual of Environmental Microbiology*. ASM Press, Washington, DC, pp. 102–114.

Stahl, D.A. and Amann, R. (1991) Development and application of nucleic acid probes. In: Stackebrandt, E. and Goodfellow, M. (eds) *Nucleic Acid Techniques in Bacterial Systematics*. John Wiley & Sons, Chichester, UK, pp. 205–248.

Stahl, D.A., Flesher, B., Mansfield, H.R. and Montgomery, L. (1988) Use of phylogenetically based hybridization probes for studies of ruminal microbial ecology. *Applied and Environmental Microbiology* 54, 1079–1084.

Stewart, C.S., Flint, H.J. and Bryant, M.P. (1997) The rumen bacteria. In: Hobson, P.N. and Stewart, C.S. (eds) *The Rumen Microbial Ecosystem*, 2nd edn. Chapman and Hall, New York, pp. 10–72.

Stumm, C.K., Gizjen, H.J. and Vogels, G.D. (1982) Association of methanogenic bacteria with ovine rumen ciliates. *British Journal of Nutrition* 47, 95–99.

Sues, H.-D. and Reisz, R.R. (1998) Origins and early evolution of herbivory in tetrapods. *Trends in Ecology and Evolution* 13, 141–145.

Tajima, K., Aminov, R.I., Nagamine, T., Ogata, K., Nakamura, M., Matsui, H. and Benno, Y. (1999) Molecular bacterial diversity of the rumen as determined by sequence analysis of 16S rDNA libraries. *FEMS Microbiology Ecology* 27, 159–169.

Tannock, G.W. (1997) Influences of the normal microbiota on the animal host. In: Mackie, R.I.,

White, B.A. and Isaacson, R.E. (eds) *Gastrointestinal Microbiology*, Vol. 2. Chapman and Hall, New York, pp. 466–497.

Thill, L.A., Kocherginskaya, S.A., Naylor, G.E., Joblin, K.N. and White, B.A. (1997) Small subunit ribosomal RNA sequence analysis of anaerobic ruminal fungi isolated in New Zealand. *Proceedings of the 24th Biennial Conference on Rumen Function*, p. 28.

Thomasson, J.R. and Voorhies, M.R. (1990) Grasslands and grazers. In: Briggs, D.E.G. and Crowther, P.R. (eds) *Paleobiology: a Synthesis*. Blackwell Scientific Publishers, Oxford, pp. 84–87.

Vogels, G.D., Hoppe, W.F. and Wheelis, M.L. (1980) Association of methanogenic bacteria with rumen ciliates. *Applied and Environmental Microbiology* 40, 608–612.

Ward, D.M. (1989) Molecular probes for analysis of microbial communities. In: Characklis, W.G. and Wilderer, P.A. (eds) *Structure and Function of Biofilms*. John Wiley & Sons, Inc., New York, pp. 145–163.

Ward, D.M., Bateson, M.M., Weller, R. and Ruff-Roberts, A.L. (1992) Ribosomal RNA analysis of microorganisms as they occur in nature. *Advances in Microbial Ecology* 12, 219–286.

Whitehead, T.R. and Cotta, M.A. (1993) Development of a DNA probe for *Streptococcus bovis* by using a cloned amylase gene. *Journal of Clinical Microbiology* 31, 2387–2391.

Whitford, M.F., Forster, R.J., Beard, C.E., Gong, J. and Teather, R.M. (1998) Phylogenetic analysis of rumen bacteria by comparative sequence analysis of cloned 16S rRNA genes. *Anaerobe* 4, 153–163.

Whitman, W.B., Coleman, D.C. and Wiebe, W.J. (1998) Prokaryotes: the unseen majority. *Proceedings of the National Academy of Sciences USA* 95, 6578–6583.

Williams, A.G. and Coleman, G.S. (1997) The rumen protozoa. In: Hobson, P.N. and Stewart, C.S. (eds) *The Rumen Microbial Ecosystem*, 2nd edn. Chapman and Hall, New York, pp. 73–139.

Wilson, K.H. and Blitchington, R.B. (1996) Human colonic biota studied by ribosomal DNA sequence analysis. *Applied and Environmental Microbiology* 62, 2273–2278.

Wintzingerode, F.V., Gobel, U.B. and Stackebrandt, E. (1997) Determination of microbial diversity in environmental samples: pitfalls of PCR-based rRNA analysis. *FEMS Microbiology Reviews* 21, 213–229.

Woese, C.R. (1994) There must be a prokaryote somewhere: microbiology's search for itself. *Microbiology Reviews* 58, 1–9.

Woese, C.R., Kandler, O. and Wheelis, M.L. (1990) Towards a natural system of organisms: proposal for the domains Archaea, Bacteria, and Eucarya. *Proceedings of the National Academy of Sciences USA* 87, 4576–4579.

Wright, A.-D.G. and Lynn, D.H. (1997a) Phylogenetic analysis of the rumen ciliate family Ophryoscolecidae based on 18S ribosomal RNA sequences, with new sequences from *Diplodinium*, *Eudiplodinium*, and *Ophryoscolex*. *Canadian Journal of Zoology* 75, 963–970.

Wright, A.-D.G. and Lynn, D.H. (1997b) Monophyly of the Trichostome ciliates (phylum Ciliophora: class Litostomatea) tested using new 18S rRNA sequences from the Vestibuliferids, *Isotrichia intestinalis* and *Dasytricha ruminantium*, and the Haptorian, *Didinium nasutum*. *European Journal of Protistology* 33, 305–315.

Wright, A.-D.G., Dehority, B.A. and Lynn, D.H. (1997) Phylogeny of the rumen ciliates *Entodinium*, *Epidinium*, and *Polyplastron* (Litostomatea: Entodiniomorphida) inferred from small subunit ribosomal RNA sequences. *Journal of Eukaryotic Microbiology* 44, 61–67.

Zoetendahl, E.G., Akkermans, A.D.L. and de Vos, W.M. (1998) Temperature gradient gel electophoresis analysis of 16S rRNA from human fecal samples reveals stable and host-specific communities of active bacteria. *Applied and Environmental Microbiology* 64, 3854–3859.

# 5 Microbial Adherence to the Plant Cell Wall and Enzymatic Hydrolysis

C.W. Forsberg[1], E. Forano[2] and A. Chesson[3]

[1]Department of Microbiology, University of Guelph, Guelph, Ontario, Canada; [2]Laboratoire de Microbiologie, INRA CR de Clermont-Ferrand–Theix, Saint-Genès-Champanelle, France; [3]Rowett Research Institute, Bucksburn, Aberdeen, UK

## Interactions of ruminal microorganisms with plant cell walls

The ruminal environment is a highly complex ecosystem in which the plant cell wall digesting microorganisms constitute a minor proportion of the total population, but make available the carbon and energy for the ecosystem when animals are fed a forage diet. The major cellulolytic bacteria, *Ruminococcus albus*, *Ruminococcus flavefaciens* and *Fibrobacter succinogenes*, may account for 0.3–4% of the bacterial population (Krause *et al.*, 1999; Weimer *et al.*, 1999). The fungi may account for 8% of the microbial biomass (Orpin, 1983), and only a portion of these produce highly active cellulases and hemicellulases (Trinci *et al.*, 1994). A limited number of protozoal genera as well have an important role in digesting plant cell walls, and they may digest 5–21% of the cellulosic materials dependent upon the diet (Dijkstra and Tamminga, 1995). The primary factor limiting cellulose digestion appears to be the availability of cellulose digestion sites in the plant material rather than low cellulolytic activity (Dehority and Tirabasso, 1998). This conclusion is supported by similar studies by Weimer *et al.* (1999). However, Weimer *et al.* (1999) reported that differences in cellulolytic populations in individual cows were larger than those attributable to diet, suggesting that each cow maintains a unique assemblage of cellulolytic species. This could result from differing efficiencies in masticating plant cell walls since the bulk of microbial digestion occurs in the secondary cell wall (Wilson and Mertens, 1995). The fascinating aspect of these studies is that they present a general overview of population dynamics *in vivo* which begs the question of the factors that regulate adhesion of microorganisms to fibrous substrates, the mechanism of hydrolysis, and the cellular and extracellular factors influencing degradation of individual but highly integrated polymers making up the plant cell wall.

## The 'cellulosome' paradigm and the molecular basis for adhesion

Many studies have been undertaken to elucidate the molecular basis of adhesion and to further improve adhesion abilities of rumen microorganisms to the plant particles. Researchers have focused on the isolation of cellulose-binding proteins (CBPs) in order to examine their role in adhesion of microorganisms, and have searched for dockerin domains on polysaccharidase sequences, and scaffoldins in order to identify enzymatic complexes similar to cellulosomes. To complement this research, mutant strains exhibiting impaired adhesion to cellulose have been isolated.

### *The cellulosome paradigm*

The cellulosome is a multi-functional, multi-enzyme complex of high molecular mass, which is attached to the cell surface and mediates both cellular adhesion to cellulose and efficient solubilization of crystalline cellulose. The cellulosome structure was first discovered and described in the anaerobic thermophilic bacterium *Clostridium thermocellum*, and is considered today as one of the major molecular paradigms of bacterial cellulolysis and adhesion to cellulose (Bayer *et al.*, 1998). Biochemical and ultrastructural staining techniques showed that this complex (about 2 MDa) aggregates into polycellulosomes (up to 100 MDa) that form protuberances on the cell surface. The *C. thermocellum* cellulosome was shown to be composed of a 250 kDa non-catalytic glycoprotein, named CipA, that binds to cellulose, and of a variety of glycosyl hydrolases (endoglucanase, cellobiohydrolase, xylanase or lichenase). CipA was further shown to act as a scaffoldin protein that anchors the cellulosomal enzymes through receptor domains (Fig. 5.1). Indeed, all of the enzymes that assemble in a cellulosome structure have at their C-terminus, a highly conserved, non-catalytic domain called a dockerin, which is composed of two similar segments of 22 amino acids each (Bayer *et al.*, 1998) that specifically bind to nine receptor domains called cohesins present in the CipA scaffold structure (Fig. 5.1). CipA also contains a cellulose-binding domain (CBD) mediating binding of the entire cellulosome to cellulose, and a dockerin domain (type II) that tethers the cellulosome to the cell through another type of cohesin–dockerin interaction and a complex set of proteins (Beguin and Lemaire, 1996; Bayer *et al.*, 1998).

Complexes similar to *C. thermocellum* cellulosomes have been described for mesophilic cellulolytic clostridia such as *Clostridium cellulovorans*, *Clostridium cellulolyticum*, *Clostridium josui*, *Clostridium papyrosolvens*; the non-clostridial anaerobic bacteria *Bacteroides cellulosolvens* and *Acetivibrio cellulolyticus*; and also for the aerobic bacterium *Vibrio* sp. (Bayer *et al.*, 1998). The evidence in favour of cellulosomes in these bacteria include: (i) the production of high molecular weight complexes; (ii) the presence of cell-surface protuberances; (iii) cross-reactivity with antibodies against the *C. thermocellum* cellulosome; and (iv) isolation and sequencing of genes encoding the scaffoldin and cellulosomal enzymes. The various scaffoldins may differ in the number of cohesins, the position of the CBD (internal or terminal) and the organization of the dockerin on the polypeptide. The cellulosomal structure seems widespread and thus would seem to have been selected through evolution as a very efficient cellulolytic system.

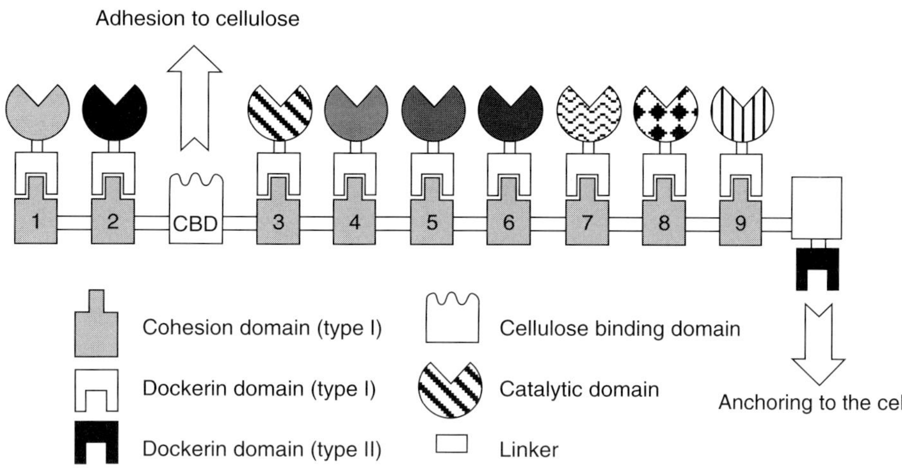


**Fig. 5.1.** Schematic representation of the cellulosome of *Clostridium thermocellum*.

## *The cellulosome equivalent in rumen microorganisms*

Three types of evidence indicate the presence of cellulosome-like structures in rumen cellulolytic microorganisms. First, protuberances similar to those found on *C. thermocellum* were observed on *R. albus*, *R. flavefaciens* and *F. succinogenes* using electron microscopy and staining with cationized ferritin (Leatherwood, 1973; Lamed *et al.*, 1987; Miron *et al.*, 1989). However, it was recently shown that cationized ferritin stains protuberances on the cell surface of a non-cellulolytic *Clostridium* species, and that the presence of these protuberances was correlated with growth rate rather than with the presence of cellulosomes (Blair and Anderson, 1998). Consequently, protuberances are not a reliable indicator of cellulosomes.

Second, high molecular mass complexes with cellulolytic and/or xylanolytic activity were shown in rumen bacteria and fungi. *R. albus* 8 was shown to produce large multi-enzyme complexes at the cell surface when cultivated in the presence of rumen fluid or phenylpropanoic acid (Stack and Hungate, 1984); similarly, *R. albus* SY3 was found to elaborate high molecular mass enzyme aggregates (Wood *et al.*, 1982). More recently, a cellulase complex of 2 MDa was isolated from the culture supernatant of *R. albus* F40, and was shown to include about ten proteins with endoglucanase or xylanase activity together with a protein of 250 kDa with no catalytic activity that might be a scaffoldin (Ohara *et al.*, 1998). *R. flavefaciens* FD1 and *Butyrivibrio fibrisolvens* H17C also produce multi-enzyme aggregates, composed either of several endoglucanases and xylanases, or only of several endoglucanases (Doerner and White, 1990; Lin and Thompson, 1991). The *R. flavefaciens* complexes were able to degrade crystalline cellulose (Doerner and White, 1990). High molecular mass complexes (4 MDa) with endoglucanase activity were also isolated from membrane fractions and culture fluid of *F. succinogenes* S85 cultures on cellulose, but they were not able to solubilize crystalline cellulose (Groleau and Forsberg, 1983). The ability of these bacterial cellulase complexes to bind to cellulose was not examined.

*Neocallimastix frontalis* produces a multi-enzyme complex (750–1000 kDa) that binds to and hydrolyses cellulose, and includes endoglucanase and exoglucanase (in majority) activities (Wood *et al.*, 1995). Similarly, 80% of the cellulase, xylanase and mannanase activity of *Piromyces* are extracellular and associated with a complex that binds to cellulose (Ali *et al.*, 1995). A multi-enzyme complex of at least ten endoglucanases was also found in the strain E2 of *Piromyces* (Dijkerman *et al.*, 1997).

Third, reiterated sequences resembling dockerins of the clostridial cellulosomal enzymes were found in the great majority of the sequences of polysaccharidases from the rumen fungi and in some of the *Ruminococci* enzymes (Table 5.1). These sequences are longer than the reiterated dockerin sequences of *C. thermocellum* or related clostridia, are sometimes internal, and may be found in one to three copies (Table 5.1). The reiterated sequences of the fungi *Neocallimastix*, *Piromyces* and *Orpinomyces* show strong homology to each other. Antibodies raised against the reiterated sequence of xylanase XYNA from *Orpinomyces* spp. reacted with numerous extracellular proteins of this fungus and also from *N. frontalis*, suggesting such sequences are widespread and share common epitopes (Li *et al.*, 1997a). Furthermore, the reiterated sequences (fused to glutathione-S-transferase) were shown to bind specifically to proteins of 116 and 97 kDa in *Neocallimastix patriciarum* and *Piromyces*, respectively, suggesting that they act as dockerins (Fanutti *et al.*, 1995). A dockerin-like sequence was found in one cellulase from *R. albus* (Karita *et al.*, 1997) while dockerin-like sequences were found in at least three glycanase enzymes from *R. flavefaciens* 17 (Kirby *et al.*, 1997). More recently, Ding *et al.* (1999) identified a 350 kDa glycosylated cellulose-binding protein in *R. flavefaciens* 17 that was a strong candidate for a scaffoldin/cellulosome-like integrating protein (CIP). The protein was subjected to trypsin digestion and fragments sequenced by Edman degradation. PCR primers designed from the amino acid sequences were used to clone fragments of the encoding gene. Sequencing of the recovered clone resulted in the identification of a series of cohesin-like repeats of 140 residues suggesting that a scaffoldin had indeed been cloned. The *Ruminococcus* reiterated dockerin sequences showed a significant, although distant, relationship with the dockerins of clostridia (Kirby *et al.*, 1997; Karita *et al.*, 1997); however, they did not show any sequence similarity with those from ruminal fungi.

C-terminal reiterated sequences of 164 and 156 amino acids in the CelA endoglucanase of *Orpinomyces joyonii* exhibited notable similarity to an N-terminal sequence of the EG3 enzyme from *F. succinogenes* and *Fibrobacter intestinalis*, but not with enzymes from other rumen organisms. The significance of this observation should become apparent with further studies.

### Isolation of cellulose-binding proteins from rumen bacteria

CBPs were identified in the three main cellulolytic rumen bacteria, *F. succinogenes, R. flavefaciens* and *R. albus* (Mitsumori and Minato, 1995; Pegden *et al.*, 1998), and in the other fibrolytic species, but also in non-cellulolytic species such as *Prevotella ruminicola, Megasphaera elsdenii* and *Eubacterium ruminantium* that do not attach to cellulose (Mitsumori and Minato, 1995).

Several CBPs ranging from 36 to 240 kDa were found in *F. succinogenes* S85 and in *F. intestinalis* DR7 (Gong *et al.*, 1996; Miron and Forsberg, 1998), and some of the

**Table 5.1.** Recently characterized glycoside hydrolases and esterases from ruminal bacteria, fungi and protozoa.

| Organism | Gene product | Description[a] | Modular arrangement[b] | Amino acid residues | $K_m/V_{max}$ [c] | Gen Bank # |
|---|---|---|---|---|---|---|
| *Butyrivibrio fibrisolvens* E14 | CinA | Cinnamoyl ester hydrolase | CD | 246 | –/76* | U44893 |
| *B. fibrisolvens* E14 | CinB | Cinnamoyl ester hydrolase | CD | 285 | | U44893 |
| *Epidium caudatum* | Epi3 | Endoglucanase | CD5 | 341 | | AB011273 |
| *Fibrobacter succinogenes* S85 | CedA | Cellobiosidase | CD5 | 357 | | U07419 |
| *F. succinogenes* S85 | CelG | Glucanase | CD 5A/SL/BTD | 519 | –/16.5* | U33887 |
| *F. succinogenes* S85 | EGB | Endoglucanase | ?/CD9 | 555 | | L14436 |
| *F. succinogenes* BL21 | EGC | Endoglucanase | CD9/BTD | 620 | | L48039 |
| *F. succinogenes* S85 | EGD | Endoglucanase | CD9/SL/BTD | 668 | –/11.5* | U05897 |
| *F. succinogenes* S85 | EGE | Endoglucanase | CD9/BTD | 467 | | U05897 |
| *F. succinogenes* S85 | EGF | Endoglucanase | CBD?/CD? | 1053 | –/57* | U39070 |
| *Neocallimastix frontalis* MCH3 | CelA | Endoglucanase | CD5/EL/TR2–38–36 | ? | | U38843 |
| *N. frontalis* | XYN3 | Xylanase | CD11/SL/CD11/TL/TR2 –40,40 | 607 | | X82266 |
| *Neocallimastix patriciarum* | BnaA | AXE | CD/TL/TR2–37,39 | 393 | | U66251 |
| *N. patriciarum* | BnaB | AXE | ?/CD | 392 | | U66252 |
| *N. patriciarum* | BnaC | AXE | CD/TR | 389 | | U66253 |
| *N. patriciarum* | CelA | Cellobiohydrolase | CBD/NL/CD6 | 428 | | U29872 |
| *N. patriciarum* | CelB | Glucanase | CD5/STL/TR2–36,37 | 473 | 0.24/- | Z31364 |
| *N. patriciarum* | XynA | Xylanase | CD11/SL/CD11/TL/TR2 –40,40 | 607 | –/5,980 | X65526 |
| *N. patriciarum* | XynB | Xylanase | CD10/L/TR12–8 | 860 | | S71569 |
| *Orpinomyces joyonii* 26 | CelA | Endoglucanase | CD5/TR2–164,156 | 910 | | U59432 |
| *O. joyonii* SG4 | CelB 2 | Cellulase | CD5/XL/$T_{doc}$1–38 | 585 | | AF015249 |
| *O. joyonii* SG4 | CelB 29 | Cellulase | CD5/XL/TR2–36,37 | 513 | | AF015248 |
| *Orpinomyces* PC-2 | AxeA | AXE | CD | 313 | 0.9/785 | AF001178 |
| *Orpinomyces* PC-2 | CelA | Endoglucanase | TR2–26,34/PL/CD6 | 459 | | U63837 |
| *Orpinomyces* PC-2 | CelB | Endoglucanase | CD5/L/TR2–36,37 | 471 | –/9.7* | U57818 |
| *Orpinomyces* PC-2 | CelC | Endoglucanase | TR2–26,29/TL/CD6 | 449 | | U63838 |
| *Orpinomyces* PC-2 | CelE | Endoglucanase | CD10/L/TR2–35–37 | 477 | 14.5* | U97153 |

*(Continued)*

**Table 5.1.** *Continued*

| Organism | Gene product | Description[a] | Modular arrangement[b] | Amino acid residues | $K_m/V_{max}$ [c] | Gen Bank # |
|---|---|---|---|---|---|---|
| *Orpinomyces* PC-2 | LICA | Lichenase | CD16 | 245 | 0.91/5,320 | U63813 |
| *Orpinomyces* PC-2 | XynA | Xylanase | CD11/L/TR2–40,40 | 362 | | U57819 |
| *Prevotella ruminicola* $B_1$4 | CdxA | β-Glucosidase | CD3 | 789 | 5.0/3.5 | U35425 |
| *P. ruminicola* $B_1$4 | XynA | Xylanase | CD10 | 369 | | Z49241 |
| *P. ruminicola* $B_1$4 | XynB | Glycosidase | CD ? | 319 | | Z49241 |
| *P. ruminicola* $B_1$4 | XynC | Xylanase | CD10/?/CD10 | 560 | | Z79595 |
| *P. ruminicola* 23 | D31d | Xylanase | CD10/TR/CD10 | | | U53926 |
| *Piromyces* spp. | MANA | Mannanase | CD26/TR3–36,36,36 | 606 | | X91858 |
| *Piromyces* spp. | MANB | Mannanase | CD26/NL/TR2–36,37 | | | X97408 |
| *Piromyces* spp. | MANC | Mannanase | CD26/NL/TR2–36,37 | | | X97520 |
| *Piromyces* spp. | XYLA | Xylanase | CD11/TR2–39,36/CD11 | 623 | | X91857 |
| *Polyplastron multivesiculatum* | PolyX | Xylanase | CD11 | 175 | | AB011274 |
| *P. multivesiculatum* | XYNA | Xylanase | CD11 | 218 | | AJ009828 |
| *Ruminococcus albus 7* | XynA | Xylanase | CD | 680 | | U43089 |
| *R. albus* F-40 | EGIV | Endoglucanase | CD5/TR2–32,31 | 312 | | D16315 |
| *Ruminococcus flavefaciens* 17 | XynB | Xylanase/AXE | CD11/TSD/$CD_{AXE}$/TL/ TR 2–32,33 | 781 | | Z35226 |
| *R. flavefaciens* 17 | XynD | Xylanase/ β glucanase | CD11/TSD/TR2- 32,33/TL/CD16 | 802 | | S61204 |
| *R. flavefaciens* 17 | EndA | Endoglucanase | CD5/?/TL/TL/TR2–32 | 759 | | Z83304 |
| *R. flavafaciens* FD-1 | CelD | Endoglucanase | CD9 | 405 | | L05368 |
| *Streptococcus bovis* JB1 | | β-(1,3–1,4) glucanase | CD16 | 237 | 2.8/338 | Z92911 [1] |

[a] AXE, acetyl xylan esterase.
[b] Modular arrangement: CD#, catalytic domain followed by the number of the glycosyl family or a ? if there is no similarity to a known family; CBD, cellulose binding domain; BTD, basic terminal domain; L, GL, SL, TL, STL, PL, Linker regions with designation, or rich in glutamic acid, serine, threonine or proline; TR, dockerin with a tandem repeat, the first number indicates the number of repeats and the subsequent numbers indicate the amino acid residues in each; $T_{doc}$, dockerin in a single sequence; X, not known; TSD, thermostabilizing domain.
[c] The number before the slash is the $K_m$ (mM) and the number after it is the $V_{max}$ (μmol $min^{-1}$ $mg^{-1}$ protein) where values are available. Where an asterisk is shown, the specific activity was listed using the relevant substrate.

proteins of 120 kDa and lower had endoglucanase activity. Two CBPs of 120 and 225 kDa were purified from *F. succinogenes* S85, and the gene coding for the 120 kDa protein was cloned in *Escherichia coli* and sequenced (Mitsumori *et al.*, 1996). This protein was later shown to be an endoglucanase bearing a CBD (Malburg *et al.*, 1997). In parallel, Gong *et al.* (1996) recognized two other CBPs, of 180 and 240 kDa. The 180 kDa protein is a glycosylated xylanase (E.E. Egbosimba and C.W. Forsberg, unpublished data) that shares common epitopes with numerous proteins from the outer-membrane. Furthermore, polyclonal antibodies raised against this CBP decreased the adhesion of *F. succinogenes* to cellulose by 60%, suggesting that it and related proteins have an important role in adhesion.

To identify proteins important in adhesion of *F. intestinalis* to cellulose, polyclonal antibodies prepared against whole cells were adsorbed with cells of the non-adherent mutant (Miron and Forsberg, 1998, Section 2.4) to remove all antibodies reacting with surface epitopes not involved in the adhesion process (Miron and Forsberg, 1999). These adsorbed antibodies reacted strongly with six of the major CBPs of strain DR7, but reacted very weakly with a non-adherent mutant. The antiserum was shown to react with glycosyl residues rather than protein. Since the non-adherent mutant was not missing any of the CBPs, glycosylation appears to have an important role in the adhesion process. Monosaccharide analysis of the CBPs showed that they contained mainly galactosamine, glucosamine, galacturonic acid and glucuronic acid. Of these compounds, glucosamine and galacturonic acid, each at 10%, blocked binding. These data were interpreted to indicate that the compounds interfered with binding by occupying adhesion sites on the cellulose substrate; however, the precise role of the glycosyl components of the CBPs involved in adhesion is not known.

Recently, two low-molecular mass cellulose-binding polypeptides (16 and 21 kDa) were isolated from *R. albus* 8 (Pegden *et al.*, 1998). One of them (CbpC) possessed structural motifs typical of the Pil protein family, which is comprised mostly of type 4 fimbrial proteins produced by Gram-negative pathogenic bacteria. Inclusion of either ruminal fluid or phenyl propionic acid in the growth medium, a treatment that was shown to increase *R. albus* adherence to cellulose, also increased the concentration of CbpC transcripts (Pegden *et al.*, 1998). DNA sequences homologous to the CbpC gene were found in other strains of *R. albus*. The authors suggest that fimbrial-type adhesion proteins may represent a novel strategy (in addition to the cellulosome?) for the adhesion of Gram-positive bacteria to cellulose.

### Isolation of non-adherent mutants

Adherence deficient mutants of *F. succinogenes*, *F. intestinalis* and *R. albus* have been obtained (Gong and Forsberg, 1989; Miron and Forsberg, 1998; Miron *et al.*, 1998). The mutants were spontaneous and obtained by enrichment of the non-adherent fraction of bacterial cultures, and the genes involved in the mutations have not been identified. Mutagenesis with transposons, which allows easy identification of the mutated gene(s), was also applied to *F. succinogenes* and *R. albus*, but was not successful (Mosoni, 1999).

An adhesion-defective mutant of *F. intestinalis* DR7, isolated by Miron and Forsberg (1998), showed limited growth on insoluble cellulosic substrates. The mutant displayed lower cellulose-degrading activities than the wild type, and these activities were located in the extracellular fluid, while they were mostly cell-bound in the wild type. This may suggest that the mutant is impaired in assembling multi-enzyme complexes involved in both adhesion and hydrolysis of cellulose. Four mutants isolated from the strain S85 of *F. succinogenes* displayed different phenotypes regarding their capability to utilize cellulose (Gong and Forsberg, 1989), suggesting that the adhesion mechanism may require several factors. An adhesion-defective derivative isolated from *R. albus* SY3 was also impaired in its capacity to grow on cellulose (Miron *et al.*, 1998). Two adhesion-defective mutants were also isolated from *R. albus* 20; they showed both a decreased rate and reduced extent of cellulose degradation (Mosoni, 1999). In addition to these studies with mutant strains, a great heterogeneity in adhesion and cellulolytic performances was observed among different wild strains of *R. albus* (Morris and Cole, 1987). Altogether, these results suggest that adhesion of *R. albus* to cellulose is a complex process that may be mediated by several factors.

In conclusion, there is evidence that the rumen fungi and the ruminococci produce multi-enzyme complexes that may reorganize to form cellulosomes. These complexes are likely to be involved in both adhesion of the cell to cellulose and degradation of this substrate. Indeed, for all the fibrolytic microorganisms, adhesion and cellulolytic activity appear closely, although not always strictly, related. However, there is no evidence at this time for a cellulosome-type organization for *F. succinogenes*. Complete understanding of the molecular basis of adhesion of the main rumen fibrolytic species will rely on further work, such as isolation of genes encoding the putative scaffoldins, and the molecular characterization of the isolated mutants.

## Diversity and catalytic properties of bacterial, fungal and protozoan glycoside hydrolases

In excess of 100 plant cell wall degrading enzymes have been cloned from ruminal organisms. Recently cloned genes are listed in Table 5.1 while previously cloned genes and their characterized native gene products are described in several reviews (Chesson and Forsberg, 1997; Forsberg *et al.*, 1997). Of great interest is the recent cloning of cellulase and xylanase genes from the ruminal protozoa *Epidinium caudatum* and *Polyplastron multivesiculatum* (Devillard *et al.*, 1999; Takenaka *et al.*, 1999). Three glucanases, one cellobiosidase and two xylanases were cloned from *E. caudatum* while one glucanase, one cellobiosidase and several xylanases were cloned from *P. multivesiculatum*. Thus we can say that all major rumen fibre-digesting organisms contain multiple hemicellulase and glucanase genes which code for related or different enzymes that can be separated into glycosyl hydrolase families based on the amino acid sequence of the catalytic domain (CD). A current overview of the glycoside hydrolase families (Henrissat, 1998) can be found at the URL: http://afmb.cnrs-mrs.fr/~pedro/CAZY/. Although no crystal structure of a ruminal glycoside hydrolase has been reported, structures of cellulases from non-rumen organisms are available for more than half of the families.

Enzymatic hydrolysis of the glycosidic bond takes place via general acid catalysis that involves a pair of carboxylic acids (Glu or Asp) in essentially all cases investigated (McIntosh *et al.*, 1996). This hydrolysis occurs via two major mechanisms giving rise to either retention or inversion of the anomeric configuration (Henrissat *et al.*, 1998).

In the past, considerable emphasis was placed on characterization of the catalytic properties of glycosidases; however, recently the focus has been on establishing the cellulosome-like structure in ruminal fungi and bacteria. As documented earlier (see pp. 80–82), most enzymes are an integral part of the cellulosome-like structure, but some are not. These include the *Orpinomyces* PC-2 LICA lichenase and AxeA acetyl xylan esterase, the *N. patriciarum* CelA cellobiohydrolase and the *N. patriciarum* BnaB acetyl xylan esterase (Table 5.1). The CelA cellobiohydrolase CD, in contrast to other fungal glycoside hydrolases, exhibited no similarity to those of other ruminal organisms and instead resembled CBHII of *Trichoderma reesei*. The CelA CBD resembled the type 1 family that is exclusive to aerobic fungal cellulases (Tomme *et al.*, 1995). Knowledge of the mechanism by which this enzyme interacts with the cellulosome-like complex in cellulose hydrolysis will be important to our understanding of cellulolysis.

The number and organization of CDs in enzymes of ruminal microorganisms is to a degree characteristic of the genus. *Neocallimastix* and *Piromyces* species often have two CDs in a single protein (Gilbert *et al.*, 1992; Fanutti *et al.*, 1995), while Xue *et al.* (1992) reported three in a broad spectrum glucanase of *N. patriciarum*. However, Fanutti *et al.* (1995) reported that the second xylanase gene did not code for a catalytically active protein. In the bacteria, the XynB and XynD xylanases from *R. flavefaciens* (Kirby *et al.*, 1997, 1998) and the XynC xylanase from *F. succinogenes* S85 each possess two catalytically active CDs. In contrast to the enzymes with a more complex structure, those produced by *R. albus* (Kirby *et al.*, 1997) and the polycentric *Orpinomyces* usually have single CDs.

The glycanase genes of the anaerobic ruminal fungi exhibit sequence conservation with their respective counterparts among the rumen bacteria, except for the *celA* gene of *N. patriciarum* as stated above (Chen *et al.*, 1998; Denman *et al.*, 1996). The similarity has led to the hypothesis that many of the fungal genes were acquired via horizontal gene transfer from the ruminal bacteria (Gilbert *et al.*, 1992; Chen *et al.*, 1997). Similarly, the protozoan cellulases and xylanases resemble those of the bacteria and fungi, for example, the *P. multivesiculatum* xylanase exhibited higher homology with family 11 xylanases from Gram-positive bacteria than with enzymes of eucaryotic origin (Devillard *et al.*, 1999). This raises the possibility that the protozoan gene might have been acquired by horizontal gene transfer, perhaps from ingested bacteria. Superimposed upon gene transfer, evidence has been presented suggesting that genes encoding three mannanases of a *Piromyces* spp. (Millward-Sadler *et al.*, 1996), two cellobiohydrolases of *Orpinomyces* PC-2 (Li *et al.*, 1997b) and two glucanases from *Orpinomyces* (Chen *et al.*, 1998) have originated via a gene duplication process. Gene duplication has also been observed in *F. succinogenes* as documented by the endoglucanases CelD and CelE (Malburg *et al.*, 1996) and three new family 10 xylanase genes from *F. succinogenes* that are in series on the chromosome (J. Ha and C.W. Forsberg, unpublished data). It was hypothesized by Fanutti *et al.* (1995) that fungal mannanases and endoglucanases could have a common progenitor sequence. This is an interesting suggestion and perhaps is validated by endoglucanases and xylanases exhibiting both activities to varying degrees (Gilkes *et al.*, 1991), and a xylanase exhibiting arabinose-debranching activity and a xylanase exhibiting arabinose-debranching activity (Matte and Forsberg, 1992).

## Structure and action of CBDs

Few genes coding for glycosyl hydrolases with CBDs have been cloned from ruminal organisms. Those containing CBD sequences include the *celA* from *N. patriciarum*, *celF* from *F. succinogenes* S85 (Malburg *et al.*, 1997), *end1* from *B. fibrisolvens* H17C (Berger *et al.*, 1989), and *celA* from *Clostridium longisporum* (Mittendorf and Thompson, 1993). One glucanase with a CBD has been cloned from each of *E. caudatum* and *P. multivesiculatum* as well. CBDs have one or more roles to play in the mechanism of cellulose hydrolysis. The CBD concentrates the attached enzyme at the cellulose surface. This was demonstrated by deletion of the CBD from CelA which reduced the Avicel cellulose hydrolysing activity by ninefold (Denman *et al.*, 1996). In a separate study, attaching a CBD from *Clostridium stercorarium* to the *R. albus* endoglucanase IV was demonstrated to enhance cellulolysis at low cellulose concentrations, but not at high concentrations where enzyme molecules lacking the CBD had ready access to cleavage sites on the cellulose (Karita *et al.*, 1996). The endoglucanase A (CenA) CBD from the non-rumen bacterium *Cellulomonas fimi* reportedly has the ability to disrupt noncovalent interactions between adjacent glucan chains in the cellulose microfibril (Din *et al.*, 1991). A similar property has been reported for a scaffolding protein from *C. cellulolyticum* (Pagès *et al.*, 1997). Thus some CBDs have the capacity to convert crystalline cellulose to amorphous strands at the surface. This may explain the ability of the *Thermomonospora fusca* E2 family II CBD to enhance the activity of the attached *P. ruminicola* glucanase against insoluble cellulose (Maglione *et al.*, 1992) whereas a hybrid enzyme constructed from a *Pseudomonas* CBD and a *Ruminococcus* endoglucanase did not show enhanced catalytic activity (Poole *et al.*, 1990). The cellulosome structure circumvents the need for individual glycosyl hydrolases to have CBDs.

CBDs from different organisms are grouped into distinctive families on the basis of amino acid sequence similarity (Tomme *et al.*, 1995). There are at least ten different families, with Type I and II being the most common domains in aerobic fungi and bacteria, respectively. A feature of many of these domains is a planar hydrophobic region in the surface, in which ligand binding is mediated by hydrophobic stacking interactions between solvent-exposed aromatic residues and the glucose-pyranose rings. There is a strong preference for insoluble forms of cellulose suggesting that the binding surface interacts with multiple cellulose chains. Binding by CBPs may in some cases seem practically irreversible. However, Jervis *et al.* (1997) tested the rate of surface diffusion of the *C. fimi* cellulases, Cex and CenA, and their respective CBDs using fluorescence recovery after photobleaching analysis. They reported that greater than 70% of bound molecules were mobile on the cellulose surface. A comparison with kinetics of crystalline cellulose hydrolysis suggested that surface diffusion rates did not limit cellulase activity. One can therefore visualize a model for the hydrolysis of plant cell wall by a cellulase with a CBD in which the CBD tethers the CD via a linker making available multiple cleavage sites. As the CBD diffuses, fresh sites become available along the path of diffusion, which presumably follow the orientation of the cellulose strands. No CBD from a ruminal organism has been fully characterized to date, but we can anticipate behaviour similar to their non-ruminal counterparts.

## Regulation of hydrolase synthesis and activity

It is commonly observed that glycosyl hydrolase synthesis by ruminal bacteria and fungi is enhanced by growth on cellulose and hemicellulose, and synthesis is reduced when cells are grown on glucose. This is well illustrated in a recent study by Bera *et al.* (1998) where they observed that seven cellulase genes, a cellodextrinase gene, a xylanase gene and a lichenase gene were over-expressed when *F. succinogenes* was grown on cellulose as compared with sugar-grown cultures. The expression of glycosyl-hydrolase genes of *F. succinogenes* therefore appears to be tightly regulated. The regulation of glycosyl hydrolases is less tightly controlled in *P. ruminicola* (Gardner *et al.*, 1995). The xylanase was induced by xylan, but not by other sugars while the β-1,4-endoglucanases and mannanases were induced by xylan, but even higher activities were observed with mannose, cellobiose, or xylose. The endoglucanases and mannanases were repressed by sucrose, arabinose or rhamnose. Gardner *et al.* (1997) provided evidence that two endoglucanases and a mannanase were present in an operon and perhaps under the same control.

Cotta *et al.* (1994) demonstrated the presence of cAMP in *P. ruminicola*, but it was practically absent from other anaerobes tested. They have since cloned a novel adenylate cyclase from *P. ruminicola* D31d (Cotta *et al.*, 1998). However, the concentration of cAMP did not alter in organisms under conditions where catabolite repression-like phenomena occurred; therefore it may not have a role in regulation of hydrolase synthesis.

Dalrymple and Swadling (1997) reported the presence of the *cinR* negative regulator of the *cinB* gene which codes for cinamoyl ester hydrolase in *B. fibrisolvens*. The addition of FAXX (*O*-[5-*O*-(*trans*-feruloyl)-α-L-arabinofuranosyl]-(1,3)-*O*-β-D-xylopyranosyl-(1,4)-D-xylopyranose) and Fara [5-*O*-(*trans*-feruloyl)-arabinofuranose], but not xylobiose or ferulic acid inhibited binding of CinR to DNA. This is the first characterized regulatory system in ruminal bacteria.

A novel enhancer of cellulase action is produced by some ruminal organisms. *O. joyonii* produces a yellow affinity substance (YAS) that increases adsorption of cellulases to microcrystalline cellulose (Hodrová *et al.*, 1998). A similar YAS is reportedly produced by *R. flavefaciens* (Kopecny and Hodrová, 1997). These substances have not been isolated, but a similar substance from *C. thermocellum* has a mass of 1036 Da and is comprised of a ring structure with a long hydrocarbon tail that may contain carbonyl and amino moieties (Ljungdahl *et al.*, 1988).

Although not previously recognized as a limitation on microbial digestion in the rumen, wheat plants produce a novel 20 kDa xylanase inhibitor that binds close to the active site of xylanase enzymes (McLauchlan *et al.*, 1999). Whether other plants produce related inhibitor molecules remains to be determined.

## Strategies to overcome limitations to the microbial degradation of plant cell walls

From studies of plant cell wall structure it has become evident that a number of organizational features determine the nature of the plant cell wall biodegradation process. The most important of these features is the pore size distribution. This is essentially a measure of the spacing between the individual polymers contributing to the wall structure and is remarkably similar in all crop vegetation used for feed purposes. Direct measurement by a variety of probe-based methods has shown that most pores have a diameter between 2 and 4 nm (Chesson *et al.*, 1997; Gardner *et al.*, 1999). These dimensions are not sufficient to allow free diffusion into the wall by simple globular enzymes with masses greater than ~20 kDa. The porosity of the wall (Gardner *et al.*, 1999) and its composition (Chesson *et al.*, 1986) changes little during the course of degradation even when ~70% of dry matter has been eroded. The data are consistent with a pattern of degradation based on surface erosion imposed on the rumen bacteria by the impervious mixed polymer nature of the cell wall.

Recent work showed that enzyme (cellulase and xylanase) supplementation of the diet can increase ruminal digestibility and milk production in cattle (Yang *et al.*, 1999). This result is surprising considering the extensive array of potent endogenous fibrolytic enzymes produced by the rumen microflora. The most likely explanation for the beneficial effect is that addition of enriched extracellular polysaccharidases results in an immediate attack on freshly ingested plant material thereby providing additional available carbohydrate that encourages more rapid microbial growth, shortening the lag time required for microbial colonization. The net effect may be equivalent to a longer retention time within the rumen. However, one cannot preclude the possibility that the exogenous cellulases have more efficacious binding and catalytic properties.

Evolution has not produced a solution that overcomes this limitation. Instead, most of the rumen bacteria and fungi have optimized a system of cell-wall degrading activities in the form of the cellulosome which recognizes the diffusion limitations imposed by the plant cell wall and which is highly adapted to a superficial mode of action. This makes it unlikely that the introduction of genes coding for single activities, as has been done in the past (see Forsberg *et al.*, 1997), will contribute significantly to the degradation process. There remains the option of engineering the cellulosome itself (Bayer *et al.*, 1998). However, more needs to be known about its structure and, in particular, the structure–function relationships of its component parts in the organism of choice, before this is likely to be effective. One possible option is the introduction of CBDs that have a greater avidity for cellulose, which disrupt the crystalline structure of the surface cellulose as shown by Din *et al.* (1991). Another is the introduction into the cellulosome structure of CDs with higher catalytic efficiency and, depending upon the organism of choice, ensuring that essential hemicellulases, for example, feruloly and coumaroyl esterase, arabinofuranosidase and acetylxylan esterase, are included as illustrated in Fig. 5.2. There may be more efficacious CBDs and CDs in non-ruminal organisms or those organisms, for example the ruminal fungi, with highly efficient cellulases, but which usually are a minor component of the population.

With surface erosion as the predominant mechanism of bacterial (and rumen fungal?) cell wall breakdown, two factors are particularly important. Firstly, the amount of surface area available for colonization and, secondly, the chemistry of the available surface. Surface is created by feed processing and then by mastication/rumination, each opening plant cells and exposing the inner surface to colonization. Subsequent removal of polysaccharide from that surface can, in more lignified cell types, lead to the development of an inert surface in which any remaining polysaccharide is 'protected' from attack by the presence of phenolic compounds. Consequently, available surface area reaches a maximum and then diminishes with time resulting in the slowing of the rate

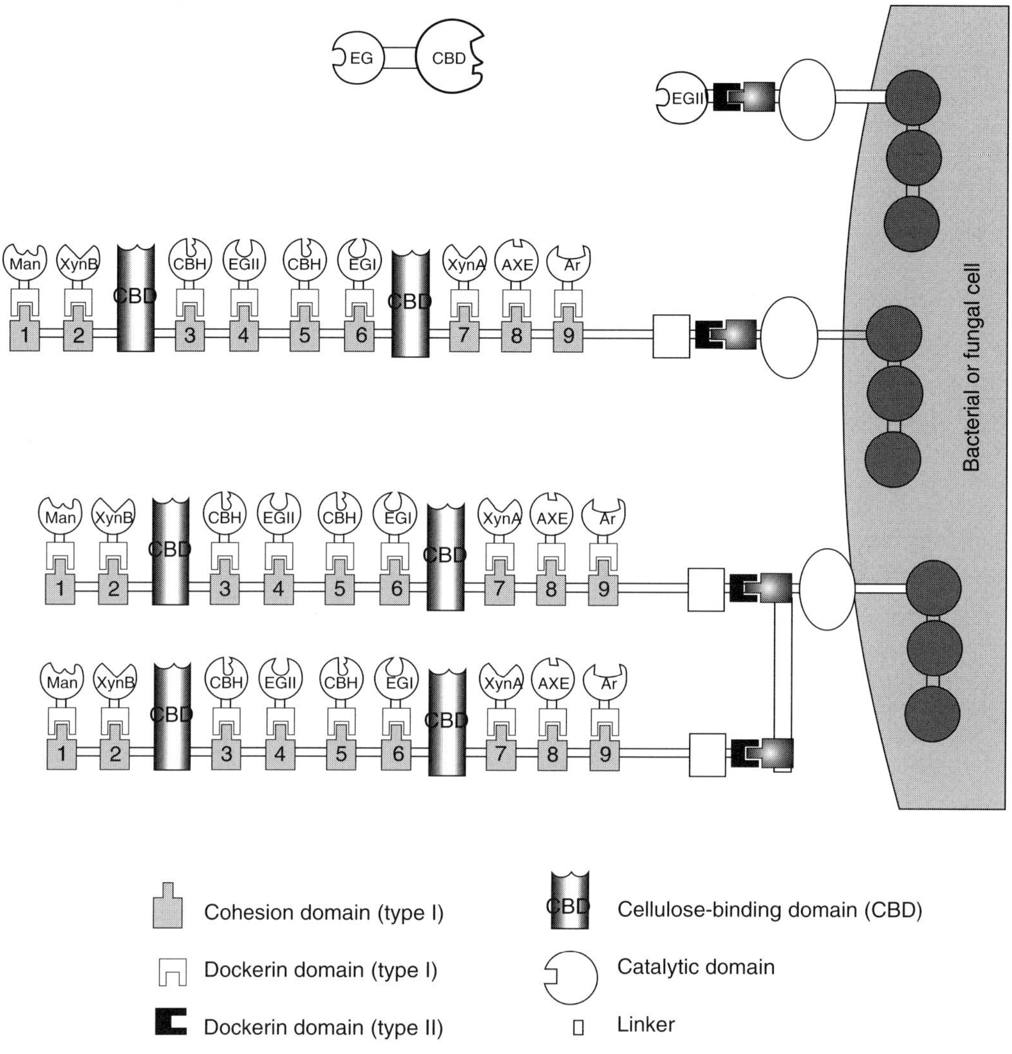


**Fig. 5.2.** Schematic representation of a cellulosome with high catalytic activity on plant cell walls. Symbols: Man, Mannanase; Xyn, xylanase; CBH, cellobiohydrolase; EG, endoglucanase; AXE, acetyl xylan esterase; Ar, arabinofuranosidase.

of degradation. The amount of surface that is denied to attacking organisms by this mechanism is a product of the amount of lignin present and the degree of its cross-linking to other wall polymers and defines the extent of degradation.

The development of ruminal organisms able to digest lignin does not appear to be an option at present because of the aerobic requirements for the process to proceed at a rate compatible with the rate of passage. A strategy more likely to succeed is the modification of lignification in plants used as ruminant feeds. Attempts to achieve this by various pre-treatment methods, although partially successful, have not proven cost-effective.

More permanent solutions through selective breeding and the use of recombinant DNA technology are under active investigation.

Enzymes involved in the biosynthesis of lignin and tannin precursors have been targeted for genetic modification, and crop plants have been produced in which one or more of these enzymes have been down-regulated. One general observation made is that phenolic polymers appear much more plastic than was originally thought (Ralph, 1997). Modifying the nature of the precursors was found rarely to decrease the total amount of lignin formed but did have significant effects on its composition and properties (Boudet, 1998). Blocking a biochemical pathway at one point can also lead to a redirection of the flux of precursor molecules. Down-regulation of cinnamyl CoA reductase, an enzyme which catalyses the reduction of phenolic acids to the corresponding aldehyde, led to greater amounts of free ferulic acid present in the cell and increased significantly the number of diferulate cross-links within the wall (Piquemal *et al.*, 1998). Because of the plasticity of lignin, targeting the extent of cross-linking between polymers within the wall may be a more effective way of altering cell wall degradation characteristics (Grabber *et al.*, 1998).

There is also an increasing appreciation of the importance to nutritional value of the spatial distribution of cell wall material within the plant. Unfortunately, the ability to apply this new understanding to the breeding of new crop varieties was, until recently, hampered by the lack of quantitative tools for the routine measurement of anatomy. Analysis of images from microscope sections now has revolutionized approaches for the measurement of anatomical features enabling the automated recognition of cell types and tissues to be used as selection criteria (Travis *et al.*, 1996). However, the lack of knowledge of the genetic basis to anatomy has meant that development of improved crop plants using anatomical features as selection criteria is restricted to conventional breeding at present.

## Concluding remarks

As we look to the future there would appear to be no short-term opportunities to develop genetically modified ruminal organisms that will radically improve the rate and extent of plant cell wall digestion. However, furthering our understanding of the structure of the cellulosome and related cellulase systems of ruminal organisms and comparative studies on the CBDs and CDs of ruminal and other organisms may reveal new opportunities to improve the catalytic properties of the ruminal cellulases. This strategy, in conjunction with continued exploration of genetic methods to reduce the lignin content of fibrous plants used as ruminant feeds, should eventually give rise to improved digestion of plant components that at present are considered to be largely recalcitrant to ruminal digestion.

## Acknowledgements

The authors thank E.A. Bayer, P. Beguin, J.-P. Belaich, J. Ha, P. Mosoni and J.H.D. Wu for communicating data prior to publication, and L.B. Selinger for providing journal articles not available at the authors' institutions.

## References

Ali, B.R.S., Zhou, L.Q., Graves, F.M., Freedman, R.B., Black, G.W., Gilbert, H.J. and Hazlewood, G.P. (1995) Cellulases and hemicellulases of the anaerobic fungus *Piromyces* constitute a multiprotein cellulose-binding complex and are encoded by multigene families. *FEMS Microbiology Letters* 125, 15–21.

Bayer, E.A., Chanzy, H., Lamed, R. and Shoham, Y. (1998) Cellulose, cellulases and cellulosomes. *Current Opinion in Structural Biology* 8, 548–557.

Beguin, P. and Lemaire, M. (1996) The cellulosome: an exocellular, multiprotein complex specialized in cellulose degradation. *Critical Review of Biochemistry and Molecular Biology* 31, 201–236.

Bera, C., Gaudet, G. and Forano, E. (1998) Regulation of glycosyl-hydrolase genes expression in *Fibrobacter succinogenes* S85. In: Ohmiya, K., Hayashi, K., Sakka, K., Kobayashi, Y., Karita, S. and Kimura, T. (eds) *Genetics, Biochemistry and Ecology of Cellulose Degradation*. Uni Publishers Co., Ltd, Japan, pp. 541–544.

Berger, E., Jones, W.A., Jones, D.T. and Woods, D.R. (1989) Cloning and sequencing of an endoglucanase (*end1*) gene from *Butyrivibrio fibrisolvens* H17c. *Molecular and General Genetics* 219, 193–198.

Blair, B.J. and Anderson, K. (1998) Comparison of staining techniques for scanning electron microscopic detection of ultrastructural protuberances on cellulolytic bacteria. *Biotechnic and Histochemistry* 73, 107–113.

Boudet, A.-M. (1998) A new view of lignification. *Trends in Plant Science* 3, 67–71.

Chen, H., Li, X.-L. and Ljungdahl, L.G. (1997) Sequencing of a 1,3–1,4-β-D-glucanase (lichenase) from the anaerobic fungus *Orpinomyces* strain PC-2: Properties of the enzyme expressed in *Escherichia coli* and evidence that the gene has a bacterial origin. *Journal of Bacteriology* 179, 6028–6034.

Chen, H., Li, X.-L., Blum, D.L. and Ljungdahl, L.G. (1998) Two genes of the anaerobic fungus *Orpinomyces* sp. strain PC2 encoding cellulases with endoglucanase activities may have arisen by gene duplication. *FEMS Microbiology Letters* 159, 63–68.

Chesson, A. and Forsberg, C.W. (1997) Polysaccharide degradation by rumen microorganisms. In: Hobson, P. and Stewart, C. (eds) *The Rumen Microbial Ecosystem*, 2nd edn. Chapman & Hall Ltd, Andover, UK, pp. 329–381.

Chesson, A., Stewart, C.S., Dalgarno, K. and King, T.P. (1986) Degradation of isolated grass mesophyll, epidermis and fibre cell walls in the rumen and by cellulolytic rumen bacteria in axenic culture. *Journal of Applied Bacteriology* 60, 327–336.

Chesson, A., Gardner, P.T. and Wood, T.J. (1997) Cell wall porosity and available surface area of wheat straw and wheat grain fractions. *Journal of the Science of Food and Agriculture* 75, 289–295.

Cotta, M.A., Wheeler, M.B. and Whitehead, T.R. (1994) Cyclic AMP in ruminal and other anaerobic bacteria. *FEMS Microbiology Letters* 124, 355–359.

Cotta, M.A., Whitehead, T.R. and Wheeler, M.B. (1998) Identification of a novel adenylate cyclase in the ruminal anaerobe, *Prevotella ruminicola* D31d. *FEMS Microbiology Letters* 164, 257–260.

Dalrymple, B.P. and Swadling, Y. (1997) Expression of a *Butyrivibrio fibrisolvens* E14 gene (*cinB*) encoding an enzyme with cinnamoyl ester hydrolase activity is negatively regulated by the product of an adjacent gene (*cinR*). *Microbiology* 143, 1203–1210.

Dehority, B.A. and Tirabasso, P.A. (1998) Effect of ruminal cellulolytic bacterial concentrations on in situ digestion of forage cellulose. *Journal of Animal Science* 76, 2905–2911.

Denman, S., Xue, G.P. and Patel, B. (1996) Characterization of a *Neocallimastix patriciarum* cellulase cDNA (celA) homologous to *Trichoderma reesei* cellobiohydrolase II. *Applied and Environmental Microbiology* 62, 1889–1896.

Devillard, E., Newbold, C.J., Scott, K.P., Forano, E., Wallace, R.J., Jouany, J.-P. and Flint, H.J.

(1999) A xylanase produced by the rumen anaerobic protozoan *Polyplastron multivesiculatum* shows close sequence similarity with family 11 xylanases from Gram-positive bacteria. *FEMS Microbiology Letters* 181, 145–152.

Dijkerman, R., Opdencamp, H.J.M., Vanderdrift, C. and Vogels, G.D. (1997) The role of the cellulolytic high molecular mass (HMM) complex of the anaerobic fungus *Piromyces* sp. strain E2 in the hydrolysis of microcrystalline cellulose. *Archives of Microbiology* 167, 137–142.

Dijkstra, J. and Tamminga, S. (1995) Simulation of the effects of diet on the contribution of rumen protozoa to degradation of fibre in the rumen. *British Journal of Nutrition* 74, 617–634.

Din, N., Gilkes, N.R., Tekant, B., Miller, R.C., Jr, Warren, R.A.J. and Kilburn, D.G. (1991) Non-hydrolytic disruption of cellulose fibres by the binding domain of a bacterial cellulase. *Bio/Technology* 9, 1096–1099.

Ding, S.-Y., Bayer, E.A., Shoham, Y. and Lamed, R. (1999) Preliminary evidence of high-molecular-weight scaffodin-like proteins from *Ruminococcus flavefaciens*. P5.12, 3rd Carbohydrate Bioengineering Meeting, University of Newcastle Upon Tyne 11–14 April.

Doerner, K.C. and White, B.A. (1990) Assessment of the endo-ß-1,4-glucanase components of *Ruminococcus flavefaciens* FD1. *Applied and Environmental Microbiology* 56, 1844–1850.

Fanutti, C., Ponyi, T., Black, G.W., Hazlewood, G.P. and Gilbert, H.J. (1995) The conserved noncatalytic 40-residue sequence in cellulases and hemicellulases from anaerobic fungi functions as a protein docking domain. *Journal of Biological Chemistry* 270, 29314–29322.

Forsberg, C.W., Cheng, K.-J. and White, B.A. (1997) Polysaccharide degradation in the rumen and large intestine. In: Mackie, R.I. and White, B.A. (eds) *Gastrointestinal Microbiology*. Chapman & Hall, New York, pp. 319–379.

Gardner, R.G., Wells, J.E., Russell, J.B. and Wilson, D.B. (1995) The effect of carbohydrates on the expression of the *Prevotella ruminicola* β-1,4-D-endoglucanase. *FEMS Microbiology Letters* 125, 305–310.

Gardner, R.G., Wells, J.E., Fields, M.W., Wilson, D.B. and Russell, J.B. (1997) A *Prevotella ruminicola* $B_14$ operon encoding extracellular polysaccharide hydrolases. *Current Microbiology* 35, 274–277.

Gardner, P.T., Wood, T.J., Chesson, A. and Stuchbury, T. (1999) Effect of degradation on the porosity and surface area of forage cell walls of differing lignin content. *Journal of the Science of Food and Agriculture* 79, 11–18.

Gilbert, H.J., Hazlewood, G.P., Laurie, J.I., Orpin, C.G. and Xue, G.P. (1992) Homologous catalytic domains in a rumen fungal xylanase: evidence for gene duplication and prokaryotic origin. *Molecular Microbiology* 6, 2065–2072.

Gilkes, N.R., Henrissat, B., Kilburn, D.G., Miller, J.R.C. and Warren, R.A.J. (1991) Domains in microbial β-1,4-glycanases: sequence conservation, function, and enzyme families. *Microbiological Reviews* 55, 303–315.

Gong, J. and Forsberg, C.W. (1989) Factors affecting adhesion of *Fibrobacter succinogenes* subsp. *succinogenes* and adherence-defective mutants to cellulose. *Applied and Environmental Microbiology* 55, 3039–3044.

Gong, J., Egbosimba, E.E. and Forsberg, C.W. (1996) Cellulose binding proteins of *Fibrobacter succinogenes* and the possible role of a 180-kDa cellulose-binding glycoprotein in adhesion to cellulose. *Canadian Journal of Microbiology* 42, 453–460.

Grabber, J.H., Hatfield, R.D. and Ralph, J. (1998) Diferulate cross-links impede the enzymatic degradation of non-lignified maize walls. *Journal of the Science of Food and Agriculture* 77, 193–200.

Groleau, D. and Forsberg, C.W. (1983) Solubilization of the extracellular membraneous carboxymethylcellulase activity produced by the rumen bacterium *Bacteroides succinogenes*. *Canadian Journal of Microbiology* 29, 710–717.

Henrissat, B. (1998) Enzymology of cell-wall degradation. *Biochemical Society Transactions* 26, 153–156.

Henrissat, B., Teere, T.T. and Warren, R.A.J. (1998) A scheme for designating enzymes that hydrolyse the polysaccharides in the cell walls of plants. *FEBS Letters* 425, 352–354.

Hodrová, B., Kopecny, J. and Kas, J. (1998) Cellulolytic enzymes of rumen anaerobic fungi *Orpinomyces joyonii* and *Caecomyces communis*. *Research in Microbiology* 149, 417–427.

Jervis, E., Haynes, C. and Kilburn, D. (1997) Surface diffusion of cellulases and their isolated binding domains on cellulose. *Journal of Biological Chemistry* 272, 24016–24023.

Karita, S., Sakka, K. and Ohmiya, K. (1996) CBDs confer an enhanced activity against insoluble cellulose to *Ruminococcus albus* endoglucanase IV. *Journal of Fermentation Biotechnology* 81, 553–556.

Karita, S., Sakka, K. and Ohmiya, K. (1997) Cellulosomes, cellulase complexes of anaerobic microbes: their structure models and functions. In: Onodera, R., Itabashi, H., Ushida, K., Yano, H. and Sasuki, Y. (eds), *Rumen Microbes and Digestive Physiology in Ruminants*. Japan Scientific Societies Press, Tokyo/S. Karger, Basel, pp. 47–57.

Kirby, J., Martin, J.C., Daniel, A.S. and Flint, H.J. (1997) Dockerin-like sequences in cellulases and xylanases from the rumen cellulolytic bacterium *Ruminococcus flavefaciens*. *FEMS Microbiology Letters* 149, 213–219.

Kirby, J., Aurilia, V., McCrae, S.I., Martin, J.C. and Flint, H.J. (1998) Plant cell wall degrading enzyme complexes from the cellulolytic rumen bacterium *Ruminococcus flavefaciens*. *Biochemical Society Transactions* 26, S169.

Kopecny, J. and Hodrová, B. (1997) The effect of yellow affinity substance on cellulases of *Ruminococcus flavefaciens*. *Letters in Applied Microbiology* 25, 191–196.

Krause, D.O., Dalrymple, B.P., Smith, W.J., Mackie, R.I. and McSweeney, C.S. (1999) 16S rDNA sequencing of *Ruminococcus albus* and *Ruminococcus flavefaciens*: design of a signature probe and its application in adult sheep. *Microbiology* 145, 1797–1807.

Lamed, R., Naimark, J., Morgenstern, E. and Bayer, E.A. (1987) Specialized cell surfaces structures in cellulolytic bacteria. *Journal of Bacteriology* 169, 3792–3800.

Leatherwood, J.M. (1973) Cellulose degradation by *Ruminococcus*. *Federation Proceedings* 32, 1815–1819.

Li, X.-L., Chen, H. and Ljungdahl, L.G. (1997a) Monocentric and polycentric anaerobic fungi produce structurally related cellulases and xylanases. *Applied and Environmental Microbiology* 63, 628–635.

Li, X.-L., Chen, H. and Ljungdahl, L.G. (1997b) Two cellulases, CelA and CelC, from the polycentric anaerobic fungus *Orpinomyces* strain PC2 contain N-terminal docking domains for a cellulase–hemicellulase complex. *Applied and Environmental Microbiology* 63, 4721–4728.

Lin, L.L. and Thomson, J.A. (1991) An analysis of the extracellular xylanases and cellulases of *Butyrivibrio fibrisolvens* H17c. *FEMS Microbiology Letters* 84, 197–204.

Ljungdahl, L.G., Coughlan, M.P., Mayer, F., Mori, Y., Hon-Nami, H. and Hon-Nami, K. (1988) Macrocellulase complexes and yellow affinity substance from *Clostridium thermocellum*. *Methods in Enzymology* 160, 483–500.

McIntosh, L.P., Hand, G., Johnson, P.E., Joshi, M.D., Körner, M., Plesniak, L.A., Ziser, L., Wakarchuk, W.W. and Withers, S.G. (1996) The p$K_a$ of the general acid/base carboxyl group of a glycosidase cycles during catalysis: A $^{13}$C-NMR study of *Bacillus circulans* xylanase. *Biochemistry* 35, 9958–9966.

McLauchlan, W.R., Garcia-Conesa, M.T., Williamson, G., Rosa, M., Ravestein, P. and Maat, J. (1999) A novel class of protein from wheat which inhibits xylanases 1. *Biochemical Journal* 338, 441–446.

Maglione, G., Matsushita, O., Russell, J.B. and Wilson, D.B. (1992) Properties of a genetically reconstructed *Prevotella ruminicola* endoglucanase. *Applied and Environmental Microbiology* 58, 3593–3597.

Malburg, L.M., Iyo, A.H. and Forsberg, C.W. (1996) A novel family 9 endoglucanase gene (*cel*D), whose product cleaves substrates mainly to glucose, and its adjacent upstream homolog (*cel*E) from *Fibrobacter succinogenes* S85. *Applied and Environmental Microbiology* 62, 898–906.

Malburg, S.R.C., Malburg, L.M., Liu, T., Iyo, A.H. and Forsberg, C.W. (1997) Catalytic properties of the cellulose binding endoglucanase F from *Fibrobacter succinogenes* S85. *Applied and Environmental Microbiology* 63, 2449–2453.

Matte, A. and Forsberg, C.W. (1992) Purification, characterization, and mode of action of endoxylanases 1 and 2 from *Fibrobacter succinogenes* S85. *Applied and Environmental Microbiology* 58, 157–168.

Millward-Sadler, S.J., Hall, J., Black, G.W., Hazlewood, G.P. and Gilbert, H.J. (1996) Evidence that the *Piromyces* gene family encoding end-1,4-mannanases arose through gene duplication. *FEMS Microbiology Letters* 141, 183–188.

Miron, J. and Forsberg, C.W. (1998) Features of *Fibrobacter intestinalis* DR7 mutant which is impaired with its ability to adhere to cellulose. *Anaerobe* 7, 35–48.

Miron, J. and Forsberg, C.W. (1999) Characterisation of cellulose binding proteins which are involved in adhesion mechanism of *Fibrobacter intestinalis* DR7. *Applied Microbiology and Biotechnology* 51, 491–497.

Miron, J., Yokoyama, M.T. and Lamed, R. (1989) Bacterial cell surface structures involved in lucerne cell wall degradation by pure cultures of cellulolytic rumen bacteria. *Applied Microbiology and Biotechnology* 32, 218–222.

Miron, J., Morag, E., Bayer, E.A., Lamed, R. and Ben-Ghedalia, D. (1998) An adhesion defective mutant of *Ruminococcus albus* SY3 is impaired in its capability to degrade cellulose. *Journal of Applied Microbiology* 84, 249–254.

Mitsumori, M. and Minato, H. (1995) Distribution of cellulose-binding proteins among the representative strains of rumen bacteria. *Journal of General and Applied Microbiology* 41, 297–306.

Mitsumori, M., Minato, H., Sekizaki, T., Uchida, I. and Ito, H. (1996) Cloning, nucleotide sequence and expression of the gene encoding the cellulose-binding protein 1 (CBP1) of *Fibrobacter succinogenes* S85. *FEMS Microbiology Letters* 139, 43–50.

Mittendorf, V. and Thomson, J.A. (1993) Cloning of an endo-(1–4)-β-glucanase gene, *celA*, from the rumen bacterium *Clostridium* sp. (*C. longisporum*) and characterization of its product, CelA, in *Escherichia coli*. *Journal of General Microbiology* 139, 3233–3242.

Morris, E.J. and Cole, O.J. (1987) Relationship between cellulolytic activity and adhesion to cellulose in *Ruminococcus albus*. *Journal of General Microbiology* 133, 1023–1032.

Mosoni, P. (1999) Characterization of two spontaneous adherence defective mutants of *Ruminococcus albus* strains. *South African Journal of Animal Science* 29, 78–79.

Ohara, H., Karita, S., Kimura, T., Sakka, K. and Ohmiya, K. (1998) Multicellulase complex of *Ruminococcus albus* F40. Proceedings of the MIE Bioforum 1998 *Genetics, Biochemistry and Ecology of Cellulose Degradation*, Suzuka, Japan, p. 140.

Orpin, C.G. (1983) The role of ciliate protozoa and fungi in the rumen digestion of plant cell walls. *Animal Feed Science and Technology* 10, 121–143.

Pagès, S., Gal, L., Bélaïch, A., Gaudin, C., Tardif, C. and Bélaïch, J.-P. (1997) Role of scaffolding protein CipC of *Clostridium cellulolyticum* in cellulose degradation. *Journal of Bacteriology* 179, 2810–2816.

Pegden, R.S., Larson, M.A., Grant, R.J. and Morrison, M. (1998) Adherence of the gram positive bacterium *Ruminococcus albus* to cellulose and identification of a novel form of cellulose-binding protein which belong to the Pil family of proteins. *Applied and Environmental Microbiology* 180, 5921–5927.

Piquemal, J., Lapierre, C., Myton, K., O'Connell, A., Schuch, W., Grima-Pettenati, J. and Boudet, A.-M. (1998) Down-regulation of cinnamoyl-CoA reductase induces significant changes of lignin profiles in transgenic tobacco plants. *Plant Journal* 13, 71–83.

Poole, D.H., Hazlewood, G.P., Laurie, J.I., Barker, P.J. and Gilbert, H.J. (1990) Nucleotide sequence of the *Ruminococcus albus* SY3 endoglucanase genes celA and celB. *Molecular and General Genetics* 223, 217–223.

Ralph, J. (1997) Fake plastic trees. *Chemistry and Industry*, p. 708.

Stack, R.J. and Hungate, R.E. (1984) Effect of 3-phenylpropionic acid on capsule and cellulases of *Ruminococcus albus* 8. *Applied and Environmental Microbiology* 48, 218–223.

Takenaka, A., D'Silva, C.G., Kudo, H., Itabashi, H. and Cheng, K.-J. (1999) Molecular cloning, expression, and characterization of an endo-β-1,4-glucanase cDNA from *Epidinium caudatum*. *Journal of General and Applied Microbiology* 45, 57–61.

Tomme, P., Warren, R.A.J. and Gilkes, N.R. (1995) Cellulose hydrolysis by bacteria and fungi. *Advances in Microbial Physiology* 37, 1–81.

Travis, A.J., Hirst, D.J. and Chesson, A. (1996) Automatic classification of plant cells according to tissue type using anatomical features obtained by the distance transform. *Annals of Botany* 78, 325–331.

Trinci, A.P.J., Davies, D.R., Gull, K., Lawrence, M.I., Nielsen, B.B., Rickers, A. and Theodorou, M.K. (1994) Anaerobic fungi in herbivorous animals. *Mycological Research* 98, 129–152.

Weimer, P.J., Waghorn, G.C., Odt, C.L. and Mertens, D.R. (1999) Effect of diet on populations of three species of ruminal cellulolytic bacteria in lactating dairy cows. *Journal of Dairy Science* 82, 122–134.

Wilson, J.R. and Mertens, D.R. (1995) Cell wall accessibility and cell structure limitations to microbial digestion of forage. *Crop Science* 35, 251–259.

Wood, T.M., Wilson, C.A. and Stewart, C.S. (1982) Preparation of the cellulase from the cellulolytic anaerobic rumen bacterium *Ruminococcus albus* and its release from the bacterial cell wall. *Biochemical Journal* 205, 129–137.

Wood, T.M., Wilson, C.A. and McCrae, S.I. (1995) The cellulase system of the anaerobic rumen fungus *Neocallimastix frontalis*: studies on the properties of fractions rich in endo (1,4)-ß-D-glucanase activity. *Applied Microbiology and Biotechnology* 44, 177–184.

Xue, G.P., Orpin, C.G., Gobius, K.S. and Orpin, C.G. (1992) A novel polysaccharide hydrolase cDNA (*celD*) from *Neocallimastix patriciarum* encoding three multi-functional catalytic domains with high endoglucanase, cellobiohydrolase and xylanase activities. *Journal of General Microbiology* 138, 2397–2403.

Yang, W.Z., Beauchemin, K.A. and Rode, L.M. (1999) Effects of an enzyme feed additive on extent of digestion and milk production of lactating dairy cows. *Journal of Dairy Science* 82, 391–403.

# 6 The Microbial Ecology and Physiology of Ruminal Nitrogen Metabolism

M. Morrison

*Department of Animal Science, School of Biological Sciences and Center for Biotechnology, University of Nebraska, Lincoln, Nebraska, USA*

## Introduction

The relationships between the ruminant animal and its resident microflora, and the ensuing impacts on protein nutrition, have been examined in detail for more than 40 years. The foundational studies such as those by Hogan and co-workers demonstrated the interface between host and microbial metabolism of non-protein nitrogen sources; and the practical ramifications of this interface were clearly demonstrated in the widely referenced study by Virtanen (1966). Egan (1965) helped establish another precedent in our understanding of ruminant protein nutrition. By showing that positive responses in animal physiology coincided with the intra-abomasal infusion of casein, their studies serve as the underpinning of the 'by-pass protein' concept. To this day, much of the practical interest in ruminal nitrogen metabolism relates to the optimization of microbial growth with different nitrogen sources, as well as the optimization of post-ruminal supply of microbial and feed proteins. Additionally, as the relative amount and/or biological value of feed protein increases, there is a shift in research interest, away from the biosynthetic processes of ruminal microbes, and more towards their degradative processes. For these reasons, the ruminal microflora continues to provide a unique and varied set of challenges for nutritionists and microbiologists intent on improving the protein nutrition of animals, in both extensive and intensive production scenarios.

Hungate (1960) considered that a meaningful analysis of any microbial habitat requires an understanding of: (i) the types of kinds of microorganisms present (ecology); (ii) the activities possessed by these microorganisms (enzymology); and (iii) factors affecting the expression of these activities (regulation). There is no shortage of recent reviews that provide a detailed accounting of our current understanding of these aspects, from an organismal and biochemical perspective (Morrison and Mackie, 1996; Cotta and Russell, 1997; Wallace *et al.*, 1998 ). This review will attempt to take a different approach to the topic, by focusing on recent and pending advances contributed via genetics and molecular biology. In addition to providing new information relating to the ecology and enzymology of ruminal nitrogen metabolism, these techniques are likely to help elucidate how nitrogen metabolism is coordinated, and what other physiological process(es) ensue. Ultimately, a more precise identification of the rate-limiting parameter(s) of ruminal nitrogen metabolism, from an ecological, enzymatic, or regulatory perspective, should expedite advances in ruminant protein nutrition.

## A molecular-based perspective of microbial ecology and ruminal nitrogen metabolism: cultivation-dependent and cultivation-independent studies

The value from examination of ruminal microorganisms in pure or defined mixed cultures can not be overstated. For instance, the principle of interspecies hydrogen transfer, foundational to our understanding of the anaerobic metabolism of organic matter, arose from such studies. However, most microbiologists and nutritionists interested in the ruminal habitat have also wondered how representative laboratory isolates are of the microbial diversity inherent to the rumen. Additionally, how representative is the growth and metabolism of ruminal microorganisms under laboratory conditions, relative to their behaviour in their 'real world'? Various molecular-based technologies are now available and are being gradually adapted for use with ruminal microorganisms. Although an inherent advantage of these technologies will be their utilization for *in situ* examination of ruminal microbiology and ecology, their greatest impact in ruminal microbiology is currently arising from their integration with other microbiological methods. Some examples with specific relevance to ruminal nitrogen metabolism are outlined below.

### *Proteolytic and obligate peptide- and amino-acid-fermenting bacteria*

Probably the most salient development in ruminal microbiology since the isolation and characterization of the anaerobic fungi has been the identification of obligate peptide- and amino-acid-fermenting bacteria. Chen and Russell (1988, 1989) obtained three isolates of Gram-positive bacteria not only with very high specific activities of ammonia production, but able to grow rapidly with peptides or amino acids as their sole energy source. Although two of the isolates were ultimately identified by taxonomic characteristics as *Peptostreptococcus anaerobius* and *Clostridium sticklandii,* the third isolate possessed taxonomic characteristics which did not clearly match the criteria necessary for assignment to established bacterial species. This discrepancy was resolved by the use of molecular techniques; in this case, 16S rRNA samples were used as templates for primer elongation by reverse transcriptase, and nearly complete sequences were aligned with other sequences available in the databases. On the basis of these studies the third isolate warranted designation as a new species and, following the isolation of additional strains with similar phenotypic and phylogenetic characteristics, the strains were classified as *Clostridium aminophilum* sp. nov. (Paster *et al.*, 1993).

Although the ecological significance of these types of bacteria in ruminal nitrogen metabolism was initially questioned by some, bacteria with similar ecophysiological properties have been isolated from ruminants in other parts of the world. Attwood *et al.* (1998) recently isolated from New Zealand ruminants a variety of bacteria with high specific rates of ammonia production. Five of these strains are phenotypically similar to the isolates identified by Chen and Russell (1988, 1989), at least in terms of sensitivity to monensin and their growth being restricted primarily to small peptides or amino acids. Phylogenetic (16S rRNA) analyses placed two of the isolates in the genus *Peptostreptococcus,* one in the genus *Eubacterium,* and two more in the family *Bacteroideacea* (presumably, the first Gram-negative bacteria of this type). Wallace *et al.* (1999) have also described the biochemical and phylogenetic characteristics of a Gram-positive rod named *Eubacterium pyruvovorans* sp. nov., which is capable of growth with either peptides, amino acids, or organic acids such as pyruvate or oxaloacetate. In addition to acetate, this bacterium also produced long- and branched-chain organic acids such as valerate and caproate during growth. It is now widely accepted that bacteria with high specific rates of ammonia production are not uncommon in ruminant animals and, in some instances, might comprise a significant percentage of the culturable bacteria in the rumen.

Despite the breakthroughs in our understanding of ruminal ammonium production arising from these cultivation-dependent studies, there are some inherent difficulties with such an approach. The culturable number of these bacteria can be relatively low ($\sim 10^7$ per gram of ruminal contents; Chen and Russell, 1989) and the lack of completely selective media can make it difficult to accurately quantify their numbers (Attwood *et al.*, 1998). To overcome these difficulties, molecular techniques have been developed and utilized effectively for population analysis, and with a variety of production/nutritional scenarios. Krause and Russell (1996) were the first to use a combination of *in vivo* sampling and a continuous culture system inoculated with predominant ruminal bacteria (PRB), to examine how monensin influences the persistence and abundance of the obligate amino-acid-fermenting bacteria. Despite all three strains being monensin sensitive in pure culture under batch conditions, *C. aminophilum* was found to persist and increase in relative abundance, both *in vivo* and within *in vitro* continuous cultures containing PRB, in the presence of monensin. Although other reasons have been forwarded to explain the 'protein sparing' effect of monensin, the conclusions reached by Krause and Russell (1996) also seem valid: monensin does not have a greater impact on ammonia production kinetics because of the persistence of *C. aminophilum.* It will be interesting to see if similar patterns of persistence exist for any of the other types of obligate amino-acid-fermenting bacteria isolated elsewhere. Studies such as these also highlight that while much can be learned from pure culture studies, microorganisms can behave quite differently when cultivated as part of a more complex consortia.

Another molecular-based technology which circumvents cultivation-dependent analysis of ruminal microbiology is competitive PCR (cPCR). The technique has been used to enumerate the abundance of the proteolytic bacterium *Clostridium proteoclasticum* in New Zealand ruminants (Reilly and Attwood, 1998). Under laboratory conditions, this bacterium produces extremely high levels of proteinase activity, and 16S rRNA analysis showed it is very similar to several bacterial strains currently classified within the genus *Butyrivibrio fibrisolvens.* The cPCR technique is dependent on the development of a 'competitor DNA' template, which is usually generated by removal of a restriction enzyme fragment within a clone of a 16S rRNA gene isolated from the bacterium being studied. Primers specific only for the intact gene ('target' DNA) and the competitor template are then used in PCR reactions, the products are resolved by agarose gel electrophoresis and the log ratio of band intensities for the target and competitor PCR products are determined. A standard curve is first generated by combining serial dilutions of cells containing the target DNA with fixed amounts of competitor DNA, and the log ratio of PCR band intensities is plotted relative to cell numbers. DNA from environmental samples are then 'spiked' with a known quantity of the competitor template and subjected to PCR. In this manner, the quantity of target DNA in an environmental sample can be quantified, and in the case of *C. proteoclasticum,* as few as 2500 cells $g^{-1}$ of ruminal contents can be detected. Reilly and Attwood (1998) showed the numbers of *C. proteoclasticum* and closely related species did not vary greatly among animals consuming rations with varying levels of carbohydrate and/or protein. These type of studies have now been expanded to include a greater range of proteolytic bacteria, including *Streptococcus, Butyrivibrio* and *Eubacterium* spp. (Attwood and Klieve, 2000). Interestingly, although the numbers of *Streptococcus* spp. were unchanged in response to diet, the numbers of *Butyrivibrio* were stimulated, and the numbers of *Eubacterium* spp. suppressed, by carbohydrate supplementation. Similar studies are now underway for populations of obligate amino-acid-fermenting bacteria (G.T. Attwood, personal communication), and it will be interesting to see whether the fluctuations in proteolytic bacteria in response to plane of nutrition extend also to this group of ruminal bacteria.

## Examination of intraruminal nitrogen recycling

Since Nolan and Leng (1972) showed that 30–50% of rumen microbial protein may recycle through the ammonium-N pool, ecological and physiological explanations for this intraruminal N-recycling have been sought. Attention in the past has been directed largely towards quantifying the impact of ruminal protozoa on this phenomenom, primarily via the use of defaunation agents. In general terms, elimination of ruminal protozoa corresponds with increased bacterial yield and improved protein outflow from the rumen, provided there is limited effect from defaunation on the ruminal fermentation of organic matter (Jouany *et al.*, 1988). Additionally, Ffoulkes and Leng (1989) estimated that as much as 74% of the protozoal biomass never leaves the rumen-reticulum. However, although this may be true for forage-fed animals, evidence suggests that ruminal protozoa not only slow down the rate of ruminal starch digestion in grain-fed animals, but shift the site of its digestion to the small intestine, as starch-laden protozoa are removed from the rumen (Mendoza *et al.*, 1993). Given the inherent difficulties associated with microscopic examinations to quantify and differentiate ruminal protozoa, as well as with the use of biochemical markers (such as diaminopimelic and aminoethylphosphonic acids) to quantify flow rates of protozoal and bacterial nitrogen from the rumen, molecular methods perhaps offer alternative approaches. The genes encoding small subunit rRNAs, and one or two other highly conserved structural proteins, have now been isolated from cDNA libraries of several predominant rumen protozoa (e.g. Wright *et al.*, 1997; Eschenlauer *et al.*, 1998). It should be possible to extend the application of these clones beyond phylogenetic analyses, and develop probes to quantify the relative abundance of particular groups of ruminal protozoa. One interesting example would be to examine the degree to which protozoal lysis contributes to intraruminal N-recycling, and whether the elevated level of ammonium present in faunated animals is attributable more to protozoal predation of bacteria, or protozoal lysis as suggested by Wells and Russell (1996).

More recently, the relationship(s) between intraruminal N-recycling and bacteriophage populations has received some attention, but unfortunately, efforts to understand this component of ruminal ecology are not widespread. Largely due to the efforts of Klieve and co-workers, molecular technologies have provided and may continue to provide relevant breakthroughs in our understanding of bacteriophage biology and population dynamics. Bacteriophages capable of infecting some of the better-known ruminal bacteria have been isolated and partially characterized (Klieve *et al.*, 1989) and both lytic and temperate bacteriophages have been identified (Lockington *et al.*, 1988; Kleive *et al.*, 1989, 1991). All known ruminal bacteriophages possess a single copy of a double-stranded DNA molecule, which has facilitated the development of techniques to examine genetic diversity and bacteriophage abundance in response to diet and feeding behaviour. Klieve and Swain (1993) developed a method of harvesting bacteriophages from ruminal fluid coupled with the lysis of the bacteriophage capsid and separation of bacteriophage DNA by pulsed-field gel electrophoresis (PFGE). Genome size can be estimated by the migration distance of the DNA molecules in PFGE gels, relative to a set of DNA standards of known size. Additionally, changes in the abundance of DNA within a particular region of the gel can be quantified by laser densitometry, and the diurnal variation in bacteriophage numbers can be estimated from these changes in DNA staining intensity. Using these methods, Klieve and Swain (1993) found bacteriophage DNA could be subdivided into two main components. The bulk of the DNA, postulated to have been derived from temperate bacteriophages, was present as a broad band ranging from 30 to 200 kb. Distinct bands of DNA, both small (~10 kb) and very large (~850 kb) in size, were also apparent and postulated to represent blooms of lytic bacteriophage. Using these same methods, Swain *et al.* (1996) also showed there is a marked diurnal fluctuation in total bacteriophage numbers, the lowest concentrations occurring within two hours of feeding, and maximal concentrations occuring 8–10 h later. In addition to this daily fluctuation in numbers, Klieve *et al.* (1998) found that animals consuming green pasture tend to have the highest concentration of bacteriophages, and the concentrations were twofold and tenfold lower in animals consuming dry forages, and grain-fed animals, respectively. However, whether these differences are the result of sampling time relative to feeding, or other factors such as bacterial resistance to bacteriophage infection, are not clear.

With the development of these techniques the next logical step appears to be to superimpose measurements of the intraruminal turnover of nitrogen among ammonium and non-ammonium pools upon quantitative measurements of fluctuations in protozoal, bacterial and bacteriophage populations. Accordingly, the relative contributions from each of these biotic components of the ruminal habitat to intraruminal nitrogen recycling, and the physicochemical factor(s) underpinning the phenomenon, should be elucidated. Such information is most likely necessary for the development of new, productive methods of curbing intraruminal nitrogen recycling.

In summation, the molecular methods outlined above are providing the means necessary to examine rumen microbial diversity and phylogeny in more detail. The transition of these technologies from providing information of fundamental microbiological significance to the realm of providing information of a more pragmatic nature is gradually occurring. These technologies also provide the opportunity for detailed, quantitative examination of microorganisms *in situ* and their response to fluctuations in nutrient supply.

## A genetics-based approach to the examination of microbial physiology and ruminal nitrogen metabolism

New insights into ruminal microbiology, especially in relation to fibre degradation, have been forthcoming from the use of recombinant DNA technology (Flint, 1997). However, application of the technology has also been one-sided, limited largely to the characterization of clones stably maintained and expressed in *Escherichia coli*. Although heterologous gene expression in ruminal bacteria still remains limited and evasive, other genetics-based methods of analysis are being used to generate a clearer understanding of ruminal nitrogen metabolism. Two methods finding recent application in ruminal microbiology will be described here: mutational analysis and differential display reverse transcriptase PCR (DD RT-PCR).

### *Mutational analysis of nitrogen metabolism in* Prevotella bryantii

Mutagenesis procedures, despite being used extensively in microbiological studies, have found little application to-date in the field of ruminal microbiology. Newbold *et al.* (1992) described the properties of ionophore-resistant *Prevotella* (*Bacteroides*) *ruminicola* cells enriched by cultivation in the presence of increasing concentrations of tetronasin; and spontaneous mutants of *P. ruminicola* strain $B_1$4 resistant to rifampicin have also been isolated in a similar manner (Shoemaker *et al.*, 1991). However, the rate of spontaneous mutagenesis is sufficiently low that without a strong selection strategy (e.g. antibiotic resistance) the isolation of mutants defective in certain enzymes or metabolic pathways is impractical. Although advances have been made in the development of gene transfer protocols for use with ruminal bacteria (for example, Shoemaker *et al.*, 1991; Gardner *et al.*, 1996), there are still limitations in terms of transfer frequency and available selective markers, to faciltiate the use of transposon mutagenesis or gene displacement strategies with these bacteria. For these reasons, we have successfully adapted the mutagenesis procedures described by Miller (1992) for use under anaerobic conditions and utilizing ethyl methane sulphonate (EMS) as a mutagen. Maximal mutation frequencies were achieved by exposure to EMS for periods of 30 min or longer, and were approximately 200-fold higher than the frequency of spontaneous mutation (Madeira *et al.*, 1997). However, the EMS-induced frequency was still much lower than that typically observed with enteric bacteria, and we chose to also use an ampicillin enrichment procedure before screening for mutants (Wen, 1998).

These mutagenesis procedures have not only allowed us to ask fundamental questions relating to bacterial physiology, but also to quantitatively evaluate the implications of changes/loss of enzyme activity on ruminal nitrogen metabolism. Our efforts to date have focused on the peptidase and glutamate dehydrogenase activities of *P. bryantii*. Wallace and co-workers have shown that the predominant forms of peptidase activity measurable in ruminal contents appear to be dipeptidyl aminopeptidases, most active against neutral oligopeptides such as penta-alanine, as well as the synthetic substrate L-glycyl-L-argininyl-methylnapthylamide (Gly-Arg-MNA; Wallace and McKain, 1989; Wallace, 1994). Using pure cultures of ruminal bacteria, Wallace *et al.* (1990) and Wallace and McKain (1991) showed that *P. ruminicola* and *B. fibrisolvens* both exhibited a dipeptidyl aminopeptidase-like activity against penta-alanine, but *P. ruminicola* isolates were the only bacteria with measurable Gly-Arg-MNAse (hereafter referred to as PrtA) activity. McKain *et al.* (1992) extended these studies by isolating bacteria from ruminal contents and screening for PrtA activity and of the twelve positive isolates obtained, ten were consistent with the subgroupings of *Prevotella* spp. outlined by Avgustin *et al.* (1994).

Wallace's group has since attempted to use chromatographic procedures to enrich and purify the peptidases produced by *Prevotella* spp. Main peaks of activity following size exclusion chromatography of *P. bryantii* and *Prevotella brevis* extracts were estimated to contain proteins in excess of 100 kDa (Wallace *et al.*, 1995); ion-exchange chromatography of *Prevotella albensis* cell extracts separated peptidase activity into four distinct peaks with different substrate specificities (Wallace *et al.*, 1997). Madeira *et al.* (1997) found the inhibition profile of the PrtA was strikingly similar to that of gingipain, an extracellular, trypsin-like enzyme isolated from *Porphyromonas* (*Bacteroides*) *gingivalis,* that requires cysteine for activation and calcium for stabilization (Chen *et al.*, 1992). Despite these efforts, none of these peptidases appear to have been purified to homogeneity and the relative contribution of individual enzymes to proteolysis and ammonium production by ruminal bacteria were not quantified. Within this context, Madeira *et al.* (1997) chose to generate mutants of *P. bryantii* strain $B_14$ defective in PrtA activity for three primary reasons: (i) to assess the physiological role of this enzyme activity on growth of *P. bryantii*; (ii) to quantify the contribution from this enzyme to ruminal proteolysis, and ammonium production by the obligate amino-acid- and peptide-fermenting bacteria isolated by Russell and co-workers; and (iii) to facilitate comparative examination of wild type and mutant strains to identify polypeptide(s) responsible for this activity. Two independently derived mutants were isolated by Madeira *et al.* (1997) and in addition to the virtual elimination of measurable PrtA activity in both mutants, activity towards Arg-Arg-MNA was also lost. Interestingly, a second cysteine proteinase from *P. gingivalis*, termed argingipain, which presented a narrow specificity for synthetic substrates containing Arg in the P1 site and hydrophobic amino acids in the P2 and P3 sites, has also been described (Kadowaki *et al.*, 1994) and subsequent studies showed gingipain and argingipain to be the same enzyme (Okamoto *et al.*, 1995). Therefore, the *P. bryantii* mutant strains may lack Arg-Arg-MNAse activity because the same enzyme is responsible for both activities, rather than the mutation(s) giving rise to polar or pleiotropic effects on the expression of multiple genes. That the mutation(s) in the mutant strains does not result in pleiotropic effects is further supported by the proteome profiles of the wild type and mutant strains, which are virtually indistinguishable from each other. The similarity of the results obtained with *P. gingivalis*, and *P. bryantii,* also suggests that this family of cysteine proteinases, although not identified in other eubacterial lines of descent, is widespread amongst bacteria belonging to the family *Bacteroidacea.*

Although the physiological role of Gly-Arg-MNAse enzyme activity in *P. bryantii* is still not clear, an ecological role for this enzyme activity was clearly demonstrated by co-culture experiments with the obligate amino-acid-fermenting bacteria isolated by

Russell's group. Gelatin hydrolysate was chosen as the substrate for these experiments, to ensure that the results were directly comparable with earlier studies that established the prominent roles of *C. aminophilum* and *P. anaerobius* in controlling ruminal ammonia production (Chen and Russell, 1989). Both the rate and extent of ammonia production was decreased approximately 25% when the *P. bryantii* mutants replaced the wild type strain in co-cultures. Future strategies which specifically inhibit PrtA activity should therefore result in productive alterations in ruminal ammonia production. We also have isolated mutant strains of *P. bryantii* defective in their growth with peptides as sole N-source (Peng and Morrison, 1995) and Wallace's group also have a set of mutant *P. albensis* strains which appear to possess reduced levels of all measurable peptidases. Further characterization of these mutants is currently underway and will probably provide further insights into the genetics and molecular biology controlling rate-limiting steps in proteolysis and peptide uptake in this numerically dominant group of ruminal bacteria. With such information, new productive strategies to control ruminal proteolysis should be forthcoming.

We have also used these mutagenesis procedures to examine ammonia assimilation by *P. bryantii.* A number of *Prevotella* isolates have been found to lose cell viability following growth in a defined medium containing a relatively high concentration of glucose and a relatively low concentration of ammonia; this glucose toxicity was correlated with methylglyoxal production and its accumulation in the growth medium (Russell, 1992). More detailed studies with *P. bryantii* showed that, unlike enteric bacteria, methylglyoxal production occurred independently of phosphate limitation, perhaps instead related to aspects of carbon and/or nitrogen metabolism (Russell, 1993). Given that glutamate dehydrogenase (GDH) serves as a key enzyme linking carbon and nitrogen metabolism, we were interested in determining the role of this enzyme under different growth conditions. A mutant lacking GDH activity was isolated following EMS mutagenesis and ampicillin enrichment procedures (Z. Wen and M. Morrison, unpublished data) and nucleotide sequence analysis of the *gdhA* gene from the mutant strain confirmed a missense mutation had been created within one of the highly conserved motifs of the GdhA protein subunit. A clone of the mutant *gdhA* gene also did not complement the *E. coli* glutamate auxotroph used to successfully isolate clones of the wild type *gdhA* gene from *P. bryantii.* A number of growth experiments have been conducted with this mutant strain, and the only measurable phenotype associated with the GDH mutant strain was a slower growth rate when ammonium was used as the sole nitrogen source (108 min as compared with 75 min for the mutant and wild type strains, respectively). The mutant strain was also still capable of growth with both limiting (1 mM) and excess (10 mM) ammonium. The mutant strain was also incapable of growth in the presence of the glutamine synthetase inhibitor methiosulphoxamine, unless trypticase or a supplement of glutamine was provided. Although GDH activity appears relevant for optimal growth of *P. bryantii,* these results confirm the bacterium possesses more than one route of ammonium assimilation and glutamate biosynthesis, which also appears to be functionally independent of the prevailing ammonium concentration in the growth medium.

In summation, mutagenesis strategies offer a valuable and productive means of examining bacterial physiology, and to evaluate the implications on ruminal function that may arise from the inhibition or elimination of specific enzyme activities. Although the strategy outlined here has been productive, other methods of mutational analysis will need to be developed, primarily to facilitate gene identification. Although progess has been slow, advances in rumen bacterial genetics are being made (Flint 1997) including the successful use of gene displacement strategies (Brooker and McCarthy, 1997). Such strategies must be broadened in scope and application if further advances in our understanding, and opportunities for manipulating ruminal function, are to be achieved.

### *Differential display RT-PCR*

The completed microbial genome sequencing projects have revealed that as many as 30% of the presumptive coding sequences in prokaryote genomes have no current match in databases and, therefore, are of unidentified function (Hinton, 1997). Recent advances in molecular biology facilitate new opportunities to identify and ascribe function to these so-called 'FUN' genes. Both *in vivo* expression technology and DD RT-PCR have permitted the identification of stress-, acid-, starvation- and colonization-inducible genes in pathogenic bacteria (Wong and McClelland, 1994; Valdivia and Falkow, 1997). While the former is applicable with bacteria for which relatively sophisticated genetic techniques are available, DD RT-PCR can be applied to microorganisms for which little or no techniques in bacterial genetics have been developed. However, there currently appears to have been limited application of this approach other than in the study of bacterial pathogenesis. We have begun to utilize DD RT-PCR as part of our studies of *Ruminococcus albus,* a Gram-positive anaerobe best recognized for its role in cellulose degradation in ruminant animals. Although our efforts to date are not directly related to nitrogen metabolism, our findings do reflect the potential power associated with this technique. The DD RT-PCR procedure used with *R. albus* RNA was essentially the same as described by Liang *et al.* (1993) except RT of the DNA-free total RNA was performed with random hexamers, and [$\alpha$-$^{33}$P] dATP was used in PCR amplification of cDNA. Ten-mer oligonucleotides were used as primers for second DNA strand synthesis, and were chosen on the basis of limited sequence homology with the 16S rDNA gene of *R. albus* strain 8. To confirm that the RNA was not contaminated with DNA, PCR reactions were carried out with RT samples which contained no reverse transcriptase. Both the RT and PCR reactions were done in duplicate, and the cDNA products resolved on 6% (wt/vol) polyacrylamide gels under denaturing conditions. Putative differentially displayed bands were excised from polyacrylamide gels, and the DNA was reamplified using the same primer and PCR conditions. The DNA was ethanol precipitated and cloned into TA-cloning vectors and the nucleotide sequence of the cloned DNA was obtained using universal T7 and M13 reverse primers. The cloned DNA was then used as a template to produce [$\alpha$-$^{32}$P] dCTP-labelled probes by random-primer labelling and northern blot analysis, to confirm differential expression of specific transcripts. Of the 18 putative dESTs identified in these initial experiments, six were confirmed to be differentially expressed. Alignments of these nucleotide sequences with other entries in the databases revealed significant homology with genes involved with protein export and solute uptake (Larson and Morrison, 2000).

Our use of DD RT-PCR has contributed valuable information about the response of *R. albus* to biochemical and physical cues, and will facilitate our efforts to develop a comparative model with other cellulolytic and/or Gram-positive bacteria. We are currently adapting these procedures to examine differential gene expression by *P. bryantii* in response to nitrogen source and availability, and similar efforts are already underway in other labs with *Ruminococcus flavefaciens*. The challenge will be to ensure that other genetic technologies are also advanced, to ensure that these efforts extend beyond an exercise in gene identification and influence our understanding of rumen physiological ecology.

## Nitrogen control in ruminal bacteria

Because microbial protein is generally the predominant source of amino acids for ruminants, animal scientists have invested considerable effort in trying to determine the 'optimal' ruminal nitrogen requirements to maximize microbial protein synthesis. However, there are still large variations in bacterial growth rates and cell yields, even when animals are fed similar diets (Owens and Goetsch, 1986; Cotta and Russell, 1997). Since the biochemical examination of ammonium assimilation and nitrogen regulation by Smith and Hespell in the early 1980s, nitrogen control in ruminal microorganisms has received scant attention. Perhaps this is due to the widely held belief that the proteolytic enzymes of ruminal microorganisms are not tightly regulated, and ruminal ammonium concentrations are unlikely to be growth-limiting under most production scenarios (Wallace *et al.*, 1998). However, as outlined above, *Prevotella* spp. support toxic schemes of carbon metabolism (methylglyoxal production) when an energy source is in great excess relative to ammonium (Russell 1992, 1993). Profound effects on bacterial yield are also seen if *Prevotella* spp. are cultivated first on peptides, then changed to a growth environment containing ammonium as sole nitrogen source. Russell (1983) found that a relatively high peptide concentration (15 g $l^{-1}$) blocked ammonium assimilation in the bacterium now classified as *P. bryantii*. While this concentration of peptides might seem high, Russell (1983) calculated that less than 5% of the peptide nitrogen available was actually utilized. Therefore, the 'effective concentration' of peptide nitrogen that blocked ammonia assimilation may actually be much lower. Russell (1983) also found that *P. bryantii* demonstrated a prolonged lag in ammonia assimilation and cell growth, even though glucose fermentation was significant, upon transfer to a medium lacking peptides. These findings show some similarity to earlier observations made by Pittman and Bryant (1964) with *P. ruminicola*. It is also interesting to note that Cotta and Russell (1982) could not establish *P. bryantii* in continuous culture with a medium lacking peptides, presumably using an inoculum previously grown with peptides. It appears that peptide nitrogen not only blocks ammonium assimilation in *Prevotella* spp., but fermentation of energy proceeds uncoupled from protein synthesis (energy spilling), once peptides are depleted. Such a situation might arise in cattle fed low protein forages, where the rumen receives some protein nitrogen, but the microbiota is primarily reliant upon recycled, non-protein nitrogen sources. However, not all ruminal bacteria are capable of energy spilling, but are instead subject to rapid rates of death and lysis once ammonium is depleted. *Fibrobacter succinogenes,* a predominant cellulolytic bacterium, is one such example (Wells and Russell, 1995). The rate of autolysis appears to be modulated by the expression of substance(s) which inactivate autolysins, the expression of which appears to be influenced by nitrogen availability (Wells and Russell, 1996).

Therefore, while the conceptual focus of nitrogen-control in ruminal bacteria has often been in relation to shifts in assimilatory and biosynthetic pathways, fluctuations in nitrogen availability also result in periods of toxic schemes of carbon metabolism, energy spilling, and elevated rates of bacterial autolysis. With the exception of Russell and co-workers, there has been relatively little attention paid to how fluctuations in ruminal peptide *and* ammonia concentrations, as well as other physicochemical factors, might affect the balance between anabolic and catabolic pathways, and thereby influence microbial yield via cell lysis and/or death. For these reasons, the study of nitrogen control in ruminal bacteria needs to be given greater emphasis. Given their numerical predominance in the rumen, combined with their somewhat unusual characteristics in terms of nitrogen sources used for growth and response to nitrogen availability, we have decided to examine nitrogen control in *Prevotella* spp. Unlike most other prokaryotes, but similar to other members of the family *Bacteroidaceae*, NADPH-utilizing GDH activity is greatest in all *Prevotella* species tested following growth with low exogenous concentrations of ammonium, and the greatest relative change in GDH activity appears to be down-regulation in response to growth in the presence of peptides. Addition of trypticase to cultures of all three *Prevotella* spp. already growing on ammonium resulted in substantial reductions in NADPH-utilizing GDH activity, although the time required for the decrease in enzyme activity was different among strains. In strain $B_1$4, significant reductions were not observed until one doubling time after the addition of trypticase (72 min), consistent with regulation at the level of GDH synthesis. In contrast, NADPH-utilizing GDH specific activity decreased 60% and 90% in *P. ruminicola* strain 23 and *P. brevis* strain GA33, respectively, within 20 min of the addition of peptides, which is more consistent with the regulation of enzyme activity. Interestingly, the NADH-utilizing GDH specific activity in strains 23 and GA33 appeared to increase once peptides had been added to the growth medium, and the cells were allowed to grow for one doubling time. We subsequently confirmed by non-denaturing PAGE that *P. brevis* actually produces a second, NAD-dependent GDH in response to growth on peptides (Wen and Morrison, 1997).

Although the underlying mechanisms of this nitrogen control remain to be determined in *Prevotella* spp., some clues have been provided from analogous studies conducted with *Bacteroides thetaiotaomicron.* Similar to *P. brevis,* this bacterium possesses two glutamate dehydrogenases, an NAD(P)H-utilizing enzyme (GdhA) as well as an NAD(H)-utilizing enzyme (GdhB) and their regulation in response to nitrogen source has been examined (Baggio and Morrison, 1996; W. Xing, L. Peng, L. Baggio, T. Chen and M. Morrison, unpublished data). In the wild type strain, GdhA activity was maximal when ammonium was provided as sole nitrogen source, and GdhB activity was only detectable when peptides were provided as the sole nitrogen source. To examine whether *gdhA* and/or *gdhB* gene expression was nitrogen regulated, a reporter gene shuttle vector was constructed, which contained a promoterless copy of a bifunctional xylosidase/arabinosidase originally cloned from *Bacteroides ovatus* (*xyaA*; W. Xing, L. Peng, L. Baggio, T. Chen and M. Morrison, unpublished data). In the wild type strain carrying a *gdhA–xyaA* fusion, xylosidase activity was maximal following growth in ammonium-limited medium, and decreased in response to the inclusion of peptides in the growth medium. In wild type strains carrying a *gdhB–xyaA* fusion, xylosidase activity was measurable only when peptides were available. Primer extension analysis and examination of the nucleotide sequence upstream of the ORF encoded by *gdhA* identified putative *cis*-acting sequences, and a motif similar to the consensus NtcA (nitrogen control protein A) binding motif, the global nitrogen regulatory system of cyanobacteria was identified. A role for this element in the regulation of *gdhA* expression in response to nitrogen source was confirmed by PCR-mediated deletion of the motif, and using the reporter gene vector described above. Despite the predicted changes in GdhA enzyme activity in response to nitrogen source, xylosidase activity remained low in those transconjugants containing the reporter gene fused with the mutagenized *gdhA* promoter. From these results, we conclude the NtcA-like motif is actually a *cis*-acting element, and that it facilitates a positive regulation of *gdhA* expression in response to ammonium. In cyanobacteria, the NtcA protein is required for the transcriptional activation of *glnN,* the structural gene encoding the glutamine synthetase (GSIII) enzyme subunit, which to date, has only been identified in the cyanobacteria, *Bacteroides, Prevotella, Butyrivibrio and Ruminococcus* . The NtcA protein has also recently been shown to repress gene expression in *Anabaena* PCC 7120, notably, the *rbcLS* operon, which encodes the large and small subunits of ribulose-1,5-bisphosphate carboxylase (Rubisco; Jiang *et al.*, 1997). It appears that in cyanobacteria, the NtcA protein has a role to play in balancing rates of nitrogen and carbon assimilation. Although putative NtcA binding motifs can be identified upstream of various nitrogen assimilation genes from the ruminal bacteria mentioned above, further studies are required to confirm conservation of this regulatory protein, in addition to the *glnN* gene.

## Concluding remarks and future perspectives

The ruminal habitat has provided microbiologists with a rich bounty of information pertaining to microbial schemes of anaerobic metabolism, and the interactions among ruminal microorganisms and the host animal have been well described. In the past, much of the information has been derived by cultivation of a particular microorganism either axenically, or as part of a restrictive subset of the rumen microbial consortia. This research has led to a number of strategies which offer the potential to manipulate ruminal nitrogen metabolism, although most of these strategies involve either feed additives (e.g. monensin or other ionophores) or 'protected' supplements of amino acids, peptides or protein. Further options for the productive alteration of ruminal nitrogen metabolism are still desired, especially for grazing and forage-fed ruminants and/or intensive production scenarios where nitrogen loss into the environment is a societal concern.

Molecular technologies now provide the potential to more completely examine a microbe's response to its growth environment, through the rapidly emerging fields of molecular ecology, genomics and proteomics. In the future, how might these technologies assist with the overall goal of optimizing ruminant production? With reference to nitrogen metabolism, there appears to be some value associated with identifying precise way(s) to inhibit specific enzymes, such as the gingipain-like activity produced by *Prevotella* spp. Although current information is limited, these peptidases appear to be functionally and structurally distinctive, relative to mammalian and other microbial enzymes, which should facilitate inhibitor specificity. However, it is also possible that we may not be fully aware of the complete suite of ruminal proteolytic enzymes, or their origin. For this reason a proteomics-based approach also seems valuable: to identify predominant proteolytic activities in ruminal samples and by reverse genetics, 'track' down the origin of this activity. Second, can we selectively control populations of proteolytic microorganisms using bacteriophages or other antimicrobial compounds such as bacteriocins? Third, and in addition to taxonomic probes, can gene-specific probes be used to evaluate the physiological state of rumen microorganisms and the physicochemical conditions necessary to ensure optimal growth and yield? Fourth, in terms of nitrogen control of physiological processes, what are the 'signals' which ruminal microorganisms actually respond to: ammonium, peptides, cell density, or other factors? Genomics-based approaches such as DD RT-PCR, which aim to examine how a microbe responds to changes in its growth environment, offer the potential to identify novel physiological processes, and their impact on ruminal function.

There will be no shortage of practical ideas for ruminal microbiologists to examine in the future and a plethora of useful approaches will be available. The greatest challenge facing ruminal microbiologists will probably be to identify and procure the financial resources necessary to execute their research. However, those research endeavours which offer the greatest potential to provide novel information concerning microbial ecology, genetics, and physiology also possess the greatest likelihood of being supported. Fortunately, acquisition of such fundamental information is also likely to maximize the probability of achieving productive alterations in nitrogen metabolism, and nitrogen retention in ruminant production systems.

## References


Attwood, G.T. and Klieve, A.V. (2000) Molecular techniques for monitoring bacteria and bacteriophage populations in the rumen. *Proceedings of the 8th International Symposium on Microbial Ecology* (in press).

Attwood, G.T., Klieve, A.V., Ouwerkerk, D. and Patel, B.K. (1998) Ammonia-hyperproducing bacteria from New Zealand ruminants. *Applied and Environmental Microbiology* 64, 1796–1804.

Avgustin, G., Wright, F. and Flint, H.J. (1994) Genetic diversity and phylogenetic relationships among strains of *Prevotella* (*Bacteroides*) *ruminicola* from the rumen. *International Journal of Systematic Bacteriology* 44, 246–255.

Baggio, L. and Morrison, M. (1996) The NAD(P)H-utilizing glutamate dehydrogenase of *Bacteroides thetaiotaomicron* belongs to enzyme family I, and its activity is affected by *trans*-acting gene(s) positioned downstream of *gdhA*. *Journal of Bacteriology* 178, 7212–7220.

Brooker, J.D. and McCarthy, D.M. (1997) Gene knockout of the intracellular amylase gene by homologous recombination in *Streptococcus bovis*. *Current Microbiology* 35, 133–138.

Chen, G. and Russell, J.B. (1988) Fermentation of peptides and amino acids by a monensin-sensitive ruminal *Peptostreptococcus*. *Applied and Environmental Microbiology* 54, 2742–2749.

Chen, G. and Russell, J.B. (1989) More monensin-sensitive, ammonia-producing bacteria from the rumen. *Applied and Environmental Microbiology* 55, 1052–1057.

Chen, Z., Potempa, J., Polanowski, A., Wikstrom, M. and Travis, J. (1992) Purification and characterization of a 50-kDa cysteine proteinase (gingipain) from *Porphyromonas gingivalis*. *Journal of Biological Chemistry* 267, 18896–18901.

Cotta, M.A. and Russell, J.B. (1982) Effect of peptides and amino acids on efficiency of rumen bacterial protein synthesis in continuous culture. *Journal of Dairy Science* 65, 226–234.

Cotta, M.A. and Russell, J.B. (1997) Digestion of nitrogen in the rumen: a model for metabolism of nitrogen compounds in gastrointestinal environments. In: Mackie, R.I. and White, B.A. (eds) *Gastrointestinal Ecosystems and Fermentations,* vol. 1. Chapman & Hall, New York, pp. 380–423.

Egan, A.R. (1965) The fate and effects of duodenally infused casein and urea in sheep fed on a low-protein roughage. *Australian Journal of Agricultural Research* 16, 169–175.

Eschenlauer, S.C., McEwan, N.R., Calza, R.E., Wallace, R.J., Onodera, R. and Newbold, C.J. (1998) Phylogenetic position and codon usage of two centrin genes from the rumen ciliate protozoan, *Entodinium caudatum. FEMS Microbiology Letters* 166, 147–154.

Ffoulkes, D. and Leng, R.A. (1989) Dynamics of protozoa in the rumen of cattle. *British Journal of Nutrition* 59, 429–436.

Flint, H.J. (1997) The rumen microbial ecosystem – some recent developments. *Trends in Microbiology* 5, 483–488.

Gardner, R.G., Russell, J.B., Wilson, D.B., Wang, G.-R. and Shoemaker, N.B. (1996) Use of a modified *Bacteroides–Prevotella* shuttle vector to transfer a reconstructed b-1,4-D-endoglucanase gene into *Bacteroides uniformis* and *Prevotella ruminicola* $B_14$. *Applied and Environmental Microbiology* 62, 196–202.

Hinton, J.C.D. (1997) The *Escherichia coli* genome sequence: the end of an era or the start of the FUN? *Molecular Microbiology* 26, 417–422.

Hungate, R.E. (1960) Symposium on selected topics in microbial ecology. 1. Microbial ecology of the rumen. *Bacteriological Reviews* 24, 353–364.

Jiang, F., Mannervik, B. and Bergman, B. (1997) Evidence for redox regulation of the transcription factor NtcA, acting both as an activator and a repressor, in the cyanobacterium *Anabaena* PCC 7120. *Biochemistry Journal* 327, 513–517.

Jouany, J.P., Demeyer, D.I. and Grain, J. (1988) Effect of defaunating the rumen. *Animal Feed Science and Technology* 21, 229–265.

Kadowaki, T., Yoneda, M., Okamoto, K., Maeda, K. and Yamamoto, K. (1994) Purification and characterization of a novel arginine-specific cysteine proteinase (argingipain) involved in the pathogenesis of periodontal disease from the culture supernatant of *Porphyromonas gingivalis. Journal of Biological Chemistry* 269, 21371–21378.

Kleive, A.V. and Swain, R.A. (1993) Estimation of ruminal bacteriophage numbers by pulsed-field electrophoresis and laser densitometry. *Applied and Environmental Microbiology* 59, 2299–2303.

Klieve, A.V., Hudman, J.F. and Bauchop, T. (1989) Inducible bacteriophages from ruminal bacteria. *Applied and Environmental Microbiology* 55, 1630–1634.

Klieve, A.V., Gregg, K. and Bauchop, T. (1991) Isolation and characterization of lytic bacteriophages from *Bacteroides ruminicola* subsp. *brevis. Current Microbiology* 23, 183–187.

Klieve, A.V., Turner, A.F and Heck, G.L. (1998) Dietary influences on bacteriophage numbers in the rumen. *Proceedings of the Australian Society of Animal Production* 22, 341.

Krause, D.O. and Russell, J.B. (1996) A rRNA approach for assessing the role of obligate amino acid fermenting bacteria in ruminal ammonia production. *Applied and Environmental Microbiology* 62, 815–821.

Larson, M.A. and Morrison, M. (2000) Application of the differential display RT-PCR technique to examine conditional gene expression in *Ruminococcus albus. Proceedings of the 8th International Symposium on Microbial Ecology* (in press).

Liang, P., Averboukh, L. and Pardee, A.B. (1993) Distribution and cloning of eukaryotic mRNAs by means of differential display: refinements and optimizations. *Nucleic Acids Research* 21, 3269–3275.

Lockington, R.A., Attwood, G.T. and Brooker, J.D. (1988) Isolation and characterization of a temperate bacteriophage from the ruminal anaerobe *Selenomonas ruminantium. Applied and Environmental Microbiology* 54, 1575–1580.

McKain, N., Wallace, R.J. and Watt, N.D. (1992) Selective isolation of bacteria with dipeptidyl aminopeptidase type I activity from the sheep rumen. *FEMS Microbiology Letters* 95, 169–174.

Madeira, H.M.F., Peng, L. and Morrison, M. (1997) Biochemical and mutational analysis of a gingipain-like peptidase activity from *Prevotella ruminicola* $B_14$, and its role in ammonia production by ruminal bacteria. *Applied and Environmental Microbiology* 63, 670–675.

Mendoza, G.D., Britton, R.A. and Stock, R.A. (1993) Influence of ruminal protozoa on site and extent of starch digestion and ruminal fermentation. *Journal of Animal Science* 71, 1572–1578.

Miller, J.H. (1992) *A Short Course in Bacterial Genetics. A Laboratory Manual and Handbook for* Escherichia coli *and Related Bacteria.* Cold Spring Harbor Laboratory Press, New York, pp. 135–142.

Morrison, M. and Mackie, R.I. (1996) Nitrogen metabolism by ruminal microorganisms: current understanding and future perspectives. *Australian Journal of Agricultural Research* 47, 227–246.

Newbold, C.J., Wallace, R.J. and Watt, N.J. (1992) Properties of ionophore-resistant *Bacteroides ruminicola* enriched by cultivation in the presence of tetronasin. *Journal of Applied Bacteriology* 72, 65–70.

Nolan, J.V. and Leng, R.A. (1972) Dynamic aspects of ammonia and urea metabolism in sheep. *British Journal of Nutrition* 27, 177–194.

Okamoto, K., Misumi, Y., Kadowaki, T., Yoneda, M., Yamamoto, K. and Ikehara, Y. (1995) Structural characterization of argingipain, a novel arginine-specific cysteine proteinase as a major periodontal pathogeneic factor from *Porphyromonas gingivalis. Archives of Biochemistry and Biophysics* 316, 917–925.

Owens, F.N. and Goetsch, A.L. (1986) Digesta passage and microbial protein synthesis. In: Milligan, L.P., Grovum, W.L. and Dobson, A. (eds) *Control of Digestion and Metabolism in Ruminants.* Prentice-Hall, Englewood Cliff, New Jersey, pp. 196–223.

Paster, B.J., Russell, J.B., Yang, C.M.J., Chow, J.M., Woese, C.R. and Tanner, R. (1993) Phylogeny of ammonia-producing ruminal bacteria *Peptostreptococcus anaerobius, Clostridium sticklandii,* and *Clostridium aminophilum* sp. nov. *International Journal of Systematic Bacteriology* 43, 107–110.

Peng, L. and Morrison, M. (1995) Mutagenesis of *Prevotella ruminicola* $B_14$ and selection of mutants defective in peptide utilization. *Abstracts of the 23rd Biennial Conference on Rumen Function. no. 26.*

Pittman, K.A. and Bryant, M.P. (1964) Peptides and other nitrogen sources for growth of *Bacteroides ruminicola. Journal of Bacteriology* 88, 401–410.

Reilly, K. and Attwood, G.T. (1998) Detection of *Clostridium proteoclasticum* and closely related strains in the rumen by competitive PCR. *Applied and Environmental Microbiology* 64, 907–913.

Russell, J.B. (1983) Fermentation of peptides by *Bacteroides ruminicola* $B_14$. *Applied and Environmental Microbiology* 45, 1566–1574.

Russell, J.B. (1992) Glucose toxicity and inability of *Bacteroides ruminicola* to regulate glucose transport and utilization. *Applied and Environmental Microbiology* 58, 2040–2045.

Russell, J.B. (1993) Glucose toxicity in *Prevotella ruminicola:* methylglyoxal accumulation and its effects on membrane physiology. *Applied and Environmental Microbiology* 59, 2844–2850.

Shoemaker, N.B., Anderson, K.L., Smithson, S.L., Wang, G.R. and Salyers, A.A. (1991) Conjugal transfer of a shuttle vector from the human colonic anaerobe *Bacteroides uniformis* to the ruminal anaerobe *Prevotella* (*Bacteroides*) *ruminicola. Applied and Environmental Microbiology* 57, 2114–2120.

Swain, R.A., Nolan, J.V. and Klieve, A.V. (1996) Natural variability and diurnal fluctuations within the bacteriophage population of the rumen. *Applied and Environmental Microbiology* 62, 994–997.

Valdivia, R.H. and Falkow, S. (1997) Fluorescence-based isolation of bacterial genes expressed within host cells. *Science* 277, 2007–2011.

Virtanen, A.I. (1966) Milk production of cows on protein-free feed. *Science* 153, 1603–1614.

Wallace, R.J. (1994) Amino acid and protein synthesis, turnover and breakdown by ruminal microorganisms. In: Asplund, J.M. (ed.) *Principles of Protein Nutrition of Ruminants.* CRC Press, Boca Raton, Florida, pp. 72–111.

Wallace, R.J. and McKain, N. (1989) Analysis of peptide metabolism by rumen microorganisms. *Applied and Environmental Microbiology* 55, 2372–2376.

Wallace, R.J. and McKain, N. (1991) A survey of peptidase activity in rumen bacteria. *Journal of General Microbiology* 137, 2259–2264.

Wallace, R.J., McKain, N. and Newbold, C.J. (1990) Metabolism of small peptides in rumen fluid. Accumulation of intermediates during hydrolysis of alanine oligomers, and comparison of peptidolytic activities of bacteria and protozoa. *Journal of the Science of Food and Agriculture* 50, 191–199.

Wallace, R.J., Kopecny, J., Broderick, G.A., Walker, N.D., Sichao, L., Newbold, C.J. and McKain, N. (1995) Cleavage of di- and tri-peptides by *Prevotella ruminicola. Anaerobe* 1, 335–343.

Wallace, R.J., McKain, N., Broderick, G.A., Rode, L.M., Walker, N.D., Newbold, C.J. and Kopecny, J. (1997) Peptidases of the rumen bacterium *Prevotella ruminicola. Anaerobe* 3, 35–42.

Wallace, R.J., Onodera, R. and Cotta, M.A. (1998) Metabolism of nitrogen containing compounds. In: Hobson, P.N. (ed.) *The Rumen Microbial Ecosystem*, 2nd edn. Elsevier Applied Science, New York, pp. 283–328.

Wallace, R.J., McKain, N., Miyagawa, E., Chaudary, L., McEwan, N.R. and Newbold, C.J. (1999) *Eubacterium pyruvovorans,* a new non-saccharolytic anaerobe from the rumen which ferments pyruvate and amino acids and carries out reverse β-oxidation of propionate and butyrate. *Abstracts of the 99th Meeting of the American Society for Microbiology* p. 381.

Wells, J.E. and Russell, J.B. (1995) The effect of growth and starvation on the lysis of the ruminal cellulolytic bacterium *Fibrobacter succinogenes. Applied and Environmental Microbiology* 62, 1342–1346.

Wells, J.E. and Russell, J.B. (1996) Why do ruminal bacteria die and lyse so quickly? *Journal of Dairy Science* 79, 1487–1495.

Wen, Z. (1998) A molecular-based examination of the glutamate dehydrogenase and glutamine synthetase of ruminal *Prevotella* spp. PhD thesis, University of Nebraska, Nebraska, USA.

Wen, Z. and Morrison, M. (1997) Glutamate dehydrogenase activity profiles for the type strains of ruminal *Prevotella* spp. *Applied and Environmental Microbiology* 63, 3314–3317.

Wong, K.K. and McClelland, M. (1994) Stress-inducible gene of *Salmonella typhimurium* identified by arbitrarily primed PCR of RNA. *Proceedings of the National Academy of Sciences USA* 91, 639–643.

Wright, A.D., Dehority, B.A. and Lynn, D.H. (1997) Phylogeny of the rumen ciliates *Entodinium, Epidinium* and *Polyplastron* (*Litostomatea:Entodiniomorphida*) inferred from small subunit ribosomal RNA sequences. *Eukaryotic Microbiology* 44, 61–67.

# III Nutrient Absorption and Splanchnic Metabolism

# 7 Tissue, Cellular and Molecular Aspects of Peptide Absorption and Utilization

K.E. Webb, Jr

*Department of Animal and Poultry Sciences, Virginia Polytechnic Institute and State University, Blacksburg, Virginia, USA*

## Introduction

As we enter the new millennium, we can reflect on recent efforts and know that an important body of literature, primarily from biomedical research, has accumulated over the last 10 years that is helping to define the physiological relevance of peptide absorption. No longer are we limited to just suggesting that peptide absorption occurs, now we are aware of the existence of special proteins that are responsible for transmembrane movement of peptides. The mRNA for some of these proteins has been cloned and the structural and functional characteristics of the encoded proteins are being determined. As the basis for understanding peptide absorption has grown, so also has the interest of animal scientists. Our laboratory has been conducting investigations in this area for over 20 years and now several others join us. In contrast with the biomedical community, our interest as animal scientists is to explain the appropriate dietary foundation of nutrient absorption that will result in the desired growth, development, and production in animals. Thus, we are interested in peptide absorption from the gastrointestinal tract, but our interests go beyond this. We are interested in knowing what, if any, role peptides may have as sources of amino acids for protein synthesis in different tissues and how they might be involved in controlling this process. These and other issues will be addressed in this chapter. The primary focus will be on the ruminant, but some attention will be given to the human and laboratory species, simply because of the greater availability of information regarding these species.

## Peptide transport: cellular and molecular aspects

### *Functional characteristics of an ovine peptide transporter*

Absorption of Met-Gly and carnosine (β-Ala-His) was observed to occur across ruminal and omasal epithelia (Matthews and Webb, 1995). At the substrate concentrations tested, absorption appeared to be primarily non-mediated and the omasal epithelium appeared to have a greater capacity than ruminal epithelium to absorb and translocate these peptides. In a subsequent study, poly(A)$^+$ RNA was extracted from omasal epithelium and injected into *Xenopus laevis* oocytes (Matthews *et al.*, 1996b). Oocytes injected with sheep omasal epithelium poly(A)$^+$ RNA absorbed more glycylsorcosine (Gly-Sar) than did oocytes injected with water. This expressed peptide transport was shown to be $Na^+$ independent and pH dependent (e.g. maximal induced absorption of Gly-Sar occurred at pH 5.0–6.0). Gly-Sar uptake by oocytes injected with poly(A)$^+$ RNA was shown to be saturable and have an affinity constant ($K_t$) of 0.40 ± 0.23 mM at pH 5.5. Uptake of 1 mM Gly-Sar by poly(A)$^+$ RNA injected oocytes was inhibited by 5 mM β-Ala-His, Met-Gly, and Gly-Leu but not by 5 mM glycine, suggesting that these dipeptides were transported by the same expressed transport protein(s) but glycine was not. Oocytes injected with ovine poly(A)$^+$ RNA and with a degenerate DNA oligomer that was complementary to the sequence flanking the initiation site of the cloned rabbit intestinal H+/peptide cotransport protein (Fei *et al.*, 1994) completely inhibited Gly-Ser uptake. This indicated that there is some homology between the rabbit peptide transporter and the ovine peptide transporter expressed under these circumstances.

In order to determine the structural features of peptides that might influence their affinity for the peptide transporter, *X. laevis* oocytes were injected with ovine poly(A)$^+$ RNA and uptake at pH 5.5 of several peptides was examined (Pan *et al.*, 1997). Oocytes injected with poly(A)$^+$ RNA showed transport ability for the dipeptides Gly-Sar, Gly-Leu, Gly-Pro, Phe-Leu and Leu-Leu. Oocytes showed transport ability for the tripeptides Leu-Ser-Phe, Leu-Gly-Phe, Lys-Tyr-Lys, Ala-Pro-Gly, Met-Leu-Phe and Leu-Leu-Tyr. The tetrapeptides, Met-Gly-Met-Met, Val-Gly-Asp-Glu, Ala-Gly-Ser-Glu and Val-Gly-Ser-Glu were transported but Pro-Phe-Gly-Lys and Val-Ala-Ala-Phe were not. No uptake occurred with the penta- (2), hexa- (1), septa- (1) and octapeptides (1) examined. These peptides constitute a variety of substrates varying in their molecular size, hydrophobicity and electrical charge under the experimental conditions used. For the di-, tri-, and tetrapeptides examined, the relationship between transport and molecular weight (MW) was $r = 0.02$ ($P = 0.94$), while the relationship between transport and hydrophobicity values was $r = 0.34$ ($P = 0.21$). When the relationship between transport and net charge was examined, values of $r = 0.1$ ($P = 0.72$) were obtained. Thus the transporter(s) seems to be capable of translocating substrates regardless of their MW, hydrophobicity or charge. Recent studies from other laboratories on the mammalian proton-coupled peptide transporter, PepT1, also demonstrate that peptide substrates are transported by electrogenic, $H^+$-coupled cotransport, which is independent of their physicochemical characteristics (i.e. size, charge; Daniel *et al.*, 1996; Wenzel *et al.*, 1996).

When *X. laevis* oocytes were injected with cRNA for ovine PepT1 (Pan, 1999), peptide uptake by the oocytes was observed to be pH and concentration dependent. External pH was varied by bathing the oocytes in buffers ranging from pH 5.0 to 7.0. Uptake of Glu-Glu and Met-Met was greatest at pH 5.0, uptake of Gly-Sar was greatest at pH 5.0–6.0, and uptake of Lys-Lys was greatest at pH 7.0. These results agree with previously published reports indicating that an inwardly directed $H^+$ gradient as well as the charge on the peptide both influence transport (Fei *et al.*, 1994). The $K_t$ for Gly-Sar uptake at pH 5.5 was observed to be 0.61 mM which is similar to the $K_t$ of

0.40 mM previously observed in our laboratory (Matthews *et al.*, 1996b). The $K_t$s for Met-Met, Lys-Lys and Glu-Glu uptake at pH 5.5 were 27.5, 20.4 and 28.7 μM, respectively. Affinity constants ranging from 81 μM for Gly-Leu to 2.5 mM for Gly-Gly were reported by Fei *et al.* (1994).

Peptide absorption across ruminal and omasal epithelia has been demonstrated *in vitro* as has mRNA in sheep omasal epithelial cells that encodes for a peptide transporter(s). Substrate specificity of the transporter(s) indicates that many but not all di-, tri- and tetrapeptides can be transported. Initial kinetic evaluation of ovine PepT1 indicates a range in affinity constants for substrates.

While results from our initial research indicated that absorption may be largely by non-mediated processes, it is obvious that mechanisms for mediated transport of peptides in the ruminant stomach are present. Omasal epithelium was collected from sheep and mounted in parabiotic chambers to measure the uptake of Gly-Sar in the presence of other peptides (McCollum and Webb, 1998). If multiple substrates are transported on a given transporter, as peptides are believed to be, then they compete with one another if more than one is present. To our surprise, co-incubating high levels of potentially competing peptide substrates with Gly-Sar resulted in a stimulation and not a depression in Gly-Sar uptake. Even though our previous results clearly indicate that mRNA for a peptide transporter is present in omasal epithelium, these results are certainly not consistent with mediated transport being the only mechanism involved in peptide transport across omasal epithelium. Madara and Pappenheimer (1987) discussed paracellular transport through intestinal epithelia and showed that a prerequisite for this process is the mediated uptake of the substrate. It may be that paracellular transport is the mechanism that was responsible for what appeared to be a non-mediated uptake of peptides by epithelial tissues mounted in parabiotic chambers (Matthews and Webb, 1995).

From the evidence that is accumulating, it seems reasonable to suggest that multiple mechanisms may be involved in the transport of peptides across gastrointestinal epithelia. Validation of the presence of and the clarification of the relative importance of each of these mechanisms will increase our understanding of peptide absorption and the contribution this may make to the overall nutritional status of the animal.

## *Tissue distribution of mRNA for an ovine and bovine peptide transporter*

Sheep omasal epithelium poly(A)$^+$ RNA was used as the template for reverse transcriptase PCR (RT-PCR) to develop a probe for Northern blot analysis (Chen *et al.*, 1999). The published sequences of rabbit, human and rat PepT1 (Fei *et al.*, 1994; Liang *et al.*, 1995; Miyamoto *et al.*, 1996) were used to design oligonucleotides to a highly homologous region (from transmembrane domain III to VI). After RT-PCR amplification, a 446-bp fragment from sheep omasal epithelium was cloned and sequenced. The predicted amino acid sequence of this fragment was 85.8%, 90.5% and 90.5% identical to rabbit, human and rat PepT1, respectively. The fragment has low similarity with any known PepT2 sequences. Poly(A)$^+$ RNA isolated from the rumen, omasum and small intestine of sheep and lactating dairy cows showed positive hybridization to the probe in Northern blot analysis. The size of the mRNA that hybridized to the probe was 2.8 kb for both sheep and dairy cows. From the summary of the Northern blot analysis for the five sheep and three cows tested, the pattern of distribution appeared to be consistent within the same species, although the relative abundance varied between animals. Among these tissues, the abundance of PepT1 was higher in the jejunum and ileum in comparison with omasum and duodenum, while abundance was lowest in the rumen. PepT1 was not detectable in the abomasum, liver, kidney, caecum, colon, longissimus and semitendinosus muscles in either sheep or cows, or in the mammary gland in cows.

## Structure of an ovine peptide transporter

Recently, we have cloned a full-length cDNA for sheep PepT1 (Pan, 1999). A partial sequence of this clone indicates that it has about 85% amino acid identity with human PepT1 in the non-loop regions. Amino acid identity within the large extracellular loop appears to be much lower.

The existence of poly(A)$^+$ RNA transcript(s) is strong evidence for the presence of a peptide transport protein(s). Thus, peptide absorption from the stomach and small intestine of ruminants appears to be a physiologically relevant process.

## Overview of human, rabbit and rat peptide transporters

The importance of peptide transport has been documented in recent years. By taking advantage of molecular technology, a few peptide transporters have been identified. By cloning and *in vitro* expression, the molecular structures of these peptide transporters and the mechanisms by which the systems work have begun to be revealed.

It was not until the mid 1990s that the structures of the peptide transporters became clear. In 1994, two groups simultaneously reported the identification of a rabbit oligopeptide transporter designated PepT1 (Boll *et al.*, 1994; Fei *et al.*, 1994). This was the beginning of a new era of peptide transport studies. The reported size of rabbit PepT1 cDNA was 2.7 kb (Fei *et al.*, 1994). It codes for a protein consisting of 707 amino acid residues. The amino acid sequence indicates that PepT1 is a membrane protein with 12 membrane-spanning domains. The protein has a large hydrophilic loop between domains 9 and 10 and this makes the protein different from transporters reported previously. Based on the information from *in vitro* translation, this loop is the target for N-linked glycosylation and the authors believe it to be an extracellular portion of the protein. Other features, including a protein kinase C site and a cAMP-dependent phosphorylation site, appear also to be present on the protein.

Others have cloned human (Liang *et al.*, 1995) and rat (Saito *et al.*, 1995) PepT1. Human PepT1 cDNA is 2.2 kb long with an open reading frame encoding for a protein composed of 708 amino acids (Liang *et al.*, 1995). Rat PepT1 cDNA is 2.9–3.0 kb long with an open reading frame encoding a 710 amino acid protein (Saito *et al.*, 1995; Miyamoto *et al.*, 1996).

Despite the different sizes of the PepT1 cDNA of these species obtained by the different groups, they all encode proteins that are very similar in structure. The 12 putative transmembrane domains and the long hydrophilic segment between domains 9 and 10 are the most significant features for this group of transport proteins. The predicted structural model shows a long hydrophilic loop on the extracellular side containing several N-linked glycosylation sites. With all the 12 $\alpha$-helices accommodated within the membrane, both the amino and carboxyl termini are located on the cytoplasmic side. These PepT1 proteins are all highly homologous. The amino acid sequence of the rat PepT1 is 77% and 83% identical with the rabbit and human PepT1, respectively (Saito *et al.*, 1995). The rabbit and human PepT1 are 81% identical (Liang *et al.*, 1995). The amino acid sequences of these transporters do not show strong homology with other known classes of transport proteins. It is worth mentioning that, while all of the transmembrane domains are highly conserved, the extracellular loops are much less so. Structural differences among these transport proteins indicate that there are different numbers and locations of sites for protein kinase A and C phosphorylation. The extent to which these differences may affect the function of these transporters is still not clear.

Another peptide transporter (PepT2) that is structurally and functionally different from PepT1 has been identified (Liu *et al.*, 1995; Boll *et al.*, 1996; Saito *et al.*, 1996). The human full-length PepT2 cDNA is 2.7 kb with an open reading frame of 2.2 kb (Liu *et al.*, 1995). The rat PepT2 cDNA is 3.9 kb (Saito *et al.*, 1996). Although the sizes of these PepT2 cDNAs are different, they both encode a 729 amino acid protein (Boll *et al.*, 1996). The shared features include the putative 12 transmembrane domains, the large extracellular loop, and several sites for protein kinase dependent phosphorylation.

These structural features are observed in both the PepT1 and PepT2 transporter groups. Even with these similarities, these appear to belong to two distinct transporter groups. When PepT1 and PepT2 are compared from the same species, they display lower identity than do the members in the same group. For example, human PepT2 has 83% amino acid identity with rat PepT2 whereas only 50% with human PepT1. PepT2 proteins have a greater molecular mass than PepT1.

Transport activities of these cloned peptide transporters have been characterized mainly by *in vitro* expression of the cDNA in *X. laevis* oocytes or other cell lines. Transport studies have shown that the cloned peptide transporters are capable of taking up a broad range of di- and tripeptides, regardless of whether they contain acidic, basic or hydrophobic amino acids. The peptides transported may be in either the charged or neutral form under different conditions. Results from studies designed to examine the effects of the net charge of a substrate on peptide transport suggest that neutral substrates are preferred by the peptide transporter when compared with charged peptides under physiological pH conditions (Lister *et al.*, 1997). Amasheh *et al.* (1997) suggest that, under physiological conditions, the affinity of PepT1 for zwitterionic or anionic substrates is greater than for cationic substrates. For transport of a cationic substrate, the pH must be higher.

One of the features of peptide transporters is the necessity for the proteins to cotransport a proton along with the peptide substrate. Brandsch *et al.* (1997) studied the effect of protons on the affinity and $V_{max}$ of Gly-Sar uptake by Caco-2 (PepT1) and SKPT cells (PepT2). Uptake of Gly-Sar in both cells was measured over a concentration range at an extracellular pH of either 6.0 or 7.0. In Caco-2 cells, the $K_t$ was 1.0 mM at pH 6.0 and 1.2 mM at pH 7.0. However, the $V_{max}$ was 13.7 ± 0.3 nmol 10 $min^{-1}$ $mg^{-1}$ of protein at pH 6.0 and 5.8 ± 0.3 nmol 10$min^{-1}$ $mg^{-1}$ of protein at pH 7.0. Similar results were obtained with SKPT cells. In both cell types, protons affected only $V_{max}$ but did not affect the affinity of the transporters for the substrates. Steel *et al.*

(1997) suggested that there is a proton–peptide coupling ratio of 1:1, 2:1 and 1:1 for neutral, acidic and basic dipeptides, respectively. They also showed that, at a pH of 5.5–6.0, PepT1 favoured substrates in neutral and acidic forms.

Given the role of these proteins as peptide transporters, factors that influence the protein structure, its localization on the membrane, and the specific position for substrate binding may all have an effect on the transport activity. It was reported that, among histidyl residues present, His-57 in PepT1 and His-87 in PepT2 are the most critical histidyl residues necessary for the uptake function and probably represent critical binding sites (Fei *et al.*, 1997).

It may be that protein kinases C and A are involved in the regulation of peptide transport. Muller *et al.* (1996) showed that increased intracellular levels of cAMP in Caco-2 cells that expressed PepT1 resulted in a 50% reduction of Gly-Sar uptake. Zinc is an essential trace element that plays a fundamental role in the structure and function of many proteins, e.g. stabilizing the structure of an enzyme, being an essential component of the active site of an enzyme, and as a regulatory factor. Daniel and Adibi (1995) concluded that zinc had a selective effect on peptide transport. They incubated brush-border membrane vesicles (BBMV) with zinc sulphate and observed an increased uptake of Gly-Gln and Leu-Tyr, without changing the diffusion rate of the substrates. Zinc had no effect on the uptake of either Gln or glucose by BBMV.

Information on the distribution of the peptide transporters among various tissues comes largely from the search for the messenger RNA that encodes for the protein. Along with the cloning of PepT1 and PepT2 from various species, distribution of their mRNA has been studied. The basic technique is Northern blot analysis using a specific radiolabelled DNA probe, where for most studies reported, full-length cDNA was used. A 2.9 kb mRNA was found in the small intestine of the rabbit, while much lower levels were observed in the liver and kidney and only trace amounts were found in the brain (Fei *et al.*, 1994). No mRNA was detected in the colon, skeletal muscle, heart, spleen or lung. The mRNA for PepT1 was 2.9–3.0 kb in the rat, the major location of which was in the small intestine (Saito *et al.*, 1995; Miyamoto *et al.*, 1996). Trace amounts of mRNA were found in kidney cortex, but none was observed in liver. Liang *et al.* (1995) reported the size of human PepT1 mRNA to be 3.3 kb, based on Northern blot analysis. They observed a major presence of this PepT1 mRNA in the small intestine as well as in an intestinal epithelium derived cell line (Caco-2). They also observed PepT1 mRNA in kidney, placenta, liver and pancreas. The mRNA transcript was absent in muscle, brain and heart.

A ~4 kb mRNA for PepT2 was found mainly in the kidney medulla and at lower abundance in the kidney cortex (Saito *et al.*, 1996). The mRNA was also detected in brain, lung and spleen, but was undetectable in the heart, liver and small intestine. In the rabbit, a 4.8 kb mRNA was found in the kidney cortex as well as in brain, lung, liver and heart (Boll *et al.*, 1996). PepT2 mRNA was detected in human kidney and small intestine (Liu *et al.*, 1995).

PepT1 mRNA was expressed all along the small intestine and to a much lesser extent in the colon (Freeman *et al.*, 1995). PepT1 mRNA was not detected in the stomach, sacculus rotundus or caecum. Expression was restricted to the epithelial surface of the small intestine and there was no detectable expression in deeper tissues such as the lamina propria, muscularis mucosa or serosa (Fei *et al.*, 1994). Along the crypt–villus axis, the mRNA was detected at or above the crypt–villus junction with the maximal expression occurring at about 100–200 μm above the junction. The mRNA was absent in the lower-to-mid crypt throughout the entire small intestine.

Ogihara *et al.* (1996) were the first to investigate the localization of the transporter protein itself instead of its mRNA. They developed an anti-PepT1 antibody and used it (by means of immunoblotting) to look for PepT1 protein in the rat. Their results confirmed the exclusive expression of PepT1 throughout the length of the small intestine and the absence of PepT1 in the crypt. They also found that PepT1 was specific to the differentiated absorptive epithelial cells and was located mainly on the brush-border membrane of the cell.

From these results, we can conclude that PepT1 and PepT2 are distributed differently between the tissues. PepT1 is mainly an intestinal peptide transporter whereas PepT2 is mainly a renal peptide transporter. Interestingly, PepT1 is also detectable in the kidney. The existence of PepT1 in the small intestine of animals suggests its nutritional importance. The PepT2 in the kidney may play a significant role in conserving peptide forms of amino nitrogen by means of reabsorption (Daniel and Herget, 1997).

## Peptide absorption

### In vitro *characterization of gastrointestinal absorption of peptides*

Casein, soybean meal and distillers' dried grains were incubated in a buffered ruminal fluid inoculum for 8 h (Jayawardena, 2000). Following incubation, cell-free supernatants were obtained by centrifugation and these were used as the mucosal fluids in parabiotic chambers containing either ruminal or omasal epithelium. Initially, free amino acid concentrations ranged from 7.2 to 60 mg $l^{-1}$ and peptide amino acid concentrations ranged from 100 to 270 mg $l^{-1}$ in these mucosal buffers. Serosal appearance of free and peptide amino acids was measured after 240 min. For ruminal epithelium, serosal appearance of free amino acids from casein, soybean meal, and distillers' dried grains was 242, 220 and 234 mg $l^{-1}$ $mg^{-1}$ dry tissue, respectively. Corresponding figures for peptide amino acids were 493, 329 and 453 mg $l^{-1}$ $mg^{-1}$ dry tissue. For omasal epithelium, serosal appearance of free amino acids from casein, soybean meal and distillers' dried grains was 438, 323 and 340 mg $l^{-1}$ $mg^{-1}$ dry tissue. Corresponding figures for peptide amino acids were 2249, 807 and 1191 mg $l^{-1}$ $mg^{-1}$ dry tissue. Serosal appearance of peptide amino acids was greater than serosal appearance of free amino acids in both tissues, probably reflecting the concentration effect of the substrates in the mucosal buffer. Movement through omasal epithelium was much greater than through ruminal epithelium, especially for peptides.

Uptake of Gly-Sar by sheep jejunal and ileal BBMV in a study we conducted showed that these membranes have the capability of translocating this dipeptide (Bowers, 1997). Uptake was greater in BBMV from jejunal tissue than from ileal tissue. Uptake of 0.3 mM Gly-Sar was not stimulated by an inwardly directed $H^+$ gradient (pH 6.4 outside, pH 7.5 inside) in either jejunal or ileal BBMV.

Uptake of 10 μM Gly-Pro by BBMV prepared from sheep duodenal epithelium was reported by Backwell *et al.* (1995), who found that uptake was dependent on a $H^+$ gradient (based on results obtained from an intravesicular pH of 8.4 and extravesicular pHs of 6.0 and 8.4). At first glance, the results of these studies appear to be in conflict.

A careful examination of the protocol reported by Backwell *et al.* (1995) reveals that the extravesicular pH in their study was probably closer to 7.0 than to the pH 6.0 reported. Furthermore, an extravesicular pH of 8.4 would cause peptide substrates to be more negatively charged than would occur physiologically. Because peptide molecules need to be in zwitterionic form in order to be transported (Ganapathy and Leibach, 1985), the results reported by Backwell *et al.* (1995) may be due to a reduction in uptake because of the charged nature of the peptide substrate at pH 8.4 rather than an enhancement of uptake in the presence of a $H^+$ gradient.

BBMV prepared from the proximal jejunum of dairy cows was used to examine Gly-Sar uptake (Wolffram *et al.*, 1998). Results from this study clearly indicate that an inwardly directed $H^+$ gradient stimulated a greater rate of uptake of Gly-Sar. In contrast to the two previous studies (Backwell *et al.*, 1995; Bowers, 1997), this transport was uphill. There is no clear explanation for this concentrative uptake of a peptide. The authors attributed their ability to measure concentrative uptake to the fact that they used a low substrate concentration (25 μM) while others used much higher concentrations (0.1–0.8 mM). Even though Backwell *et al.* (1995) used a concentration of 10 μM and did not observe concentrative uptake, there may be other problems with the design of this study, as previously discussed, that preclude observing a concentrative uptake. Wolffram *et al.* (1998) argue that if a transport mechanism with a low substrate affinity or transport capacity or both is responsible for peptide transport, then, under conditions of a high substrate concentration, the driving forces for uphill transport might be exhausted well before significant amounts of the substrate accumulates inside the vesicle.

## Portal flux of peptides

One of the more controversial issues regarding peptide transport revolves around the issue of the quantitative aspects of portal flux of peptides. Our suggestion many years ago that there may be a sizeable flux of peptides across portal-drained viscera (PDV) is the origin of this controversy. The fact that we suggested that the ruminant stomach may be involved in this process served only to heighten the controversy. This single issue, more than any other, has probably been responsible for stimulating laboratories other than ours to initiate investigations into peptide transport in ruminants. These aspects have been extensively reviewed previously by us (Matthews *et al.*, 1996a; Webb, 1986, 1990; Webb and Bergman, 1991; Webb *et al.*, 1992, 1993; Webb and Matthews, 1994, 1998).

In the first study we conducted, we attempted to quantify the flux of free and peptide amino acids across the PDV of calves (Koeln *et al.*, 1993). Quantifying peptides as the difference between total amino acids in the protein-free filtrate (following acid hydrolysis) and free amino acids (following sulphosalicylic acid precipitation), we observed that there was a much greater flux of peptide amino acids than free amino acids across the PDV. Subsequently, we reported that not only did peptides constitute a major fraction of PDV flux of amino acids but that non-mesenteric tissues contributed substantially to this (Webb *et al.*, 1992, 1993).

Seal and Parker (1996) reported that there was a considerable net appearance of peptide-N in both mesenteric-drained viscera (MDV) and PDV of steers. Net appearance of free amino acids in blood was greater than peptide amino acids across MDV than across PDV. Conversely, net appearance of peptide amino acids in blood was greater than free amino acids across PDV than across MDV. This is consistent with the ruminant stomach contributing significantly to the appearance of peptides in blood.

Gastrointestinal absorption of peptides was examined in sheep fed a forage-based diet (Backwell *et al.*, 1997). They failed to observe any net flux of peptide amino acids across either MDV or PDV. They also observed lower peptide concentrations in blood plasma than reported previously by Koeln *et al.* (1993) and Seal and Parker (1996). They attributed this absence of a net flux of peptide amino acids across MDV and PDV as being due to a matter of technique, more specifically a more efficient deproteinization of blood plasma, thus not creating artificially high 'peptide' concentrations.

Lactating dairy cows were fed diets based on either dry-rolled or steam-flaked sorghum and portal fluxes of free and peptide amino acids were quantified (H. Tagari *et al.*, personal communication). As would be expected, there was a positive portal flux for all amino acids. The quantity of peptide amino acids appearing in the portal vein was lower than free amino acids, but appearance was positive for 70% of the amino acids. This included fluxes differing ($P < 0.059$) from zero for Ala, Arg, Asp, Glu, Gly, His, Leu, Lys, Met, Ser, Thr and Val. The deproteinization procedure used in this study involved protein precipitation with methanol followed by ultrafiltration through 3000 mol. wt filters. Residual peptides would, therefore, be small.

Evidence suggests a net appearance of peptide amino acids in portal and possibly mesenteric blood plasma. Methodological differences are likely contributors to variations observed in the magnitude of portal appearance of peptide amino acids. The procedure employed by H. Tagari *et al.* (personal communication) appears to be the soundest yet employed. Continued efforts in this area will provide further clarification regarding the absolute magnitude of the contribution of peptide amino acids to amino acid flux across MDV and PDV.

## Utilization of circulating peptides

### *Peptide utilization by cultured myogenic cells*

L-Methionine-containing peptides were evaluated as sources of methionine to support protein accretion in $C_2C_{12}$ myogenic cells from mouse muscle (Pan *et al.*, 1996). Expressed as a percentage of the response to free methionine, growth of $C_2C_{12}$ cells differed due to the type of dipeptide (11–108%). Met-Met, Met-Val, and Leu-Met were utilized as efficiently as free methionine. Pro-Met and Gly-Met were poorly utilized by $C_2C_{12}$ cells. Met-Pro, Phe-Met, Met-Phe, Met-Leu, Met-Gly, Ala-Met and Met-Ala were utilized at a rate of about 62–86% of the rate of free methionine and Met-Ser, Ser-Met and Val-Met were utilized at 26–43% of the rate of free methionine.

Primary cultures of ovine myogenic satellite cells were evaluated for their ability to use peptide-bound methionine as a source of methionine for protein accretion and cell proliferation after isolation from skeletal muscle (Pan and Webb, 1998). The cultured myogenic cells were able to utilize all the methionine-containing dipeptides tested for protein accretion with responses ranging from about 49 to 95% of the response for free methionine. This is consistent with the concept that peptide-bound amino acids can serve as amino acid sources for protein accretion in sheep skeletal muscle. In some cases, the molecular arrangement of the dipeptides with the same amino acid composition influenced the relative ability of the dipeptides to serve as methionine sources. For all peptides studied, however, only Ala-Met was utilized to support protein accretion as well as free methionine.

### Peptide utilization by cultured mammary cells

L-Methionine-containing peptides were evaluated for their ability to be a source of methionine to support protein accretion in MAC-T, bovine mammary epithelial, cells (Pan *et al.*, 1996). All of the methionine-containing dipeptides examined were able to support protein accretion in cultured MAC-T cells with the response ranging from 35% to 122% of the free methionine growth response. Met-Val, Met-Leu, Met-Met and Leu-Met supported greater protein accretion than did free methionine. Phe-Met, Met-Phe, Ala-Met, Met-Ala, Met-Ser and Met-Gly were utilized as effectively as free methionine. Gly-Met, Pro-Met and Ser-Met were the least utilized peptides in MAC-T cells. Dipeptides with methionine at the N-terminus were preferred substrates to those with methionine at the C-terminus.

Regulation of the use of peptides as amino acid substrates for protein accretion and cell proliferation in MAC-T cells by serum factors was examined (Pan *et al.*, 1998). Results indicated that adult animal sera from humans, horses, chickens, pigs and rabbits promote the utilization of most methionine-containing peptides. By themselves, neither insulin nor serum lipids were able to facilitate peptide utilization.

Cultured MAC-T bovine mammary epithelial cells were used to study the ability of methionine-containing peptides to substitute for free methionine in the synthesis of secreted proteins (Wang, 1994). All of the methionyl peptides examined were utilized by the MAC-T cells as sources of methionine for the synthesis of both secreted and cellular proteins. Most of the methionine-containing peptides were as efficient as free methionine in promoting protein synthesis.

These results indicate that MAC-T mammary epithelial cells are able to utilize small methionine-containing peptides as sources of methionine to support cellular protein accretion and the synthesis of secreted proteins.

### Peptide utilization by mammary tissue explants

Mammary tissue explants from lactating (10–11 days) CD-1 mice were used to study the ability of methionine-containing peptides on the synthesis of secreted proteins (Wang *et al.*, 1996). Mammary tissue explants were able to utilize methionine from all peptides studied. Eleven of the peptides promoted 15–76% greater synthesis of secreted proteins than did free methionine. Dipeptides containing either valine or serine promoted the greatest synthesis. The remaining six peptides were not different from free methionine in promoting synthesis of secreted proteins.

Mammary explants were able to utilize lysine from all lysyl peptides examined for the synthesis of secreted proteins (Wang, 1994). These lysyl peptides generally were similar to free lysine in promoting synthesis of secreted proteins. The synthesis of proteins promoted by Gly-His-Lys was about 17% greater than that promoted by lysine. The other peptides were not different from lysine in promoting protein synthesis which ranged from 91 to 108% of the synthesis promoted by lysine. Within each of the three peptide pairs, Asp-Lys and Lys-Asp, Gly-Lys and Lys-Gly, and Val-Lys and Lys-Val, location of the lysyl residue at either the N- or C-terminal position did not affect protein synthesis.

These results are consistent with those previously discussed for cultured MAC-T mammary epithelial cells. Together, these studies indicate that a wide range of peptide-bound methionine and lysine substrates can support the synthesis of milk proteins by mammalian epithelial cells, at least as well as free methionine and lysine.

### *Peptide flux across the mammary gland* in vivo

Use of free and peptide amino acids by the mammary gland of lactating goats was investigated by Backwell *et al.* (1994). They observed that Phe from Gly-Phe and Leu from Gly-Leu were incorporated into casein in the mammary gland. It is obvious that the incorporation of Phe and Leu from these peptides into casein was an intracellular process. The authors were unable to determine whether there was transport of intact peptides into the cell or whether extracellular hydrolysis preceded transport of the amino acids. Shennan *et al.* (1998) concur that amino acids of peptide origin are utilized by the mammary gland (rat), but their data indicate that peptides are probably not transported but are hydrolysed prior to the constituent amino acids entering the cell. We cannot rule out the possibility that a peptide transporter other than PepT1 may be involved, but data from our laboratory supports their claims because we were unable to detect mRNA for PepT1 in the mammary gland of lactating cows (Chen *et al.*, 1999).

Estimates of the extent of incorporation of Phe, Met, Lys and Tyr arriving at the mammary gland of lactating goats in peptide form have been recently reported (Bequette *et al.*, 1999). They report that 5–11% of Phe, 8–18% of Met, 4–13% of Lys, and 13–25% of Tyr incorporated into casein is of peptide origin. This confirms earlier work from the same laboratory indicating that peptides can contribute amino acids for milk protein synthesis (Backwell *et al.*, 1996).

When fluxes of free amino acids across the mammary gland of lactating cows fed either steam-flaked or dry-rolled sorghum were examined, it was observed that, except for Gly, there was a net uptake of free amino acids (H. Tagari *et al.*, personal communication). Uptake of amino acids of peptide origin by the mammary gland was much smaller but was positive for about 65% of the amino acids including positive fluxes that differed ($P < 0.05$) from zero for Glu, Leu, Lys and Ser.

Collectively, these studies indicate that peptides contribute their constituent amino acids for the synthesis of milk proteins. In magnitude, these amino acids account for only a small portion of the total amino acids incorporated into the proteins. This source, however, may be critical and may play a role in controlling milk protein synthesis.

## References


Amasheh, S., Wenzel, U., Boll, M., Dorn, D., Weber, W.-M., Clauss, W. and Daniel, H. (1997) Transport of charged dipeptides by the intestinal H+/peptide symporter PepT1 expressed in *Xenopus laevis* oocytes. *Journal of Membrane Biology* 155, 247–256.

Backwell, F.R.C., Bequette, B.J., Wilson, D., Calder, A.G., Metcalf, J.A., Wray-Cahen, D., MacRae, J.C., Beever, D.E. and Lobley, G.E. (1994) Utilization of dipeptides by the caprine mammary gland for milk protein synthesis. *American Journal of Physiology* 267, R1–R6.

Backwell, F.R.C., Wilson, D. and Schweizer, A. (1995) Evidence for a glycyl-proline transport system in ovine enterocyte brush-border membrane vesicles. *Biochemical and Biophysical Research Communications* 215, 561–565.

Backwell, F.R.C., Bequette, B.J., Wilson, D., Metcalf, J.A., Franklin, M.F., Beever, D.E., Lobley, G.E. and MacRae, J.C. (1996) Evidence for the utilization of peptides for milk protein synthesis in the lactating dairy goat in vivo. *American Journal of Physiology* 271, R955–R960.

Backwell, F.R.C., Hipolito-Reis, M., Wilson, D., Bruce, L.A., Buchan, V. and MacRae, J.C. (1997) Quantification of circulating peptides and assessment of peptide uptake across the gastrointestinal tract of sheep. *Journal of Animal Science* 75, 3315–3322.

Bequette, B.J., Backwell, F.R.C., Kyle, C.E., Calder, A.G., Buchan, V., Crompton, L.A., France, J. and MacRae, J.C. (1999) Vascular sources of phenylalanine, tyrosine, lysine, and methionine for casein synthesis in lactating goats. *Journal of Dairy Science* 82, 362–377.

Boll, M., Markovich, D., Weber, W.-M., Korte, H., Daniel, H. and Murer, H. (1994) Expression cloning of a cDNA from rabbit small intestine related to proton-coupled transport of peptides, β-lactam antibiotics and ACE-inhibitors. *European Journal of Physiology* 429, 146–149.

Boll, M., Herget, M., Wagener, M., Weber, W.-M., Markovich, D., Biber, J., Clauss, W., Murer, H. and Daniel, H. (1996) Expression cloning and functional characterization of the kidney cortex high-affinity proton-coupled transporter. *Proceedings of the National Academy of Sciences USA* 93, 284–289.

Bowers, S.H. (1997) Characterization of glycyl-sarcosine uptake by ovine intestinal brush border membrane vesicles. MS Thesis, Virginia Polytechnic Institute and State University, Blacksburg, Virginia.

Brandsch, M.C., Ganapathy, M.E., Chew, C.S., Ganapathy, V. and Leibach, F.H. (1997) Influence of proton and essential histidyl residues on the transport kinetics of the $H^+$/peptide cotransport systems in intestine (PEPT1) and kidney (PEPT2). *Biochimica et Biophysica Acta* 1324, 251–262.

Chen, H., Wong, E.A. and Webb, K.E., Jr (1999) Tissue distribution of a peptide transporter mRNA in sheep, dairy cows, pigs, and chickens. *Journal of Animal Science* 77, 1277–1283.

Daniel, H. and Adibi, S.A. (1995) Selective effect of zinc on uphill transport of oligopeptides into kidney brush border membrane vesicles. *FASEB Journal* 9, 1112–1117.

Daniel, H. and Herget, M. (1997) Cellular and molecular mechanisms of renal peptide transport. *American Journal of Physiology* 273, F1–F8.

Daniel, H., Amasheh, S., Wenzel, U., Weber, W.-M. and Clauss, W. (1996) Transport of charged substrates by the intestinal peptide carrier PepT1 expressed in *Xenopus laevis* oocyte. *FASEB Journal* 10, A121 (Abstract).

Fei, Y.-J., Kanal, Y., Nussberger, S., Ganapathy, V., Leibach, F.H., Romero, M.F., Singh, S.K., Boron, W.F. and Hediger, M.A. (1994) Expression cloning of a mammalian proton-coupled oligopeptide transporter. *Nature* 368, 563–566.

Fei, Y.-J., Liu, W., Prasad, P.D., Kekuda, R., Oblak, T.G., Ganapathy, V. and Leibach, F.H. (1997) Identification of the histidyl residue obligatory for the catalytic activity of the human H+/peptide cotransporters PEPT1 and PEPT2. *Biochemistry* 36, 452–460.

Freeman, T.C., Bentsen, B.S., Thwaites, D.T. and Simmons, N.L. (1995) $H^+$/di-tripeptide transporter (PepT1) expression in the rabbit intestine. *European Journal of Physiology* 430, 394–400.

Ganapathy, V. and Leibach, F.H. (1985) Is intestinal peptide transport energized by a proton gradient? *American Journal of Physiology* 249, G153–G160.

Jayawardena, V. (2000) Ruminal and omasal absorption of ruminally derived amino acids and peptides *in vitro*. PhD Dissertation, Virginia Polytechnic Institute and State University, Blacksburg, Virginia.

Koeln, L.L., Schlagheck, T.S. and Webb, K.E., Jr (1993) Amino acid flux across the gastrointestinal tract and liver of calves. *Journal of Dairy Science* 76, 2275–2286.

Liang, R., Fei, Y.-J., Prasad, P.D., Ramamoorthy, S., Han, H., Yang-Feng, T.L., Hediger, M.A., Ganapathy, V. and Leibach, F.H. (1995) Human intestinal H+/peptide cotransporter cloning, functional expression, and chromosomal localization. *Journal of Biological Chemistry* 270, 6456–6463.

Lister, N., Bailey, P.D., Collier, I.D., Boyd, C.A.R. and Bronk, J.R. (1997) The influence of luminal pH on transport of neutral and charged dipeptides by rat small intestine, *in vitro. Biochimica et Biophysica Acta* 1324, 245–250.

Liu, W., Liang, R., Ramamoorthy, S., Fei, Y.-J., Ganapathy, M.E., Hediger, M.A., Ganapathy, V. and Leibach, F.H. (1995) Molecular cloning of PEPT 2, a new member of the H+/ peptide cotransporter family, from human kidney. *Biochimica et Biophysica Acta* 1235, 461–466.

McCollum, M.Q. and Webb, K.E., Jr (1998) Glycyl-L-sarcosine absorption across ovine omasal epithelium during coincubation with other peptide substrates and volatile fatty acids. *Journal of Animal Science* 76, 2706–2711.

Madara, J.L. and Pappenheimer, J.R. (1987) Structural basis for physiological regulation of paracellular pathways in intestinal epithelia. *Journal of Membrane Biology* 100, 149.

Matthews, J.C. and Webb, K.E., Jr (1995) Absorption of L-carnosine, L-methionine, and L-methionylglycine by isolated sheep ruminal and omasal epithelial tissue. *Journal of Animal Science* 73, 3464–3475.

Matthews, J.C., Pan, Y.L., Wang, S., McCollum, M.Q. and Webb, K.E., Jr (1996a) Characterization of gastrointestinal amino acid and peptide transport proteins and the utilization of peptides as amino acid substrates by cultured cells (myogenic and mammary) and mammary tissue explants. In: Kornegay, E.T. (ed.) *Nutrient Management of Food Animals to Enhance the Environment*, CRC Press, Inc., Boca Raton, Florida.

Matthews, J.C., Wong, E.A., Bender, P.K., Bloomquist, J.R. and Webb, K.E., Jr (1996b) Demonstration and characterization of dipeptide transport system activity in sheep omasal epithelium by expression of mRNA in *Xenopus laevis* oocytes. *Journal of Animal Science* 74, 1720–1727.

Miyamoto, K.-I., Shiraga, T., Morita, K., Yamamoto, H., Haga, H., Taketani, Y., Tamai, I., Sai, Y., Tsuji, A. and Takeda, E. (1996) Sequence, tissue distribution and developmental changes in rat intestinal oligopeptide transporter. *Biochimica et Biophysica Acta* 1305, 34–38.

Muller, U., Brandsch, M., Prasad, P.D., Fei, Y.-J., Ganapathy, V. and Leibach, F.H. (1996) Inhibition of the $H^+$/peptide cotransporter in the human intestinal cell line Caco-2 by cyclic AMP. *Biochemical and Biophysical Research Communications* 218, 461–465.

Ogihara, H., Saito, H., Shin, B.-C., Terada, T., Takenoshita, S., Nagamachi, Y., Inui, K.-I. and Takata, K. (1996) Immuno-localization of H+/peptide cotransporter in rat digestive tract. *Biochemical and Biophysical Research Communications* 220, 848–852.

Pan, Y.X. (1999) Structural and functional characteristics of ovine PepT1. PhD Dissertation, Virginia Polytechnic Institute and State University, Blacksburg, Virginia.

Pan, Y. and Webb, K.E., Jr (1998) Peptide-bound methionine as methionine sources for protein accretion and cell proliferation in primary cultures of ovine skeletal muscle. *Journal of Nutrition* 128, 251–256.

Pan, Y., Bender, P.K., Akers, R.M. and Webb, K.E., Jr (1996) Methionine-containing peptides can be used as methionine sources for protein accretion in cultured $C_2C_{12}$ and MAC-T cells. *Journal of Nutrition* 126, 232–241.

Pan, Y.-X., Wong, E.A., Bloomquist, J.R. and Webb, K.E., Jr (1997) Poly(A)$^+$ RNA from sheep omasal epithelium induces expression of a peptide transport protein(s) in *Xenopus laevis* oocytes. *Journal of Animal Science* 75, 3323–3330.

Pan, Y., Bender, P.K., Akers, R.M. and. Webb, K.E., Jr (1998) One or more serum factors promote peptide utilization in cultured animal cells. *Journal of Nutrition* 128, 744–750.

Saito, H., Okuda, M., Terada, T., Sasaki, S. and Inui, K.-I. (1995) Cloning and characterization of a rat H$^+$/peptide cotransporter mediating absorption of beta-lactam antibiotics in the intestine and kidney. *Journal of Pharmacology and Experimental Therapeutics* 275, 1631–1637.

Saito, H., Terada, T., Okuda, M., Sasaki, S. and Inui, K.-I. (1996) Molecular cloning and tissue distribution of rat peptide transporter PEPT2. *Biochimica et Biophysica Acta* 1280, 173–177.

Seal, C.J. and Parker, D.S. (1996.) Effect of intraruminal propionic acid infusion on metabolism of mesenteric- and portal-drained viscera in growing steers fed a forage diet: II. Ammonia, urea, amino acids, and peptides. *Journal of Animal Science* 74, 245–256.

Shennan, D.B., Calvert, D.T., Backwell, F.R.C. and Boyd, C.A.R. (1998) Peptide aminonitrogen transport by the lactating rat mammary gland. *Biochimica et Biophysica Acta* 1373, 252–260.

Steel, A., Nussberger, S., Romero, M.F., Boron, W.F., Boyd, C.A.R. and Hediger, M.A. (1997) Stoichiometry and pH dependence of the rabbit proton-dependent oligopeptide transporter PepT1. *Journal of Physiology* 498, 563–569.

Wang, S. (1994) Peptides as amino acid sources for the synthesis of secreted proteins by mammary tissue explants and cultured mammary epithelial cells. PhD Dissertation, Virginia Polytechnic Institute and State University, Blacksburg, Virginia.

Wang, S., Webb, K.E., Jr and Akers, M.R. (1996) Peptide-bound methionine can be a source of methionine for the synthesis of secreted proteins by mammary tissue explants from lactating mice. *Journal of Nutrition* 126, 1662–1672.

Webb, K.E., Jr (1986) Amino acid and peptide absorption from the gastrointestinal tract. *Federal Proceedings* 45, 2268–2271.

Webb, K.E., Jr (1990) Intestinal absorption of protein hydrolysis products: A review. *Journal of Animal Science* 68, 3011–3022.

Webb, K.E., Jr and Bergman, E.N. (1991) Amino acid and peptide absorption and transport across the intestine. *Proceedings of the 7th International Symposium on Ruminant Physiology*, pp. 111–128.

Webb, K.E., Jr and Matthews, J.M. (1994) Absorption of amino acids and peptides In: Asplund, J.M. (ed.) *Amino Acid Nutrition of Ruminants*. CRC Press, Boca Raton, Florida.

Webb, K.E., Jr and Matthews, J.C. (1998) Peptide absorption and its significance in ruminant protein nutrition. In: Grimble, G. and Backwell, C. (eds) *Peptides in Mammalian Protein Metabolism: Tissue Utilization and Clinical Targeting*. Portland Press Ltd, London, pp. 1–10.

Webb, K.E., Jr, Matthews, J.C. and DiRienzo, D.B. (1992) Peptide absorption: A review of current concepts and future perspectives. *Journal of Animal Science* 70, 3248–3257.

Webb, K.E., Jr, DiRienzo, D.B. and Matthews, J.C. (1993) Recent developments in gastrointestinal absorption and tissue utilization of peptides – A review. *Journal of Dairy Science* 76, 351–361.

Wenzel, U., Gebert, I. and Daniel, H. (1996) Characteristics of uptake of neutral and anionic cephalosporin antibiotics by the cloned intestinal peptide transporter PePT1. *FASEB Journal* 10, A283 (abstract).

Wolffram, S., Grenacher, B. and Scharrer, E. (1998) H$^+$-coupled uphill transport of the dipeptide glycylsarcosine by bovine intestinal brush-border membrane visicles. *Journal of Dairy Science* 81, 2595–2603.

# 8 Influence of Gastrointestinal Metabolism on Substrate Supply to the Liver

C.J. SEAL AND D.S. PARKER

*Department of Biological and Nutritional Sciences, Faculty of Agriculture and Biological Science, University of Newcastle upon Tyne, Newcastle upon Tyne, UK*

## Introduction

Tissues of the splanchnic bed include the gastrointestinal tract (GIT), liver, spleen, pancreas and mesenteric fat depots. Cumulatively these organs, together with the associated connective tissue and blood vessels, contribute approximately 15–20% of total body mass. During pregnancy and lactation, hypertrophy of the GIT can increase this still further. Collectively the splanchnic bed plays a central role in moderating nutrient supply to peripheral tissues for maintenance and productive processes such as muscle deposition, wool growth and milk production. Liver output is affected by the pattern and quantity of nutrients delivered to the portal vein from the GIT and this process is discussed separately in Chapter 9. Intestinal tissues maintain a high rate of metabolic activity which is sustained by both the luminal supply of nutrients available during the process of absorption and also from metabolites derived from the arterial blood supply. This chapter will review the effects of GIT metabolism on the supply of energy-yielding substrates to the liver, and will examine the effects of changing diet on the supply of nutrients such as short-chain volatile fatty acids (SCVFA), glucose and amino acids to the liver.

## Tissues of the splanchnic bed

### *Anatomy and vasculature of the gastrointestinal tract and liver*

The common blood vessel supplying the liver is the portal vein which divides into the lobes of the liver at the porta hepatis plexus (Seal and Reynolds, 1993). Blood arriving at the liver in the portal vein represents the cumulative drainage of venous blood from the entire GIT and its associated organs including the pancreas, spleen and mesenteric fat. Blood sampled at the porta hepatis and compared with arterial blood, therefore, represents the cumulative effects of metabolism across a range of diverse metabolically active tissues. For the GIT, the additional effects of compartmentalization of the digestive process along the tract cannot be identified with reference to portal blood alone.

Technically, the biggest advance in understanding splanchnic metabolism has been the development of *in vivo*, multi-catheterization procedures to sample portal blood and blood draining other areas of the GIT. Elegantly described by Huntington *et al.* (1989) for cattle, this has provided the benchmark for much of the work in this field over the past decade, following the early work of Bergman's group with sheep (Katz and Bergman, 1969). Anatomically, however, there are significant differences between cattle and sheep, which has made direct comparison of catheterization techniques difficult. The common portal vein of cattle, for example, is much shorter than in sheep making the implantation of transonic flow probes complex (see later). The proximity of the gastro-splenic vein also makes location of the catheter tip difficult (Huntington, 1982). Additionally, the venous drainage of sheep is not associated with discrete gut components as observed with the closed loop draining into a 'mesenteric' vein in cattle (Seal and Reynolds, 1993). In an attempt to resolve this problem in sheep, Neutze *et al.* (1994) have described a complex procedure in which they have measured α-amino–N uptake in the caudal mesenteric vein upstream and downstream of the ileocaecocolic vein junction. The surgery involved in this procedure is technically demanding, requiring the precise location of catheters in small blood vessels. Their results, however, were very comparable with similar data from mesenteric vein samples obtained from cattle, suggesting that this may be a reasonable approach for future studies.

### *Measurement of blood flow*

Measuring blood flow through catheterized vessels is a key parameter for the determination of net nutrient fluxes across the splanchnic bed. Errors in the determination of blood flow, and variability in its measurement, are critical factors in all studies of this type. For many metabolites, small venous–arterial (V–A) differences are amplified by large blood flows to give widely variable (and perhaps unreliable) flux rates. This problem is further exacerbated by the increasing use of more 'sensitive' analytical procedures coupled with labelled metabolites leading to V–A differences that are at the limit of analytical precision.

The majority of blood flow data reported in the literature have been obtained using dye-dilution procedures, the most common being dilution of *p*-amino hippuric acid (PAH). The method was initially described in sheep by Katz and Bergman (1969) and involves the use of a primed, continuous infusion of PAH upstream of the sampling catheter through a second infusion catheter, usually implanted in a small mesenteric vein. Alternative dyes used in other animal species and man include indocyanine green (ICG; Shoemaker, 1964; Wangsness and McGilliard, 1972; Ferrannini *et al.*, 1988) and antipyrine (Marconi *et al.*, 1989). Blood flow is calculated by determining the difference in indicator dye concentration between the target vessel and the peripheral circulation relative to the dye infusion rate. Corrections for water absorption/resorption across the gut tissues can be made using differences in haematocrit between splanchnic blood and peripheral blood. In our experience, however, this correction was small and inconsistent and for the majority of our studies was not significant. Accurate determination of blood flow, regardless of the dye used, is dependent on complete mixing of the dye with the blood before it reaches the tip of the sampling catheter. This is usually achieved by implanting the infusion catheter several centimetres distal to the sampling tip. However, 'streaming' of blood through vessels may still be a major factor in causing variability in this data and can be reduced by placing the catheter tip in the lateral lobe of the liver beyond the turbulent flow of the porta hepatis (Seal and Reynolds, 1993).

Direct measurement of portal and mesenteric vein blood flow is possible through the use of transonic flow probes. This procedure has been in use for more than 30 years (Carr and Jacobson, 1968), but has been limited by technical difficulties during surgical implantation and in calibration. This is a particular problem for large cattle, where the size and accessibility of the portal vein severely restricts the positioning of the probe (Huntington *et al.*, 1990). The size of the probe necessary to encase the portal vein in these animals may also influence the transmission of the Doppler signal across the vein, giving false measurements of blood flow. The procedure has been used with greater success in sheep (Neutze *et al.*, 1994; Remond *et al.*, 1998) and in small growing steers (Kim *et al.*, 1998). In some of these studies it has been reported that blood flow determined by ultrasonic transit-time flow meters underestimates blood flow compared with indicator dilution techniques (Huntington *et al.*, 1990; Kristensen *et al.*, 1996; Remond *et al.*, 1998), but this observation is not consistent, and may be improved with the use of newer types of probe (Remond *et al.*, 1998). This underestimate is presumed to be due to turbulence in the blood flow through the probe. The principal advantage of the transonic probes is the ability to measure blood flow continuously over long time periods using automated data-handling systems. This reduces the frequency with which blood samples are taken, and gives an 'integrated' blood flow profile which can be used to describe and model diurnal variations. Measurements of this type in steers show that blood flow in twice-daily-fed animals varies in a sinusoidal manner (Fig. 8.1; Kim *et al.*, 1998) which is similar to data obtained using frequent blood sampling and complex non-parametric curve-fitting methods (Whitt *et al.*, 1996). Future developments in transonic probe technology will further increase the value of this methodology.

### Use of labelled substrates

Measurements of V–A concentration differences and blood flow across tissues of the GIT and liver have been invaluable in providing information linking the digestion and absorption of feed in the gut lumen and the subsequent appearance of nutrients in the venous drainage. However, flux rates thus determined represent the net effect of transfers across the apical and basolateral surfaces of the intestinal cells and metabolism within the cell. Positive V–A differences result from a net flux of a nutrient from lumen to blood that exceeds the sum of extraction from blood and utilization within the intestinal tissues. Conversely, a negative V–A indicates that extraction from arterial blood and utilization within the intestinal tissues is greater than the flux of the nutrient from the gut lumen. For some nutrients, for example glucose, V–A may fluctuate from positive to negative, depending on the diet (see, for example, Seal *et al.*, 1992, 1994; Seal and Parker, 1994; Balcells *et al.*, 1995; Piccioli Cappelli *et al.*, 1997). For others,

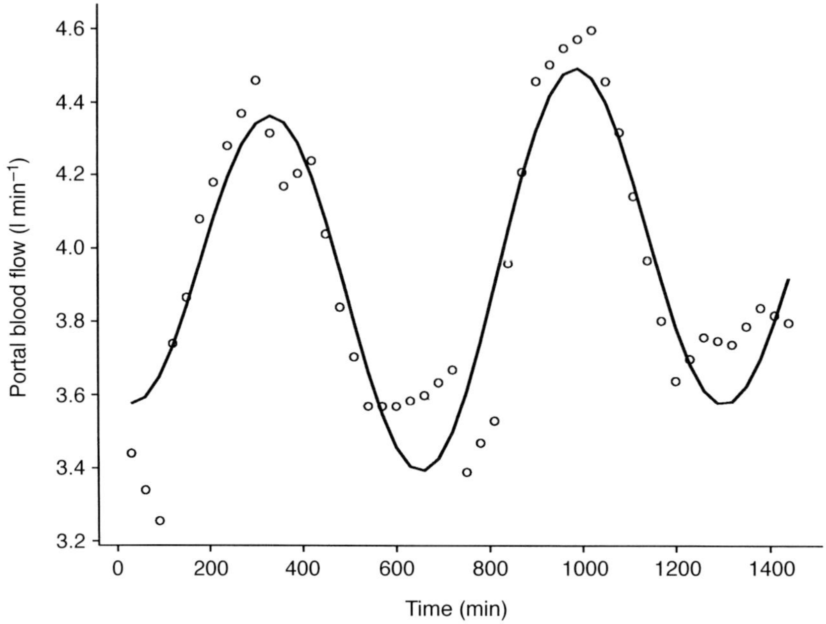


**Fig. 8.1.** A typical circadian pattern for portal vein blood flow in a 150 kg Holstein–Friesian steer (feed offered at 60 and 780 min). Reproduced with permission from Kim *et al.* (1998).

for example acetate, net fluxes are positive across ruminal tissues, but may be negative across mesenteric-drained tissues (Seal *et al.*, 1992; Seal and Parker, 1994). Hind-gut fermentation may also result in positive fluxes of SCVFA into the caecal vein (DeGregorio *et al.*, 1984).

Isotopic labelling of nutrients in combination with classic V–A techniques provides the opportunity to estimate true rates of absorption from the gut and can be used to determine rates of utilization of nutrients within gut tissues as well as sequestration of nutrients from the blood supplying the gut. Combinations of luminal and vascular infusions of labelled substrates can be used to quantify whole-body irreversible loss rates for individual nutrients, inter-conversion of metabolites and metabolism across tissues in single experiments, thus maximizing the data produced from a multi-faceted approach. If these studies are further combined with surgical cannulation of the gut tissues, direct measurements of intestinal disappearance of nutrients can be compared with simultaneous measurements of nutrient flux into the venous drainage. Improvements in analytical procedures and developments in stable isotope technologies will undoubtedly further increase the potential for this type of study.

## Energy metabolism in the gastrointestinal tract

Estimates of the proportion of total oxygen consumption attributed to the tissues of the whole splanchnic bed range from 35 to 60% (Seal and Reynolds, 1993) and approximately 20% for the GIT alone (Cant *et al.*, 1996). These tissues account for about 10–13% of total body mass, and thus, on a unit mass basis, their metabolic activity is considerable. Quantitatively, the major components of this energy expenditure are $Na^+$, $K^+$-ATPase-linked ion transport systems (approximately 30–60% of the expenditure) and protein turnover (20–23% for protein synthesis and 4% for protein degradation; McBride and Kelly, 1990). Factors which influence energy expenditure include meal consumption (Christopherson and Brockman, 1989; Kelly *et al.*, 1989) and energy intake expressed as either metabolizable energy (see Seal and Reynolds, 1993) or digestible energy (Goetsch and Patil, 1997; Goetsch, 1998). Physiological state also has a major impact on intestinal energy expenditure. For example, pregnancy and lactation are associated with increased intestinal tissue mass (Fell *et al.*, 1972; Hammond and Diamond, 1994). This latter response, observed in both ruminant and non-ruminant species, results in increased mucosal mass and concomitant increases in protein synthetic rates in GIT tissues. This increased intestinal transport capacity also results in the synchronous upregulation of all intestinal transporters (Bird *et al.*, 1996). The relationships between digestible energy intake, fibre digestion and urea-N flux have been described in some detail by Goetsch (1998) in a series of regression equations based on 11 separate experiments with sheep. These data show that, in forage-fed animals, there is a complex interaction between portal-drained viscera energy use and digestible energy intake and neutral detergent fibre (NDF) digestion; these factors and hepatic urea net flux also impact on total splanchnic energy use and increase with increased forage consumed. The combined effects of these variables, especially the increased energy required in hepatic ureagenesis to dispose of ruminal ammonia (Lobley *et al.*, 1996b), may account for the observed inefficiency of growth observed in animals fed forage-based diets. The processes involved in changing GIT energy expenditure, and the consequences on animal performance, have been the focus of several recent reviews (Lobley, 1994; Cant *et al.*, 1996; Goetsch, 1998).

## Short-chain volatile fatty acid metabolism

Quantitatively, the major source of absorbed energy in ruminants is in the form of SCVFAs (acetate, propionate and butyrate). The net flux of these SCVFAs across the portal-drained viscera is highly correlated with both the composition of the diet and the level of feed intake (Huntington, 1983; Huntington and Prior, 1983). For any single diet at different levels of intake, the proportions of SCVFA appearing in portal blood remains constant, reflecting ruminal production rates for the individual acids. Changing ruminal fermentation pattern, by altering the forage to concentrate ratio of the diet, results in a concomitant change in the pattern of SCVFA in ruminal fluid and portal blood, but the extent of metabolism of the individual acids remains constant (Seal and Reynolds, 1993). Studies in which direct comparison between measured production rates of SCVFAs in the rumen and net flux into the portal vein have been made are not common. Those that are available, however, confirm *in vitro* studies which show that there is extensive metabolism of all the acids during absorption across the gut wall.

## Absorption of SCVFA

SCVFA may be absorbed across the rumen epithelium in both undissociated and dissociated forms. Undissociated SCVFAs are lipid soluble and cross the epithelial membrane more easily than the dissociated form; absorption rates of the undissociated acids are thus affected by the pH of ruminal fluid although decreasing pH only results in a small increase in SCVFA clearance (Dijkstra *et al.*, 1993). In contrast to non-ruminant colonic and caecal tissue (Holtug *et al.*, 1992), $Na^+/H^+$ exchange is not involved in SCVFA absorption (López *et al.*, 1996). Dissociated SCVFAs are absorbed via anion exchange with bicarbonate, in a mechanism which appears similar for non-ruminant large intestinal and ruminal tissues and involves an additional electroneutral anion exchange between bicarbonate and chloride.

## SCVFA metabolism by gut tissues

Within ruminal tissues, acetate is primarily metabolized for energy; propionate contributes to energy supply, lactate and $CO_2$; and butyrate is extensively metabolized to 3-hydroxy butyrate, $CO_2$ and energy (Bergman, 1990; Britton and Krehbiel, 1993). The initial stage of SCVFA metabolism involves the formation of activated CoA-esters of the acids, and the extent of their metabolism depends on the activities of two acyl-CoA-synthetases present in the epithelial cells: one which is non-specific for all three major SCVFAs and one which is specific for butyrate. This specificity ensures that butyrate is preferentially metabolized in the rumen epithelium. Although the extent of acetate metabolism is lower than that for the other SCVFAs, quantitatively it may account for the majority of energy expended by ruminal tissues due to the high ruminal production rates of acetate compared with propionate and butyrate (Britton and Krehbiel, 1993). Metabolism of SCVFA during absorption increases with increasing chain length with between 40 and 69% of acetate, 30 and 78% of propionate and 8 and 66% of butyrate ruminal production appearing in portal blood (Table 8.1). Care should be taken in interpreting measured rates of SCVFA appearance into portal blood which may underestimate true rates of absorption from the rumen. For example, acetate is utilized by the mesenteric drained viscera (Bergman and Wolff, 1971; Seal *et al.*, 1992; Seal and Parker, 1994; C.J. Seal, D.S. Parker, J.C. MacRae and G.E. Lobley, unpublished data) as shown by negative V–A in the mesenteric vein which must be balanced, and then exceeded by acetate absorbed from the rumen. In addition for some diets, microbial fermentation of residual carbohydrate in the hind-gut may contribute SCVFA to portal blood (DeGregorio *et al.*, 1984; Owens *et al.*, 1986), which cannot be distinguished from SCVFA derived from the rumen. In our studies in steers with mesenteric vein catheters, we have been unable to detect positive V–A for propionate across post-ruminal tissues. This suggests that although approximately 30% of propionate produced in the rumen may escape into the abomasum and omasum (Peters *et al.*, 1990), this is completely metabolized within the tissues of these gut compartments. Further studies using labelled SCVFA are needed to elucidate the processes of absorption and metabolism of these nutrients during passage through, and absorption from, different compartments of the GIT.

**Table 8.1.** Net portal flux of acetate, propionate and butyrate in sheep and cattle expressed as a proportion of measured ruminal production rate. (Updated from Seal and Reynolds, 1993.)

| | | Proportion of rumen production rate appearing in portal vein | | | |
|---|---|---|---|---|---|
| Animal | Diet | Acetate | Propionate | Butyrate | Reference[a] |
| Lambs | Intragastric infusion | 0.45 | 0.77 | 0.25 | a |
| | | 0.40 | 0.52 | 0.14 | |
| | | 0.53 | 0.78 | 0.15 | |
| Lambs | Intragastric infusion | 0.53 | 0.66 | 0.28 | b |
| | | 0.60 | 0.66 | 0.28 | |
| | | 0.44 | 0.67 | 0.26 | |
| Sheep | Intragastric infusion | | 0.57 | | c |
| | | | 0.59 | | |
| Sheep | Forage | 0.69 | 0.49 | 0.08 | d |
| Sheep | High fibre | 0.62 | 0.51 | | e |
| | Low fibre | 0.57 | 0.58 | | e |
| Cattle | Concentrate | 0.69 | 0.49 | 0.08 | f |
| Cattle | Concentrate | 0.43 | 0.40 | 0.66 | g |
| Steers | Forage | 0.48 | 0.30 | | h |
| | Concentrate | 0.52 | 0.30 | | |
| Steers | Forage | 0.96 | 0.52 | | i |
| | + 0.5 mol propionate $day^{-1}$ | 0.74 | 0.56 | | |
| | + 1 mol propionate $day^{-1}$ | 0.76 | 0.58 | | |
| Steers | Concentrate | | | 0.47 | j |
| | + butyrate | | | 0.31 | |
| | | | | 0.34 | |
| | | | | 0.32 | |
| | | | | 0.30 | |
| | | | | 0.29 | |

[a]Data from: a, Gross *et al.*, 1990b; b, Gross *et al.*, 1990a; c, Weekes and Webster, 1975; d, Bergman, 1990; e, Calculated from Linington *et al.*, 1998a, and Linington *et al.*, 1998b; f, Harmon *et al.*, 1988; g, Calculated from Huntington and Reynolds, 1983 and Sharp *et al.*, 1982); h, Seal *et al.*, 1992; i, Seal and Parker, 1994; j, Krehbiel *et al.*, 1992.

## Glucose metabolism

### *Glucose metabolism by gut tissues*

Glucose metabolism in ruminant species is dominated by the requirement for suitable precursors for gluconeogenesis, reflecting the lack of glucose absorbed from the digestive tract of forage-fed animals. Glucose requirements for tissue metabolism are similar to those of other species (Weekes, 1991), although glucose sparing from cellular activity such as fatty acid synthesis limits carbon flux through key metabolic pathways. Experiments which measure overall glucose utilization by the animal and the contribution of different tissues to that flux, show that the gut plays a major role in these processes. Glucose has been shown to be an energy substrate for the intestinal mucosa

(Stangassinger and Giesecke, 1986; Britton and Krehbiel, 1993; Okine *et al.*, 1993) and calculations of net glucose utilization by ruminants fed a range of diets (Parker, 1990) demonstrated that, in most dietary situations, the gut is a net consumer of glucose (negative V–A difference across the tissue). In studies in pregnant and lactating sheep (Van der Walt *et al.*, 1983; Perry *et al.*, 1994), glucose utilization by both the mesenteric-drained viscera (MDV) and portal-drained viscera (PDV) increased in animals post partum, with a concomitant rise in lactate production. In studies with sheep and steers where glucose irreversible loss (GIL) has been manipulated by either increasing glucose or propionate supply, a similar relationship between portal glucose utilization and GIL has been shown (Seal and Parker, 1994; Balcells *et al.*, 1995; Piccioli Cappelli *et al.*, 1997); the results of these studies are summarized in Fig. 8.2. These data clearly show that the use of glucose by gut tissues increases in line with whole body turnover rate. Further work in our laboratory has also shown that glucose utilization by gut tissues is responsive to changes in both vascular and luminal glucose supply (Piccioli Cappelli *et al.*, 1997), underlining the ability of the tissue to capitalize on both sources of the nutrient.

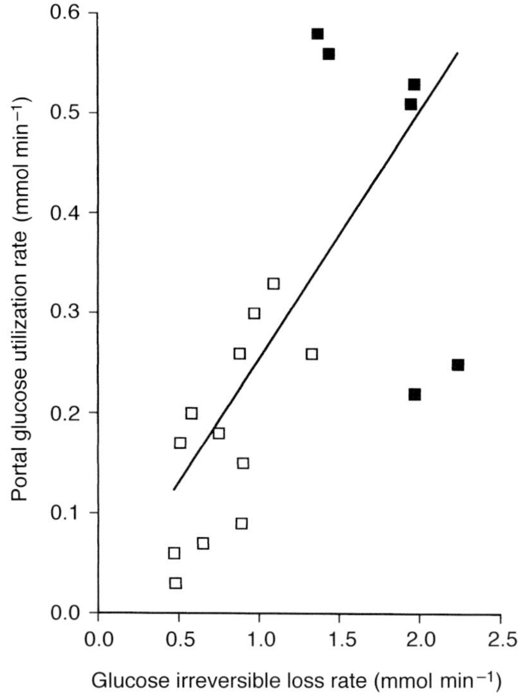


**Fig. 8.2.** Relationship between whole body glucose irreversible loss rate (mmol min$^{-1}$) and portal glucose utilization rate (mmol min$^{-1}$) in sheep and steers (data from Balcells *et al.*, 1995, and C.J. Seal, D.S. Parker, J.C. MacRae and G.E. Lobley, unpublished data; $y = 0.25x + 0.007$, $r = 0.642$).

### *Starch digestion, glucose release and absorption*

In most dietary situations, the availability of glucose in the gut lumen is low (Armstrong and Smithard, 1979) and gut tissue requirements are met from the arterial supply. The inclusion of significant amounts of raw materials such as maize, either as ensiled forage or as grain, in ruminant diets, however, has been shown to result in the flow of undegraded starch to the small intestine. These observations have provoked considerable discussion as to the ability of the ruminant gut to digest starch and absorb the glucose monomers (Nocek and Tamminga, 1991). A review of digestibility studies investigating the extent and site of digestion of starch within the ruminant gut (Huntington, 1997) gives total tract values of between 87 and 99% with, on average, 5–20% of starch consumed digested post-ruminally (Fig. 8.3). A number of experiments in this area have provided evidence that there may be a limitation in the ability of the ruminant small intestine to digest high levels of the polymer. It has been suggested that this restriction is linked to low levels of secretion of pancreatic amylase (Hill *et al.*, 1991; Kreikemeier *et al.*, 1991; Harmon, 1992). This hypothesis has been analysed in some detail by Huntington (1997) in a series of simulations of starch flow and digestion incorporating digesta flow measurements with net glucose absorption data and values for glucose transport kinetics across the small intestine. These calculations suggest that gut glucose transporter capacity is unlikely to limit glucose uptake

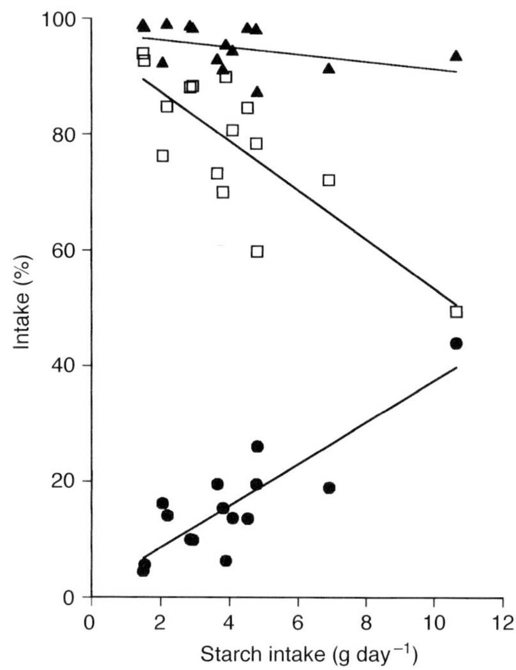


**Fig. 8.3.** Relationship between starch intake (g day$^{-1}$) and digestibility (expressed as a percentage of intake) in the rumen (□; $y = -4.24x + 95.83$, $r = 0.793$), postrumen (●; $y = 3.61x + 1.35$, $r = 0.859$) and total tract (▲; $y = -0.64x + 97.66$, $r = 0.404$) of beef and dairy cattle. Data from Huntington (1997).

and that in many cases the capacity of the ruminant small intestine to absorb glucose is in excess of supply. These data also suggest that any reduction in starch digestibility is a result of insufficient amylase activity and that stimulation of amylase synthesis/ release would overcome this limitation.

Recent experiments investigating the expression of SGLT1 $Na^+$/glucose co-transporter gene in gut tissues from lactating cows (Zhao *et al.*, 1998) confirm that the epithelial cells throughout the digestive tract, including the rumen and omasum, have the capacity to actively transport glucose. The kinetics of the transport process show that the system has a high affinity for D-glucose with a $K_m$ of about 0.1 mM confirming earlier studies by Bauer *et al.* (1995). In these experiments, ruminants unadapted to digesting starch were shown to have an ability to transport glucose which was blocked by the presence of phlorizin. In contrast to earlier work with sheep (Shirazi-Beechey *et al.*, 1991), these later experiments demonstrated that both cattle and sheep maintained a capacity for glucose uptake by the small intestine in the absence of a flow of dietary starch to the small intestine, an observation confirmed in work in our own laboratory (Balcells *et al.*, 1995; Piccioli Cappelli *et al.*, 1997). The relationship between transporter number at the apical and basolateral membranes and gut capacity to transport glucose has been reviewed (Ferraris *et al.*, 1992; Croom *et al.*, 1998) and endocrine factors involved in transport regulation identified (Bird *et al.*, 1996). Almost all these studies have been carried out on monogastric species, apart from one in which sheep treated with bST showed up-regulation of $Na^+$-dependent glucose transport in the duodenum (Bird *et al.*, 1994). This effect was not apparent in jejunal or ileal tissue indicating a site-specific response, possibly promoted by release of factor(s) such as epidermal growth factor.

### *Effects of glucose availability on nutrient flux*

The site and extent of starch digestion in the ruminant gut has an influence on the absorbed end products of digestion and also the flow of other nutrients within the tract. Studies such as that by Reynolds *et al.* (1991) provide clear evidence of the impact that forage versus concentrate diets have upon the pattern of nutrient uptake and subsequent animal performance. Greater energy retention in concentrate-fed heifers was linked to changes in the metabolism of visceral tissues, including an increase in net portal-drained visceral glucose flux as a result of increased starch flow to the small intestine. A similar effect was reported by Goetsch and Ferrell (1995), who fed different forage sources with increasing levels of maize and monitored nutrient flux across the gut. Increasing dietary maize level had no effect on PDV glucose flux. However, for two of the forage sources, propionate release increased linearly with maize inclusion, whereas for the third, a linear response in PDV release of lactate was reported. These data underline the impact that diet composition has upon the site of digestion and the subsequent availability of substrates for gut tissue metabolism. These ideas have been extended in a number of interesting studies in which combinations of glucose and protein infusions into the rumen and small intestine have been undertaken to attempt to define the interactions between site and nature of nutrients on animal responses. In the work reported by Taniguchi *et al.* (1995), starch and casein were infused into either the rumen or the abomasum and net absorption of the end products of digestion monitored across the gut. Energy release by the PDV per unit of digestible energy intake was the same, whichever site was used for starch infusion. Net release of glucose by the PDV, however, was higher during simultaneous infusion of starch and casein into the abomasum, when compared with abomasal starch infusion and intraruminal casein infusion. These data support an earlier hypothesis that increased protein flow to the duodenum stimulated pancreatic enzyme release, resulting in increased digestibility of starch escaping fermentation in the rumen. Analysis by Huntington (1997) on the impact of maize inclusion in the diet of feedlot steers and high-yielding dairy cows on glucose kinetics and the use of glucose by the gut tissues identifies the contribution of post-ruminal starch digestion to splanchnic tissue metabolism. Although this is significant in both production situations, it is apparent that for the growing steer increased fermentation of the available starch in the rumen, resulting in energy substrate release and an increase in microbial protein flow might improve the overall efficiency of nutrient use for growth.

## Amino acid metabolism

### *Amino acid requirements of gut tissues*

Although the tissues of the gut represent only a small proportion of body protein mass (0.05 in cattle), they contribute a disproportionate amount to whole body protein synthesis (0.32–0.45, Lobley *et al.*, 1980). In addition, when compared to muscle, gut tissues are characterized by having a high rate of protein turnover and a low level of protein accretion, reflecting the ability to adapt rapidly to changing nutritional situations so as to maintain a supply of nutrients to the whole animal. The impact of changes in gut function during the transition from milk-fed to weaning in lambs, for example, has been shown to result in enhanced rates of protein synthesis throughout the tract (Attaix *et al.*, 1992), reflecting the increased intake of a more fibrous diet of lower digestibility. This adaptive process has been suggested as one of the reasons why ruminant animals show a low apparent efficiency of utilization of absorbed amino acids when compared to monogastric species. This hypothesis was supported by the pioneering work on amino acid uptake in sheep (Tagari and Bergman, 1978) in which apparent disappearance of amino acids from the gut was compared to the net absorption measured in the portal vein. These studies suggested that, for a number of the essential amino acids, less than 50% of the measured loss from the gut could be accounted for in terms of net absorption into the portal vein. Since that time, there has been a continued interest in the extent to which amino acid metabolism by the tissues of the gut impacts upon whole body protein accretion and the factors which influence this effect. Our understanding of the processes involved has been considerably enhanced by the development of infusion and tracer techniques using mixed amino acid substrates such as $^{13}C$ algal protein hydrolysates (Lobley *et al.*, 1996a). This technique has been applied in an experiment to investigate the relationship between luminal and arterial supply of amino acids to the gut tissues (MacRae *et al.*, 1997) by sequential infusion of labelled amino acids into either the lumen of the gut or intravascularly. These data have shown that, in contrast to our previous understanding, luminal amino acids are efficiently transported into portal blood and that the majority (80%) of the amino acids sequestered in the gut tissue arise from the arterial supply to the gut. These data are shown in Fig. 8.4, which includes data for the proportion of whole body flux of selected amino acids which is utilized by the gut. These are important observations because they indicate that, far from capitalizing on the availability of dietary amino acids, the GIT is effectively competing with the other tissues of the body for arterially supplied amino acids, similar to the situation with glucose (see above). Recent work from Lobley *et al.* (1998) in which increased levels of a mixture of amino acids were infused into the mesenteric vein tends to support this hypothesis. During these infusions, net appearance of amino acids across the PDV was below the theoretical level, suggesting that either there was an inhibition of amino acid uptake from the gut or there was increased uptake of amino acids from the systemic circulation. Sequestration of amino acids in gut proteins represents a significant element in overall amino acid flux, whether it is derived from arterial or luminal supply. The extent to which this process can be manipulated in ruminant species is unclear. Studies in which steers were treated with recombinant somatotrophin during infusion of casein into the abomasum (Bruckental *et al.*, 1997) demonstrate that hormone treatment can alter the partition of amino acids into gut tissues resulting, in this case, in reduced PDV flux. Similarly, Yu *et*

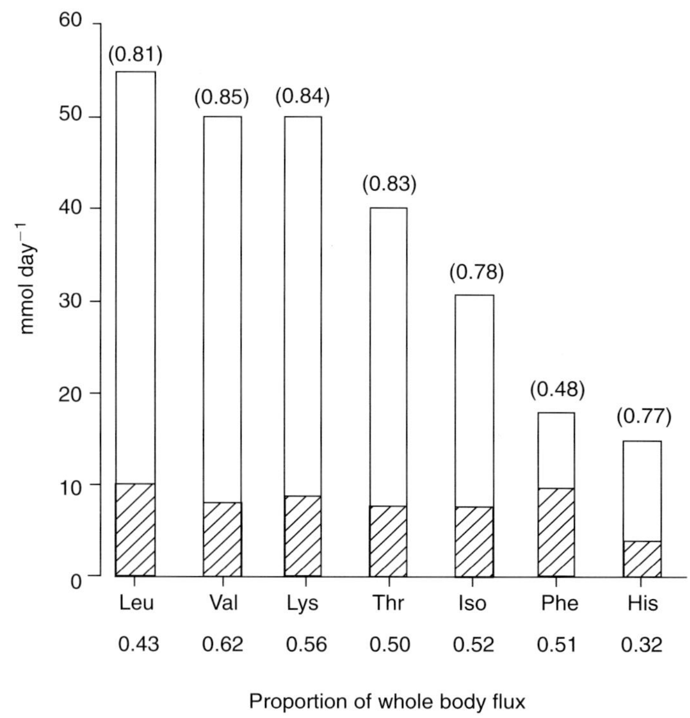


**Fig. 8.4.** Rates of amino acid sequestration (mmol day$^{-1}$) in gut tissues from arterial (open portion of bars) and luminal (hatched portion of bars) precursors in sheep fed 800 g day$^{-1}$ lucerne pellets (reproduced with permission from MacRae *et al.*, 1997).

*al.* (1998) showed that subclinical infection of *Trichostongylus colubriformis* in sheep increased leucine sequestration in gut tissues and reduced net release of absorbed leucine to the peripheral tissues; gut protein accretion in response to the parasite infection limiting synthetic processes elsewhere in the body. The extent to which the results from the study of MacRae *et al.* (1997) can be applied across a range of physiological and nutritional states has yet to be assessed. The use of a similar stable isotope infusion technique in baby pigs has provided somewhat conflicting data showing extensive oxidation of dietary amino acids during first pass through the gut tissues (Reeds, 1998). These latter observations may reflect differences between ruminant and non-ruminant species or changes associated with gut development. In this respect, the earlier work of Alpers (1972) in which $^{14}$C-leucine was infused either intraluminally or intravenously into the rat allowed the pattern of incorporation into enterocyte cells along the villus to be determined. The results showed that cells in the crypts and the base of the villus incorporated leucine from the blood whereas those at the tip of the villus and in the upper section utilized the labelled amino acid provided in the gut lumen. It is not known how such a differential use of amino acids might be influenced by situations in which enterocyte migration rate is altered during, for example, administration of gut active growth promoters (Parker, 1990).

### Amino acid absorption and metabolism

Amino acid use by the tissues of the ruminant gut has been studied using a number of models in which the pattern of nutrients has been manipulated or the demands of the tissue altered. In many of these studies, ruminant tissue oxidation of amino acids contributes only a limited extent to amino acid use, and the metabolism of glutamine, for example, may be different from that observed in non-ruminant species. Recent data from fasted sheep (Gate *et al.*, 1999) showed that conversion of this amino acid to ammonia was much lower in sheep than in rats (Windmeuller and Spaeth, 1978) and that the fraction of glutamine extracted by the digestive tract to be catabolized is also lower in ruminants. Similar observations have been made for the branch-chain amino acids (Pell *et al.*, 1986). Leucine utilization by gut tissues was shown to be directed primarily towards protein synthesis and these data are similar to our own obtained in sheep in which glucose flux, manipulated by intravenous or intraduodenal infusion of glucose, had no effect on overall leucine oxidation by gut tissues (Piccioli Cappelli *et al.*, 1997). In earlier experiments in our laboratory, manipulation of glucose availability through intravenous infusion (Balcells *et al.*, 1995) was shown to decrease net absorption of amino acids across the PDV in sheep, whereas propionate infusion into the rumen of steers resulted in an increase in amino acid flux across the MDV (Seal and Parker, 1996). These differences could reflect the impact that changes in the pattern of nutrient supply may have on metabolism by different parts of the gut, as determined by measurements made at different sites, i.e. PDV versus MDV (Reynolds and Huntington, 1988; Reynolds *et al.*, 1994). The application of new isotope infusion techniques, and the recognition that the amino acid 'nutrition' of the gut represents flux from both arterial and luminal supplies, has advanced our appreciation of the impact that this active tissue has upon partition of amino acids within the whole animal. Quantification of these processes under different physiological and nutritional conditions will provide a framework for improving the efficiency of protein utilization by ruminant species.

## References


Alpers, D.H. (1972) Protein synthesis in intestinal mucosa: the effect of route of administration of precursor amino acids. *Journal of Clinical Investigation* 51, 167–173.

Armstrong, D.G. and Smithard, R.R. (1979) The fate of carbohydrates in the small and large intestines of the ruminant. *Proceedings of the Nutrition Society* 38, 283–294.

Attaix, D., Aurousseau, E., Rosolowska-Huszcz, D., Bayle, G. and Arnal, M. (1992) In vivo longitudinal variations in protein synthesis in developing ovine intestines. *American Journal of Physiology* 263, R1318–R1323.

Balcells, J., Seal, C.J. and Parker, D.S. (1995) Effect of intravenous glucose infusion on portal-drained visceral metabolism in sheep fed a cereal/straw-based diet. *Journal of Animal Science* 73, 2146–2155.

Bauer, M.L., Harmon, D.L., McLeod, K.R. and Huntington, G.B. (1995) Adaptation to small intestinal starch assimilation and glucose transport in ruminants. *Journal of Animal Science* 73, 1828–1838.

Bergman, E.N. (1990) Energy contributions of volatile fatty acids from the gastrointestinal tract in various species. *Physiological Reviews* 70, 567–590.

Bergman, E.N. and Wolff, J.E. (1971) Metabolism of volatile fatty acids by liver and portal-drained viscera in sheep. *American Journal of Physiology* 221, 586–592.

Bird, A.R., Croom, W.J., Fan, Y.K., Daniel, L.R., McBride, B.W. and Taylor, I.L. (1994) The effects of recombinant bovine somatotrophin on intestinal absorption in sheep. *Journal of Animal Science* 72 (Suppl 1), 70A.

Bird, A.R., Croom, W.J., Fan, Y.K., Black, B.L., McBride, B.W. and Taylor, I.L. (1996) Peptide regulation of intestinal glucose absorption. *Journal of Animal Science* 74, 2523–2540.

Britton, R. and Krehbiel, C. (1993) Nutrient metabolism by gut tissues. *Journal of Dairy Science* 76, 2125–2131.

Bruckental, I., Huntington, G.B., Baer, C.K. and Erdman, R.A. (1997) The effect of abomasal infusion of casein and recombinant somatotropin hormone injection on nitrogen balance and amino acid fluxes in portal-drained viscera and net hepatic and total splanchnic blood in Holstein steers. *Journal of Animal Science* 75, 1119–1129.

Cant, J.P., McBride, B.W. and Croom, W.J. (1996) The regulation of intestinal metabolism and its impact on whole animal energetics. *Journal of Animal Science* 74, 2541–2553.

Carr, S.B. and Jacobson, D.R. (1968) Method for measurement of gastrointestinal absorption in normal animals, combining portal-carotid differences and telemetered portal blood flow by Doppler shift. *Journal of Dairy Science* 51, 721–729.

Christopherson, R.J. and Brockman, R.P. (1989) Effect of feeding and a β-adrenergic blocking agent on oxygen consumption by the portal-drained viscera, liver and hind quarters of the sheep. In: Van der Honing, Y. and Close, W.H. (eds) *Energy Metabolism of Farm Animals* (EAAP Publication No 43). Wageningen, Netherlands: Centre for Agricultural Publishing and Documentation (Pudoc), pp. 147–150.

Croom, W.J., McBride, B., Bird, A.R., Fan, Y.-K., Odle, J., Froetschel, M. and Taylor, I.L. (1998) Regulation of intestinal glucose absorption: a new issue in animal science. *Canadian Journal of Animal Science* 78, 1–13.

DeGregorio, R.M., Tucker, R.E., Mitchell, G.E. and Gill, W.W. (1984) Acetate and propionate production in the caecum and proximal colon of lambs. *Journal of Animal Science* 58, 203–207.

Dijkstra, J., Boer, H., Van Bruchem, J., Bruining, M. and Taminga, S. (1993) Absorption of volatile fatty acids from the rumen of lactating dairy cows as influenced by volatile fatty acid concentration, pH and rumen liquid volume. *British Journal of Nutrition* 69, 385–396.

Fell, B.F., Campbell, R.M., Mackie, W.S. and Weekes, T.E.C. (1972) Changes associated with pregnancy and lactation in some extra-reproductive organs in the ewe. *Journal of Agricultural Science, Cambridge* 79, 397–407.

Ferrannini, E., DeFronzo, R.A., Gusberg, R., Tepler, J., Jacob, R., Aaron, M., Smith, D. and Barrett, E.J. (1988) Splanchnic amino acid and glucose metabolism during amino acid infusion in dogs. *Diabetes* 37, 237–245.

Ferraris, R.P., Villenas, S.A. and Diamond, J. (1992) Regulation of brush-border enzyme activities and enterocyte migration rates in mouse small intestine. *American Journal of Physiology* 262, G1047–G1059.

Gate, J.J., Parker, D.S. and Lobley, G.E. (1999) The metabolic fate of the amido-N group of glutamine in the tissues of the gastrointestinal tract in 24 h fasted sheep. *British Journal of Nutrition* 81, 297–306.

Goetsch, A.L. (1998) Splanchnic tissue energy use in ruminants that consume forage-based diets ad libitum. *Journal of Animal Science* 76, 2737–2746.

Goetsch, A.L. and Ferrell, C.L. (1995) Effects of dietary level of maize on net flux of oxygen and nutrients across splanchnic tissues in sheep consuming ad libitum different forages. *Animal Science* 61, 43–55.

Goetsch, A.L. and Patil, A.R. (1997) Relationships among splanchnic tissue energy consumption and net flux of nutrients, feed intake and digestibility in wethers consuming forage-based diets ad libitum. *Journal of Applied Animal Research* 11, 1–18.

Gross, K.L., Harmon, D.L. and Avery, T.B. (1990a) Portal-drained visceral flux of nutrients in lambs fed alfalfa or maintained by intragastric infusion. *Journal of Animal Science* 68, 214–221.

Gross, K.L., Harmon, D.L., Minton, J.E. and Avery, T.B. (1990b) Effects of isoenergetic infusions of propionate and glucose on portal-drained visceral nutrient flux and concentrations of hormones in lambs maintained by total intragastric infusion. *Journal of Animal Science* 68, 2566–2574.

Hammond, K.A. and Diamond, J.M. (1994) Limits to dietary nutrient intake and intestinal nutrient uptake in lactating mice. *Phyiological Zoology* 67, 282–303.

Harmon, D.L. (1992) Dietary influences on carbohydrases and small intestinal starch hydrolysis capacity in ruminants. *Journal of Nutrition* 122, 203–210.

Harmon, D.L., Avery, T.B., Huntington, G.B. and Reynolds, P.J. (1988) Influence of ionophore addition to roughage and high-concentrate diets on portal blood flow and net nutrient flux in cattle. *Canadian Journal of Animal Science* 122, 203–210.

Hill, T.M., Schmidt, S.P., Russell, R.W., Thomas, E.E. and Wolfe, D.F. (1991) Comparison of urea treatment with established methods of sorghum grain preservation and processing on site and extent of starch digestion by cattle. *Journal of Animal Science* 69, 4570–4576.

Holtug, K., Clausen, M.R., Hove, H., Christiansen, J. and Mortensen, P.B. (1992) The colon in carbohydrate malabsorption – short-chain fatty acids, pH and osmotic diarrhoea. *Scandanavian Journal of Gastroenterology* 27, 545–552.

Huntington, G.B. (1982) Portal blood flow and net absorption of ammonia-nitrogen, urea-nitrogen, and glucose in nonlactating Holstein cows. *Journal of Dairy Science* 65, 1155–1162.

Huntington, G.B. (1983) Net nutrient absorption in beef steers fed silage or high concentrate diets containing four levels of limestone. *Journal of Nutrition* 113, 1157–1164.

Huntington, G.B. (1997) Starch utilisation by ruminants: from basics to the bunk. *Journal of Animal Science* 75, 852–867.

Huntington, G.B. and Prior, R.L. (1983) Digestion and absorption of nutrients by beef heifers fed a high concentrate diet. *Journal of Nutrition* 113, 2280–2288.

Huntington, G.B. and Reynolds, P.J. (1983) Net volatile fatty acid absorption in non-lactating Holstein cows. *Journal of Dairy Science* 66, 86–92.

Huntington, G.B., Reynolds, C.K. and Stroud, B.H. (1989) Techniques for measuring blood flow in the splanchnic tissues of cattle. *Journal of Dairy Science* 66, 1901–1908.

Huntington, G.B., Eismann, J.H. and Whitt, J.M. (1990) Portal blood flow in beef steers: comparison of techniques and relation to hepatic portal blood flow, cardiac output and oxygen uptake. *Journal of Animal Science* 68, 1666–1673.

Katz, M.L. and Bergman, E.N. (1969) A method for simultaneous catheterisation of the major splanchnic vessels of the sheep. *American Journal of Veterinary Research* 30, 655–661.

Kelly, J.M., Vaage, A.S., McBride, B.W. and Milligan, L.P. (1989) Oxygen consumption and the energy costs of $Na^+$, $K^+$-ATPase in rumen epithelial papillae from Hereford steers. *Journal of Dairy Science* 72 (Suppl. 1), 560.

Kim, E.J., Dhanoa, M.S., Gooden, J.M., Dawson, J.M., Buttery, P.J. and Scollan, N.D. (1998) The effect of dietary asynchrony on circadian changes in net nutrient fluxes across the portal-drained viscera in calves. In: McCracken, K., Unsworth, E.F. and Wylie, A.R.G. (eds) *Proceedings of the 14th Symposium on Energy Metabolism, Newcastle, Co. Down, Northern Ireland.* CAB International, Wallingford, UK, pp. 15–18.

Krehbiel, C.R., Harmon, D.L. and Schneider, J.E. (1992) Effects of increasing ruminal butyrate on portal and hepatic nutrient flux in steers. *Journal of Animal Science* 70, 904–914.

Kreikemeier, K.K., Harmon, D.L., Brandt, R.T.J., Avery, T.B. and Johnson, D.E. (1991) Effects of various levels of abomasal glucose, corn starch and corn dextrin on small intestinal disappearance and net glucose utilisation. *Journal of Animal Science* 69, 328–338.

Kristensen, N.B., Danfaer, A., Tetens, V. and Agergaard, N. (1996) Portal recovery of intraruminally infused short-chain fatty acids in sheep. *Acta Agriculturae Scandanavica. Section A. Animal Science* 46, 26–38.

Linington, M.J., Meyer, J.H.F. and Van der Walt, J.G. (1998a) Ruminal VFA production rates, whole body metabolic kinetics and blood hormone concentrations in sheep fed high- and low-fibre diets. *South African Journal of Animal Science* 28, 82–98.

Linington, M.J., Meyer, J.H.F. and Van der Walt, J.G. (1998b) The use of energy by the splanchnic tissues and their metabolism of VFA, glucose and FFA in sheep fed high- and low-fibre diets. *South African Journal of Animal Science* 28, 99–119.

Lobley, G.E. (1994) Nutritional and endocrine regulation of energy metabolism. In: Aguilera, J.F. (ed.) *13th Symposium on Energy Metabolism of Farm Animals, Mojacar, Spain.* CSIC Publishing Service, Spain, pp. 139–149.

Lobley, G.E., Milne, V., Lovie, J.M., Reeds, P.J. and Pennie, K. (1980) Whole body and tissue protein synthesis in cattle. *British Journal of Nutrition* 43, 491–502.

Lobley, G.E., Connell, A., Revell, D.K., Bequette, B.J., Brown, D.S. and Calder, A.G. (1996a) Splanchnic-bed transfers of amino acids in sheep blood and plasma, as monitored through the use of a multiple U-13C-labelled amino acid mixture. *British Journal of Nutrition* 75, 217–235.

Lobley, G.E., Weijs, P.J.M., Connell, A., Calder, A.G., Brown, D.S. and Milne, E. (1996b) The fate of absorbed and exogenous ammonia as influenced by forage or forage-concentrate diets in growing sheep. *British Journal of Nutrition* 76, 231–248.

Lobley, G.E., Bremner, D.M., Nieto, R., Obitsu, T., Hotson Moore, A. and Brown, D.S. (1998) Transfers of N metabolites across the ovine liver in response to short-term infusions of an amino acid mixture into the mesenteric vein. *British Journal of Nutrition* 80, 371–379.

López, S., Deb Hovell, F.D. and MacLeod, N.A. (1996) Osmotic pressure, water kinetics and volatile fatty acid absorption in the rumen of sheep, sustained by intragastric infusions. *British Journal of Nutrition* 71, 153–168.

McBride, B.W. and Kelly, J.M. (1990) Energy costs of absorption and metabolism in the ruminant gastrointestinal tract and liver: a review. *Journal of Animal Science* 68, 2997–3010.

MacRae, J.C., Bruce, L.A., Brown, B.S. and Calder, G. (1997) Amino acid use by the gastrointestinal tract of sheep given lucerne forage. *American Journal of Physiology* 273, G1200–G1207.

Marconi, A.M., Battaglia, F.C., Meschia, G. and Sparks, J.W. (1989) A comparison of amino acid arteriovenous differences across the liver and placenta of the fetal lamb. *American Journal of Physiology* 257, E909–E915.

Neutze, S.A., Oddy, V.H. and Gooden, J.M. (1994) A cranial mesenteric vein preparation for measurement of amino acid uptake by lambs. *Journal of Agricultural Science, Cambridge* 122, 309–314.

Nocek, J.E. and Tamminga, S. (1991) Site of digestion of starch in the gastrointestinal tract of dairy cows and its effect on milk yield and composition. *Journal of Dairy Science* 74, 3598–3629.

Okine, E.K., Cherry, R. and Kennelly, J.J. (1993) Glucose and amino acid transport and metabolism in duodenal flat sheets of dairy cattle at three stages of lactation. *Comparative Biochemistry and Physiology* 107, 719–726.

Owens, F.N., Zinn, R.A. and Kim, Y.K. (1986) Limits to starch digestion in the ruminant small intestine. *Journal of Animal Science* 63, 1634–1648.

Parker, D.S. (1990) Manipulation of the functional activity of the gut by dietary and other means (antibiotics/probiotics) in ruminants. *Journal of Nutrition* 120, 639–648.

Pell, J.M., Caldarone, E.M. and Bergman, E.N. (1986) Leucine and α-ketoisocaproate metabolism and interconversions in fed and fasted sheep. *Metabolism* 35, 1005–1016.

Perry, K.W., Janes, A.N., Weekes, T.E., Parker, D.S. and Armstrong, D.G. (1994) Glucose and L-lactate metabolism in pregnant and lactating ewes fed barley- or ground maize-based diets. *Experimental Physiology* 79, 35–46.

Peters, J.P., Shen, R.Y.W., Robinson, J.A. and Chesters, S.T. (1990) Disappearance and passage of propionic acid from the rumen of the beef steer. *Journal of Animal Science* 68, 3337–3349.

Piccioli Cappelli, F., Seal, C.J. and Parker, D.S. (1997) Glucose and [13C]leucine metabolism by the portal-drained viscera of sheep fed on dried grass with acute intravenous and intraduodenal infusions of glucose. *British Journal of Nutrition* 78, 931–946.

Reeds, P.J. (1998) In: McCracken, K., Unsworth, E.F. and Wylie, A.R.G. (eds) *Proceedings of the 14th Symposium on Energy Metabolism, Newcastle, Co. Down, Northern Ireland.* CAB International, Wallingford, UK.

Remond, D., Ortigues-Marty, I., Isserty, A. and Lefaivre, J. (1998) Technical note: measuring portal blood flow in sheep using an ultrasonic transit time flow probe. *Journal of Animal Science* 76, 2712–2716.

Reynolds, C.K. and Huntington, G.B. (1988) Partition of portal-drained visceral net flux in beef steers. 1. Blood flow and net flux of oxygen, glucose and nitrogenous compounds across stomach and post-stomach tissues. *British Journal of Nutrition* 60, 539–551.

Reynolds, C.K., Tyrell, H.F. and Reynolds, P.J. (1991) The effect of diet forage-to-concentrate ratio and intake on energy metabolism in growing beef heifers: net nutrient metabolism by gut tissues. *Journal of Nutrition* 121, 1004–1015.

Reynolds, C.K., Harmon, D.L. and Cecava, M.J. (1994) Absorption and delivery of nutrients for milk protein synthesis by portal-drained viscera. *Journal of Dairy Science* 77, 2787–2808.

Seal, C.J. and Parker, D.S. (1994) Effect of intraruminal propionic acid infusion on metabolism of mesenteric- and portal-drained viscera in growing steers fed a forage diet. 1. Volatile fatty acid, glucose and lactate. *Journal of Animal Science* 72, 1325–1334.

Seal, C.J. and Parker, D.S. (1996) Effect of intraruminal propionic acid infusion on metabolism of mesenteric- and portal-drained viscera in growing steers fed a forage diet. 2. Urea, ammonia, amino acids and peptides. *Journal of Animal Science* 74, 245–255.

Seal, C.J. and Reynolds, C.K. (1993) Nutritional implications of gastrointestinal and liver metabolism in ruminants. *Nutrition Research Reviews* 6, 185–208.

Seal, C.J., Parker, D.S. and Avery, P.J. (1992) The effect of forage and forage-concentrate diets on rumen fermentation and metabolism of nutrients by the mesenteric and portal-drained viscera in growing steers. *British Journal of Nutrition* 67, 355–370.

Seal, C.J., Parker, D.S., MacRae, J.C. and Lobley, G.E. (1994) Effect of intraduodenal starch infusion on net portal absorption of amino acids in growing steers fed lucerne. *Animal Production* 58, 434A.

Sharp, W.M., Johnson, R.R. and Owens, F.N. (1982) Ruminal VFA production with steers fed whole or ground corn grain. *Journal of Animal Science* 55, 1505–1514.

Shirazi-Beechey, S.P., Smith, M.W., Wang, Y. and James, P.S. (1991) Postnatal development of lamb intestinal digestive enzymes is not regulated by diet. *Journal of Physiology* 437, 691–698.

Shoemaker, W.C. (1964) Methods and techniques for measurement of hepatic physiology and metabolism. In: Rouiller, C. (ed.) *The Liver: Morphology, Biochemistry, Physiology*. Academic Press, New York, pp. 243–266.

Stangassinger, M. and Giesecke, D. (1986) Splanchnic metabolism of glucose and related energy substrates. In: Milligan, L.P., Grovum, W.L. and Dobson, A. (eds) *Proceedings of the 6th International Symposium on Ruminant Physiology*. Prentice Hall, Englewood Cliffs, New Jersey, pp. 347–366.

Tagari, H. and Bergman, E.N. (1978) Intestinal disappearance and portal blood appearance of amino acids in sheep. *Journal of Nutrition* 108, 790–803.

Taniguchi, K., Huntington, G.B. and Glenn, B.P. (1995) Net nutrient flux by visceral tissues of beef steers given abomasal and ruminal infusions of casein and starch. *Journal of Animal Science* 73, 236–249.

Van der Walt, J.G., Baird, G.D. and Bergman, E.N. (1983) Tissue glucose and lactate metabolism and interconversions in pregnant and lactating sheep. *British Journal of Nutrition* 50, 267–280.

Wangsness, P.J. and McGilliard, A.D. (1972) Measurement of portal blood flow in calves by dye dilution. *Journal of Dairy Science* 55, 1439–1446.

Weekes, T.E.C. (1991) Hormonal control of glucose metabolism. In: Tsuda, T., Sasaki, Y. and Kawashima, R. (eds) *Physiological Aspects of Digestion and Metabolism in Ruminants*. Academic Press, New York, pp. 183–200.

Weekes, T.E.C. and Webster, A.J.F. (1975) Metabolism of propionate in the tissues of the sheep gut. *British Journal of Nutrition* 33, 425–468.

Whitt, J., Huntington, G., Zetina, E., Casse, E., Taniguchi, K. and Potts, W. (1996) Plasma flow and net nutrient flux across gut and liver of cattle fed twice daily. *Journal of Animal Science* 74, 2450–2461.

Windmeuller, H.G. and Spaeth, A.E. (1978) Identification of ketone bodies and glutamine as the major respiratory fuels *in vivo* for postabsorptive rat small intestine. *Journal of Biological Chemistry* 253, 69–76.

Yu, F., Bruce, L.A., Coop, R.L. and MacRae, J.C. (1998) Leucine metabolism across the gastrointestinal tract of sheep infected with *Trichostrongylus colubriformis*. *Proceedings of the British Society of Animal Science* p. 1.

Zhao, F.-Q., Okine, E.K., Cheeseman, C.I., Shirazi-Beechey, S.P. and Kennelly, J.J. (1998) Glucose transporter gene expression in lactating bovine gastrointestinal tract. *Journal of Animal Science,* 76, 2921–2929.

# 9 The Liver: Integrator of Nitrogen Metabolism

G.E. LOBLEY[1], G.D. MILANO[2] AND J.G. VAN DER WALT[3]

[1]*Rowett Research Institute, Bucksburn, Aberdeen, UK;* [2]*Departamento de Fisiopatologia, Facultad de Ciencias Veterinarias UNCPBA, Campus Universitario, Paraje Arroyo Seco (7000) Tandil, Argentina;* [3]*Department of Physiology, Faculty of Veterinary Sciences, University of Pretoria, Onderstepoort, South Africa*

## Introduction

The liver is positioned at the anatomic and metabolic crossroads of the body and, as such, controls the amount and nature of nutrients available to the peripheral tissues from digestive tract absorption. This role in nutrient partitioning encompasses all of the major macronutrients and may be influenced by the involvement of the liver in the production (insulin-like growth factor-I, IGF-I) and removal (insulin, glucagon) of key hormones. For the sake of brevity, this review will focus on the metabolic products of protein-N, particularly ammonia and amino acids. Hepatic detoxification of the former, to prevent deleterious peripheral hyperammonaemia, is probably a more vital function in ruminants than non-ruminants, particularly for grazing animals. Because peripheral hyperaminoacidaemia may also produce adverse effects, it is evident that the regulation of systemic plasma amino acid concentrations is also important. This is achieved by modifying the quantity of individual absorbed amino acids removed by the liver and, in particular, controlling the amount oxidized versus the proportion converted to other metabolites, such as the plasma export proteins. Although the biochemical pathways involved in these partitions are well documented and a number of recent reviews have described the quantitative events (e.g. van der Walt, 1993; Reynolds, 1995; Lobley and Milano, 1997), the regulatory aspects still remain obscure. Furthermore, the limited data available to provide a conceptual framework for metabolic control mechanisms are often obtained from non-ruminants, or *in vitro* studies, and may not apply directly to ruminants. The shortcomings of these approaches should always be kept in mind.

## Liver architecture

The liver is the most richly vascularized organ in the body, with blood comprising approximately 25% of its mass. Despite constituting less than 2% of body weight in

 *Ruminant Physiology: Digestion, Metabolism, Growth and Reproduction* (ed. P.B. Cronjé)

adult ruminants, the liver also receives approximately 25% of cardiac output. Total blood flow through the liver is sensitive to nutrition, plus other regulatory factors, and can reach values of 3 l $min^{-1}$ $kg^{-1}$ liver weight (or 60 ml $min^{-1}$ $kg^{-1}$ body weight). Blood supply to the liver arises from two sources, the hepatic portal vein and the hepatic artery, although the latter makes only a small contribution to total hepatic blood flow in cattle and sheep (8–12%; Reynolds, 1995; Isserty *et al.*, 1998).

Regulation of hepatic blood flow is unique. The main blood supply, through the hepatic portal vein, is not controlled by the liver; even if vascular resistance is increased to a maximum (for example, by stimulation of the hepatic sympathetic nerves), the flow remains unaltered, although the pressure rises. Furthermore, the well-known autoregulation of arterial blood flow through most other organs (driven by the level of tissue oxygenation) is absent from the liver. Instead, the arterial flow appears to be inversely linked to the hepatic portal venous flow, providing the liver with a constant total flow (Lautt, 1996). The mechanism may involve steady release of adenosine into the Space of Mall, followed by removal through both arterial and hepatic portal inflows. If the portal supply is compromised, the local concentration of adenosine increases, stimulating arterial vasodilation and flow to remove the metabolite. During periods of maximal hepatic portal flow, e.g. at peak absorption from the rumen, the inverse should occur.

While most information on spatial organization within the mammalian liver is derived from rodents, ruminants may differ somewhat. The liver is comprised of four cell types: hepatocytes (by far the largest group); macrophages (Kupffer cells); endothelial cells lining the Space of Disse; and fat storage (Ito) cells (see Gebhardt, 1992; Jungermann and Kietzmann, 1996). The hepatic acinus, accepted as the smallest functional unit, comprises parenchymal cells in berry-like clusters, approximately 2 mm in radius, on a vascular stalk (portal triad) containing the finest branches of the hepatic artery, hepatic portal vein and the bile duct. This arrangement means that blood flows past both sides of not more than 16 hepatocytes, before exiting via the central drainage system (Lautt, 1996). The hepatocytes exhibit heterogeneity in both spatial and biochemical characteristics and may be divided simplistically into two populations, periportal and perivenous (also known as pericentral), based on their ability to synthesize urea or glutamine (Fig. 9.1). Although other metabolic activities of these cells show a smooth transition along the acinus from periportal to perivenous (Jungermann and Kietzmann, 1996), the enzymes of the ornithine cycle and glutamine synthesis are sharply separated, at least in the rodent.

Periportal cells are the first to receive blood from the afferent hepatic portal vein and contain characteristically all five enzymes of the ornithine (urea) cycle, plus glutaminase in the mitochondria. Furthermore, mRNA for a variety of export proteins, including albumin, are also present (Gebhardt, 1992; Jungermann and Kietzmann, 1996), as are transporters for most of the amino acids, with the notable exceptions of aspartate and glutamate. The perivenous cells are clustered around the efferent hepatic vein and lack the ornithine cycle enzymes, albumin mRNA and mitochondrial glutaminase. Instead, they possess cytosolic glutamine synthetase, the $X_{ag}$ transporter for glutamate and aspartate entry (Häussinger and Gerok, 1983), plus ornithine aminotransferase (necessary for the catabolism of arginine; Kuo *et al.*, 1991). A similar perivenous localization of glutamine synthetase in sheep is suggested by the data obtained from antegrade and retrograde perfusion of isolated liver, with and without inhibition of glutamine synthesis by sulphoximine (Rossouw *et al.*, 1997).

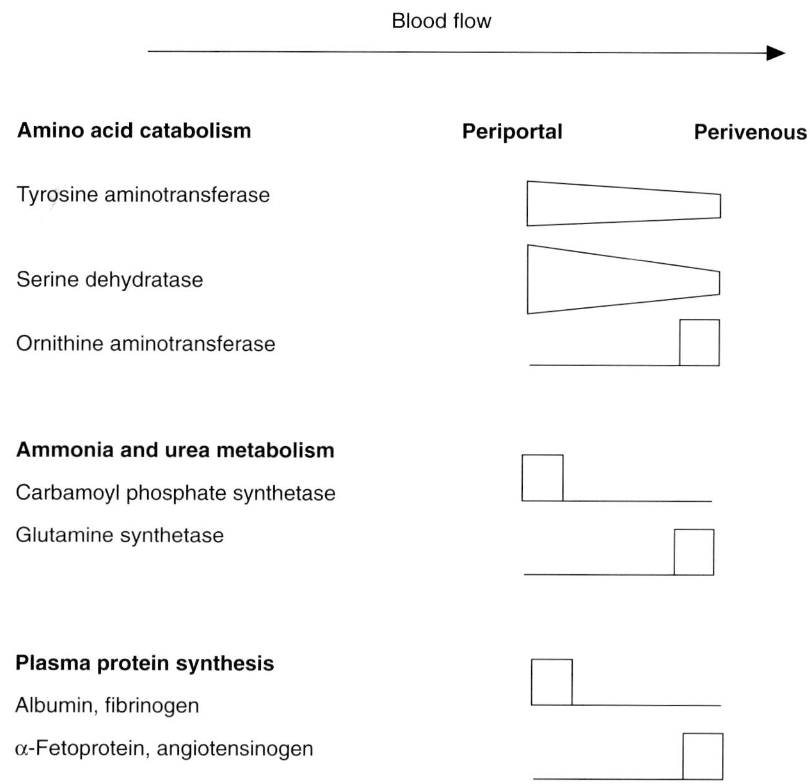


**Fig. 9.1.** Metabolic zonation and pattern of enzyme distribution between periportal and perivenous hepatocytes in non-ruminants (data from Jungermann and Kietzmann, 1996; O'Sullivan *et al.*, 1998). The magnitude of the symbols indicates the relative contribution within each zone, e.g. serine dehydratase activity declines between periportal and perivenous hepatocytes, but still occurs in the latter, while glutamine synthetase is restricted to the perivenous region.

## Ammonia metabolism and ureagenesis

The spatial organization of hepatic cells confers a 'metabolic zonation' (Häussinger *et al.*, 1992a), particularly with reference to ammonia metabolism and ureagenesis (Fig. 9.1). Based on the high $K_m$ of carbamoyl phosphate synthetase 1 (CPS1) for ammonia (approximately 1–2 mM; Lusty, 1978), the ornithine cycle provides a primary high capacity, low affinity system for the hepatic removal of ammonia. In contrast, limited amounts of perivenous glutamine synthetase, with a low $K_m$ for ammonia (0.11 mM in humans, Kaiser *et al.*, 1988; 0.33 mM in rats, Deuel *et al.*, 1978), offer a secondary, low capacity, high affinity system for capture of ammonia. The effectiveness of this two-stage system ensures that up to 98% of [$^{15}$N]ammonia in the portal vein is removed during each pass across the liver (Nieto *et al.*, 1996). This system is vital to maintain peripheral plasma ammonia concentrations at very low concentrations. Failure to do so may lead to a range of deleterious consequences, including reduced fertility

(McEvoy *et al.*, 1997), lowered appetite and, in extreme cases, coma and death (Summerskill and Wolpert, 1970).

## The ornithine cycle

The ornithine cycle requires equal and concomitant inflows of N into carbamoyl phosphate and aspartate and, in theory at least, an imbalance at either entry point could lead to the potentially deleterious accumulation of metabolites (e.g. an inability to remove ammonia). Specific mechanisms exist to reduce the risk of such occurrences. For example, under normal cell redox conditions in periportal hepatocytes, the $K_{eq}$ of mitochondrial glutamate dehydrogenase (GDH) strongly favours synthesis of glutamate from ammonia and oxoglutarate. This action, when coupled with that of aspartate : 2-oxoglutarate transaminase (AOT), transfers ammonia-N to mitochondrial aspartate. This would be important when ammonia extraction by the liver exceeds hepatic catabolism of amino acids, probably the normal circumstance with ruminants fed fresh or conserved forages (e.g. Whitt *et al.*, 1996; Lobley *et al.*, 1998). Indeed, in studies involving [$^{15}$N]ammonia across the sheep liver (Lobley *et al.*, 1995; Milano *et al.*, 1996) and with ovine hepatocytes (Luo *et al.*, 1995; Mutsvangwa *et al.*, 1997), [$^{15}$N$^{15}$N]urea was produced, proving that ammonia-N can be transferred to aspartate (Table 9.1). Furthermore, in ovine hepatocytes incubated with [$^{15}$N]ammonia, the $^{15}$N enrichments of glutamate and aspartate were similar (Luo *et al.*, 1995), confirming the involvement of GDH and AOT.

The contribution of ammonia-$^{15}$N to both carbamoyl phosphate and aspartate-N inflows can be obtained quite simply from isotopomer analysis of the relative proportions of $^{14}$N$^{15}$N and $^{15}$N$^{15}$N urea formed (Milano *et al.*, 1996; Milano, 1997). For example, under conditions where rapid equilibrium occurs via GDH-AOT (Luo *et al.*, 1995), the N precursor enrichment, *a*, is obtained from

$$a = 2/[2 + (^{14}N^{15}N : {}^{15}N^{15}N)]$$

Thus, under these circumstances, only a sample of urea is required to determine this contribution. A similar approach has been applied to studies with the perfused rat liver (Brosnan *et al.*, 1996). More sophisticated equations have been developed (Milano, 1997) to accommodate situations where equilibrium between ammonia and aspartate-N is not achieved, as is the probable case *in vivo* (Lobley *et al.*, 1995). This approach has shown that in fasted sheep, subjected to an ammonia overload, approximately 32% of aspartate-N was derived from ammonia (Milano *et al.*, 1996). When coupled with net transfers of ureagenic metabolites across the liver, this technique offers a powerful means to determine how much of the urea truly arises from ammonia- or amino acid-N.

In non-ruminants, or ruminants offered grain-based rations with additional rumen undegradable protein, supply of amino acid-N may exceed that of ammonia (Burrin *et al.*, 1991; Goetsch and Ferrell, 1995), and extra ammonia must be generated to maintain ornithine cycle activity and prevent peripheral hyperaminoacidaemia. This is achieved through mitochondrial glutaminase, which generates ammonia (and glutamate) from glutamine. Indeed, glutaminase may act as a continuous amplification mechanism to maintain high concentrations of ammonia within the mitochondria (Meijer *et al.*, 1985), to compensate for the high $K_m$ for ammonia of CPS1. In sheep, with intra-abomasal infusions of either ammonium bicarbonate or amino acids, coupled with intravascular infusion of 5-$^{15}$N (amido) glutamine, between 28 and 61% of the [$^{15}$N]urea produced across the liver was derived directly from glutamine (R.M. Nieto, T. Obitsu, A. Fernandez and G.E. Lobley, unpublished results).

Net hepatic removal of glutamine in pregnant, dry cows was proportional to the amounts of an amino acid mixture infused into the mesenteric vein (Wray-Cahen *et al.*, 1997). In contrast, complementary studies involving amino acid infusions into sheep (Lobley *et al.*, 1998) failed to alter liver glutamine uptake. These observed differences probably relate to the proportions of ammonia and amino acids extracted by the liver (Lobley *et al.*, 1998). Amino acid-N may also transfer to ammonia more directly, because ovine hepatocytes, incubated with unlabelled ammonia and [$^{15}$N]alanine, also produced [$^{15}$N$^{15}$N]urea, although this was smaller than the amount of [$^{14}$N$^{15}$N]urea synthesized (Mutsvangwa *et al.*, 1997; Table 9.1).

Whether these various mechanisms operate effectively enough may have important implications for ruminant productivity. Before the paradigm of the ornithine/glutamine cycles was proposed, it was thought that, under conditions of ammonia excess, either peripheral hyperammonaemia would occur or amino acid catabolism would be stimulated to provide additional aspartate-N. Indeed, from retrospective analysis of trans-hepatic ruminant data, a scenario was proposed that removal of ammonia-N by the liver was accompanied by a similar input from other N sources, i.e. amino acid catabolism would match ammonia detoxification (Parker *et al.*, 1995). This hypothesis would

**Table 9.1.** Effect of propionate on urea production in ovine hepatocytes and sources of ornithine cycle-N.

| | nmol mg$^{-1}$ wet cells h$^{-1\dagger}$ | | | |
|---|---|---|---|---|
| Propionate* (mM) | $NH_3$ uptake | Urea production | Glucose production | Urea ([$^{15}$N$^{15}$N]:[$^{14}$N$^{15}$N]) |
| 0 | 17$^a$ | 28 | 6$^a$ | 1.0 |
| 0.2 | 58$^b$ | 63$^b$ | 16$^b$ | 1.1 |
| 0.4 | 54$^b$ | 62$^b$ | 22$^b$ | 1.1 |
| 0.8 | 43$^c$ | 54$^c$ | 29$^c$ | 1.2 |
| 1.6 | 36$^c$ | 43$^c$ | 31$^c$ | 1.1 |

| Propionate$^\ddagger$ (mM) | nmol mg$^{-1}$ DM h$^{-1\dagger}$ Urea production | Urea ([$^{15}$N$^{15}$N]:[$^{14}$N$^{15}$N]) |
|---|---|---|
| 0 | 7.1$^a$ | 0.29 |
| 0.31 | 6.9$^a$ | 0.31 |
| 0.63 | 7.2$^a$ | 0.26 |
| 1.25 | 3.9$^b$ | 0.27 |

* Medium also contained 0.67 mM $^{15}NH_4Cl$, plus a physiological mixture of amino acids; data from M.A. Lomax and G.E. Lobley (unpublished data).
† Within study different superscripts in a column indicate $P < 0.05$ or better.
‡ Medium also contained 0.63 mM $NH_4Cl$, plus 1.25 mM [$^{15}$N]alanine; data from Mutsvangwa *et al.* (1997).

apparently account for the poorer utilization of dietary amino acids by ruminants, particularly those fed fresh or conserved forages. In the event, controlled perturbation of ammonia supply to the liver, by intra-mesenteric vein infusion of ammonium bicarbonate, showed that the extra urea produced exceeded ammonia uptake by between 13 and 17% (Lobley *et al.*, 1996b; Lobley and Milano 1997). This is substantially less than the doubling predicted by the hypothesis but, none the less, does represent increased amino acid catabolism – thereby losing the equivalent of 0.6–1.3 g amino acid-N for daily anabolism. With ovine hepatic preparations *in vitro*, addition of ammonia led to similar amounts of N released as urea plus glutamine (Luo *et al.*, 1995; Rossouw *et al.*, 1997, 1999), i.e. the removal of ammonia did not appear to stimulate amino acid catabolism. Overall, therefore, the importance of ammonia detoxification may well have resulted in the process being effectively uncoupled from amino acid oxidation.

### Capacity and substrate priority for the ornithine cycle

There is a finite capacity for the rate of ureagenesis and a number of studies in both cattle and sheep have yielded values of 2 mmol urea-N $kg^{-1}$ fresh liver weight $min^{-1}$ (see Lobley *et al.*, 1998). This amounts to approximately 29 and 435 g urea-N $day^{-1}$, for a 40 kg sheep or a 600 kg bovine respectively, within the range of N intake by these animals on high levels of nutrition (Sarraseca *et al.*, 1998). At higher intakes, unless ornithine cycle enzyme content or activity increases adaptively, amino acids and ammonia will escape beyond the liver. Furthermore, even at lower mean daily urea production rates, diurnal fluctuations in ammonia and amino acid absorption (Whitt *et al.*, 1996) may result in periods when these capacities are exceeded. This raises the important question: when the capacity of the ornithine cycle is reached does the liver prioritize removal of ammonia or of amino acids?

When ammonia inflow to the liver of young sheep was raised from 0.58 to 2.01 mM $min^{-1}$, through acute (31 min) infusions of ammonium bicarbonate into the mesenteric vein, hepatic extraction was elevated from 0.44 to 1.21 mM $min^{-1}$ (Milano, 1997; see Lobley and Milano, 1997). This incomplete removal led to ammonia overspill into the peripheral circulation and elevated arterial concentrations (from 0.08 to 0.31 mM), a finding compatible with the low capacity of glutamine synthetase in the ovine perivenous hepatocytes (Rossouw *et al.*, 1999). Interestingly, the non-ammonia N contribution to ureagenesis declined (from 0.36 to 0.14 mM $min^{-1}$; $P$ = 0.07), suggesting that amino acid catabolism was reduced in order to allow more ammonia to be detoxified. These acute ammonia infusions might relate to situations of diurnal feeding, e.g. morning or evening intakes of fresh grass at pasture, where it is possible that absorption of large quantities of ammonia might actually reduce, temporarily, the hepatic oxidation of amino acids. Interestingly, this is the converse of the original ammonia hypothesis (Parker *et al.*, 1995). This priority in favour of ammonia removal was confirmed in a converse study when 1.84 mM $min^{-1}$ of mixed amino acid-N was infused into the mesenteric vein of sheep fed a basal supra-maintenance grass pellet ration (Lobley *et al.*, 1998). Even though the potential ureagenic supply to the liver exceeded ornithine cycle activity, hepatic ammonia removal was unchanged and severe arterial blood hyperaminoacidaemia resulted. Although hepatic extraction favours ammonia removal, neither substrate was able to completely suppress *in vivo* the ureagenic capacity of the other. This is despite the observations with ovine hepatocytes *in vitro* that ureagenesis can proceed with ($^{15}$N) ammonia as the sole exogenous N substrate and with [$^{15}$N$^{15}$N]urea as the primary product (Luo *et al.*, 1995); a finding confirmed with the perfused sheep liver (Rossouw *et al.*, 1997, 1999).

### Control of ureagenesis

A major aim of any high-production system is to increase the amount of anabolic metabolites (e.g. amino acids) available to the peripheral tissues. Clearly, any treatment that changes anabolism without changing intake or absorption of nutrients must alter ureagenesis (e.g. bovine-somatotropin; Bruckentahl *et al.*, 1997). Despite many years of research, it is still unclear whether ureagenesis is actively or passively controlled. Simple consideration of the quantities and maximal activities of the ornithine cycle enzymes would suggest that arginosuccinate synthetase (ASS) is limiting (Rattenbury *et al.*, 1980). In practice, however, CPS1 is a more probable candidate because this is positively regulated by the mitochondrial concentration of *N*-acetylglutamate (N-AG; see Meijer *et al.*, 1990). The activity of N-AG synthetase is regulated *in vitro* by several factors, including insulin (an inhibitor) and glucagon (an activator). This would fit with the competing roles of these hormones; glucagon leads to poorer N retention (thus more urea synthesis), while insulin diverts amino acids towards peripheral tissue anabolism. Direct, or immediate, actions of insulin on the ornithine cycle, however, are probably doubtful *in vivo*. For example, although the liver of cattle, which were acutely (3 h) hyper-insulinaemic, extracted substantial amounts of insulin, urea production remained unchanged (Eisemann and Huntington, 1994). In other bovine studies, chronic administration of growth hormone (or its releasing factor) decreased hepatic amino acid extraction and ureagenesis (Bruckentahl *et al.*, 1997), although this has not been a universal finding (McLeod *et al.*, 1997). The long-term nature of these latter studies does not allow resolution of direct hormonal effects on the ornithine cycle enzymes.

Another putative regulator of ureagenesis is propionate, which has been shown at supra-physiological concentrations to be a potent *in vitro* inhibitor of ureagenesis (e.g. Stewart and Walser, 1980). Such regulation through propionate would provide an attractive explanation for the superior performance of animals on concentrate versus forage diets. This role has been reinvestigated recently with ovine hepatocytes and, while a minimal amount of propionate (or other 3-carbon unit) is needed to activate the ornithine cycle, ureagenesis is suppressed at propionate concentrations above 0.63 mM (Mutsvangwa *et al.*, 1997; see Table 9.1). At these higher propionate concentrations, gluconeogenesis continues to be stimulated, thus breaking the normal link between glucose synthesis and urea production. Inhibition of ureagenesis by propionate probably does not affect the relative inflows into the ornithine cycle of ammonia and amino acids, as shown by incubating hepatocytes in the presence of either [$^{15}$N]ammonia or [$^{15}$N]alanine. Indeed, the relative ratios of [$^{15}$N$^{15}$N]:[$^{14}$N$^{15}$N] urea produced were uninfluenced by the presence of propionate (Table 9.1). While these effects are only apparent towards the upper end of the physiological range *in vitro*, the elevated arterial ammonia concentrations, which result from episodic infusion of propionate into the rumen (Choung and Chamberlain, 1995), may indicate that the ornithine cycle may also be inhibited *in vivo*.

## *Fate of urea synthesized*

In ruminants, urea-N synthesis often exceeds the amount of apparently digestible N and, at submaintenance intakes, even total ingested N (e.g. Goetsch and Ferrell, 1995; Sarraseca *et al.*, 1998). This apparent anomaly is due to substantial entry of urea into the gastrointestinal tract (GIT), where it may either provide an N source for microbial protein synthesis (and thus contribute to potential N anabolism) or be re-absorbed as ammonia across the portal-drained viscera (PDV) (with augmentation of hepatic ureagenesis). Hitherto, resolution of these two fates has been difficult but can now be overcome by the use of the [$^{15}N^{15}N$]urea isotopomer technique (see p. 152, Fig. 9.2). This method estimates in normal humans that approximately 80–90% of the urea-N which enters the GIT may be returned for anabolic purposes. This proportion may be markedly influenced by dietary treatments (Jackson, 1998).

Hitherto, this approach has failed to account for multiple re-entries of [$^{14}N^{15}N$]urea into the GIT, leading to the formation of daughter [$^{14}N^{15}N$]molecules, indistinguishable chemically from the parent labelled urea. This leads to over-estimates of the [$^{14}N^{15}N$]urea transfers and, for ruminants, the error introduced can be as large as 60%. Application of this isotopomer technique to sheep, but with correction for multiple entries of [$^{14}N^{15}N$]urea to the GIT, revealed that although absolute rates of

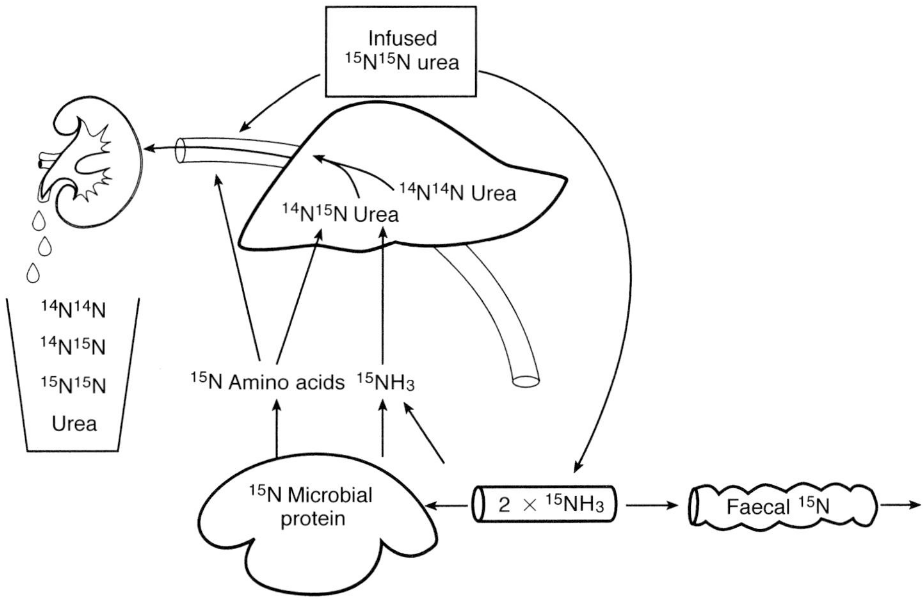


**Fig. 9.2.** Re-utilization of hepatic urea from the digestive tract quantified by use of infusion of [$^{15}N^{15}N$]urea. Urea-N is either lost in faecal-N, converted to microbial protein (then digested and absorbed as amino acids), or reabsorbed as $^{15}NH_3$ when it can either be reconverted to urea (as $^{14}N^{15}N$) or used to aminate (transaminate) amino acids.

ureagenesis and urea-N entry to the GIT doubled as intake of grass pellets was increased from 0.6 to 1.8 × maintenance, the proportion recycled to the ornithine cycle remained constant and represented 22–30% of urea production (Sarraseca *et al.*, 1998). Similar values (30–31%) were obtained with rations of either a mixed forage-concentrate or chopped hay plus grass pellets (unpublished results). In practice, this means that, under these dietary conditions (Lobley *et al.*, 1996b, 1998), as much as 50% of the ammonia absorbed from the GIT and detoxified by the liver arises from urea synthesized earlier and thus represents part of N recycling in the ruminant.

## Amino acid metabolism

### *Liver N balance*

Studies that have examined trans-hepatic nitrogen metabolism are confounded by a number of apparent anomalies. Logically, there must be a sensible balance between N substrate inflow and outflow, with due allowance made for synthesis of export (plasma) proteins and accretion (or depletion) of liver protein mass. In practice, several recent studies show larger outflow of urea-N than the combined inflow of ammonia- and amino acid-N, even at high intakes, and before allowance is even made for net protein synthesis (Burrin *et al.*, 1991; Reynolds and Tyrrell, 1991; Goetsch and Ferrell, 1995). Some studies found both negative and positive balances (e.g. Reynolds *et al.*, 1994; Bruckentahl *et al.*, 1997; Wray-Cahen *et al.*, 1997) while others reported only net positive N uptake (e.g. Lobley *et al.*, 1995, 1996b, 1998; Krehbiel *et al.*, 1998). To blame may be unsatisfactory methodology, i.e. 'α-amino N' (ninhydrin based), 'free amino N' (trinitrobenzene sulphonate) techniques, since these data may be up to threefold higher than values obtained from a full amino acid analysis (Burrin *et al.*, 1991; compare Reynolds and Tyrrell, 1991 with Reynolds *et al.*, 1994). Of course, some of the discrepancies may be accounted for by other N inputs, notably peptides (Webb, 1999). In many situations, however, physiologically-sensible balances can be achieved using improved techniques to measure the main N substrates (ammonia, urea and free amino acids), without involving peptide uptake. As a result, peptide exchanges are probably minor (but not necessarily unimportant). Further precision can be gained by a gravimetric approach (Lobley *et al.*, 1998) or by a stable-isotope dilution technique (Calder *et al.*, 1999).

### *Plasma versus blood transport*

Amino acids may be transported in either the plasma and/or the red blood cells. In sheep (Heitmann and Bergman, 1980) and cattle (Houlier *et al.*, 1991), earlier studies on mass exchanges suggested that carriage to the liver was via plasma, but that hepatic transfer to peripheral tissues involved the erythrocytes. The former concept was supported, in sheep, by data with [$^{13}$C] amino acids infused systemically and where, across the PDV, absorption caused a dilution of enrichments in the plasma but not of those in the red blood cell (Lobley *et al.*, 1996a). Across the liver, however, there was only a small dilution of plasma enrichments (0–6%), suggesting that transfers were primarily unidirectional, i.e. into the liver cells. Furthermore, because liver intracellular enrichments usually exceed those in the red blood cell (Lobley *et al.*, 1996a; Connell *et al.*, 1997), then any outward transfer from hepatocytes to erythrocytes should increase enrichment in the latter. In practice, enrichment within the red blood cells was either unchanged or decreased slightly (0 to −6%). These data suggest strongly there is little net transfer of amino acids from the liver via the erythrocytes. This issue needs to be resolved, because of the marked consequences for both net and gross (isotopic) transfers.

### *Fates of individual amino acids*

For each individual amino acid, the proportion of that absorbed from the PDV, which is then removed across the liver, varies widely (see Table 9.2), although there are some consistencies. For example, the branch-chain amino acids (BCAA) have the lowest rates of fractional removal, under a wide range of physiological and nutritional conditions, in both cattle and sheep. Post-hepatic supply of BCAA will thus closely reflect changes in intake and absorption. The capacity of the liver to degrade BCAA is limited and, once this is exceeded, additional inflows are not extracted and catabolism proceeds in the extra-hepatic tissues (Lobley, 1992).

In a number of studies with cattle and sheep, the net splanchnic appearance of phenylalanine has either been negative (Burrin *et al.*, 1991; Koeln *et al.*, 1993), close to zero (Reynolds *et al.*, 1994), or only slightly positive (Lobley *et al.*, 1995, 1996b; Bruckentahl *et al.*, 1997; Table 9.2). In sheep, a similar situation exists for histidine (see Table 9.2). Some of these variations may again relate to analytical difficulties, because the proportion of total hepatic inflow that is removed may be as low as 1% (Table 9.3). Indeed, it has been argued that expression of the hepatic extraction against total amino acid inflow (hepatic portal plus hepatic artery flows) has greater physiological validity than comparison with net absorption (e.g. Reynolds *et al.*, 1994; Hanigan *et al.*, 1998). This approach does 'dampen' the differences between studies (compare lysine in Tables 9.2 and 9.3, for example) and may have particular use in the development of modelling equations (Hanigan *et al.*, 1997).

Notwithstanding analytical uncertainties, hepatic fractional extractions of specific amino acids (Tables 9.2 and 9.3) do vary in response to differences in total intake (Burrin *et al.*, 1991), protein and b-ST supply (Bruckentahl *et al.*, 1997) and amino acid infusions (Reynolds *et al.*, 1994; Wray-Cahen *et al.*, 1997; G.E. Lobley, unpublished data). This variation may relate, in part, to the physiological role of the liver as the site of synthesis of vital metabolites, e.g. glucose, neurotransmitters, xenobiotic conjugates, transmethylation products, etc., all of which involve the use of carbon skeletons derived from amino acids. For example, removal of glycine by the liver is associated with one-carbon metabolism and the synthesis of hippuric acid. There are finite demands for both of these and, when satisfied, glycine uptake by the liver is reduced. In addition, hepatic export proteins are relatively rich in phenylalanine, histidine and cysteine (64, 41 and 66 g $kg^{-1}$ bovine albumin respectively versus 36, 23 and 26 g $kg^{-1}$ for non-wool whole body sheep protein; MacRae *et al.*, 1993). The pattern of amino acids extracted by the liver to fulfil these various functions does not match that absorbed from the PDV. Consequently, post-splanchnic supply of amino acids to the peripheral tissues differs considerably from that either absorbed from the PDV or

**Table 9.2.** Fractional extraction by the liver of absorbed amino acids.

| | Cattle | | | | | | | | Sheep | | | |
|---|---|---|---|---|---|---|---|---|---|---|---|---|
| | Lapierre *et al.* (2000) | Koeln *et al.* (1993)[a] | Reynolds *et al.* (1994)[b] | | Wray-Cahen *et al.* (1997)[c] | | Bruckentahl *et al.* (1997)[d] | | Burrin *et al.* (1991)[e] | | Lobley[f] | |
| | | | Control | +Ala | Pre | +AA | Control | +Casein | M | AL | Control | +AA |
| Alanine | 0.41 | 0.58 | 1.17 | 0.75 | 0.97 | 0.54 | 0.70 | 0.74 | 1.30 | 3.18 | 0.92 | 0.66 |
| Arginine | 0.25 | 2.66 | – | – | 0.20 | 0.70 | 0.42 | 0.80 | −25.5 | 7.29 | 0.38 | 0.39 |
| Glycine | 0.67 | 1.18 | 1.00 | 0.82 | 1.26 | 0.73 | 0.70 | 0.60 | 1.29 | 2.01 | 2.43 | 0.88 |
| Serine | 0.52 | 0.80 | 1.10 | 0.85 | 0.73 | 0.37 | 1.12 | 0.46 | 0.49 | 1.65 | 0.61 | 0.44 |
| Tyrosine | 0.50 | 1.16 | 0.90 | 0.80 | – | – | 0.57 | 0.78 | 0.89 | 1.46 | 0.89 | – |
| Histidine | 0.39 | −0.13 | 0.57 | 0.50 | 0.57 | 0.50 | 0.61 | 0.78 | −0.85 | 2.01 | 1.64 | 0.83 |
| Isoleucine | 0.01 | −0.01 | 0.45 | −0.08 | 0.41 | 0.49 | 0.40 | 0.32 | −0.23 | 0.50 | 0.28 | 0.02 |
| Leucine | 0.20 | 0.23 | 0.45 | 0.13 | 0.01 | 0.30 | 0.35 | 0.28 | −0.14 | 0.42 | 0.43 | 0.08 |
| Lysine | 0.14 | 0.53 | 0.71 | 0.41 | 0.16 | 0.31 | 0.42 | 0.45 | −1.94 | 0.44 | 0.68 | 0.38 |
| Methionine | 0.29 | 1.12 | – | – | 0.70 | 0.83 | 0.68 | 0.73 | −0.36 | 1.37 | 0.83 | 0.61 |
| Phenylalanine | 0.64 | 1.11 | 1.00 | 0.82 | 0.67 | 0.87 | 0.94 | 0.84 | 0.49 | 1.39 | 1.03 | 0.63 |
| Threonine | 0.23 | 0.56 | 0.84 | 0.48 | 0.72 | 0.49 | 1.79 | 1.32 | 0.43 | 1.04 | 0.24 | 0.37 |
| Valine | −0.04 | 0.32 | 0.50 | 0.17 | 0.12 | 0.25 | 0.45 | 0.41 | −0.27 | 0.63 | 0.25 | −0.08 |

[a] Fed calves.
[b] Fed 75% maize:soybean meal, with (+ala) or without (control) infusion of L-alanine into the mesenteric vein.
[c] Pregnant dry cows (plasma data), basal ration (pre) plus intra-mesenteric vein infusion of an amino acid mixture (+AA).
[d] Fed a high concentrate diet without (control) or with (+casein) 300 g $day^{-1}$ casein infused into the abomasum.
[e] Sheep at maintenance (M), or fed *ad libitum* (AL).
[f] Sheep fed at 1.2 × maintenance with (+AA) or without (control) intra-mesenteric vein infusion of amino acid mixture (equivalent to 4 × basal intake); unpublished data.

**Table 9.3.** Per cent hepatic extraction of total amino acid inflows to the liver[a].

| | Cattle | | | Sheep | |
|---|---|---|---|---|---|
| | Control[b] | +ala[b] | Control[c] | Control[d] | +AA[d] |
| Alanine | 19.7 | 19.6 | 6.7 | 9.7 | 13.0 |
| Arginine | – | – | 1.8 | 6.4 | 6.0 |
| Glycine | 8.8 | 6.4 | 7.2 | 3.4 | 8.7 |
| Serine | 16.7 | 15.1 | 9.4 | 9.3 | 18.3 |
| Tyrosine | 11.2 | 12.4 | 8.2 | 7.5 | – |
| Histidine | 3.9 | 2.5 | 3.0 | 5.7 | 11.0 |
| Isoleucine | 3.1 | 0.8 | 1.7 | 2.3 | 1.4 |
| Leucine | 3.3 | 1.5 | 0.1 | 2.4 | 1.4 |
| Lysine | 6.3 | 5.2 | 1.7 | 4.3 | 5.4 |
| Methionine | – | – | 8.8 | 9.9 | 14.4 |
| Phenylalanine | 13.3 | 13.6 | 5.9 | 11.5 | 12.3 |
| Threonine | 9.1 | 7.4 | 7.0 | 1.2 | 7.3 |
| Valine | 0.7 | 1.0 | 0.5 | 1.2 | 1.0 |

[a] Based on portal vein + hepatic artery inflows and hepatic vein outflows.
[b] Reynolds et al. (1994); fed 75 maize-soybean meal with (+ala) and without (control) alanine infusion (3 days).
[c] Wray-Cahen *et al.* (1997); pre-infusion data only.
[d] G.E. Lobley; unpublished results, with (+AA) and without (control) mesenteric vein infusion (4.5 h) of amino acid (AA) mixture (equivalent to 4 × basal intake).

required by peripheral tissues. In sheep, for example, this may result in a relative under-provision of histidine and phenylalanine, with a lesser deficit of methionine. Their availability may limit peripheral tissue growth, but leave the other essential amino acids in excess and subject to catabolism.

If, however, plasma export proteins can provide amino acids for peripheral tissue needs, then, based on known rates of synthesis (Connell *et al.*, 1997), the limitation changes to methionine and, to a lesser extent, phenylalanine. This fits with the generally perceived view that methionine is often the first-limiting amino acid for many ruminant diets. Extra-hepatic use of hepatic export proteins has been shown in rodents, by adding serum proteins tagged with residualizing labels (Thorpe *et al.*, 1993), where as much as 20% of albumin is degraded in each of skeletal muscle and skin (Maxwell *et al.*, 1990). Interestingly, hepatic removal and degradation of albumin appears to be dominated by endothelial cells (Eskild *et al.*, 1989), adding another aspect to liver metabolic zonation.

Because the balance between the synthesis of metabolites and export protein synthesis will vary with physiological state, then clearly neither the magnitude of hepatic fractional extractions nor the pattern of amino acids available to the periphery will be constant. In sheep, protein intake and parasitic infestation (Abbott *et al.*, 1985) markedly influence both albumin pool size and biological half-life (degradation and clearance). Ovine albumin synthesis is sustained even under fasting conditions (Koeln *et al.*, 1993) and is sensitive to intake (Connell *et al.*, 1997), at least up to a certain limit of amino acid supply (Lobley *et al.*, 1998).

### Glutamine

Aside from the major metabolic role of glutamine, it is now used as a therapeutic aid in certain clinical situations, particularly those where cells proliferate, such as in the GIT epithelia and the immune system (Heitmann and Bergman 1978; Calder 1995). Besides a key role in maintaining ureagenic flux and ammonia detoxification, glutamine may also impact on hepatic mechanisms related to acid–base balance and regulation of protein turnover (see next section).

Maintenance of acid–base balance is a major role for the liver in non-ruminants. In rodents, chronic metabolic acidosis, induced by either HCl or ammonium chloride, results in suppression of ureagenesis and urinary urea elimination, but with concomitant stimulation of amino acid oxidation, net glutamine synthesis and urinary $NH_4^+$ release (Welbourne *et al.*, 1986; May *et al.*, 1992). Production of urea yields a neutral product (involving utilization of both $NH_4^+$ and $HCO_3^-$), whereas synthesis of glutamine spares bicarbonate. This bicarbonate is then available to neutralize protons, and, on transfer to the kidney, the amido group is released into urine as ammonium ($NH_4^+$) ion (Meijer *et al.*, 1990). In sheep, acidosis induced by ammonium chloride had little effect on whole body and liver flux of [U-$^{14}$C]glutamine (Heitmann and Bergman, 1978), but the complication of acidosis and the additional supply of free ammonia confounded interpretation. Certainly, mild acidosis, induced by [$^{15}$N]ammonium chloride infusion into sheep, resulted in elevated enrichments of hepatic [$^{15}$N]glutamine, comparable to those observed for urea (Lobley *et al.*, 1995). Net movements of glutamine across the liver are little affected by ammonia but, under both neutral and acidotic conditions, there is a large change in rates of both utilization and synthesis (Table 9.4), i.e. transfer of ammonia through the glutamine amido group is stimulated.

For sheep infused with HCl, urea production and urinary elimination were elevated, with a smaller increase in hepatic glutamine synthesis (Milano, 1997). The latter

**Table 9.4.** Effect of ammonia supply and acidosis on ovine hepatic glutamine kinetics.

| | Glutamine kinetics (μmol min$^{-1}$) | | |
|---|---|---|---|
| | Net utilization | Hepatic synthesis | Hepatic utilization |
| Study 1 | | | |
| Control | −30[a] | 47[a] | −77[a] |
| +$NH_4Cl$ | −11[b] | 87[b] | −98[b] |
| Study 2 | | | |
| Control | −43 | 26 | −70 |
| +HCl | −37 | 42 | −79 |
| Study 3 | | | |
| Control | −21 | 37[a] | −58 |
| +$NH_4HCO_3$ | −33 | 51[b] | −84 |

Data from: Study 1, [U-$^{14}$C]glutamine, acidosis, Heitmann and Bergman (1980); Study 2, [5-$^{15}$N]glutamine, acidosis, Milano (1997); Study 3, [5-$^{15}$N]glutamine, ammonia overload (R.M. Nieto, T. Obitsu, A. Fernandez, G.E. Lobley, unpublished results).
For each study, values within a column but with different superscripts are significantly different by $P < 0.05$ or lower.

accounted for a minor fraction of the extra $NH_4^+$ appearing in the urine, with most arising from extra-hepatic sources. This suggests that the liver is probably less important in regulation of acid–base balance in sheep, compared with the rat (Fig. 9.3). For example, the activity of the ovine hepatic glutamine synthetase (50–100 μmol min$^{-1}$ kg$^{-1}$ liver) is only 15% that of the rat (Häussinger, 1990). Thus, the maximal rate of urea production in sheep exceeds hepatic glutamine synthesis by 5–10-fold (Rossouw *et al.*, 1999; Table 9.5), compared with equal activities in the rodent. Furthermore, both the ornithine cycle and glutamine synthesis are sensitive to pH in the rat but not in the sheep, as shown by perfusion of the ovine dorsal lobe with buffers varying in pH from 7.2 to 7.6 (Rossouw *et al.*, 1999; Table 9.5). There are other metabolic differences in acid–base responses between species. Notably, net changes in protein-N retention appear to be regulated by decreased protein synthesis in sheep (Milano, 1997) but by elevated protein degradation in rodents (Manier *et al.*, 1994; Fig. 9.5).

## Hepatic protein turnover

Liver protein metabolism has two major components – production, or replacement, of constitutive material and synthesis of proteins destined for export. The liver responds rapidly to alterations in intake with short-term, but limited, fluctuations in protein mass (Lobley, 1990). These short-term changes may form part of a small, but mobile, protein reserve, sensitive to acute variations in protein supply (MacRae *et al.*, 1991). Liver protein mass also alters, both in absolute terms and relative to body weight, in response to longer-term changes in nutrition (Burrin *et al.*, 1992; Lobley *et al.*, 1994). The rate of change is relatively slow, however, even when intake is altered from

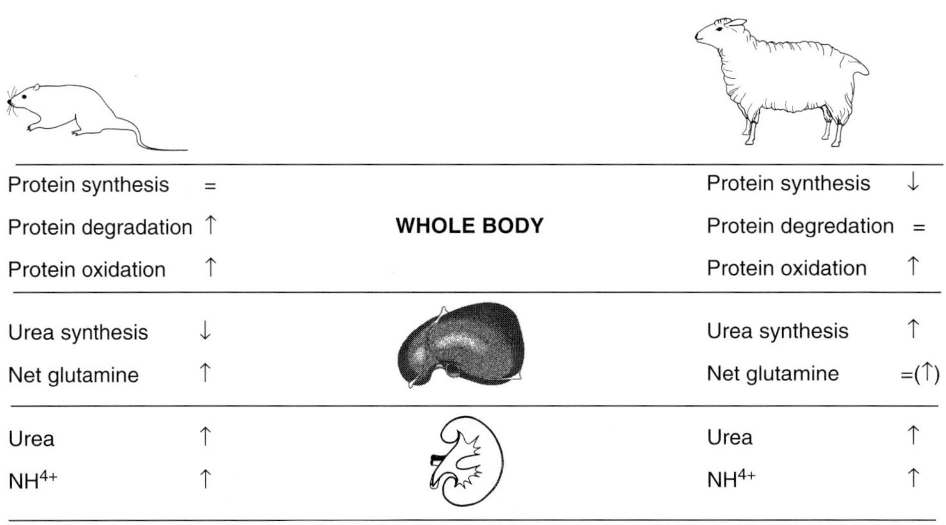


**Fig. 9.3.** Responses in whole body protein metabolism, plus liver and kidney urea and glutamine kinetics to acidosis in rodents and sheep (from Welbourne *et al.*, 1986; May *et al.*, 1992; Manier *et al.*, 1994; Milano, 1997).

**Table 9.5.** Effect of ammonia supply and acidosis on ovine hepatic glutamine kinetics *in vitro*.

| Condition | $NH_3$ uptake | Urea production | Glutamine production |
|---|---|---|---|
| *Antegrade* | | | |
| $+NH_4Cl$ (0.3 mM), pH 7.2 | 827 ± 134 | 596 ± 66 | 176 ± 27 |
| $+NH_4Cl$ (0.3 mM), pH 7.4 | 770 ± 118 | 590 ± 71 | 172 ± 40 |
| $+NH_4Cl$ (0.3 mM), pH 7.6 | 791 ± 96 | 594 ± 60 | 184 ± 34 |
| $-NH_4Cl$, pH 7.4 | −24 ± 40[a] | 34 ± 10[a] | 40 ± 10[a] |
| *Retrograde* | | | |
| $+NH_4Cl$ (0.3 mM), pH 7.2 | 606 ± 83 | 530 ± 52[b] | 192 ± 19 |
| $+NH_4Cl$ (0.3 mM), pH 7.4 | 707 ± 108 | 590 ± 58 | 222 ± 26[c] |
| $+NH_4Cl$ (0.3 mM), pH 7.6 | 730 ± 83 | 556 ± 60 | 186 ± 28 |
| $-NH_4Cl$, pH 7.2 | −19 ± 20[a] | 38 ± 40[a] | 20 ± 4[a] |

Values are expressed as nmol N $g^{-1}$ wet weight liver $min^{-1}$ and are the means of seven perfusions (H. Rossouw, J.G. van der Walt, unpublished results).
All analyses conducted *within* direction of flow.
[a] Differs from all values obtained with $NH_4Cl$, at all pH values ($P < 0.05$).
[b] Differs from the value obtained with $NH_4Cl$, at pH 7.4 ($P < 0.10$).
[c] Differs from the value obtained with $NH_4Cl$, at pH 7.6 ($P < 0.10$).

30 g $kg^{-1}$ weight $day^{-1}$ to either *ad libitum* or just sufficient to maintain body weight. Under these conditions, fractional rates of protein gain or loss were, at maximum, 5 and −1.5% $day^{-1}$ (Burrin *et al.*, 1992).

In contrast, fractional rates of hepatic protein synthesis are much greater, but are also very variable (from 3 to 98% $day^{-1}$; see Lescoat *et al.*, 1997). This wide disparity makes interpretation of metabolic regulation difficult, with the problem exacerbated by technical difficulties. These include the under-estimation of synthesis of export proteins, as well as those constitutive proteins with high rates of turnover, using the continuous infusions of labelled amino acids. This problem is further complicated by the presence of different precursor pools within the liver. Mixed intracellular sources provide for the synthesis of constitutive proteins, while polypeptides destined for secretion are derived from amino acids that have newly entered the cell (see Connell *et al.*, 1997).

Notwithstanding these technical concerns, the large difference between fractional rates of protein synthesis and gain indicates that constitutive proteins are extensively degraded within the ruminant liver. If these liberated amino acids were to be channeled into a pool for catabolism, it would significantly affect the overall N economy of the animal. There is also strong evidence that the regulation of constitutive protein production differs from that of proteins destined for export. For example, in the transition from fasting to feeding, synthesis of albumin is markedly stimulated, in both absolute and fractional terms (Connell *et al.*, 1997). Regulation through synthesis probably also largely accounts for changes in plasma protein concentration and content commonly observed under conditions of stress (e.g. Abbott *et al.*, 1985) and under-nutrition (e.g. Liu *et al.*, 1995). In contrast, liver constitutive protein synthesis is less susceptible to acute changes in amino acid supply (Lobley *et al.*, 1998) or general intake (Connell *et al.*, 1997). Consequently, changes in ovine hepatic protein mass probably occur through alteration in protein catabolism, possibly involving up- and down-regulation of lysosomal activity as proposed for the perfused rodent liver (Mortimore *et al.*, 1989).

How then is such control exerted? Recent studies with rodent hepatocytes and perfused livers have indicated that the mechanism probably involves changes in cell volume (Häussinger *et al.*, 1994). Cellular uptake of a number of amino acids, mediated by $Na^+$-dependent transporters (e.g. the A system for glycine and alanine, or the N-system in the case of glutamine), is accompanied by concomitant inflow of $Na^+$. This is then expelled by exchange with $K^+$ ion, the energy being provided by the membrane $Na^+$-$K^+$ ATPase. This flow of ions leads to increased uptake of intracellular water and, therefore, cell swelling. As cell volume expands, protein degradation is reduced, probably related to a decrease in the rate of formation and acidification of lysosomes (Luiken *et al.*, 1996). Under the converse conditions, i.e. efflux of these amino acids, the hepatocytes shrink, degradation is enhanced and protein synthesis inhibited (Häussinger *et al.*, 1994). Net production of export protein appears to be regulated primarily by changes in synthesis, whereas constitutive protein mass is modulated more by alterations in the rate of proteolysis (Connell *et al.*, 1997). These contrasting mechanisms may have a common step involving intracellular targetting of the mRNAs for the export proteins to the endoplasmic reticulum (Hesketh *et al.*, 1998) and phosphorylation of ribosomal protein S6 (Blommart *et al.*, 1995). The binding there of additional ribosomes will augment synthesis of export proteins and, at the same time, reduce the release of reticulum membrane needed for formation of the autophagosomes. Thus, proteolysis would be inhibited. If targetting and binding to the endoplasmic reticulum is inhibited (as at low intakes) then export protein synthesis decreases while lysosome formation and activity is enhanced.

Interestingly, two major protein anabolic hormones, insulin and IGF-I, also lead to increased cell hydration, while the catabolic hormone, glucagon, causes shrinkage. Both insulin and glucagon are extracted by the ruminant liver (Lapierre *et al.*, 1992) and play competing roles in determining the fates of amino acids. Cell volume increases also lead to enhancements of amino acid catabolism (Häussinger *et al.*, 1992b), ureagenesis, possibly through increased ASS activity (Quillard *et al.*, 1996), and glutaminase activity (Häussinger *et al.*, 1990). Change in cell hydration, therefore, provides a link for the responses often seen with improved nutrition, i.e. a co-stimulation of protein anabolism and amino acid catabolism.

## References


Abbott, E.M., Parkins, J.J. and Holmes, P.H. (1985) Influence of dietary protein on the pathophysiology of ovine haemonchosis in Finn Dorset and Scottish Blackface lambs given a single moderate infection. *Research Veterinary Science* 38, 54–60.

Blommart, E.F.C., Luiken, J.J.F.P., Blommart, P.J.E., van Woerkom, G.M. and Meijer, A.J. (1995) Phosphorylation of ribosomal protein S6 is inhibitory for autophagy in isolated rat hepatocytes. *Journal of Biological Chemistry* 270, 2320–2326.

Brosnan, J., Brosnan, M.E., Charron, R. and Nissim, I. (1996) A mass isotopomer study of urea and glutamine synthesis from $^{15}N$-labelled ammonia in the perfused rat liver. *Journal of Biological Chemistry* 271, 16199–16207.

Bruckentahl, I., Huntington, G.B., Baer, C.K. and Erdman, R.A. (1997) The effect of abomasal infusion of casein and recombinant somatotropin hormone injection on nitrogen balance and amino acid fluxes in portal-drained viscera and net hepatic and total splanchnic blood in Holstein steers. *Journal of Animal Science* 75, 1119–1129.

Burrin, D.G., Ferrell, C.L., Eisemann, J.H. and Britton, R.A. (1991) Level of nutrition and splanchic metabolite flux in young lambs. *Journal of Animal Science* 69, 1082–1091.

Burrin, D.G., Britton, R.A., Ferrell, C.L. and Bauer, M.L. (1992) Level of nutrition and visceral organ protein synthetic capacity and nucleic-acid content in sheep. *Journal of Animal Science* 70, 1137–1145.

Calder, P.C. (1995) Glutamine action in cells of the immune system. In: Cynober, L., Fürst, P. and Lawin, P.W. (eds) *Pharmacological Nutrition – Immune Nutrition*. Zuckschwerdt Verlag, München, pp. 10–20.

Calder, A.G., Garden, K.E., Anderson, S.E. and Lobley, G.E. (1999) Quantitation of blood and plasma amino acids using isotope dilution mass spectrometry impact gas chromatography/mass spectrometry with U-$^{13}$C amino acids as internal standards. *Rapid Communications in Mass Spectrometry* 13, 2080–2083.

Choung, J.-J. and Chamberlain, D.G. (1995) Effects of intra-ruminal infusion of propionate on the concentrations of ammonia and insulin in peripheral blood of cows receiving intraruminal infusion of urea. *Journal of Dairy Research* 62, 549–557.

Connell, A., Calder, A.G., Anderson, S.E. and Lobley, G.E. (1997) Hepatic protein synthesis in the sheep: effect of intake as monitored by use of stable-isotope-labelled glycine, leucine and phenylalanine. *British Journal of Nutrition* 77, 255–271.

Deuel, T.F., Lovie, M. and Lerner, A. (1978) Glutamine synthetase from rat liver. *Journal of Biological Chemistry* 253, 6111–6118.

Eisemann, J.H. and Huntington, G.B. (1994) Metabolite flux across portal-drained viscera, liver, and hindquarters of hyperinsulinemic, euglycemic beef steers. *Journal of Animal Science* 72, 2919–2929.

Eskild, W., Kindberg, G.M., Smedsrod, B., Blomhoff, R., Norum, K.R. and Berg, T. (1989) Intracellular transport of formaldehyde-treated serum albumin in liver endothelial cells after uptake via scavenger receptors. *Biochemical Journal* 258, 511–520.

Gebhardt, R. (1992) Metabolic zonation of the liver – regulation and implications for liver function. *Pharmacological Therapeutics* 53, 275–354.

Goetsch, A.L. and Ferrell, C.L. (1995) Effects of dietary maize level on net flux across splanchnic tissues of oxygen and nutrients in wethers consuming *ad libitum* different forages. *Animal Science* 61, 43–55.

Hanigan, M.D., Dijkstra, J., Gerrits, W.J.J. and France, J. (1997) Modelling post-absorptive protein and amino acid metabolism in the ruminant. *Proceedings of the Nutritional Society* 56, 631–643.

Hanigan, M.D., France, J., Wray-Cahen, D., Beever, D.E., Lobley, G.E., Reutzel, L. and Smith, N. W. (1998) Alternative models for analyses of liver and mammary transorgan metabolite extraction data. *British Journal of Nutrition* 79, 63–78.

Häussinger, D. (1990) Nitrogen metabolism in liver: structural and functional organization and physiological relevance. *Biochemical Journal* 267, 281–290.

Häussinger, D. and Gerok, W. (1983) Hepatocyte heterogeneity in glutamate uptake in isolated perfused rat liver. *European Journal of Biochemistry* 136, 421–425.

Häussinger, D., Lang, F., Bauers, K. and Gerok, W. (1990) Interactions between glutamine metabolism and cell volume regulation in perfused rat liver. *European Journal of Biochemistry* 188, 689–695.

Häussinger, D., Lamers, W.H. and Moorman, A.F.M. (1992a) Hepatocyte heterogeneity in the metabolism of amino acids and ammonia. *Enzyme* 46, 72–93.

Häussinger, D., Stoll, B., Morimoto, Y., Lang, F. and Gerok, W. (1992b) Anisoosmotic liver perfusion: redox shifts and modulation of α-ketoisocaproate and glycine metabolism. *Biological Chemistry Hoppe-Seyler* 373, 723–734.

Häussinger, D., Lang, F. and Gerok, W. (1994) Regulation of cell function by the cellular hydration state. *American Journal of Physiology* 267 (*Endocrinol. Metab.* 30), E343–E355.

Heitmann, R.N. and Bergman, E.N. (1978) Glutamine metabolism, inter-organ transport and glucogenicity in the sheep. *American Journal of Physiology* 234, E197–E203.

Heitmann, R.N. and Bergman, E.N. (1980) Transport of amino acids in whole blood and plasma of sheep. *American Physiological Society* 239, E242–E247.

Hesketh, J.E., Vasconcelos, M.H. and Bermano, G. (1998) Regulatory signals in messenger RNA: determinants of nutrient–gene interaction and metabolic compartmentation. *British Journal of Nutrition* 80, 307–321.

Houlier, M.L., Mirand, P.P., Durand, D., Bauchart, D., Lefaivre, J. and Bayle, G. (1991) Transport des acides aminés dans l'aire splanchnique par le plasma sanguin et le sang chez le veau préruminant. *Reproduction Nutrition Development* 31, 399–410.

Isserty, A., Ortigues, I. and Remond, D. (1998) Measure of splanchnic blood flow by marker dilution: comparison of 4 methods of para-aminohippuric acid determination. *Reproduction Nutrition Development* 38, 93–106.

Jackson, A.A. (1998) Salvage of urea-nitrogen in the large bowel: functional significance in metabolic control and adaptation. *Biochemistry Society Transactions* 26, 231–236.

Jungermann, K. and Kietzmann, T. (1996) Zonation of parenchymal and nonparenchymal metabolism in liver. *Annual Review of Nutrition* 16, 179–203.

Kaiser, S., Gerok, W. and Häussinger, D. (1988) Ammonia and glutamine metabolism in human liver slices – new aspects on the pathogenesis of hyperammonemia in chronic liver disease. *European Journal of Clinical Investigation* 18, 535–542.

Koeln, L.L., Schlagheck, T.G. and Webb, K.E., Jr (1993) Amino acid flux across the gastrointestinal tract and liver of calves. *Journal of Dairy Science* 76, 2275–2285.

Krehbiel, C.R., Ferrell, C.L. and Freetly, H.C. (1998) Effects of frequency of supplementation on dry matter intake and net portal and hepatic flux of nutrients in mature ewes that consume low-quality forage. *Journal of Animal Science* 76, 2464–2473.

Kuo, F.C., Hwu, W.L., Valle, D. and Darnell, Jr (1991) Co-localisation in pericentral hepatocytes in adult mice and similarity in developmental expression of ornithine aminotransferase and glutamine synthetase mRNA. *Proceedings of the National Academy of Sciences USA* 88, 9468–9472.

Lapierre, H., Reynolds, C.K., Elsasser, T.H., Gaudreau, P., Brazeau, P. and Tyrrell, H.F. (1992) Effects of growth hormone-releasing factor and feed intake on energy metabolism in growing beef steers: Net hormone metabolism by portal-drained viscera and liver. *Journal of Animal Science* 70, 742–751.

Lapierre, H., Bernier, J.F., Dubreuil, P., Reynolds, C.K., Farmer, C., Ouellet, D.R. and Lobley, G.E. (2000) The effect of feed intake level on splanchnic metabolism in growing beef steers. *Journal of Animal Science* 78, 1084–1099.

Lautt, W.W. (1996) Intrinsic regulation of hepatic blood flow. *Canadian Journal of Physiology and Pharmacology* 74, 223–233.

Lescoat, P., Sauvant, D. and Danfaer, A. (1997) Quantitative aspects of protein fractional synthesis rates in ruminants. *Reproduction Nutrition Development* 37, 493–515.

Liu, S.M., Lobley, G.E., MacLeod, N.A., Kyle, D.J., Chen, X.B. and Ørskov, E.R. (1995) Effects of long-term protein excess or deficiency on whole-body protein turnover in sheep nourished by intragastric infusion of nutrients. *British Journal of Nutrition* 73, 829–839.

Lobley, G.E. (1990) Energy metabolism reactions in muscle: responses to age, nutrition and hormonal status. *Reproduction Nutrition Development* 30, 13–34.

Lobley, G.E. (1992) Control of the metabolic fate of amino acids in ruminants: a review. *Journal of Animal Science* 70, 3264–3275.

Lobley, G.E. and Milano, G.D. (1997) Regulation of hepatic nitrogen metabolism in ruminants. *Proceedings of the Nutrition Society* 56, 547–563.

Lobley, G.E., Connell, A., Milne, E., Newman, A.M. and Ewing, T.A. (1994) Protein synthesis in splanchnic tissues of sheep offered two levels of intake. *British Journal of Nutrition* 71, 3–12.

Lobley, G.E. Connell, A., Lomax, M.A., Brown, D.S., Milne, E., Calder, A.G. and Farningham,

D.A.H. (1995) Hepatic detoxification of ammonia in the ovine liver: possible consequences for amino acid catabolism. *British Journal of Nutrition* 73, 667–685.

Lobley, G.E., Connell, A., Revell, D.K., Bequette, B.J., Brown, D.S. and Calder, A.G. (1996a) Splanchnic-bed transfers of amino acids in sheep blood and plasma, as monitored through use of a multiple U-$^{13}$C-labelled amino acid mixture. *British Journal of Nutrition* 75, 217–235.

Lobley, G.E., Weijs, P.J.M., Connell, A., Calder, A.G., Brown, D.S. and Milne, E. (1996b) The fate of absorbed and exogenous ammonia as influenced by forage or forage-concentrate diets in growing sheep. *British Journal of Nutrition* 76, 231–248.

Lobley, G.E., Bremner, D., Nieto, R., Obitsu, T., Hotston Moore, A. and Brown, D.S. (1998) Transfers of N-metabolites across the ovine liver in response to short-term infusions of an amino acid mixture into the mesenteric vein. *British Journal of Nutrition* 80, 371–379.

Luiken, J.J.F.P., Aerts, J.M.F.G. and Meijer, A.J. (1996) The role of the intralysosomal pH in the control of autophagic proteolytic flux in rat hepatocytes. *European Journal of Biochemistry* 235, 564–573.

Luo, Q.J., Maltby, S.A., Lobley, G.E., Calder, A.G. and Lomax, M.A. (1995) The effect of amino acids on the metabolic fate of $^{15}NH_4Cl$ in isolated sheep hepatocytes. *European Journal of Biochemistry* 228, 912–917.

Lusty, C.J. (1978) Carbamoylphosphate synthetase 1 of rat-liver mitochondria. Purification, properties, and polypeptide molecular weight. *European Journal of Biochemistry* 85, 373–383.

McEvoy, T.G., Robinson, J.J., Aitken, R.P., Findlay, P.A. and Robertson, I.S. (1997) Dietary excesses of urea influence the viability and metabolism of preimplantation sheep embryos and may affect fetal growth among survivors. *Animal Reproduction Science* 47, 71–90.

McLeod, K.R., Bauer, M.L., Harmon, D.L., Reynolds, C.K. and Mitchell, G.E., Jr (1997) Effects of exogenous somatostatin and cysteamine on net nutrient flux across the portal-drained viscera and liver of sheep during intraduodenal infusion of starch hydrolysate and casein. *Journal of Animal Science* 75, 3026–3037.

MacRae, J.C., Bruce, L.A., Hovell, F.D. deB., Inkster, J., Walker, A. and Atkinson, T. (1991) Influence of protein nutrition on the response of growing lambs to exogenous bovine growth hormone. *Journal of Endocrinology* 130, 53–61.

MacRae, J.C., Walker, A., Brown, D. and Lobley, G.E. (1993) Accretion of total protein and individual amino acids by organs and tissues of growing lambs and the ability of nitrogen balance techniques to quantitate protein retention. *Animal Production* 57, 237–245.

Manier, S., Laouari, D., Dechaux, M., Motel, V., Yvert, J.P., Mathan, B. and Kleinknecht, C. (1994) *In vivo* unaltered muscle protein synthesis in experimental chronic metabolic acidosis. *Kidney International* 46, 1705–1712.

Maxwell, J.L., Terracio, L., Borg, T.K., Baynes, J.W. and Thorpe, S.R. (1990) A fluorescent residualizing label for studies on protein uptake and catabolism *in vivo* and *in vitro*. *Biochemistry Journal* 267, 155–162.

May, R.C., Masud, T., Logue, B., Bailey, J. and England, B. (1992) Chronic metabolic acidosis accelerates whole body proteolysis and oxidation in awake rats. *Kidney International* 41, 1535–1542.

Meijer, A.J., Lof, C., Ramos, I.C. and Verhoeven, A.J. (1985) Control of ureagenesis. *European Journal of Biochemistry* 148, 189–196.

Meijer, A.J., Lamers, W.H. and Chamuleau, R.A.F.M. (1990) Nitrogen metabolism and ornithine cycle function. *Physiological Reviews* 70, 701–748.

Milano, G.D. (1997). Liver nitrogen transactions in sheep (*Ovis aries*). PhD Thesis, University of Aberdeen, UK.

Milano, G.D., Lomax, M.A. and Lobley, G.E. (1996) Estimation of the enrichment of urea N precursors. *Proceedings of the Nutrition Society* 55, 42A.

Mortimore, G.E., Poso, A.R. and Lardeux, B.R. (1989) Mechanism and regulation of protein degradation in liver. *Diabetes Metabolism Reviews* 5, 49–70.

Mutsvangwa, T., Buchanan-Smith, J.G. and McBride, B.W. (1997) Effects of ruminally degradable nitrogen intake and in vitro addition of ammonia and propionate on the metabolic fate of L-[1-$^{14}$C]alanine and L-[$^{15}$N]alanine in isolated sheep hepatocytes. *Journal of Animal Science* 75, 1149–1159.

Nieto, R., Calder, A.G., Anderson, S.E. and Lobley, G.E. (1996) A new method for determination of $^{15}NH_3$ enrichment in biological samples by gas chromatography/electron ionisation mass spectrometry. *Journal of Mass Spectrometry* 31, 289–294.

O'Sullivan, D., Brosnan, J.T. and Brosnan, M.E. (1998) Hepatic zonation of the catabolism of arginine and ornithine in the perfused rat liver. *Biochemical Journal* 330, 627–632.

Parker, D.S., Lomax, M.A., Seal, C.J. and Wilton, J.C. (1995) Metabolic implications of ammonia production in the ruminant. *Proceedings of the Nutrition Society* 54, 549–563.

Quillard, M., Husson, A. and Lavoinne, A. (1996) Glutamine increases argininosuccinate synthetase mRNA levels in rat hepatocytes. The involvement of cell swelling. *European Journal of Biochemistry* 236, 56–59.

Rattenbury, J.M., Jeacock, M.K. and Shepherd, D.A.L. (1980) Urea synthesis in the liver and kidney of developing sheep. *Biochimica et Biophysica Acta* 630, 210–219.

Reynolds, C.K. (1995) Quantitative aspects of liver metabolism in ruminants. In: Engelhardt, W. von, Leonhard-Marek, S., Breves, G. and Giesecke, D. (eds) *Ruminant Physiology: Digestion, Metabolism, Growth and Reproduction: Proceedings of the Eighth International Symposium on Ruminant Physiology*. Ferdinand Enke Verlag, Stuttgart, pp. 351–371.

Reynolds, C.K. and Tyrrell, H.F. (1991) Effects of mesenteric vein L-alanine infusion on liver metabolism in beef heifers fed on diets differing in forage:concentrate ratio. *British Journal of Nutrition* 66, 437–450.

Reynolds, C.K., Harmon, D.L., Prior, R.L. and Tyrrell, H.F. (1994) Effects of mesenteric vein L-alanine infusion on liver metabolism of organic acids by beef heifers fed diets differing in forage:concentrate ratio. *Journal of Animal Science* 72, 3196–3206.

Rossouw, H.C., van der Walt, J.G. and Nell, M. (1997) Effect of pH and osmolarity on urea metabolism of the liver in small ruminants. *Proceedings of the 25th Annual Congress of Physiology Society South Africa, Durban, South Africa, OP47.*

Rossouw, H.C., Nell, M.J., Mohamed Ali, A. and van der Walt, J.G. (1999) Ammonia partitioning between urea and glutamine in the perfused sheep liver: role of extracellular pH. *South African Journal of Animal Science* 29, 246–247.

Sarraseca, A., Milne, E., Metcalf, M.J. and Lobley, G.E. (1998) Urea recycling in sheep: effects of intake. *British Journal of Nutrition* 79, 79–88.

Stewart, P.M. and Walser, M. (1980) Failure of normal ureagenic response to amino acids in the organic-acid loaded rat: a proposed mechanism for the hyperammonemia of propionic and methylmalonicacidemia. *Journal of Clinical Investigation* 66, 484–492.

Summerskill, W.H.J. and Wolpert, E. (1970) Ammonia metabolism in the gut. *American Journal of Clinical Nutrition* 23, 633–639.

Thorpe, S.R., Baynes, J.W. and Chroneos, Z.C. (1993) The design and application of residualizing labels for studies of protein catabolism. *FASEB Journal* 7, 399–405.

van der Walt, J.G. (1993) Nitrogen metabolism of the ruminant liver. *Australian Journal of Agricultural Research* 44, 381–403.

Webb, K.E., Jr (1999) Tissue, cellular and molecular aspects of peptide absorption and utilization. *IX International Symposium on Ruminant Physiology, Pretoria, S. Africa.*

Welbourne, T.C., Childress, D. and Givens, G. (1986) Renal regulation of interorgan glutamine flow in metabolic acidosis. *American Journal of Physiology* 226, R858–R866.

Whitt, J., Huntington, G., Zetina, E., Casse, E., Taniguchi, K. and Potts, W. (1996) Plasma flow and net nutrient flux across gut and liver of cattle fed twice daily. *Journal of Animal Science* 74, 2450–2461.

Wray-Cahen, D., Roberts, S., Metcalf, J.A., Backwell, F.R.C., Bequette, B.J., Brown, D. S., Sutton, J.D. and Lobley, G.E. (1997) Hepatic response to increased exogenous supply of plasma amino acids by infusion into the mesenteric vein of Holstein-Friesian cows in late gestation. *British Journal of Nutrition* 78, 913–930.

# IV Tissue Maintenance and Utilization of Endogenous Body Reserves

# 10 Adipose Tissue: Beyond an Energy Reserve

R.G. VERNON[1] AND K.L. HOUSEKNECHT[2]

[1]*Hannah Research Institute, Ayr, UK;* [2]*Animal Health Drug Discovery, Pfizer Inc., Groton, Connecticut, USA*

## Introduction

Adipose tissue is a much maligned tissue! The growing prevalence of obesity in the USA and Western Europe in particular has led to a search for means to reduce the amount of adiposity both in ourselves and also in the animals we eat. A number of agents (e.g. growth hormone, β-agonists) which can markedly reduce adiposity have been produced and immunological approaches to destroying adipocytes, and hence decreasing adiposity, have been developed. However, in this rush to reduce adiposity, it is worth remembering that adipose tissue has played a key role in mammalian evolution and still has important physiological functions today, both as an energy reserve and as a source of paracrine and endocrine factors.

## Role in mammalian evolution and physiology

Survival away from an immediate source of food requires a reserve of energy; for terrestrial animals this ideally needs to be light and compact. Triacylgycerol meets these requirements well, being energy dense and also hydrophobic. A gram of stored triacylglycerol in a cell contains virtually no water; in contrast, a gram of stored glycogen contains about 75% water. Triacylglycerol can be stored, in theory, in any cell of the body, but vertebrates evolved a specialist cell type, the adipocyte, for this purpose (Pond, 1992). Poikilotherms store triacylglycerol in skeletal muscle and liver, but in addition have some mesenteric adipose tissue (arguably an extension of the liver as its venous drainage enters the hepatic portal vein); the relatively limited capacity of these species to store energy reflects their much lower energy requirements than homeotherms (Pond, 1992; Sheridan, 1994). The evolution of homeothermy freed animals from some constraints imposed by environmental temperature and allowed exploitation of colder regions, but has a cost as metabolic rate, and hence energy requirements are considerably increased. This increases the need for reserves of energy and is reflected in the much more complex array of adipose tissue depots in mammals and birds than in poikilotherms (Pond, 1992). Detailed anatomical studies by Pond (1986) on a wide variety of mammals has shown that there are some 16 depots, some in the abdominal cavity, some within the musculature and some under the skin, which are present in essentially all mammals including marsupials. The relative size of different depots may vary between species, and in fat animals, depots may effectively merge; such changes are considered to be adaptations to meet the needs of that particular species (Pond, 1992). While adipocytes from different depots appear superficially to be identical, there are subtle differences in their properties, and there is some evidence to suggest that the small depots of the musculature are more metabolically active than the larger abdominal and subcutaneous depots which are usually studied (Pond, 1992).

Evolution of adipose tissue has allowed mammals to inhabit some remarkably inhospitable environments; it also facilitates migration, allowing animals to exploit seasonally available food in different locations. Furthermore, adipose tissue has permitted development of a variety of reproductive strategies (Vernon and Pond, 1997). Pregnancy increases the energy requirements of mammals, which are further increased by lactation. Additional demands are mostly met by increasing food intake, but in addition there is often an accumulation of adipose tissue lipid during early pregnancy for use during the later stages of pregnancy, at parturition and during lactation (Vernon and Pond, 1997). Indeed, some species of seal, whale and also boreal bears actually fast during lactation, so deriving all the nutrients required for milk production from tissue reserves (Oftedahl, 1992). Dairy cows, especially those producing large amounts of milk, usually draw heavily on adipose tissue reserves during early lactation, when for reasons as yet unknown, milk output increases more rapidly than appetite. As a result, for a period, mobilization of adipose tissue lipid may account for more than 50% of milk fat production (Vernon and Flint, 1984).

Thus adipose tissue has played a central role in mammalian evolution (indeed our evolution would have been impossible without it), and continues to play a key role in mammalian physiology in both wild and domestic species. However, there is a potential cost, at least for wild species, in that too much adipose tissue can reduce mobility and turn an animal into a succulent meal for a predator. Thus the amount of adipose tissue in an animal will usually reflect the relative threats to survival of predation and starvation. For example, African gazelles with a plentiful supply of food have relatively little adipose tissue; in contrast, reindeer in the Arctic have a layer of subcutaneous adipose tissue about 5 cm thick at the beginning of the winter (Larsen *et al.*, 1985). Predators also have this problem as obesity can decrease their chance of obtaining a meal. European foxes are normally thin, but in the Arctic they are markedly fatter (Lindström, 1983). Barnacle geese before migrating from the Arctic to Scotland for the winter, accumulate just enough adipose tissue lipid to meet the needs of the flight (Butler *et al.*, 1998). Animals can thus regulate their levels of adiposity very closely to meet their needs; this means that there must be signals which indicate the size of the body reserves of lipid. Indeed, it is now apparent that adipocytes secrete a variety of substances (Box 10.1) which have autocrine, paracrine and endocrine functions.

**Box 10.1.** Some substances secreted by adipose tissue cells.

Metabolic modulators
- Lipoprotein lipase
- Acylation-stimulating protein
- Apoprotein E
- Fatty acids
- Prostaglandin $E_2$
- Adenosine

Vasoactive factors
- Prostacyclin (Prostaglandin $I_2$)
- Monobutyrin
- Angiotensinogen/Angiotensin II
- Atrial natriuretic peptide

Growth factors
- IGF-I
- Transforming growth factors α and β
- Macrophage colony-stimulating factor

Complement system
- Factors B, C3 and D (adipsin)

Binding proteins
- IGF-binding proteins
- Retinol binding protein
- Cholesterol ester transfer protein

Hormones
- Leptin
- Oestrone, oestradiol
- Testosterone

Cytokines
- Tumour necrosis factor α (TNF-α)
- Interleukin-6

Other
- Plasminogen activator inhibitor-1
- Acrp30/Adipo Q

## Factors secreted by adipocytes

Adipose tissue secretes a surprisingly large number of substances; they have been roughly categorized in Box 10.1, but several have more than one role (Ailhaud, 1998; Gregoire *et al.*, 1998; Mohammed-Ali *et al.*, 1998). Adipose tissue contains a variety of different types of cell including endothelial cells, undifferentiated adipocyte precursor cells, macrophages, as well as adipocytes. Indeed, in adult animals, adipocytes are a minor component in terms of cell number (10% or less in sheep) (Travers *et al.*, 1997), but being very large cells, adipocytes comprise the bulk of the mass of adipose tissue. While most of the substances listed in Box 10.1 are produced by adipocytes, stromal-vascular cells are thought to be the source of steroid hormones (Mohammed-Ali *et al.*, 1998) and interleukin-6 (Fried *et al.*, 1998) produced by adipose tissue. Adipocytes and stromal-vascular cells may also cooperate; thus while adipocytes can produce and release small amounts of prostaglandin $E_2$ and prostacyclin, production is markedly enhanced by the presence of endothelial cells of the stromal-vascular fraction, adipocytes releasing arachidonic acid which is converted to prostaglandin by the endothelial cells (Richelsen, 1992).

The fact that a substance is secreted by adipose tissue cells does not necessarily mean that a physiologically significant quantity will enter the general circulation. Arterio-venous difference studies in humans have shown that there is a net release of interleukin-6 but not tumour necrosis factor α (TNF-α) by subcutaneous adipose tissue (Mohamed-Ali *et al.*, 1997). Arterio-venous difference studies across adipose tissue have also shown a net release of oestradiol and oestrone in women; indeed, adipose tissue is the major source of oestrogens in the blood of post-menopausal women (Mohamed-Ali *et al.*, 1998), and adipocytes are the major source of leptin in the general circulation in all states. However, for some substances (e.g. prostaglandins, insulin-like growth factor-I (IGF-I), acylation-stimulating protein), secretion appears to be for autocrine or paracrine effects in adipose or adjacent tissues.

For many of the substances listed in Box 10.1, there is no direct evidence for their being produced by adipose tissue in ruminant animals. Prostaglandin $E_2$ has been shown, using microdialysis, to be released by sheep adipose tissue *in vivo* (Doris *et al.*, 1996). IGF-I, IGF-II (Hovey *et al.*, 1998), IGF-binding protein (Beattie and Vernon, 1995), leptin, as well as lipoprotein lipase have all been shown to be produced by adipose tissue in ruminants.

## Autocrine and paracrine effects of locally produced factors within adipose tissue

Adipocytes do not divide and are derived from precursor cells within the tissue. The mechanisms and factors which control the rate of proliferation and differentiation of these precursors (which vary with age and adipose tissue depot) are only partially resolved, but include locally produced factors (e.g. IGF-I, prostaglandins), as well as hormones such as glucocorticoids and insulin (Flint and Vernon, 1993; Gregoire *et al.*, 1998; Soret *et al.*, 1999). The production of retinol-binding protein (Tsutsumi *et al.*, 1992) (and also IGF-binding protein) by adipose tissue may also be involved in control of adipogenesis. Retinol itself inhibits adipogenesis (Ohyama *et al.*, 1998), so production of a binding protein may help to attenuate this effect. Interestingly, it seems that retinol may have depot-specific effects on adipogenesis in ruminants (Torii *et al.*, 1996); whether this relates to differences in the ability to produce retinol-binding protein is unknown. Once formed, adipocytes accumulate lipid and can become very large, achieving a size of 2–3 nl in ruminants (Vernon, 1986). However, adipocytes do appear to have a limit to hypertrophic expansion and there is evidence that as they become very large, they induce formation of new adipocytes from precursor cells (Faust *et al.*, 1978). The factors involved in this have not been resolved, but paradoxically, increasing adipocyte size leads to increased production of TNF-$\alpha$, which inhibits adipogenesis (Hotamisligil and Spiegelman, 1994).

Adipose tissue contains an extensive capillary network (Crandall *et al.*, 1997). As adipocytes enlarge, the ratio of adipocytes to other cells in the tissue diminishes. The identity of the increased number of stromal-vascular cells has not been determined, but it would seem likely that increased numbers of endothelial cells are involved. Adipocytes produce a number of angiogenic factors including monobutyrin and IGF-I (Crandall *et al.*, 1997).

The best-known metabolic factor is lipoprotein lipase, which is synthesized and secreted by adipocytes, and then migrates to the luminal surface of endothelial cells where it hydrolyses triacylglycerols of very low density lipoproteins (VLDLs) and chylomicrons (Vernon and Clegg, 1985). Most of the fatty acids so released are taken up by the adipocytes, but some escape and are used elsewhere in the body (Vernon and Clegg, 1985). A second key autocrine factor which promotes fatty acid esterification and glucose uptake in adipocytes is acylation-stimulating protein (Cianflone, 1997).

The curious, chylomicron-dependent synthesis of this protein from a component of the complement system is discussed later (p. 177). Angiotensin II, which is derived from angiotensinogen, also has autocrine effects, increasing lipogenesis in adipocytes (Jones *et al.*, 1997). IGF-I can enhance lipogenesis in adipose tissue *in vitro*, but this requires a high concentration of the hormone and probably arises from IGF-I interacting with the insulin receptor. As adipocytes (Vernon and Flint, 1989), including sheep adipocytes (S. Wastie, P.J. Buttery and R.G. Vernon, unpublished observations), lack IGF-I receptors this effect is unlikely to be physiologically significant.

Obesity results in increased production of TNF-α by adipocytes which induces insulin resistance, in part at least due to a decrease in insulin receptor kinase activity (Hotamisligil and Spiegelman, 1994; Gregoire *et al.*, 1998). This effect of TNF-α has excited a lot of attention as a possible cause of the non-insulin-dependent diabetes associated with obesity, but the importance of adipocyte derived TNF-α in this remains controversial (Schreyer *et al.*, 1998). Indeed, TNF-α may not be released from adipose tissue *in vivo* (Mohammed-Ali *et al.*, 1997), and so may act as an autocrine regulator of adipocyte metabolism. TNF-α decreases glucose transport, lipoprotein lipase activity and adipsin production by adipocytes (Hotamisligil and Spiegelman, 1994; Gregoire *et al.*, 1998) and enhances lipolysis. As discussed previously, while animals need a reserve of fat, obesity is usually a disadvantage, so signals which attenuate further fat accumulation as adipocytes enlarge are clearly important. Adipocyte TNF-α, rather than having a pathological role, may indeed be a key signal of normal physiology for adipocytes. As well as a putative physiological role, production of TNF-α and other cytokines may also have pathological roles in some disease states leading to cachexia (Hotamisilgil and Spiegelman, 1994). TNF-α production by adipocytes has not been investigated as yet in ruminants, but preliminary studies show that TNF-α attenuates insulin stimulation of lipogenesis in sheep adipose tissue *in vitro*, but is markedly less effective than growth hormone (GH) in this respect (S.E. Melrose, M.D. Houslay and R.G. Vernon, unpublished observations). Leptin, which shares some properties with cytokines such as TNF-α, may also decrease lipogenesis and enhance lipolysis in rodent adipocytes (Houseknecht and Portocarrero, 1998), but another study found no such effects (Mick *et al.*, 1998). Physiological concentrations of ovine leptin had no effect on either lipogenesis or lipolysis in sheep adipose tissue *in vitro* (D. Newby, A. Gertler and R.G. Vernon, unpublished observations).

Adipose tissue produces prostaglandin $E_2$ and it also produces adenosine (again this is synthesized primarily by stromal-vascular cells by an ectocellular 5′-nucleotidase) (Vernon *et al.*, 1991). Both substances inhibit catecholamine-stimulated lipolysis (Vernon *et al.*, 1991). Interestingly, catecholamines increase prostaglandin $E_2$ production *in vivo* and this coincides with a fall in the rate of lipolysis (Doris *et al.*, 1996).

Fatty acids released by adipocytes are transported in blood bound to albumin, and albumin supply to adipose tissue can become a limiting factor at high rates of lipolysis (Vernon and Clegg, 1985). Albumin supply is a function of blood flow and blood flow through adipose tissue is higher before a meal (lipolytic state) than immediately after a meal (lipogenic state) (Barnes *et al.*, 1983). Adipose tissue produces monobutyrin, prostacyclin, prostaglandin $E_2$, adenosine, atrial natriuretic peptide and angiotensinogen (converted to angiotensin II within adipose tissue by angiotensinogen-converting enzyme) which are all vasoactive; most are vasodilators but angiotensin is a vasoconstrictor (Crandall *et al.*, 1997). The system is complex, for angiotensin (vasoconstrictor) promotes production of prostacyclin (vasodilator) by adipocytes (Crandall *et al.*, 1997). Catecholamines, produced by sympathetic nervous activity in the adipose tissue are also vasoactive (Vernon and Clegg, 1985; Crandall *et al.*, 1997). Indeed, IGF-I may also stimulate blood flow, as there is recent evidence for it stimulating nitric oxide production in endothelial cells, and nitric oxide is again a vasodilator (Böger, 1998). Why so many vasoactive substances are produced by adipose tissue and their relative importance in different physiological situations is unknown. Interestingly, several factors (catecholamines, adenosine, prostaglandin $E_2$) modulate both lipolysis and blood flow in adipose tissue, so control of fatty acid release from adipose tissue *in vivo* may be more complex than perceived at present. Furthermore, the interaction is not unidirectional, for endothelin-1, which is produced by endothelial cells, inhibits insulin-stimulated glucose uptake in adipocytes (Lee *et al.*, 1998).

## Paracrine interactions with neighbouring tissues

The widespread but conserved distribution of adipose tissue throughout the mammalian body, and its intimate association with several other tissues, has suggested that adipose tissue may have paracrine effects on these tissues.

In young non-pregnant animals, rudimentary mammary glands are found embedded in adipose tissue (Cunha and Hom, 1996; Sheffield, 1988). During pregnancy in rats and mice, the developing mammary tissue spreads through the adjacent inguinal adipose tissue depot and histological examination of mammary tissue during involution reveals a mosaic of clumps of mammary acini and adipocytes. That this association is obligatory rather than coincidental was shown by removing the adipose tissue depot from around the rudimentary mammary gland of young animals and showing that development was severely retarded in the absence of adipose tissue; transplantation of rudimentary mammary glands into adipose tissue resulted in normal development (Cunha and Hom, 1996; Sheffield, 1998). The search for factors involved in this interaction between adipocytes and mammary cells has involved the use of co-culture systems, or culture of mammary tissue with so-called 'conditioned' media from adipocyte cultures. Initial studies (Rudland *et al.*, 1984) suggested a role for prostaglandin $E_2$; further studies have suggested an involvement of IGF-I which is both mitogenic and also anti-apoptotic with respect to mammary cells (Hovey *et al.*, 1998; Forsyth *et al.*, 1999). There may well be other factors, and all are not necessarily stimulatory (adipose tissue secretes IGF-binding proteins and transforming growth factor-β (TGF-β) which could act to attenuate mammary development) (Rahimi *et al.*, 1998). As well as growth factors, adipocytes may contribute to the extracellular matrix required for mammary development (Wiens *et al.*, 1987). A further aspect is that mammary growth requires angiogenesis, which could also be facilitated by adipocytes.

In ruminants, the developing gland does not grow into the adjacent fat pad during pregnancy as in rodents, and involution does not reveal a mosaic of mammary acini and adipocytes (Akers *et al.*, 1990). The mammary gland is, however, associated with adipose tissue and this appears to be critical for prepubertal development (Hovey *et al.*, 1998; Forsyth *et al.*, 1999).

The heart has a cap of adipose tissue, the so-called epicardial adipose tissue (Marchington *et al.*, 1989). This depot is present in fetal animals, including fetal ruminants, and the mass of the depot does not correlate with the total mass of adipose tissue in the animal (Marchington *et al.*, 1989). Futhermore, the epicardial adipose tissue abuts directly onto the myocardium without any intervening fascia as is found in skeletal muscle (Marchington *et al.*, 1989). Although circumstantial, these various observations suggest a role for epicardial adipose tissue in the development and functioning of the heart. The former could again relate to the ability of adipose tissue to produce mitogenic factors, while it is possible that cardiac adipocytes provide a direct supply of fatty acids for use by cardiocytes. There is as yet no direct evidence for either possibility and for the latter, there is always a question of whether the rate of supply of fatty acids (presumably transported by diffusion, for there is no obvious portal system in the tissue) would be sufficient for lipolysis in cardiac adipocytes to be a quantitatively significant direct source of fatty acids for heart tissue.

The possibilities that inter- and intramuscular adipocytes may influence the development of surrounding skeletal muscle (Hossner *et al.*, 1997), and that they may provide a direct source of fatty acids for adjacent skeletal muscle have also been proposed, but the presence of a fascia between intermuscular adipose tissue and surrounding muscles (Marchington *et al.*, 1989) would suggest that diffusion of fatty acids and other factors is likely to be restricted.

The liver receives a direct supply of fatty acids from an adipose tissue depot, for venous drainage from mesenteric adipose tissues leads into the hepatic portal vein. The physiological importance of this is uncertain, but it may have pathological consequences (Barzilai *et al.*, 1999).

## Adipose tissue interactions with the immune system

Adipose tissue interacts with the immune system in several ways: by production of cytokines, proteins of the complement system and leptin and by interactions between adipocytes and cells of the lymph nodes.

Complement acts to lyse, and hence destroy, targeted cells during infection. Adipocytes secrete three components of the alternative system for complement production, factors B, C3 and D; factor D is also known as adipsin (Flier, 1995; Cianflone, 1997). Adipsin production is decreased by insulin and TNF-$\alpha$ and is decreased during obesity in rodents but apparently not in obese humans (Flier, 1995). The physiological significance of this adipocyte-based system for complement production is still unclear. Curiously, factors C3 and B bind to chylomicrons and this promotes cleavage of factor C3 by adipsin to produce factor C3A which, in turn, is cleaved by carboxypeptidase to produce factor C3A desarg (acylation-stimulating protein) (Cianflone, 1997). This raises the possibility that the primary function of factor B, C3 and D production by adipocytes is autocrine and concerned with adipocyte metabolism, rather than paracrine modulation of complement formation.

Lymph nodes are embedded in adipose tissue, mostly in minor depots that are too small to make a significant contribution to whole body lipid storage and which are selectively conserved in starvation (Pond, 1996). Studies with an intermuscular (popliteal) adipose tissue depot have shown that both the lipolytic response and the fatty acid composition of triacylglycerols of adipocytes varies with distance from the lymph nodes in guinea pigs (Pond, 1996). It appears that lymphocytes secrete cytokines (e.g. TNF-α) which accentuate lipolysis (Feingold *et al.*, 1992; Green *et al.*, 1994), while the fatty acids released from adipocytes stimulate lymphocyte proliferation (Pond, 1996). Stimulation of the immune system by injection of an endotoxin (lipopolysaccharide) results in enhanced lipolysis and the appearance of TNF-α type I receptors on adipocytes close to the lymph node but not on adipocytes remote from the lymph node (MacQueen and Pond, 1998; Pond and Mattacks, 1998). However, studies with ovine popliteal adipose tissue failed to detect any variation in fatty acid composition with distance from the lymph node (Vernon and Pond, 1997), and TNF-α had no effect on basal or catecholamine-stimulated lipolysis in adipose tissue near to (5 mm or less) or distant from the lymph node (S.E. Melrose, M.D. Houslay and R.G. Vernon, unpublished observations). However, preliminary studies suggest that the ability of the adenosine analogue, $N^6$-phenylisopropyladenosine, to inhibit lipolysis appears to be diminished in adipocytes close to the lymph nodes, and culture with growth hormone diminished the ability of PIA to inhibit lipolysis in adipocytes distant from, but not in those close to, the lymph node (S.E. Melrose, M.D. Houslay and R.G. Vernon, unpublished observations). Thus there may be differences in the lipolytic system in different parts of popliteal adipose tissue in sheep, but, if so, they appear to be much smaller than in guinea pigs.

Leptin also influences the immune system. Impaired T-lymphocyte immunity is found in ob/ob mice which produce a defective form of leptin (Chandra, 1980), and very recent studies show that leptin enhances proliferation of T-lymphocytes and their subsequent ability to respond to antigens (Lord *et al.*, 1998). In addition, preadipocytes may also contribute to the body's defence mechanisms as there is evidence for their having macrophage-like properties (Cousin *et al.,* 1999); this may explain the production of macrophage colony-stimulatory factor by adipose tissue and its promotion of adipose tissue growth (Levine *et al.*, 1998). Thus there are several ways in which adipose tissue can modulate functioning of the immune system, but their relative importance is uncertain. That such interactions are physiologically relevant is indicated by the fact that anorexia, which results in greatly diminished adiposity, causes diminished functioning of the immune system (Chandra, 1991). High-yielding dairy cows tend to be more susceptible to disease (e.g. mastitis, laminitis) and this is exacerbated by poor nutrition (Sinclair *et al.*, 1999); adiposity is likely to markedly diminish in such animals. From an individual's point of view, for starvation, and other conditions which lead to a loss of adipose tissue reserves, to result in a failure of the immune system seems singly unfortunate. However, when considered from the perspective of the species as a whole, it does make sense, for when food is in short supply it increases the likelihood that the weakest animals, with the least reserves, die leaving more of the limited food available for the stronger. A case of the survival of the fattest!

## Leptin

The discovery of leptin provided the long-postulated link between adipose tissue stores and brain centres which regulate food intake and energy homeostasis. Since this seminal discovery, our understanding of the role of leptin has expanded from that of a 'sensor' of body fat mass to include participation in, and regulation of, multiple physiological systems including reproduction, inflammation and cell-mediated immunity, as well as the coordination of whole body energy homeostasis (Friedman and Halaas, 1998; Houseknecht and Portocarrero, 1998; Houseknecht *et al.*, 1998a).

The leptin (*ob*) gene has been cloned in several species including the mouse, pig, chicken and human (Houseknecht and Portocarrero, 1998) and also the cow (Ji *et al.*, 1998); sequence homology among species is high. Leptin is primarily expressed in white adipocytes; however, in rodents and humans, leptin expression has also been reported in the placenta (Hoggard *et al.*, 1998) and stomach (Bado *et al.*, 1998). In cattle and swine, leptin is reported to be expressed exclusively in adipose tissue (Houseknecht and Portocarrero, 1998). Furthermore, Ji *et al.* (1998) reported no difference in the level of leptin gene expression among adipose tissue depots (subcutaneous, renal and omental depots) in cattle, which differs from reports in humans (Montague *et al.*, 1998; Russell *et al.*, 1998; Vanharmelen *et al.*, 1998).

The leptin receptor is a member of the cytokine family of receptors and is expressed as either a long form, or as multiple short forms due to alternative splicing (Tartaglia, 1997). The long form of the leptin receptor predominates in the hypothalamus and is credited with the central regulation of appetite; the short isoforms, and to a limited extent, the long form, are expressed in multiple peripheral tissues as well and may be involved in the regulation of tissue metabolism (Tartaglia, 1997). Soluble forms of the leptin receptor have been reported (Houseknecht *et al.*, 1996); their role is uncertain, but leptin binding to proteins in blood complicates the accurate assay of circulating leptin in many species. Dyer *et al.* (1997a) cloned a partial ovine long-form leptin receptor cDNA and reported that mRNA expression for the long-form receptor is present in the hypothalamus, anterior pituitary and adipose tissue of sheep. Furthermore, they reported that expression was highest in the arcuate nucleus and the ventromedial hypothalamus, and that hypothalamic expression was up-regulated by undernutrition as found in rodents (Houseknecht and Portocarrero, 1998).

Leptin is a powerful regulator of food intake in rodents; effects are most dramatic in ob/ob mice which lack leptin (Friedman and Halaas, 1998; Houseknecht *et al.*, 1998a; Houseknecht and Portocarrero, 1998). Intracerebro ventricular administration of recombinant ovine leptin to sheep also caused a reduction in food intake (Morrison *et al.*, 1998). Large-scale preparation of recombinant ruminant leptin is now underway (Gertler *et al.*, 1998) which will facilitate further studies in ruminants. Studies on the mechanism whereby leptin decreases appetite and increases whole-body energy expenditure have, in the past, focused mostly on the neuropeptide Y (NPY) system of the hypothalamus, but other neuropeptides are most probably involved. Leptin acts centrally to prevent NPY-induced appetite stimulation and suppression of thermogenesis by inhibiting its synthesis in the arcuate nucleus of the hypothalamus (Friedman and Halaas, 1998; Houseknecht *et al.*, 1998a). Dyer *et al.* (1997b) showed that peripheral NPY administration up-regulated the expression of leptin and NPY receptor subtype 1 in ovine adipose tissue. These data are indicative of a feedback loop between leptin and NPY expression which has been shown for rodents by others (Houseknecht *et al.*, 1998a; Houseknecht and Portocarrero, 1998).

Leptin expression is under complex control by both hormones (Friedman and Halaas, 1998; Houseknecht *et al.*, 1998a) and the sympathetic nervous system (Trayhurn *et al.*, 1998); modulatory hormones include insulin, glucocorticoids, thyroid hormones and oestrogen. Effects appear to be chronic and due to changes in leptin gene expression rather than to acute stimulation of leptin secretion from intracellular storage pools. *In vitro* studies from cattle (Houseknecht *et al.*, 1998b) show that leptin mRNA abundance in bovine subcutaneous adipose tissue is increased by insulin and glucocorticoids. *In vivo,* growth hormone increased leptin gene expression in bovine (Houseknecht *et al.*, 1998b) and ovine adipose tissue (Raymond *et al.*, 1997). However, effects of GH may be indirect, as leptin expression in bovine adipose tissue was strongly, positively correlated with IGF-I gene expression while GH had no effect on leptin expression *in vitro* (Houseknecht *et al.*, 1998b).

Serum leptin concentration in the fed state is highly, positively correlated with body fat mass in rodents, pigs and humans (Friedman and Halaas, 1998; Houseknecht *et al.*, 1998a) and in cattle (Ji *et al.*, 1997; Chilliard *et al.*, 1998; Minton *et al.*, 1998) and sheep (Kumar *et al.,* 1998). This led to the hypothesis that leptin allows the body to 'sense' the size of the body's energy storage pool and adjust appetite and whole-body energy metabolism accordingly. Fasting causes a rapid, profound down-regulation of leptin gene expression in rodents and humans (Friedman and Halaas, 1998). Fasting (2 days) also down-regulates leptin gene expression in subcutaneous adipose tissue of cattle (Tsuchiya *et al.*, 1998) and swine (Spurlock *et al.*, 1998a). However, the effects of fasting on leptin expression in pigs are modest compared with rodents, and there appears to be no effect of maintenance intake on leptin gene expression (Spurlock *et al.*, 1998a), suggesting that this species may be less sensitive to this form of regulation of the leptin gene. Fasting also results in changes in the secretion of pituitary hormones, including thyroid stimulating hormone (TSH), GH and the gonadotropins, and these changes are partly prevented by administration of leptin (Friedman and Halaas, 1998). These various observations indicate that leptin plays a central role in the adaptations to fasting and suggest that the primary evolutionary role for leptin is to prevent death by starvation rather than to prevent obesity (Spiegelman and Flier, 1996).

Leptin is an important regulator of reproduction in rodents and humans, providing a link between energy reserves and an energetically demanding function (Hoggard *et al.*, 1998; Houseknecht *et al.*, 1998a; Macut *et al.*, 1998). There is potential for leptin to act both centrally and peripherally in both males and females, as the leptin receptor is expressed not only in the hypothalamus and pituitary but in the ovary and testis as well (Houseknecht *et al.*, 1998a). Leptin stimulates gonadotrophin secretion in ob/ob mice and undernourished animals, and advances the onset of puberty (Houseknecht *et al.*, 1998a). These data have obvious, important implications for productive efficiency of livestock.

Links between leptin and the immune system have been discussed in a previous section. In addition, leptin expression is up-regulated by inflammatory cytokines such as TNF-$\alpha$ and interleukins, and an increase in leptin secretion may cause the anorexia following administration of endotoxin (Houseknecht *et al.*, 1998a). However, Spurlock *et al.* (1998b) found no effect of acute endotoxin treatment on leptin expression in fasted pigs, despite physiological indications of a robust inflammatory response. Additionally, Leininger *et al.* (1998) found that endotoxin treatment down-regulated leptin gene expression in fully fed pigs, and that the endotoxin-induced fall in leptin was highly correlated with changes in blood insulin, IGF-I, glucose and free fatty acid concentrations. These data (Leininger *et al.*, 1998) suggest that changes in energy metabolism and associated hormones can overcome, at least in the pig, the positive effects of inflammatory cytokines on leptin gene expression. Obviously, further work is necessary to elucidate the role of leptin (if any) in the anorexia and altered energy metabolism associated with acute and chronic inflammation in livestock.

## Conclusions

Ideas on adipose tissue have progressed a long way from the days when it was perceived as a relatively inert tissue. Not only do we now know that it is very active metabolically, it is also a source of a plethora of factors which influence events within adipose tissue, within adjacent tissues or indeed with endocrine effects on distant parts of the body – a veritable 'node on the information superhighway' (Flier, 1995). The importance of adipose tissue in vertebrate evolution, and mammalian evolution in particular, is now apparent, and its central role in mammalian physiology is beginning to be appreciated. This is emphasized by two very recent studies with transgenic mice essentially devoid of adipose tissue; many mice died before reaching adulthood and those that survived showed symptoms of acute diabetes (McKnight 1998); the reason for the diabetes is uncertain but may be due to an ability to remove fatty acids from the blood. Curiously, both too little and too much adipose tissue can lead to the same pathological condition! Animals have affected means of adjusting their amount of adipose tissue to meet their needs, although the mechanisms may be attenuated by domestication. Some of the substances produced by adipocytes and their associated stromal-vascular cells must act as signals to help achieve this. Leptin, initially perceived as the factor which could solve the problem of obesity, may, in fact, be more important for signalling inadequate or declining adipose tissue reserves; leptin clearly does not prevent the development of

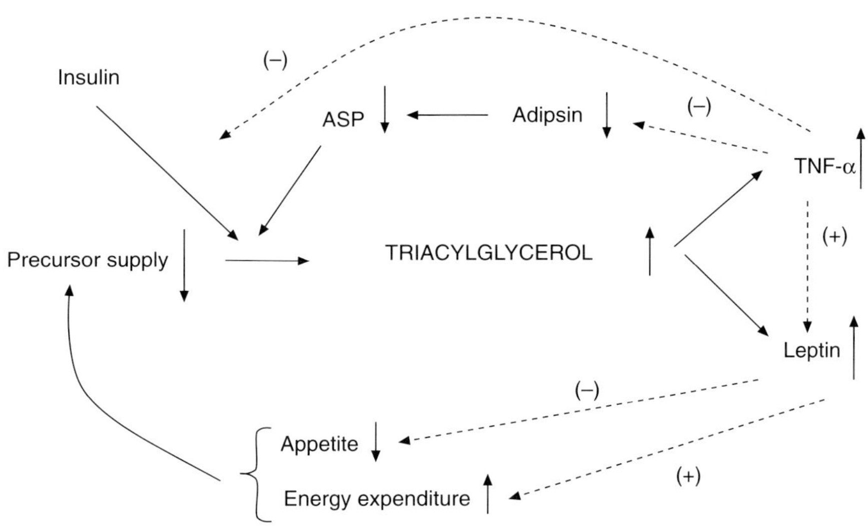


**Fig. 10.1**. Mechanisms whereby leptin and TNF-α attenuate triacylglycerol accumulation. ASP, acylation-stimulating protein; (+) stimulates, (–) inhibits, ↑ increased, ↓ decreased.

obesity, but it may retard it. Leptin also has an important role in the adaptations to negative energy balance. As adipocytes enlarge, they increase production of other substances (e.g. TNF-$\alpha$) as well as leptin, and decrease production of others (e.g. adipsin, acylation-stimulation protein) which may act to limit further expansion. Adipose tissue mass thus appears to be modulated both directly by autocrine factors and indirectly by endocrine factors such as leptin (Fig. 10.1). When one chooses to ignore these signals, obesity follows and thence undesirable pathological conditions (e.g. diabetes).

The implication of this for domestic ruminants (and indeed other food animals) is that we need to know more precisely the amounts of adipose tissue required for optimal function, for while too much adipose tissue is undesirable both for the animal and as a food for us, equally too little is deleterious for the animal's well-being.

## References


Ailhaud, G. (1998) L'adipocyte, cellule sécrétrice et endocrine. *Médecine/sciences* 14, 858–864.

Akers, R.M., Beal, W.E., McFadden, T.B. and Capuco, A.V. (1990) Morphometric analysis of involuting bovine mammary tissue after 21 or 42 days of non-suckling. *Journal of Animal Science* 68, 3604–3613.

Bado, A., Levasseur, S., Attoub, S., Kermorgant, S., Laigneau, J.P., Bortoluzzi, M.N., Moizo, L., Lehy, T., Guerre-Millo, M., Marchand-Brustel, Y. le and Lewin, M.J.M. (1998) The stomach is a source of leptin. *Nature* 394, 790–793.

Barnes, R.J., Comline R.S. and Dobson, A. (1983) Changes in the blood flow to the digestive organs of sheep induced by feeding. *Quarterly Journal of Experimental Physiology* 68, 77–88.

Barzilai, N., She, L., Liu, B.-Q., Vuguin, P., Cohen, P., Wang, J. and Rossetti, L. (1999) Surgical removal of visceral fat reverses hepatic insulin resistance. *Diabetes* 48, 94–98.

Beattie, J. and Vernon, R.G. (1995) Glucocorticoids regulate the secretion of a 21kDa-IGF-binding protein by sheep adipose tissue explants. *Molecular Cellular Biochemistry* 145, 151–157.

Böger, R.H. (1998) Role of nitric oxide in the haemodynamic effects of growth hormone. *Growth Hormone and IGF Research* 8, 163–165.

Butler, P.J., Woakes, A.J. and Bishop, C.M. (1998) Behaviour and physiology of Svalbard Barnacle Geese *Branta leucopsis* during their autumn migration. *Journal of Avian Biology* 29, 536–545.

Chandra, R.K. (1980) Cell-mediated immunity in genetically obese (C57BL/6J *ob/ob*) mice. *American Journal of Clinical Nutrition* 33, 13–16.

Chandra, R.K. (1991) Nutrition and immunity: lessons from the past and new insights into the future. *American Journal of Clinical Nutrition* 53, 1087–1101.

Chilliard, Y., Ferlay, A., Delavaud, C. and Bocquier, F. (1998) Plasma leptin in underfed or overfed adult holstein and charolais cows and its relationship with adipose tissue cellularity. *International Journal of Obesity* 22 (Suppl. 3), S171.

Cianflone, K. (1997) Acylation stimulating protein and the adipocyte. *Journal of Endocrinology* 155, 203–206.

Cousin, B., Munoz, O., Andre, M., Fontanilles, A.M., Dani, C., Cousin, J.L., Laharrague, P., Casteilla, L. and Penicaud, L. (1999) A role for preadipocytes as macrophage-like cells. *FASEB Journal* 13, 305–312.

Crandall, D.L., Hausman, G.J. and Kral, J.G. (1997) A review of the microcirculation of adipose tissue: anatomic, metabolic and angiogenic perspectives. *Microcirculation* 4, 211–232.

Cunha, G.R. and Hom, Y.K. (1996) Role of mesenchymal-epithelial interactions in mammary gland development. *Journal of Mammary Gland Biology and Neoplasia* 1, 21–35.

Doris, R., Thompson, G.E., Finley, E., Kilgour, E., Houslay, M.D. and Vernon, R.G. (1996) Chronic effects of somatotropin treatment on response of subcutaneous adipose tissue lipolysis to acutely acting factors in vivo and in vitro. *Journal of Animal Science* 74, 562–568.

Dyer, C.J., Simmons, J.M., Matteri, R.L. and Keisler, D.H. (1997a) Leptin receptor mRNA is expressed in ewe anterior pituitary and adipose tissues and is differentially expressed in hypothalamic regions of well-fed and feed-restricted ewes. *Domestic Animal Endocrinology* 14, 119–128.

Dyer, C.J., Simmons, J.M., Matteri, R.L. and Keisler, D.H. (1997b) Effects of an intravenous injection of NPY on leptin and NPY-Y1 receptor mRNA expression in ovine adipose tissue. *Domestic Animal Endocrinology* 14, 325–333.

Faust, I.M., Johnson, P.R., Stern, J.S. and Hirsh, J. (1978) Diet-induced adipocyte number increase in adult rats: a new model of obesity. *American Journal of Physiology* 235, E279-E286.

Feingold, K.R., Doerrler, W., Dinarello, C.A., Fiers, W. and Grunfeld, C. (1992) Stimulation of lipolysis in cultured fat cells by tumor necrosis factor, interleukin-1 and the interferons is blocked by inhibition of prostaglandin synthesis. *Endocrinology* 130, 10–16.

Flier, J.S. (1995) The adipocyte: storage depot or node on the energy information superhighway? *Cell* 80, 15–18.

Flint, D.J. and Vernon, R.G. (1993) Hormones and adipose growth. In: Schreibman, M.P., Scanes, C.G. and Pang, P.K.T. (eds) *The Endocrinology of Growth, Development and Metabolism in Vertebrates.* Academic Press, San Diego, pp. 469–499.

Forsyth, I.A., Gabai, G. and Morgan, G. (1999) Spatial and temporal expression of insulin-like growth factor-I, insulin-like growth factor-II and the insulin-like growth factor-I receptor in the sheep fetal mammary gland. *Journal of Dairy Research* 66, 35–44.

Fried, S.K., Bunkin, D.A. and Greenberg, A.S. (1998) Omental and subcutaneous adipose tissues of obese subjects release interleukin-6: depot difference and regulation by glucocorticoid. *Journal of Clinical Endocrinology and Metabolism* 83, 847–850.

Friedman, J.M. and Halaas, J.L. (1998) Leptin and the regulation of body weight in mammals. *Nature* 395, 763–770.

Gertler, A., Simmons, J. and Keisler, D.H. (1998) Large-scale preparation of biologically active recombinant ovine obese protein (leptin). *FEBS Letters* 422, 137–140.

Gregoire, F.M., Smas, C.M. and Sul, H.S. (1998) Understanding adipocyte differentiation. *Physiological Reviews* 78, 783–809.

Green, A., Dobias, S.B., Walters, D.J.A. and Brasier, A.R. (1994) Tumor necrosis factor increases the rate of lipolysis in primary cultures of adipocytes without altering levels of hormone-sensitive lipase. *Endocrinology* 134, 2581–2588.

Hoggard, N., Hunter, L., Trayhurn, P., Williams, L.M. and Mercer, J.G. (1998) Leptin and reproduction. *Proceedings of the Nutrition Society* 57, 421–427.

Hossner, K.L., Yemm, R., Vierck, J. and Dodson, M.V. (1997) Insulin-like growth factor (IGF)-I and -II and IGFBP secretion by ovine satellite cell strains grown alone or in coculture with 3T3-L1 preadipocytes. *In Vitro Cellular and Developmental Biology* 33, 791–795.

Hotamisligil, G.S. and Spiegelman, B.M. (1994) Tumor necrosis factor $\alpha$: a key component of the obesity-diabetes link. *Diabetes* 43, 1271–1278.

Houseknecht, K.L. and Portocarrero, C.P. (1998) Leptin and its receptors: Regulators of whole-body energy homeostasis. *Domestic Animal Endocrinology* 15, 457–475.

Houseknecht, K.L., Mantxoros, C.S., Kuliawat, R., Hadro, E.T., Flier, J.S. and Kahn, B.B. (1996) Evidence for leptin binding to proteins in serum of rodents and humans: Modulation with obesity. *Diabetes* 45, 1638–1643.

Houseknecht, K.L., Baile, C.A., Matteri, R.L. and Spurlock, M.E. (1998a) The biology of leptin: A review. *Journal of Animal Science* 76, 1405–1420.

Houseknecht, K.L., Portocarrero, C.P., Ji, S., Lemenager, R.P. and Spurlock, M.E. (1998b).

Growth hormone (GH) regulation of leptin gene expression in bovine adipose tissue: In vitro and in vivo studies. *International Journal of Obesity* 22 (Suppl. 3), S166 (abstract).
Hovey, R.C., Davey, H.W., Mackenzie, D.D.S. and McFadden, T.B. (1998) Ontogeny and epithelial-stromal interactions regulate IGF expression in the ovine mammary gland. *Molecular and Cellular Endocrinology* 136, 139–144.
Ji, S.Q., Willis, G.M., Scott, R.R. and Spurlock, M.E. (1997) Partial cloning of the bovine leptin gene and its expression in adipose depots and in cattle before and after finishing. *Journal of Animal Science* 75 (Suppl. 1), 167.
Ji, S.Q., Willis, G.M., Scott, R.R. and Spurlock, M.E. (1998) Partial cloning and expression of the bovine leptin gene. *Animal Biotechnology* 9, 1–14.
Jones, B.H., Standridge, M.K. and Moustaid, N. (1997) Angiotensin II increases lipogenesis in 3T3-L1 and human adipose cells. *Endocrinology* 138, 1512–1519.
Kumar, B., Francis, S.M., Suttie, J.M. and Thompson, M.P. (1998) Expression of obese mRNA in genetically lean and fat selection lines of sheep. *Comparative Biochemistry and Physiology* 120, 543–548.
Larsen, T.S., Nilsson, N.Ö. and Blix, A.S. (1985) Seasonal changes in lipogenesis and lipolysis in isolated adipocytes from Svalbard and Norwegian reindeer. *Acta Physiological Scandinavia* 123, 97–104.
Lee, Y.-C., Juan, C.-C., Fang, V.S., Hsu, Y.-P., Lin, S.-H., Kwok, C.-F. and Ho, L.-T. (1998) Evidence that endothelin-1 (ET-1) inhibits insulin-stimulated glucose uptake in rat adipocytes mainly through ETA receptors. *Metabolism* 47, 1468–1471.
Leininger, M.T., Portocarrero, C.P., Bidwell, C.A., Spurlock, M.E., Nielsen, J.N. and Houseknecht, K.L. (1998) Response to immune challenge in pigs selected for high lean gain: Role of leptin. *Journal of Animal Science* 76 (Suppl. 1), 129.
Levine, J.A., Jensen, M.D., Eberhardt, N.L. and O'Brien, T. (1998) Adipocyte macrophage colony-stimulating factor is a mediator of adipose tissue growth. *Journal of Clinical Investigation* 101, 1557–1564.
Lindström, E. (1983) Condition and growth of red foxes (*Vulpes vulpes*) in relation to food supply. *Journal of Zoology,* London 199, 117–122.
Lord, G.M., Matarese, G., Howard, J.K., Baker, R.J., Bloom, S.R. and Lechler, R.I. (1998) Leptin modulates the T-cell immune response and reverses starvation-induced immunosuppression. *Nature* 394, 897–900.
McKnight, S.L. (1998) WAT-free mice: diabetes without obesity. *Genes and Development* 12, 3145–3148.
MacQueen, H.A. and Pond, C.M. (1998) Immunofluorescent localisation of tumour necrosis factor-α receptors on the popliteal lymph node and the surrounding adipose tissue following a simulated immune challenge. *Journal of Anatomy* 192, 223–231.
Macut, D., Micic, D., Pralong, F.P., Bischof, P. and Campana, A. (1998) Is there a role for leptin in human reproduction? *Gynaecology Endocrinology* 12, 321–326.
Marchington, J.M., Mattacks, C.A. and Pond, C.M. (1989) Adipose tissue in the mammalian heart and pericardium: structure, foetal development and biochemical properties. *Comparative Biochemistry and Physiology* 94B, 225–232.
Mick, G., Vanderbloomer, T., Fu, C.L. and McCormick, K. (1998) Leptin does not affect adipocyte glucose metabolism: studies in fresh and cultured adipocytes. *Metabolism* 47, 1360–1365.
Minton, J.E., Bindel, D.J., Drouillard, J.S., Titgemeyer, E.C., Grieger, D.M. and Hill, C.M. (1998) Serum leptin is associated with carcass traits in finishing cattle. *Journal of Animal Science* 76 (Suppl. 1), 231.
Mohamed-Ali, V., Goodrick, S., Rawesh, A., Katz, D.R., Miles, J.M., Yudkin, J.S., Klein, S. and Coppack, S.W. (1997) Subcutaneous adipose tissue releases interleukin-6, but not tumor necrosis factor-α, in vivo. *Journal of Clinical Endocrinology and Metabolism* 82, 4196–4200.

Mohamed-Ali, V., Pinkney, J.H. and Coppack, S.W. (1998) Adipose tissue as an endocrine and paracrine organ. *International Journal of Obesity* 22, 1145–1158.

Montague, C.T., Prins, J.B., Sanders, L., Zhang, J.L., Sewter, C.P., Digby, J., Byrne, C.D. and O'Rahilly, S. (1998). Depot-related gene expression in human subcutaneous and omental adipocytes. *Diabetes* 47, 1384–1391.

Morrison, C.D., Daniel, J.A., Holmberg, B.J., Bolden, O.U., Raver, N., Gertler, A. and Keisler, D.H. (1998) Effects of lateral cerebroventricular infusion of leptin on ewe lambs. *Journal of Animal Science* 76 (Suppl. 1), 225.

Oftedal, O.T. (1992) The adaptation of milk secretion to the constraints of fasting in bears, seals and baleen whales. *Journal of Dairy Science* 76, 3234–3246.

Ohyama, M., Matsuda, K., Torii, S., Matsui, T., Yano, H., Kawada, T. and Ishihara, T. (1998) The interaction between vitamin A and thiazolidinedione on bovine adipocyte differentiation in primary culture. *Journal of Animal Science* 76, 61–65.

Pond, C.M. (1986) The natural history of adipocytes. *Science Progress* Oxford 70, 45–71.

Pond, C.M. (1992) An evolutionary and functional view of mammalian adipose tissue. *Proceedings of the Nutrition Society* 51, 367–377.

Pond, C.M. (1996) Interactions between adipose tissue and the immune system. *Proceedings of the Nutrition Society* 55, 111–126.

Pond, C.M. and Mattacks, C.A. (1998) In vivo evidence for the involvement of the adipose tissue surrounding lymph nodes in immune responses. *Immunology Letters* 63, 159–167.

Rahimi, N., Tremblay, E., McAdam, L., Roberts, A. and Elliott, B. (1998) Autocrine secretion of TGF-ß1 and TGF-ß2 by preadipocytes and adipocytes: a potent negative regulator of adipocyte differentiation and proliferation of mammary carcinoma cells. *In Vitro Cellular and Developmental Biology* 34, 412–420.

Raymond, S.R., Thomas, M.G., Carroll, J.A., Matteri, R.L. and Keisler, D.H. (1997) Zeranol and growth hormone treatment differentially influenced mRNA levels of the obesity protein, leptin, and the GH receptor in growing wethers. *Journal of Animal Science* 75 (Suppl. 1), 225.

Richelsen, B. (1992) Release and effects of prostaglandins in adipose tissue. *Prostaglandins Leukotrienes and Essential Fatty Acids* 47, 171–182.

Rudland, P.S., Davies, A.C.T. and Tsao, S.-W. (1984) Rat mammary preadipocytes in culture produce a trophic agent for mammary epithelia – prostaglandin $E_2$. *Journal of Cellular Physiology* 120, 364–376.

Russell, C.D., Petersen, R.N., Rao, S.P., Ricci, S.P., Prasad, A., Zhang, Y., Brolin, R.E. and Fried, S.K. (1998) Leptin expression in adipose tissue from obese humans: depot-specific regulation by insulin and dexamethasone. *American Journal of Physiology* 38, E507-E515.

Schreyer, S.A., Chua, S.C. and LeBoeuf, R.C. (1998) Obesity and diabetes in TNF-$\alpha$ receptor-deficient mice. *Journal of Clinical Investigation* 102, 402–411.

Sheffield, L.G. (1988) Organisation and growth of mammary epithelia in the mammary gland fat pad. *Journal of Dairy Science* 71, 2855–2874.

Sheridan, M.A. (1994) Regulation of lipid metabolism in poikilothermic vertebrates. *Comparative Biochemistry and Physiology* 107B, 495–508.

Sinclair, M.C., Nielsen, B.L., Oldham, J.D. and Reid, H.W. (1999) Consequences for immune function of metabolic adaptations to load. *British Society of Animal Science Occasional Publication* 24, 113–118.

Soret, B., Lee, H.-J., Finley, E., Lee, S.C. and Vernon, R.G. (1999) Regulation of differentiation of sheep subcutaneous and abdominal preadipocytes in culture. *Journal of Endocrinology* 161, 517–524.

Spiegelman, B.M. and Flier, J.S. (1996) Adipogenesis and obesity-rounding out the big picture. *Cell* 87, 377–389.

Spurlock, M.E., Frank, G.R., Cornelius, S.G., Ji, S.Q., Willis, G.M. and Bidwell, C.A. (1998a) Obese gene expression in porcine adipose tissue is reduced by food deprivation but not by maintenance or submaintenance intake. *Journal of Nutrition* 128, 677–682.

Spurlock, M.E., Ranalletta, M.A., Cornelius, S.G., Frank, G.R., Willis, G.M., Ji, S., Grant, A.L. and Bidwell, C.A. (1998b) Leptin expression in porcine adipose tissue is not increased by endotoxin but is reduced by growth hormone. *Journal of Interferon Cytokine Research* 18, 1051–1058.

Tartaglia, L.A. (1997) The leptin receptor. *Journal of Biological Chemistry* 272, 6093–6096.

Torii, S., Matsui, T. and Yano, H. (1996) Development of intramuscular fat in Wagyu beef cattle depends on adipogenic or antiadipogenic substances present in serum. *Animal Science* 63, 73–78.

Travers, M.T., Vernon, R.G. and Barber, M.C. (1997) Repression of the acetyl-coA carboxylase gene in ovine adipose tissue during lactation: the role of insulin responsiveness. *Journal of Molecular Endocrinology* 19, 99–107.

Trayhurn, P., Duncan, J.S., Hoggard, N. and Rayner, D.V. (1998) Regulation of leptin production: a dominant role for the sympathetic nervous system? *Proceedings of the Nutrition Society* 57, 413–419.

Tsuchiya, T., Nagao, Y., Ozawa, A., Matsumoto, M., Sugahara, K., Kubo, T. and Kato, H. (1998) Decrease of the obese gene expression in bovine subcutaneous adipose tissue by fasting. *Bioscience Biotechnology and Biochemistry* 62, 2068–2069.

Tsutsumi, C., Okuno, M., Tannous, L., Piantedosi, R., Allan, M., Goodman, D.S. and Blaner, W.S. (1992) Retinoids and retinoid-binding protein expression in rat adipocytes. *Journal of Biological Chemistry* 267, 1805–1810.

Vanharmelen, V., Reynisdottir, S., Eriksson, P., Thorne, A., Hoffstedt, J., Lonnqvist, F. and Arner, P. (1998) Leptin secretion from subcutaneous and visceral adipose tissue in women. *Diabetes* 47, 913–917.

Vernon, R.G. (1986) The growth and metabolism of adipocytes. In: Buttery, P.J., Haynes, N.B. and Lindsay, D.B. (eds) *Control and Manipulation of Animal Growth.* Butterworths, London, pp. 67–83.

Vernon, R.G. and Clegg, R.A. (1985) The metabolism of white adipose tissue in vivo and in vitro. In: Cryer, A. and Van, R.L.R. (eds) *New Perspectives in Adipose Tissue.* Butterworths, London, pp. 65–86.

Vernon, R.G. and Flint, D.J. (1984) Adipose tissue: metabolic adaptation during lactation. *Symposium of the Zoological Society of London* 51, 119–145.

Vernon, R.G. and Flint, D.J. (1989) Role of growth hormone in the regulation of adipocyte growth and function. In: Heap, R.G., Prosser, C.G. and Lamming, G.E. (eds.) *Biotechnology in Growth Regulation.* Butterworths, London, pp. 57–71.

Vernon, R.G. and Pond, C.M. (1997) Adaptations of maternal adipose tissue to lactation. *Journal of Mammary Gland Biology and Neoplasia* 2, 231–241.

Vernon, R.G., Finley, E. and Watt, P.W. (1991) Adenosine and the control of adrenergic regulation of adipose tissue lipolysis during lactation. *Journal of Dairy Science* 74, 695–705.

Wiens, D., Park, C.S. and Stockdale, F.E. (1987) Milk protein expression and ductal morphogenesis in the mammary gland in vitro: hormone-dependent and -independent phases of adipocyte-mammary epithelial cell interaction. *Developmental Biology* 120, 245–258.

# 11 Regulation of Growth and Metabolism During Postnatal Development

B.H. Breier, M.H. Oliver and B.W. Gallaher

*Research Centre for Developmental Medicine and Biology, Faculty of Medicine and Health Science, University of Auckland, Auckland, New Zealand*

## Introduction

This chapter will review the endocrine regulation of growth and metabolism in ruminants and focus on recent insights into the central role of the somatotropic axis. Growth and metabolism during postnatal life are regulated by multiple hormones and growth factors acting in an endocrine (systemic) and an autocrine or paracrine (local) manner. The somatotropic axis is a multi-level hormonal system, primarily consisting of growth hormone (GH; somatotropin) and insulin-like growth factor-I (IGF-I; somatomedin), their associated carrier proteins and receptors. There are a number of other hormonal axes that are involved in the regulation of intermediary metabolism and the growth process. Among these are insulin, thyroid hormones, glucocorticoids, sex steroids, the melanocortin-leptin axis and a number of locally produced growth factors.

Circulating GH is a major regulator of metabolism and growth during postnatal life. GH deficiency is associated with a reduction in growth rate and in final adult size. During postnatal development and during adult life circulating GH regulates key metabolic pathways of intermediary metabolism (Breier *et al.*, 1991). Treatment with GH is lipolytic, elevating circulating concentrations of free fatty acids (FFA), increasing the body's ability to respond to lipolytic stimuli and reducing lipogenesis. GH treatment also creates a state of positive nitrogen balance by increasing nitrogen retention and decreasing protein catabolism. Furthermore, GH administration elevates plasma insulin and glucose levels, which in the long term creates a state of insulin resistance.

IGF-I was originally described as an endocrine factor secreted from the liver in response to GH stimulation which mediated the effects of GH in peripheral tissues. However, while IGF-I treatment increases growth in normal and in GH-deficient animals, growth rates rarely equal those obtained with GH treatment. GH stimulation of IGF-I mRNA expression is not limited to the liver, suggesting that some of the effects of GH treatment on tissue growth are derived from the direct stimulation of autocrine or paracrine IGF-I production. Treatment with IGF-I stimulates protein metabolism and reduces plasma glucose, as a result of both increased glucose uptake and reduced glucose production (Douglas *et al.*, 1991). IGF-I is important for the function of a variety of organs. For example, elevated circulating IGF-I levels increase glomerular filtration rate (Hammerman and Miller, 1993) and renal growth, while local concentrations of IGF-I determine the destiny of maturing follicles in the ovary (Giudice *et al.*, 1995). IGF-I is also a potent inotropic agent in the heart and regulates immune cell numbers as well as other aspects of the immune response. While IGF-I is often considered a proliferation or differentiation factor, it is also a potent cell survival factor through its role as an inhibitor of programmed cell death. This review will elucidate the complexity of the regulation of growth and metabolism during postnatal life by the somatotropic axis using four main examples: developmental changes, effects of nutrition, GH and IGF-I treatment, and altered GH action in conditions of metabolic imbalance.

## Regulation of postnatal growth by the somatotropic axis

### *Circulating growth hormone*

Growth hormones and lactogenic hormones form a family of biochemically related hormones; they are single-chain polypeptides with two or three intrachain disulphide bridges and about 200 residues. There is considerable variation between species in the structure of GH, prolactin (PRL) and placental lactogen (PL), which in some cases can be correlated with significant variation in biological activity. Regulation of GH synthesis and secretion from the pituitary primarily reflects interactions between GH releasing factor (GRF) and inhibition by somatostatin (SRIF). GH secretion is initiated firstly by a reduction in the secretion of SRIF and an increase in the release of GRF from the hypothalamus into the hypophyseal portal system. Subsequently the pituitary enters a period of GRF-reduced responsiveness which generates the characteristic pulsatile pattern of GH secretion. The pattern of GH secretion is sexually dimorphic, with males exhibiting large pulses and low basal levels while females exhibit small pulses and high baseline levels (Gatford *et al.*, 1998). The secretion of GH is also regulated by a series of feedback mechanisms. For example, GH directly feeds back to inhibit its own secretion, while IGF-I inhibits GH gene transcription and the stimulation of GH release from the pituitary by GRF.

Circulating GH is a major regulator of postnatal growth. Subjects with acromegaly (excessive pituitary GH secretion) exhibit bone overgrowth in the form of increased height and enlarged hands and feet resulting from the stimulatory effects of GH on osteoblast activity and bone mineral deposition. GH deficiency is associated with a reduction in growth rate and in final adult size. In the adult animal circulating GH regulates diverse metabolic pathways including protein, fat and carbohydrate metabolism.

### *Growth hormone receptors*

The GH receptor (GHR) is a member of the class 1 cytokine receptor family. GHR mRNA and immunoreactivity have been detected in most tissues postnatally, with highest abundance in liver and adipose tissue (Bingham *et al.*, 1994). The GH molecule has two binding sites for the extracellular domain of the GHR and forms a 1:2 GH:GHR complex by sequential homodimerization. Ilondo *et al.* (1994) demonstrated in adipocytes that the receptor dimerization process is critical for GH-stimulated glucose incorporation into lipid. GH binding to the GHR activates a number of intracellular pathways. Transcription of the IGF-I gene is rapidly increased soon after GH administration, suggesting that it is an immediate consequence of GHR signalling. The GHR may also interact with cell membrane associated G proteins to regulate the metabolic effects of GH (Roupas and Herington, 1994).

The multiple physiological and metabolic actions of GH could partly be explained by heterogeneity of the GHR. Heterogeneity of the GHR has been demonstrated by ligand binding studies, by affinity cross-linking, by epitope mapping with monoclonal antibodies and by molecular studies. The soluble growth hormone binding protein (GHBP) can also be viewed as a variant form of the GHR. The GHBP represents the extracellular domain of the GHR, although its derivation appears to be species-specific. In rodents, a separate alternatively spliced mRNA exists for GHBP, conferring on the protein a unique hydrophilic 17-residue. In some species plasma GHBP arises from proteolysis of the full-length GHR protein (Baumann, 1994), while in humans, sheep and chickens the GHBP is thought to be mainly produced by proteolytic cleavage of the GHR extracellular domain (Bingham *et al.*, 1994). The GHBP in ruminant plasma has been investigated by Davis *et al.* (1992) using a charcoal separation assay, demonstrating the presence of GHBP in cattle and sheep. Devolder *et al.* (1993) provided further evidence for the existence of GHBP in cattle using size-exclusion chromatography and ligand and Western blotting techniques. Three major specific bands that bind GH were demonstrated; two of these bands (190 and 58 kDa) were recognized by a monoclonal antibody directed against a part of the extracellular domain of the bovine growth hormone receptor (bGHR). The function of GHBP is not clear at present, however, it is thought to prolong the half-life of GH in circulation and may play a role in tissue-specific delivery. Since the GHBP is regulated by a number of physiological variables, it is possible that the GHBP may represent an additional mechanism that regulates GH action.

### Insulin-like growth factors

The IGFs are a family of polypeptides related structurally and evolutionarily to proinsulin. The classical somatomedin hypothesis, which postulated that the effects of GH are mediated by circulating IGFs of hepatic origin (endocrine action), has been modified. It is now generally accepted that IGFs are synthesized at multiple sites and the concept of local action of GH in many tissues to promote both cell differentiation and proliferation through local production of IGF-I acting in an autocrine or paracrine manner has become firmly established (Isaksson *et al.*, 1991). Although feedback mechanisms between the two systems exist, the relative importance of each mechanism is uncertain. The significance of endocrine IGF-I is supported by the observation that the selection of mice (Blair *et al.*, 1988) or sheep (Sherlock *et al.*, 1993) on the basis of high or low plasma concentrations of IGF-I results in a marked divergence of their growth rate. While IGF-I treatment increases growth in the GH-deficient state, growth rates rarely equal those obtained with GH, which may reflect that some of the effects of

GH treatment on tissue growth are derived from the direct stimulation of autocrine or paracrine IGF-I production. As will be discussed later, the endocrine and the paracrine or autocrine mode of action are interrelated and contribute to a coordinated adaptive process.

An interesting concept in the regulation of IGF-I action relates to the IGF-I gene itself. The IGF-I gene consists of at least six exons dispersed over a region of more than 80 kb (Rotwein, 1991). Exons 3 and 4 encode the mature IGF-I peptide. The synthesis of IGF-I is a complex process involving two promoters, multiple sites of transcription initiation, alternate splicing and multiple polyadenylation sites (Adamo *et al.*, 1993). The observation that exons 1 and 2 are differentially spliced to exon 3, producing alternate class 1 and class 2 transcripts, has stimulated discussion about endocrine versus paracrine or autocrine functions of the IGF-I peptide. Changes in abundance of hepatic class 2 transcripts have been associated with concomitant changes in circulating (endocrine) IGF-I. For example, Pell *et al.* (1993) suggest that in sheep, hepatic IGF-I output may be regulated by nutrition and GH, primarily through class 2 transcripts. Thus, in conditions of optimal growth, the increase in circulating IGF-I correlates with an increased abundance of exon 2 transcripts in liver IGF-I mRNA. However, in peripheral tissues like muscle, exon 1 IGF-I transcripts predominate and do not respond markedly to a range of metabolic and endocrine stimuli (Gilmour, 1994). The switching of class 1 and class 2 transcripts in different tissues, depending on the endocrine and metabolic conditions of the animal, could represent an adaptive process to accommodate the changing needs of different tissues.

While the growth promoting properties of IGF-I are well established, the function of IGF-II in ruminants is not clear, although there is some evidence that IGF-II can impair the anabolic effects of IGF-I in sheep (Koea *et al.*, 1992a). It is known that at the cellular level, IGF-II has proliferative actions, enhancing proliferation of ovarian granulosa cells and stimulating wound repair processes such as keratinocyte proliferation. A proliferative role for IGF-II is also evident in many cancerous tissues (Isaksson *et al.*, 1991).

## *IGF binding proteins and receptors*

The IGF binding proteins (IGFBPs) are a family of peptides which exhibit high affinity for the IGFs. Six high affinity IGFBPs (IGFBP-1 to -6) have been identified in blood plasma to date. There is strong sequence homology at their N and C termini, especially with respect to the presence of 16 Cys residues, indicating the likely importance of these regions in IGF binding. There is considerably more sequence heterogeneity between the intermediate portions of IGFBPs suggesting that this region may define many of the variations in activity between the binding proteins. The IGFBPs vary considerably in size, their affinity for IGFs and other molecules, phosphorylation and glycosylation states, and protease sensitivity. The regulation and function is specific to each of the IGFBPs. Each of the IGFBPs could have a unique role as a modulator of IGF action and the different IGFBPs appear to be under independent endocrine and metabolic regulation. Changes in the production of IGFBPs may initiate tissue-specific targeting and distribution of IGFs. In addition, either systemic or locally produced

IGFBPs may alter the distribution of circulating IGFs within target tissues by modifying receptor–IGF interactions, thus enhancing or reducing biological activity.

*IGFBP-1*

The circulating IGFBP-1 peptide is approximately 26 kDa in size and has a similar affinity for IGF-I and IGF-II. IGFBP-1 is under strong metabolic control and is produced mainly by hepatocytes, although some expression is also found in the kidney and uterine decidua. Serum levels of IGFBP-1 are high in the fetus but decline postnatally. Plasma levels of IGFBP-1 are inversely proportional to those of insulin (Lee *et al.*, 1993) and rapidly increase in response to treatment with stress-related hormones such as glucocorticoids, glucagon and catecholamines (Hooper *et al.*, 1994). The regulation of IGF-I availability by IGFBP-1 plays an acute glucose counter-regulatory function, restricting the insulin-like actions of the IGFs during states of a relative insulin deficiency (Baxter, 1995).

*IGFBP-2*

The IGFBP-2 protein is approximately 32 kDa in size and has two distinct but overlapping IGF binding sites, of which the IGF-II site exhibits a fivefold greater affinity for its ligand than the IGF-I site. Circulating IGFBP-2 is mainly produced by hepatocytes (Scharf *et al.*, 1995) but it is also present at high concentrations in milk and seminal plasma. Plasma levels of IGFBP-2 in the postnatal circulation vary in response to many factors including ontogeny, nutritional status and GH status. Despite a striking inverse relationship between plasma concentrations of GH and IGFBP-2, the regulation of IGFBP-2 is unlikely to occur directly via GH action (Gallaher *et al.*, 1995). The role of plasma IGFBP-2 in the functional regulation of circulating IGFs is unclear. IGFBP-2 levels are high in catabolic states where IGFBP-3 is suppressed and it may restrict IGF-I availability in such situations. However, since the sensitivity of IGFBP-2 gene transcription to insulin is much lower compared with that of IGFBP-1, IGFBP-2 is more likely be a chronic, as opposed to an acute, regulator of IGF action.

*IGFBP-3*

IGFBP-3 is expressed in a wide range of tissues and undergoes extensive post-translational modifications. IGFBP-3 is secreted in a phosphorylated state but the effect on IGF affinity is minimal. However, since charged regions on IGFBP-3 are important in ternary complex formation in plasma and cell surface binding, phosphorylation has been implicated in serum half-life or tissue distribution (Coverly and Baxter, 1995). The predominant IGF carrier in the postnatal circulation is a 150 kDa ternary complex consisting of IGF-I or IGF-II, IGFBP-3 and an acid labile subunit (ALS). Circulating IGFBP-3 is predominantly derived from the non-parenchymal cells of the liver (Chin *et al.*, 1994) while ALS and endocrine IGF-I are produced by adjacent hepatocytes. The ternary complex is formed firstly by the association of IGF-I or -II with IGFBP-3, followed by the binding of ALS to the IGF–IGFBP-3 dimer. When IGFs are incorporated in the ternary complex they are considerably more stable than the free peptide. The binary complex IGFBP-3/IGF-I has a much shorter circulating half-life and rapidly crosses into the extravascular compartment where it is targeted to specific tissues (Arany *et al.*, 1993). Plasma IGFBP-3 correlates with GH and IGF-I, and is a chronic indicator of GH-dependent growth status. However, this relationship is indirect and reflects the GH dependency of ALS and IGF-I production.

### *IGFBP-4, -5 and -6*

The mature IGFBP-4 peptide is 24 kDa in size and is *n*-glycosylated to create a 28 kDa form. The function of IGFBP-4 glycosylation is unclear although one role may be in the regulation of vascular endothelial transport. Unlike other IGFBPs, IGFBP-4 persistently inhibits IGF action. For example, IGFBP-4 is elevated and IGF-I activity is suppressed in atretic ovarian follicles but the opposite is observed in dominant follicles (Giudice *et al.*, 1995). The IGFBP-5 gene is expressed in a tissue-specific fashion and the mature peptide is approximately 29 kDa in size, with phosphorylation and glycosylation variants (Jones and Clemmons, 1995). IGFBP-5 generally enhances the effects of IGFs on cell proliferation. IGFBP-6 mRNA is widely expressed and yields a mature peptide that is approximately 22 kDa in size. It has a strong preference for binding IGF-II over IGF-I. IGFBP-6 has been found in plasma, amniotic fluid, follicular fluid and cerebrospinal fluid (CSF) in sheep and humans. Levels in serum decline slightly from mid to late gestation but are stable after birth (Lewitt *et al.*, 1995).

### *IGF receptors*

The classical IGF-I receptor (IGF-IR) is structurally homologous to the insulin receptor, forming a heterotetramer. The α subunit, 90 kDa in size, is involved in ligand binding and forms the extracellular portion of the receptor. The affinity of the IGF-IR is highest for IGF-I, twofold lower for IGF-II and 100-fold reduced for insulin. Following the binding of IGF-I to the IGF-IR, there is an activation of the receptor tyrosine kinase activity which results in the phosphorylation of insulin receptor substrate I (IRS-I). The mitogenic actions of IGF-I appear to be exclusively mediated through the IGF-IR.

In cells where both IGF-I and insulin receptor subunits are both synthesized, hybrid receptors (IGF-HR) can form by the association of an (αβ) subunit from each of the IGF-IR and insulin receptors to create a heterotetramer. IGF-I is bound to these receptors with an affinity similar to that of the IGF-IR, and IGF-II binds with fourfold lower affinity. Insulin binds with a very low affinity to the IGF-HR relative to IGF-I and also relative to insulin binding to its own receptor (Soos *et al.*, 1994). While the physiological significance of the IGF-HR is unclear, the sequestration of IR subunits into less insulin-responsive hybrids might be a mechanism to modulate insulin sensitivity in tissues.

The IGF-II receptor (IGF-II/M6PR) exclusively binds IGF-II with high affinity and is identical to the cation-independent mannose-6-phosphate receptor, consisting of a short intracellular/membrane spanning domain and a large extracellular domain made up of multiple structural repeats (Kiess *et al.*, 1994). In rat tissues, IGF-II/M6PR immunoreactivity and mRNA abundance is high in the fetus but very low after birth, suggesting a minimal role in postnatal development and an active role in fetal growth and/or development. Also deduced from IGF gene knockout experiments (Baker *et al.*, 1993) is the existence of a further IGF-II receptor (IGF-XR) which is expected to mediate a number of the effects of IGF-II on fetal and placental development.

## Developmental changes within the somatotropic axis

There are crucial differences in the mechanisms that regulate growth and development before and after birth. The regulation of fetal growth depends on maternal phenotype and the stage of gestation. Fetal growth in late gestation is limited by maternal constraint. The primary elements ensuring adequate fetal growth are the ability of the mother to provide adequate nutritional support to the placenta and the fetus, and the ability of the fetus to manage the maternal resources appropriately. One of the key roles of endocrine factors during fetal growth and development therefore is provided by hormones which influence the partitioning of nutrients between mother, placenta and fetus, and which regulate the ability of the fetus to utilize the available substrate. Postnatal growth is predominantly determined by genotype and by nutrition. Animals of different genotype grow within predictable boundaries and the growth process is under central endocrine control of pituitary GH.

### Perinatal transition of the somatotropic axis from a fetal to a postnatal state

Fetal growth and development is rapidly adapted to conditions of maternal and/or placental substrate limitation. After birth the neonate does not need this acute link to the mother and the developmental switch to a mature somatotropic axis allows growth to be directly linked to environmental factors. In sheep, normal parturition is triggered by a rise in fetal cortisol which begins approximately 10–15 days before term and increases markedly in the last 3–5 days before delivery (Fowden *et al.*, 1996). The essential role of the rise in fetal glucocorticoids on lung maturation are well known. However, the maturational effects of the glucocorticoids may have a far more general role in switching the fetal physiological state to a postnatal state. Glucocorticoids play a major role in the induction of hepatic gluconeogenic enzymes in the perinatal period, since gluconeogenesis is a major metabolic pathway for neonatal adaptation. Glucocorticoids also have important effects on cardiac development and a variety of endocrine systems.

Major changes occur during the immediate prepartum and neonatal period at many levels of the somatotropic axis. Plasma concentrations of GH are reduced gradually during the last weeks before birth (Breier and Gluckman, 1991) and there is a major decrease in GH secretion at the time when the final cortisol surge is observed prior to parturition. The expression of GHR mRNA and GHR receptor number are very low during mid- or late gestation (Klempt *et al.*, 1993), particularly in the liver, until after birth. As described above, there is a marked increase in plasma IGF-I and a marked decline in plasma IGF-II after birth, and the respective hepatic mRNA levels are also altered accordingly (Li *et al.*, 1993). Plasma concentrations of ovine IGFBP-1 and IGFBP-2 and the circulating M6P/IGF-IIR during the perinatal and neonatal period markedly decrease while IGFBP-3 concentrations increase markedly after birth (Gallaher *et al.*, 1994). Similarly, IGF-IR concentrations in skeletal muscle decline markedly after birth. The mechanisms of these changes within the somatotropic axis are not well understood at present. However, the perinatal increase in hepatic GHR and the rise in IGF-I are not activated if the cortisol rise is abolished by lesioning of the paraventricular nucleus (Breier *et al.*, 1994) or by adrenalectomy (Li *et al.*, 1996). These studies suggest that the prepartum cortisol surge plays an important role in initiating the perinatal switch of the somatotropic axis from the fetal to the postnatal state and function generally observed during postnatal life.

### Postnatal ontogeny of the somatotrophic axis

The ontogeny of GH secretion in the perinatal period has been most extensively studied in the sheep. In the fetal lamb, GH concentrations are markedly elevated compared with postnatal concentrations (Breier and Gluckman, 1991). Within 24 h of birth, plasma GH concentrations fall tenfold with both a reduction in pulse amplitude and interpulse values to a secretory pattern not markedly different from the adult. However, sexually differentiated patterns of GH secretion are documented in a number of species, with a higher pulse amplitude and lower baseline values in the male than in the female. Recent work by Gatford *et al.* (1998) suggests that GH secretion in ram lambs shows markedly higher GH pulse amplitudes than in ewe lambs, and that this sexually dimorphic pattern of GH secretion may be regulated by differences in portal GRF secretion. It is well established that gonadal steroids and anabolic compounds enhance GH release in ruminants through a direct stimulatory effect on GH synthesis and release at the pituitary level. The increase in baseline concentrations of GH in oestradiol-treated steers independent of nutritional plane (Breier *et al.*, 1988a) is compatible with such a conclusion.

In most species there is a gradual increase in hepatic GHR expression during the first weeks or months after birth (Breier *et al.*, 1988b). This rise in hepatic GHR levels correlates with an increase in plasma concentrations of IGF-I and with the postnatal growth curve. The coordinated increase in GHR concentrations and plasma IGF-I suggests that GHR activity may be a sensitive indicator of increased tissue responsiveness to GH after birth (Breier *et al.*, 1994). A recent study by Min *et al.* (1999) investigated the endocrine responsiveness of sheep at different ages, ranging from 7 days after birth until one year of age. While the number of GHRs in the liver did not differ in the basal state, GH treatment increased the hepatic GHR number only after puberty. Similarly, the IGF-I response to GH was markedly higher in yearling sheep in comparison with prepubertal animals (Min *et al.*, 1999).

In ruminants, as in other species, plasma levels of IGF-I are low at birth and show a postnatal rise which is concomitant with the major increase of GHR in the liver (Breier *et al.*, 1994). A pubertal rise of plasma IGF-I is well documented and this rise is related to sex steroid production. However, in lambs sex differences in plasma IGF-I already exist before weaning with ram lambs showing twofold higher IGF-I levels than ewe lambs (Gatford *et al.*, 1998) and this sexual dimorphism is maintained after puberty.

The abundance of IGF-II mRNAs is high before birth and declines postnatally although this developmental pattern of regulation is subject to tissue specificity (Delhanty and Han, 1993). An ontogenic reduction is seen in fetal plasma levels of IGF-II. However, plasma IGF-II still remains higher than IGF-I postnatally in sheep. While IGF-II is integral to adequate prenatal growth in the mouse (Baker *et al.*, 1993), researchers have struggled to show a correlation between IGF-II and birth weight, and its role in postnatal growth is not clearly defined, being stimulatory in some studies but not in others. The ability of IGF-II to affect glucose clearance is significantly less than IGF-I (Douglas *et al.*, 1991) and unlike IGF-I, IGF-II does not ameliorate protein catabolism in lambs, but rather acts as an IGF-I antagonist.

Although there is some evidence that plasma IGFBP-1 concentrations are high in the fetus and decline postnatally (Liu *et al.*, 1991), there is little data on IGFBP-1 in ruminants. Given that IGFBP-1 is under strong metabolic control and that plasma levels of IGFBP-1 are inversely proportional to those of insulin, there is considerable scope for further research in this area. IGFBP-2 is the most abundant IGFBP prior to birth and, although plasma concentrations fall after birth, it is the second most abundant carrier protein in adult sheep plasma (Gallaher *et al.*, 1995). Recent work in sheep confirms data from other species which shows that plasma concentrations IGFBP-2 are high during fetal life, they fall gradually after birth and plateau after one year of age. Levels of IGFBP-3 observed in fetal sheep plasma are significantly lower than those in the adult (Gallaher *et al.*, 1998) and during postnatal life there is a distinct ontogenic increase in plasma concentrations of IGFBP-3 in sheep, similar to the ontogenic pattern observed for IGF-I (Gatford *et al.*, 1998). The marked ontogenic changes in IGF-IR concentrations serve as additional evidence for a key role of IGF-I during fetal development. There is a rapid decline of IGF-IR levels in skeletal muscle; the concentrations in fetal muscle are about tenfold higher than in postnatal muscle (Boge *et al.*, 1995).

## Influence of nutrition on the somatotropic axis

It is now widely accepted that nutritional status plays a major role in determining circulating GH concentrations. Plasma levels of GH are elevated as a result of undernutrition in ruminants, reflecting the influence of such factors as hypoglycaemia, stress and low serum free fatty acids on the pituitary secretion of GH (Breier and Gluckman, 1991). In sheep and cattle the GH response to GRF is enhanced during restricted feeding compared with *ad libitum* feeding. This suggests an explanation for the increase in GH secretion during reduced nutritional intakes, which is related to diminished negative feedback within the somatotropic axis mediated in part via a decrease of hypothalamic SRIF (Thomas *et al.*, 1991). Undernutrition is also associated with reduced hepatic GH binding capacity in lambs (Sauerwein *et al.*, 1991) and steers (Breier *et al.*, 1988c) leading to a state of relative GH resistance which may reflect an inhibition of GHR dimerization or altered signal transduction.

Plasma concentrations of IGF-I are clearly dependent on nutritional status. The ability of GH to increase plasma IGF-I is impaired under conditions of reduced nutritional intake when the number of GHRs is reduced (Breier *et al.*, 1988a). Basal concentrations of IGF-I and the response to GH are reduced during periods of insufficient protein and/or energy intake. Dietary protein supply seems to be the limiting factor for a maximal stimulation of IGF-I plasma concentrations by GH (Elsasser *et al.*, 1989). The relative GH resistance in poorly fed animals and consequently the inverse changes in plasma GH and IGF-I concentrations may represent a mechanism for preferential utilization of mobilized substrates to maintain homeostasis rather than cell growth and proliferation.

While transcription of all hepatic IGF-I mRNA variants is reduced by undernutrition in the sheep, the effect is greatest on the GH sensitive (endocrine IGF-I) exon 2 transcripts (Pell *et al.*, 1993). In the rat, fasting is associated with a widespread reduction in hepatic and most extra-hepatic IGF-I mRNA expression, although brain and heart expression is minimally affected (Olchovsky *et al.*, 1993). Preservation of IGF-I synthesis in these tissues may represent the mechanism by which critical tissues and organs are protected during catabolic states. Conversely, IGF-II is slightly increased following undernutrition in adult sheep (Hua *et al.*, 1995). This observation together with the demonstration that co-administration of IGF-II blocks the reduction in protein catabolism induced by IGF-I treatment in the fasted lamb (Koea *et al.*, 1992b) invites the hypothesis that the increase in the plasma IGF-II/IGF-I ratio during undernutrition may exacerbate the development of a catabolic state.

The effect of undernutrition on the IGFBPs is varied. Consistent with its glucose counter-regulatory function, IGFBP-1 gene transcription and serum concentrations are significantly and rapidly elevated by reduced food intake. Nutritional restriction generally results in an increase in plasma IGFBP-2 which occurs via an increase in transcription rate. This increase has been specifically linked to a reduction in protein intake (Smith *et al.*, 1995). However, there is still some controversy about the regulation of IGFBP-1 and IGFBP-2 in farm animals. Fasting of newborn pigs results in an increase in plasma levels of IGFBP-1 but a decrease in IGFBP-2 levels (McCusker *et al.*, 1991). In response to 72 h starvation in adult sheep, IGFBP-1 is not significantly changed, while IGFBP-2 increases (Gallaher *et al.*, 1992). A clear inverse relationship has been demonstrated recently in sheep between plasma insulin and glucose concentrations and plasma concentrations of IGFBP-1 and IGFBP-2, suggesting that both are inversely regulated by plasma insulin (Gallaher *et al.*, 1995). While the changes in plasma IGFBP-1 and -2 following undernutrition appear to restrict the insulin-like activities of the IGFs during catabolic states, the changes in IGFBP-3 and -4 may reflect an attempt to maximize the availability of the remaining IGFs to the tissues. Circulating levels of IGFBP-3 and the ternary complex (IGF storage capacity) are reduced (Gallaher *et al.*, 1992) while specific IGFBP-3 protease activity to reduce IGF affinity for IGFBP-3 is enhanced. In conditions of reduced nutritional intake, IGFBP-4, an inhibitor of IGF action, is decreased both at the serum and transcriptional level (Holt *et al.*, 1996).

## Effects of GH and IGF-I on protein and energy metabolism

The endocrine and metabolic effects of GH treatment are well established. Exogenous GH treatment induces re-partitioning of nutrients leading to increased lipolysis, increased protein synthesis, decreased protein degradation and increased bone growth. While it is established that GH has a dominant influence on IGF-I synthesis, the precise mode of action of GH treatment is complex. In the lamb, chronic treatment with bGH led to a dose-dependent increase in the number of high-affinity hepatic GHRs, plasma concentrations of IGF-I and growth rate, while carcass protein and fat content showed positive and negative dose-response relationships respectively (Sauerwein *et al.*, 1991). Body weight gain and protein content of the carcass showed significant positive correlations with plasma IGF-I. Treatment with GH can even improve protein metabolism during an acute catabolic state (simple 72 h starvation) (Ogawa *et al.*, 1996). Beginning GH therapy before or during the onset of starvation has comparable effects on the amelioration of protein breakdown. A number of components of the somatotropic axis are influenced by GH treatment, which contribute to the coordination of nutrient partitioning between different tissues. The up-regulation of the GHR and stimulation of post-receptor mechanisms may lead to increased synthesis of IGF-I and a rise in its plasma concentrations. Since plasma IGFBP-3 is increased while IGFBP-2 is decreased with GH treatment, this change in relative concentrations of IGFBPs may facilitate an increased availability of IGF-I at target tissues. Treatment with bGH of lambs significantly reduces plasma concentrations of IGFBP-2 while nutritional restriction increases plasma concentrations of IGFBP-2. Since these changes observed in plasma concentrations of IGFBP-2 show a negative correlation with changes in plasma IGF-I, it has been suggested that IGF-I itself may be an important regulating factor for the availability of IGFs (Gallaher *et al.*, 1995).

There is increasing evidence for ontogenic changes in responsiveness to GH treatment at identical doses. The treatment with GH (bGH at 0.3 mg $kg^{-1}$ $day^{-1}$) of well-fed healthy adult ewes leads to a marked reduction in protein breakdown and elevated circulating concentrations of IGF-I. In addition, a marked diabetogenic effect on plasma glucose and insulin is observed, and a major rise in plasma FFA levels shows significant lipolytic action in the healthy adult state (Hennies *et al.*, 1998). While the lipolytic action of GH is well recognized, it is minimal in young animals or in conditions of enhanced metabolic demand such as lactation, where high levels of energy expenditure and utilization prevent a major rise in plasma FFA. There is no increase in plasma FFA in young pre-pubertal lambs with GH (bGH at 0.3 mg $kg^{-1}$ $day^{-1}$). In addition, these lambs showed a comparatively modest increase in plasma IGF-I and a very mild diabetogenic response with only slightly elevated plasma glucose and insulin levels during the well-fed phase of the study (Ogawa *et al.*, 1996). This comparison suggests that the same dose of GH leads to somewhat different actions depending on the age of the animal.

Consistent with its growth promoting activity, IGF-I administration stimulates protein metabolism and it reduces protein catabolism in the lamb, while passive immunization against IGF-I in lambs elevates net protein catabolism (Koea *et al.*, 1992b). Treatment with IGF-I also has insulin-like effects on glucose metabolism. IGF-I reduces plasma glucose as a result of both increased glucose uptake and reduced glucose production. However, in the sheep such data reflect a pharmacological response since at physiological doses, IGF-I suppresses plasma insulin but increases glucose concentrations. Only as the dose increases does IGF-I saturate the IGFBPs, 'free IGF-I' levels are increased and a direct effect on glucose metabolism is observed (Douglas *et al.*, 1991). There is general agreement that elevation of plasma concentrations of IGF-I by treatment with high doses of IGF-I shows characteristic insulin-like effects inducing hypoglycaemia. This effect depends on saturation of the IGFBPs in circulation leading to a marked increase in 'free IGF-I' which binds to the insulin receptor.

Two different experimental paradigms were developed to investigate the metabolic effects of short-term IGF-I infusion (5 h infusion) in the 48-h fasted lamb and long-term effects of 8-hourly IGF-I injections for 8 weeks in well-fed lambs. In short-term treatment studies, young lambs were infused with IGF-I for 5 h after a 48-h period of starvation (Douglas *et al.*, 1991). Two doses of IGF-I were used: a low dose which caused no elevation of 'free IGF-I' in the circulation, and a higher dose which caused a sevenfold elevation of 'free IGF-I'. The lower dose did not alter plasma concentrations of glucose, or the rate of glucose production or clearance, while the higher dose caused mild hypoglycaemia. When this higher dose of IGF-I and a dose of insulin of equivalent hypoglycaemic potential were compared, only IGF-I increased protein synthesis in skeletal muscle, heart and liver. Both IGF-I infusions at the low and at the high dose markedly reduced protein degradation and this effect may be mediated via the IGF-I receptor (Douglas *et al.*, 1991). Thus, short-term IGF-I infusion in the lamb can both reduce protein degradation and increase protein synthesis in a number of different tissues.

In long-term studies, well-fed yearling sheep were treated for 8 weeks with IGF-I, which increased plasma concentrations of IGF-I by approximately 40–60% (Cottam *et al.*, 1992). There was no increase in body weight or food intake. The weight of the spleen increased by 40%, but there was no effect of IGF-I on carcass weight, composition and dimensions, or on long bone length, although the weight per unit length of the tibia was increased. The somatogenic effect of long-term IGF-I treatment was minimal, suggesting that in well-fed animals with an intact somatotropic axis, IGF-I treatment at doses which increase plasma IGF-I within the physiological range do not enhance somatic growth performance (Cottam *et al.*, 1992). Exogenous IGF-I may suppress tissue IGF-I production by interactive feedback between the systemic and the local IGF-I system. Of particular note in the long-term IGF-I treatment study were the effects on plasma glucose and insulin (Cottam *et al.*, 1992). There was a marked and rapid decline in circulating levels of insulin. Decreased plasma insulin was associated with, and may have been responsible for, a significant increase in blood glucose levels in IGF-I-treated sheep. It is possible that insulinopenia reduces the growth response by restricting substrate uptake. Thus, the insulinopenic effect of long-term IGF-I treatment contrasts with the effect of GH, which induces hyperinsulinaemia. The lack of a somatogenic effect with long-term IGF-I treatment may also be related to changes in IGFBPs. While bGH treatment of sheep significantly increases plasma IGFBP-3 and the ternary 150 kDa complex, IGF-I treatment of sheep was unable to induce such an increase. Since plasma IGFBP-1 was increased and IGFBP-2 levels were suppressed with IGF-I treatment, the change in the relative concentration of the different IGFBPs may have affected the availability of injected IGF-I to target tissues. An independent but related study by Min *et al.* (1996) showed that long-term administration of IGF-I in young energy-restricted sheep lead to negative feed-back of circulating IGF-I on GH secretion and the GHR number in the liver. Activation of such negative feed-back loops by IGF-I treatment explains the lack of effects on growth rate, metabolic parameters or body composition in endocrinologically normal animals.

## Role of the somatotropic axis in conditions of metabolic imbalance

The somatotropic axis plays an important role in maintaining homeostasis under a variety of stress conditions including malnutrition, starvation, microbial infection and septic shock. During critical illness body protein and energy reserves are used at a greatly accelerated rate which leads to severe muscle wasting, impaired wound healing, reduced cardiopulmonary performance and diminished immune function. Both GH and IGF-I treatment have a general protein sparing effect during mild or short-term nutritional restriction in humans and sheep, decreasing protein catabolism and increasing cardiac and skeletal muscle protein synthesis. In humans, GH and IGF-I co-treatment appears to be more beneficial since increased IGFBP-3 ternary complex formation induced by GH reduces the levels of 'free IGF-I' and subsequent hypoglycaemia associated with IGF-I treatment alone (Kupfer *et al.*, 1993). However, the ability of GH to ameliorate the effects of severe catabolism is much reduced due to the induction of GH resistance in these patients (Roth *et al.*, 1995). It has been argued that in cases such as elective surgery GH pre-treatment prior to the onset of GH resistance could minimize post-operative catabolic stress. Such an approach has been used successfully in the fasted sheep (Ogawa *et al.*, 1996). However, the interactions between the intermediary metabolism, the immune system and the somatotropic axis during infections and septic shock are even more complex. A number of studies have employed endotoxin-induced septic shock to investigate the endocrine interactions under pathophysiological conditions similar to microbial infections.

Endotoxin is a complex lipopolysaccharide molecule situated within the outer membrane of Gram-negative bacteria; it is the main mediator of septic shock (Morrison and Ryan, 1987). Endotoxin administration stimulates the immune system and modulates the secretion of several mediators and hormones. Endotoxin induces a rapid increase in plasma concentrations of tumour necrosis factor, interleukin-1, and interleukin-6. Its action is mediated, at least in part, by an endotoxin receptor, which has been identified in brain, pituitary, monocytes and plasma (Holst *et al.*, 1996). Changes in several hormonal axes have also been observed after endotoxin injection; there is an increase in activity of the hypothalamic–pituitary–adrenal axis (Dadoun *et al.*, 1998) and a decrease in thyroid-stimulating hormone and alterations in the gonadotropic axis (Battaglia *et al.*, 1997). Interactions between the immune and endocrine systems are multi-directional and the hormonal changes induced by endotoxin or cytokines can act on the immune system. The pathophysiological responses to endotoxin injection include increased heart rate, increased respiratory rate, changes in lung permeability, activation of the complement cascade, hypoglycaemia, insulinaemia, reduction in plasma concentrations of FFA and hypocalcaemia (Kinsbergen *et al.*, 1994). Septic shock is also associated with metabolic abnormalities, cardiovascular dysfunction, and multiple organ failure mainly related to ischaemia in part due to alterations in regional microcirculatory blood flow (Kinsbergen *et al.*, 1994).

Endotoxin administration induces species-specific effects on GH secretion; it increases GH secretion in sheep but decreases it in rats and cattle (Coleman *et al.*, 1993). In a recent study, Hennies *et al.* (1998) investigated the effects of GH therapy in well-nourished sheep exposed to endotoxin-induced septic shock. Endotoxin injections alone resulted in a modest fall in plasma glucose and no change in plasma insulin concentrations. However, the metabolic responses to bGH (0.3 mg $kg^{-1}$ $day^{-1}$) were markedly altered during endotoxin-induced metabolic stress, leading to a dissociation of the known effects of GH on intermediary metabolic pathways. The anticatabolic effect of GH treatment was abolished despite continuation of considerably elevated levels of plasma IGF-I. Furthermore, the GH-treated sheep showed a decline in plasma insulin and glucose levels within 48 h after endotoxin challenge. This suggests that, after observing an initial phase of diabetogenic effects of GH, endotoxin-induced septic shock altered glucose homeostasis, despite continuation of the marked lipolytic effect of GH. Studies in rats have shown that GH-treatment amplifies the adverse effects of endotoxin-induced septic shock. Rats primed with a GH infusion developed more severe hypoglycaemia, hyperlipidaemia and renal and hepatic dysfunction than endotoxin-treated control animals (Liao *et al.*, 1996). These effects were not observed when IGF-I was administered to rats given an endotoxin challenge, implying that the enhanced toxicity of endotoxin was a direct effect of GH administration. Whether the major change in GH action on energy metabolism after endotoxin-induced septic shock in the study of Hennies *et al.* (1998) reflects changes in glucose up-take and utilization or whether the reduction in plasma glucose levels reflect decreased gluconeogenesis, remains to be investigated.

Whilst few studies in different species have shown that GH treatment during conditions of metabolic imbalance may lead to changes in GH action, a recent study by Takala *et al.* (1999) reports a marked increase in mortality and morbidity among critically ill patients treated with GH. In this study a total of 532 critically ill patients were given GH (at a mean dose of 0.1 mg $kg^{-1}$ body weight $day^{-1}$) or placebo to investigate whether GH therapy can attenuate the catabolic response to injury, major surgery and sepsis. However, the mortality rate in patients receiving GH treatment was two times that in the patients receiving placebo. While the underlying mechanisms of the increased mortality rate in GH-treated critically ill patients are not understood, most of the deaths were attributed to multiple-organ failure and septic shock or uncontrolled infection.

## Acknowledgements

The authors acknowledge support from the Health Research Council of New Zealand.

## References

Adamo, M.L., Lanau, F., Neuenschwander, S., Werner, H., LeRoith, D. and Roberts, C.T. (1993) Distinct promoters in the rat insulin-like growth factor-I (IGF-I) gene are active in CHO cells. *Endocrinology* 132, 935–937.

Arany, E., Zabel, P., Freeman, D. and Hill, D.J. (1993) Elimination of radiolabelled recombinant human insulin-like growth factor binding protein-3 from the circulation, and its distribution amongst organs and tissues in adult male rats. *Regulatory Peptides* 48, 133–143.

Baker, J., Liu, J.-P., Robertson, E.J. and Efstratiadis, A. (1993) Role of insulin-like growth factors in embryonic and postnatal growth. *Cell* 75, 73–82.

Battaglia, D.F., Bowen, J.M., Krasa, H.B., Thrun, La., Viguie, C. and Karsch, F.J. (1997) Endotoxin inhibits the reproductive neuroendocrine axis while stimulating adrenal steroids: a simultaneous view from hypophyseal portal and peripheral blood. *Endocrinology* 138, 4273–4281.

Baumann, G. (1994) Growth hormone-binding proteins: state of the art. *Journal of Endocrinology* 141, 1–6.

Baxter, R.C. (1995) Insulin-like growth factor binding proteins as glucoregulators. *Metabolism* 44 (Suppl. 4), 12–17.

Bingham, B., Oldham, E.R. and Baumbach, W.R. (1994) Regulation of growth hormone receptor and binding protein expression in domestic species. *Proceedings of the Society for Experimental Biology and Medicine* 206, 195–199.

Blair, H.T., McCutcheon, S.N., Mackenzie, D.D.S., Ormsby, E., Siddiqui, R.A., Breier, B.H. and Gluckman, P.D. (1988) Genetic selection for insulin-like growth factor-I in growing mice is associated with altered growth. *Endocrinology* 123, 1690–1692.

Boge, A., Sauerwein, H. and Meyer, H.H.D. (1995) IGF-I and insulin receptors in bovine skeletal muscle: Comparisons of different developmental ages, two different genotypes and various individual muscles. *Experimental and Clinical Endocrinology and Diabetes* 103, 99–104.

Breier, B.H. and Gluckman, P.D. (1991) The regulation of postnatal growth: Nutritional influences on endocrine pathways and function of the somatotropic axis. *Livestock Production Science* 27, 77–94.

Breier, B.H., Gluckman, P.D. and Bass, J.J. (1988a) Influence of nutritional status and oestradiol-17β on plasma growth hormone, insulin-like growth factors-I and -II and the response to exogenous growth hormone in young steers. *Journal of Endocrinology* 118, 243–250.

Breier, B.H., Gluckman, P.D. and Bass, J.J. (1988b) The somatotrophic axis in young steers: influence of nutritional status and oestradiol-17B on hepatic high- and low-affinity somatotrophic binding sites. *Journal of Endocrinology* 116, 169–177.

Breier, B.H., Gluckman, P.D. and Bass, J.J. (1988c) The somatotrophic axis in young steers: influence of nutritional status and oestradiol-17β on hepatic high- and low-affinity somatotrophic binding sites. *Journal of Endocrinology* 116, 169–177.

Breier, B.H., Gluckman, P.D., McCutcheon, S.N. and Davis, S.R. (1991) Physiological responses to somatotropin in the ruminant. *Journal of Dairy Science* 74 (Suppl. 2), 20–34.

Breier, B.H., Ambler, G.R., Sauerwein, H., Surus, A. and Gluckman, P.D. (1994) The induction of hepatic somatotrophic receptors after birth in sheep is dependent on parturition associated mechanisms. *Journal of Endocrinology* 141, 101–108.

Chin, E., Zhou, J., Dai, J., Baxter, R.C. and Bondy, C.A. (1994) Cellular localization and regulation of gene expression for components of the insulin-like growth factor ternary binding protein complex. *Endocrinology* 134, 2498–2504.

Coleman, E.S., Elsasser, T.H., Kemppainen, R.J., Coleman, D.A. and Sartin, J.L. (1993) Effect of endotoxin on pituitary hormone secretion in sheep. *Neuroendocrinology* 58, 111–122.

Cottam, Y.H., Blair, H.T., Gallaher, B.W., Purchas, R.W., Breier, B.H., McCutcheon, S.N. and Gluckman, P.D. (1992) Body growth and carcass composition in lambs chronically treated with recombinantly derived insulin-like growth factor-1. *Endocrinology* 130, 2924–2930.

Coverly, J.A. and Baxter, R.C. (1995) Regulation of insulin-like growth factor (IGF) binding protein-3 phosphorylation by IGF-I. *Progress in Growth Factor Research* 6, 223–229.

Dadoun, F., Guillaume, V., Sauze, N., Farisse, J., Velut, J.G., Orsoni, J.C., Gaillard, R. and Oliver, C. (1998) Effect of endotoxin on the hypothalamus pituitary axis in sheep. *European Journal of Endocrinology* 38,193–197.

Davis, S.L., Graf, M., Morrison, C.A., Hall, T.R. and Swift, P.J. (1992) Identification and partial purification of serum growth hormone binding protein in domestic animal species. *Journal of Animal Science* 70, 773–780.

Delhanty, P.J. and Han, V.K. (1993) The expression of insulin-like growth factor (IGF)- binding protein-2 and IGF-II genes in the tissues of the developing ovine fetus. *Endocrinology* 132, 41–52.

Devolder, A., Renaville, R., Sneyers, M., Callebaut, I., Massart, S., Goffinet, A., Burny, A. and Portetelle, D. (1993) Presence of growth hormone-binding proteins in cattle plasma and milk. *Journal of Endocrinology* 138, 91–98.

Douglas, R.G., Gluckman, P.D., Ball, K., Breier, B.H. and Shaw, J.H. (1991) The effects of infusion of insulin-like growth factor (IGF) I, IGF-II, and insulin on glucose and protein metabolism in fasted lambs. *Journal of Clinical Investigation* 88, 614–622.

Elsasser, T.H., Rumsey, T.S. and Hammond, A.C. (1989) Influence of diet on basal and growth hormone-stimulated plasma of IGF-I in beef cattle. *Journal of Animal Science* 67, 128–141.

Fowden, A.L., Szemere, J., Hughes, P., Gilmour, R.S. and Forhead, A.J. (1996) The effects of cortisol on the growth rate of the sheep fetus during late gestation. *Journal of Endocrinology* 151, 97–105.

Gallaher, B.W., Breier, B.H., Oliver, M.H., Harding, J.E. and Gluckman, P.D. (1992) Ontogenic differences in the nutritional regulation of circulating IGF binding proteins in sheep plasma. *Acta Endocrinologica* 126, 49–54.

Gallaher, B.W., Oliver, M.H., Eichhorn, K., Kessler, U., Kiess, W., Harding, J.E., Gluckman,

P.D. and Breier, B.H. (1994) Circulating insulin-like growth factor binding proteins (IGFBPs) and IGF-II/mannose-6-phosphate receptor levels are regulated by glucose and insulin in the fetal sheep. *Acta Endocrinologica* 131, 398–404.

Gallaher, B.W., Breier, B.H., Blum, W.F., McCutcheon, S.N. and Gluckman, P.D. (1995) An homologous radioimmunoassay for ovine insulin-like growth factor binding protein 2 (oIGFBP-2): ontogenesis and the response to growth hormone, placental lactogen and IGF-I treatment. *Journal of Endocrinology* 144, 75–82.

Gallaher, B.W., Breier, B.H., Keven, C.L., Harding, J.E. and Gluckman, P.D. (1998) Fetal programming of insulin-like growth factor (IGF)-1 and IGF-binding protein-3: evidence for an altered response to undernutrition in late gestation following exposure to periconceptual undernutrition in the sheep. *Journal of Endocrinology* 159, 501–508.

Gatford, K.L., Egan, A.R., Clark, I.J. and Owens, P.C. (1998) Sexual dimorphism of the somatotrophic axis. *Journal of Endocrinology* 157, 373–389.

Gilmour, R.S. (1994) The implications of insulin-like growth factor mRNA heterogeneity. *Journal of Endocrinology* 140, 1–3.

Giudice, L.C., van Dessel, H.J.H.M., Cataldo, N.A., Chandrasekher, Y.A., Yap, O.W.S. and Fauser, B.C.J.M. (1995) Circulating and ovarian IGF binding proteins: potential roles in normo-ovulatory cycles and in polycystic ovarian syndrome. *Progress in Growth Factor Research* 6, 397–408.

Hammerman, M.R. and Miller, S.B. (1993) The growth hormone insulin-like growth factor axis in kidney revisited. *American Journal of Physiology* 265, F1–F14.

Hennies, M., Breier, B.H., Gallaher, B.W., Krishnamurthi, R.V. and Gluckman, P.D. (1998) Growth hormone induces electrolyte and metabolite imbalance in endotoxin challanged sheep. *Proceedings of the 80th Annual Meeting of the Endocrine Society, New Orleans, USA,* pp. 1–108.

Holst, O., Ulmer, A.J., Brade, H., Flad, H.D. and Rietschel, E.T. (1996) Biochemistry and cell biology of bacterial endotoxins. *FEMS Immunology Medicine Microbiology* 16, 83–104

Holt, R.I., Baker, A.J., Jones, J.S., Crossey, P.A., Stone, N.M., Preedy, V.R. and Miell, J.P. (1996) Differential effects of malnutrition, bile duct ligation and galactosamine injection in young rats on serum levels and gene expression of IGF-binding proteins. *Journal of Endocrinology* 149, 465–472.

Hooper, S.B., Bocking, A.D., White, S.E., Fraher, L.J., McDonald, T.J. and Han, V.K.M. (1994) Catecholamines stimulate the synthesis and release of insulin-like growth factor binding protein-1 (IGFBP-1) by fetal sheep liver in vivo. *Endocrinology* 134, 1104–1112.

Hua, K.M., Hodgkinson, S.C. and Bass, J.J. (1995) Differential regulation of plasma levels of insulin-like growth factors-I and -II by nutrition, age and growth hormone treatment in sheep. *Journal of Endocrinology* 147, 507–516.

Ilondo, M.M., Damholt, A.B., Cunningham, B.A., Wells, J.A., Demeyts, P. and Shymko, R.M. (1994) Receptor dimerization determines the effects of growth hormone in primary rat adipocytes and cultured human IM-9 lymphocytes. *Endocrinology* 134, 2397–2403.

Isaksson, O.G.P., Lindahl, A., Isgaard, J., Nilsson, A., Tornell, J. and Carlsson, B. (1991) Dual regulation of cartilage growth. In: Spencer, E.M. (ed.) *Modern Concepts of Insulin-Like Growth Factors.* Elsevier, Amsterdam, pp. 121–126.

Jones, J.I. and Clemmons, D.R. (1995) Insulin-like growth factors and their binding proteins: biological actions. *Endocrine Reviews* 16, 3–34.

Kiess, W., Yang, Y., Kessler, U. and Hoeflich, A. (1994) Insulin-like growth factor II (IGF-II) and the IGF-II/ mannose-6-phosphate receptor: The myth continues. *Hormone Research* 41, 66–73.

Kinsbergen, M., Bruckmaier, R.M. and Blum, J.W. (1994) Metabolic, endocrine and haematological responses to intravenous E. coli endotoxin administration in 1-week-old calves. *Zentralblatt fur Veterinärmedizin – Reihe* A 41, 530–547.

Klempt, M., Bingham, B., Breier, B.H., Baumbach, W.R. and Gluckman, P.D. (1993) Tissue distribution and ontogeny of growth hormone receptor mRNA and ligand binding to hepatic tissue in the midgestation sheep fetus. *Endocrinology* 132, 1071–1077.

Koea, J.B., Breier, B.H., Shaw, J.H. and Gluckman, P.D. (1992a) A possible role for IGF-II: Evidence in sheep for in vivo regulation of IGF-I mediated protein anabolism. *Endocrinology* 130, 2423–2425.

Koea, J.B., Gallaher, B.W., Breier, B.H., Douglas, R.G., Hodgkinson, S., Shaw, J.H.F. and Gluckman, P.D. (1992b) Passive immunization against circulating IGF-I increases protein catabolism in lambs: Evidence for a physiological role for circulating IGF-I. *Journal of Endocrinology* 135, 279–284.

Kupfer, S.R., Underwood, L.E., Baxter, R.C. and Clemmons, D.R. (1993) Enhancement of the anabolic effects of growth hormone and insulin-like growth factor-i by use of both agents simultaneously. *Journal of Clinical Investigation* 91, 391–396.

Lee, P.D., Conover, C.A. and Powell, D.R. (1993) Regulation and function of insulin-like growth factor-binding protein-1. *Proceedings of the Society for Experimental Biology and Medicine* 204, 4–29.

Lewitt, M.S., Scott, F.P., Clarke, N.M. and Baxter, R.C. (1995) Developmental regulation of circulating insulin-like growth factor binding proteins in normal pregnancies and in preeclampsia. *Progress in Growth Factor Research* 6, 475–480.

Li, J., Saunders, J.C., Gilmour, R.S., Silver, M. and Fowden, A.L. (1993) Insulin-like growth factor-II messenger ribonucleic acid expression in fetal tissues of the sheep during late gestation: effects of cortisol. *Endocrinology* 132, 2083–2089.

Li, J., Owens, J.A., Owens, P.C., Saunders, J.C., Fowden, A.L. and Gilmour, R.S. (1996) The ontogeny of hepatic growth hormone receptor and insulin-like growth factor-I gene expression in the sheep fetus during late gestation: developmental regulation by cortisol. *Endocrinology* 137, 1650–1657.

Liao, W., Rudling, M. and Angelin, B. (1996) Growth hormone potentiates the in vivo biological activities of endotoxin in the rat. *European Journal of Clinical Investigation* 26, 254–258.

Liu, F., Powell, D.R., Styne, D.M. and Hintz, R.L. (1991) Insulin-like growth factors (IGFs) and IGF-binding proteins in the developing rhesus monkey. *Journal of Clinical Endocrinology and Metabolism* 72, 905–911.

McCusker, R.H., Cohick, W.S., Busby, W.H. Jr. and Clemmons, D.R. (1991) Evaluation of the developmental and nutritional changes in porcine insulin-like growth factor-binding protein-1 and -2 serum levels by immunoassay. *Endocrinology* 129, 2631–2638.

Min, S.H., Mackenzie, D.D.S., Breier, B.H., McCutcheon, S.N. and Gluckman, P.D. (1996) Responses of young energy-restricted sheep to chronically administered IGF-I: Evidence that IGF-I suppresses the hepatic growth hormone and receptor. *Endocrinology* 137, 1129–1137.

Min, S.H., Mackenzie, D.D.S., Breier, B.H., McCutcheon, S.N. and Gluckman, P.D. (1999) Ontogeny of insulin-like growth factor-1 responsiveness to bovine growth hormone in young lambs. *Domestic Animal Endocrinology* 17, 409–419.

Morrison, D.C. and Ryan, J.L. (1987) Endotoxins and disease mechanisms. *Annual Reviews in Medicine* 38, 417–432.

Ogawa, E., Breier, B.H., Bauer, M.K., Gallaher, B.W., Grant, P.A., Walton, P.E., Owens, J.A. and Gluckman, P.D. (1996) Pretreatment with bovine growth hormone is as effective as treatment during metabolic stress to reduce catabolism in fasted lambs. *Endocrinology* 137, 1242–1248.

Olchovsky, D., Song, J., Gelato, M.C., Sherwood, J., Spatola, E., Bruno, J.F. and Berelowitz, M. (1993) Pituitary and hypothalamic insulin-like growth factor-I (IGF-I) and IGF-I receptor expression in food-deprived rats. *Molecular and Cellular Endocrinology* 93, 193–198.

Pell, J.M., Saunders, J.C. and Gilmour, R.S. (1993) Differential regulation of transcription initiation from insulin-like growth factor-I (IGF-I) leader exons and of tissue IGF-I expression in response to changed growth hormone and nutritional status in sheep. *Endocrinology* 132, 1797–1807.

Roth, E., Valentini, L., Semsroth, M., Holzenbein, T., Winkler, S., Blum, W.F., Ranke, M.B., Schemper, M., Hammerle, A. and Karner, J. (1995) Resistance of nitrogen metabolism to growth hormone treatment in the early phase after injury of patients with multiple injuries. *Journal of Trauma* 38, 136–141.

Rotwein, P. (1991) Structure, evolution, expression and regulation of insulin-like growth factors I and II. *Growth Factors* 5, 3–18.

Roupas, P. and Herington, A.C. (1994) Postreceptor signalling mechanisms for growth hormone. *TEM* 5, 154–158.

Sauerwein, H., Breier, B.H., Bass, J.J. and Gluckman, P.D. (1991) Chronic treatment with bovine growth hormone upregulates high-affinity hepatic somatotropic receptors in sheep. *Acta Endocrinologica* 124, 1–7.

Scharf, J.G., Ramadori, G., Braulke, T. and Hartmann, H. (1995) Cellular localisation and hormonal regulation of insulin-like growth factor binding proteins and of the acid-labile subunit within rat liver. *Progress in Growth Factor Research* 6, 175–180.

Sherlock, R.G., Blair, H.T., Kuhn, B., Mackenzie, D.D.S., Breier, B.H., Gluckman, P.D. and McCutcheon, S.N. (1993) Selection for insulin-like growth factor-1 causes hypoglycaemia in sheep. *Proceedings of the New Zealand Society of Animal Production* 53, 381–384.

Smith, W.J., Underwood, L.E. and Clemmons, D.R. (1995) Effects of caloric or protein restriction on insulin-like growth factor-I (IGF-I) and IGF-binding proteins in children and adults. *Journal of Clinical Endocrinology and Metabolism* 80, 443–449.

Soos, M.A., Field, C.E., Nave, B.T. and Siddle, K. (1994) Hybrid and atypical insulin-like growth factor receptors. In: Baxter, R.C., Gluckman, P.D. and Rosenfeld, R.G. (eds) *The Insulin-like Growth Factors and their Regulatory Proteins.* Elsevier, Amsterdam, pp. 95–106.

Takala, J., Ruokonen, E., Webster, N.R., Nielsen, M.S., Zandstra, D.F., Vundelinckx, G. and Hinds, C.J. (1999) Increased mortality associated with growth hormone treatment in critically ill adults. *The New England Journal of Medicine* 341, 785–792.

Thomas, G.B., Cummins, J.T., Sudbury, A.W., McCloud, P.I. and Clarke, I.J. (1991) Effect of restricted feeding on the relationship between hypophysial portal concentrations of growth hormone (GH)-releasing factor and somatostatin, and jugular concentrations of GH in ovariectomized ewes. *Endocrinology* 128 1151–1158.

# 12 Direct Effects of Photoperiod on Lipid Metabolism, Leptin Synthesis and Milk Secretion in Adult Sheep

Y. CHILLIARD AND F. BOCQUIER

*Adipose Tissue and Milk Lipids Team, Herbivore Research Unit, INRA-Theix, 63122 St Genès Champanelle, France*

## Introduction

Grazing ruminants have to cope with large seasonal changes in the amount and quality of herbage consumed. Their ability to survive is improved by seasonality of reproduction, which confines productive activities, namely late gestation, lactation and growth, to spring and summer when the climate is favourable and food of good quality is abundant. Furthermore, the chances of survival are increased by depositing fat reserves which are then used during periods of food shortage. Domesticated livestock, which are descended from ancestral types that were probably strongly seasonal, may have conserved some of these peculiarities. However, genetic selection pressure, applied under standardized conditions and aimed at increasing year-round production, may have had the effect of smoothing this seasonality. Sheep, however, are not able to reproduce during all seasons. Nevertheless, there is great variability between sheep breeds, the most sensitive to seasonal changes being located in high latitudes (Soay breed of the Hebrides), while tropical breeds exhibit few responses to seasonal changes. Among seasonal cues, photoperiod (i.e. daily light duration) has been clearly identified to be an effective signal which controls the period of reproduction in sheep (Ortavant *et al.*, 1988; Lincoln and Richardson, 1998). This raises the possibility that photoperiod, which varies during the year, may also have effects during subsequent physiological stages, i.e. reinforcing other physiological adaptive mechanisms.

## Food intake and energy expenditure

Voluntary food intake is generally linked to physiological status and the corresponding level of animal performance (growth rate, milk yield). Because of interactions between photoperiod and these factors, only results obtained with non-pregnant, non-lactating adult sheep will be discussed in this section. Voluntary food intake is increased (45%) by long daylength exposure in wethers (Gordon, 1964). This effect is more marked

 *Ruminant Physiology: Digestion, Metabolism, Growth and Reproduction* (ed. P.B. Cronjé)

with high-quality diets (+11% with fescue hay and +35 % with dehydrated grass; Milne *et al.*, 1978). There are, however, differences between rams of Soay (+71%) vs. Suffolk (+52%) breeds, and gonad-intact vs. castrated animals, the former being more sensitive (Kay, 1985). Although it is generally accepted that minimum food intake occurs during winter and maximal food intake during summer, few experiments have been conducted to analyse the exact timing of the effect of photoperiod on food intake. When photoperiod variations are cyclic (semestrial cycle), maximum food intake occurs between 6 and 13 weeks after the maximum duration of light is reached (Kay, 1985). However, in the case of abrupt changes of photoperiod, the food intake of Soay rams begins to change within 2 weeks (Lincoln and Richardson, 1998).

In sheep energy balance studies, neither the digestibility (Walker *et al.*, 1991) nor the metabolizability (Blaxter and Boyne, 1982) of the diet were affected by daylength. Voluntary food intake is generally positively correlated with fasting heat production, and this could suggest that intake oscillations may be driven by seasonal changes in metabolic rate. To test this hypothesis, measurements of heat production were made at a constant level of food intake (near maintenance) in intact ewes subjected to the following photoperiod treatments: constant long-day, natural, simulated natural or reverse-natural (Walker *et al.*, 1991). Oscillations of metabolic heat production observed in ewes subjected to constant illumination provide evidence of a long-term (more than one year) endogenous rhythm in energy metabolism. In all other sinusoidal photoperiodic treatments, whole animal energy metabolism was driven by the pattern of changes in daylight, even when the light cycle was reversed. The peak of metabolic heat production did not occur in phase with the longest or the shortest daylength periods, but 15 ± 3 weeks following the start of the day-length increment (i.e. 11 weeks before the maximum duration of light). The amplitude of changes in fasting heat production (+21%) is in general agreement with previous results obtained in mature wethers (+29%) kept under natural daylength (Blaxter and Boyne, 1982). It is noteworthy that the relative annual amplitude of variation in metabolic rate is far lower than that of food intake variations. Furthermore, maximum heat production under annual photoperiod conditions occurs before maximum daylength and long before maximum food intake would occur.

## Adipose tissue and muscle metabolism

Wild ruminants exhibit large adaptations to the seasonality of environmental changes. Detailed studies of seasonal changes in adipose tissue metabolism have been done in the reindeer, which inhabits arctic areas. Interestingly, seasonal body weight changes of this animal are mainly due to body fat, and the non-fat component remains fairly constant (Reimers *et al.*, 1982). This suggests a high capacity to spare body proteins, associated with extensive use of body fat during the winter period. The reindeer shows a marked reduction in voluntary food intake during winter, even when offered the same diet *ad libitum* throughout the year (Larsen *et al.*, 1985b). This results in seasonal cycles of fat deposition and adipose tissue lipogenic activity which parallel those of food intake and are accompanied by changes in plasma concentrations of glucose, acetate, triglyceride, cholesterol and insulin. There was no seasonal change in the concentration of plasma catecholamines, although plasma non-esterified fatty acids were higher during the winter period (Larsen *et al.*, 1985a,b). However, a decrease in *in vitro* adrenaline-stimulated lipolytic capacity was observed during winter. This could prevent excessive breakdown of fat stores.

In the domestic sheep in Scotland, the rate of adipose tissue fatty acid synthesis and the activity of the lipoprotein lipase, a key enzyme for tissue uptake of plasma triglycerides, increased between October and May (Vernon *et al.*, 1986). These increases were, however, probably related to increased food intake, since plasma concentrations of insulin, glucose and acetate were increased also.

In the animal models discussed previously, changes in adipose tissue metabolism paralleled changes in voluntary food intake and/or changes in plasma insulin and energy metabolites that were available for lipogenesis. For this reason it is not possible to distinguish whether the effects were due to a direct effect of photoperiod (or other seasonal factors) on adipose tissue metabolism or due to indirect effects related to changes in food intake. In order to avoid these problems Bocquier *et al.* (1998) used dry non-pregnant adult ewes which were subjected during winter to either long (16L:8D) or short (8L:16D) daylength for 4–6 weeks, and given fixed amounts of food within two feeding treatments: (i) restricted to 22% of energy requirements for 1 week; or (ii) subsequently re-fed to 190% of requirements for 2 weeks (pair-feeding across photoperiod treatments). Ewes were also ovariectomized in order to avoid any putative indirect effect of photoperiod via changes in the secretion of reproductive hormones, as these are potential modulators of tissue metabolism (Chilliard, 1987). Under these conditions, long days increased lipoprotein lipase activity in Longissimus thoracis muscle and, in overfed ewes, the activities of lipoprotein lipase and malic enzyme (an enzyme involved in NADPH generation for fatty acid synthesis) in subcutaneous adipose tissue, as well as malic enzyme activity in perirenal adipose tissue. Furthermore, long days increased the amount of lipoprotein lipase mRNA in cardiac muscle and perirenal adipose tissue (Table 12.1). There was also a non-significant trend for long days to slightly increase the activity of three other lipogenic enzymes in adipose tissues of re-fed ewes (Bocquier *et al.*, 1998; Faulconnier *et al.*, 1999). Simultaneously, long days increased plasma leptin concentrations and the amount of adipose tissue leptin mRNA independent of feeding level, and decreased plasma non-esterified fatty acids concentrations in underfed ewes (Fig. 12.1).

It was previously reported that long photoperiod also increased adipose tissue leptin mRNA in Djungarian hamsters (Klingenspor *et al.*, 1996). However, food consumption was not controlled in this trial. Furthermore, adipose tissue mass and gonadal activity, which have been shown to modulate leptin production in rats and humans, were sharply increased along with leptin mRNA (Klingenspor *et al.*, 1996). The study of Bocquier *et al.* (1998) in sheep shows, for the first time, an effect of photoperiod on leptin that is independent of food intake, adiposity or gonadal activity. Moreover, the photoperiod-driven changes in leptin, lipogenic enzyme activities and plasma non-esterified fatty acids were not related to changes in plasma insulin, glucose, acetate, lactate, 3-hydroxybutyrate, triglycerides or urea concentrations, since these parameters were not affected by photoperiod. These results suggest the existence of direct effects of photoperiod on sheep adipose tissue lipogenic potential, leptin secretion and lipomobilization, as well as on muscle lipoprotein lipase activity, which are not induced by nutrient supply.

**Table 12.1.** Effects of nutritional status and daylength on malic enzyme activity in adipose tissue and lipoprotein lipase activity and mRNA in adipose tissue and muscles (from Bocquier *et al.*, 1998; Faulconnier *et al.*, 1999).

| | Underfed | | Re-fed | | |
|---|---|---|---|---|---|
| | Short | Long | Short | Long | Statistical effects[a] |
| Subcutaneous adipose tissue | | | | | |
| Malic enzyme activity[b] | 41 | 51 | 117 | 158 | F, p |
| Lipoprotein lipase activity[c] | 34 | 35 | 264 | 415 | F, P, F×P |
| Perirenal adipose tissue | | | | | |
| Malic enzyme activity | 74 | 87 | 147 | 208 | F, P, f×p |
| Lipoprotein lipase activity[c] | 158 | 184 | 589 | 604 | F |
| Lipoprotein lipase mRNA[d] | 22 | 32 | 111 | 291 | F, p |
| Cardiac muscle | | | | | |
| Lipoprotein lipase activity[e] | 133 | 129 | 187 | 160 | F |
| Lipoprotein lipase mRNA[d] | 71 | 118 | 158 | 252 | F, p |
| Longissimus thoracis muscle | | | | | |
| Lipoprotein lipase activity[e] | 23 | 30 | 33 | 43 | F,p |

[a] 5 ewes in each group ($N = 20$). P, p: effect of photoperiod ($P$: $P < 0.05$; p: $P < 0.10$). F, f: effect of feeding level (F: $P < 0.05$; f: $P < 0.10$). f×p: = interaction ($P < 0.10$).
[b] nmol NADPH $min^{-1}$ $10^{-6}$ adipocytes.
[c] nmol fatty acids $min^{-1}$ $10^{-6}$ adipocytes.
[d] Arbitrary units.
[e] nmol fatty acids $min^{-1}$ $\mu g^{-1}$ DNA.

## Secretion of milk and nutrient partitioning

Linzell (1973) demonstrated that the rate of milk secretion oscillates in a sinusoidal fashion when goats are kept in continuous milk production for 2–4 years, with maximum milk yields occurring in summer and minimum yields in winter. This phenomenon was confirmed in non-cyclic ovariectomized goats and was not diet related. In selected dairy cows there is a limited effect of photoperiod on milk yield (+5 to +10%), which is followed by an increase in food intake (e.g. Dahl *et al.*, 1997).

In sheep there is a clear influence of long photoperiod on milk secretion during early lactation (+52% and +30% in two trials with Prealpes ewes, Bocquier *et al.*, 1986; +25% in Sardinian ewes, Bocquier *et al.*, 1997). Furthermore, the effect of long photoperiod was proportional to the time elapsed between full establishment of daylength and lambing. Large responses (52 and 30%) were obtained when photoperiod treatments started 42 days before lambing, whereas a smaller response (25%) was observed when treatments were established 25 days before lambing. In addition, no effect was detected when photoperiod was changed within the week prior to lambing (Bocquier *et al.*, 1990). A progressive reversal of long and short photoperiod treatments during established lactation altered lactation persistency and led to large differences in favour of long photoperiod (+12 to 34%; Fig. 12.2). In order to separate the effects of constant

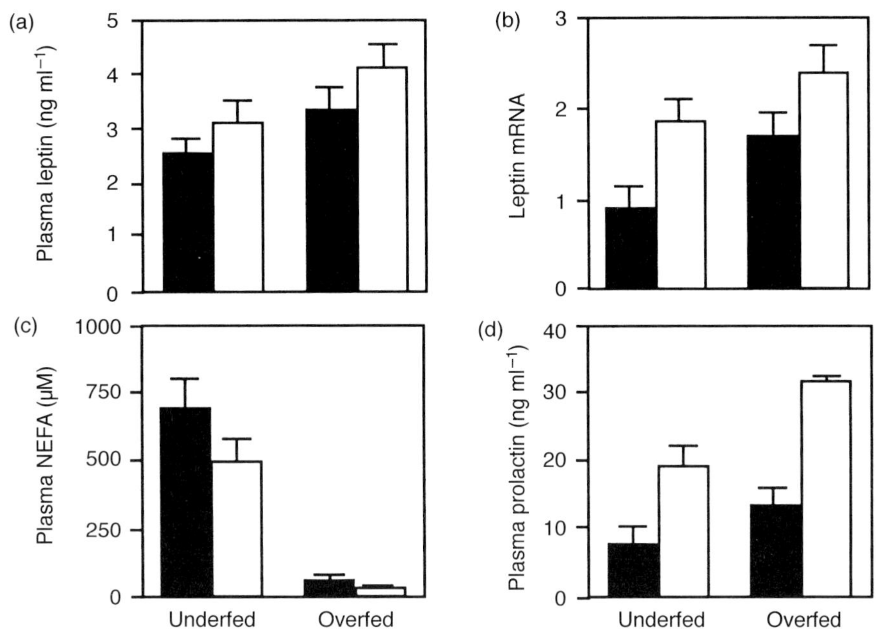


**Fig. 12.1.** Effects of daylength (shaded bars: short days (8 h day$^{-1}$) and open bars: long days (16 h day$^{-1}$)) and nutritional status (underfed (22%) or overfed (190% of energy requirements)) on (a) plasma leptin, (b) adipose tissue leptin mRNA (in arbitrary units), (c) plasma NEFA and (d) plasma prolactin (5 ewes per group) (from Bocquier *et al.*, 1998).

light duration from those of changes in light duration, light treatments were abruptly switched from constant long (15.30L:08.30D), or constant short (08.30L:15.30D) to a common treatment of equal light and dark phases (12.00L:12.00D) (Bocquier *et al.*, 1997). The shortening (−3.30 h) of light exposure induced a dramatic decline in milk yield (−38% 42 days later), while lengthened light exposure (+3.30 h) limited the normal decline of milk yield to 8% during the same period. Hence sheep milk production is affected both by changes in daylength and by the duration of light exposure itself, via multiple effects either on mammogenesis, lactogenesis and/or galactopoiesis.

Milk fat and protein concentrations are lower during long than during short days. During early lactation, the maximal difference between short and long days was observed (Bocquier, 1985; Bocquier *et al.*, 1997) 30–40 days after lambing for both fat (Prealpes: −10 g l$^{-1}$; Sardinian: −14 g l$^{-1}$) and protein content (Prealpes: −3 g l$^{-1}$; Sardinian: −11 g l$^{-1}$). After the progressive switch of light treatments during late lactation (Fig. 12.2), milk concentration of ewes exposed to long days was even more lowered (−19 g l$^{-1}$ for fat and −13 g l$^{-1}$ for protein) when compared with ewes exposed to short days. After the abrupt change in daylength to an equal dark/light exposure treatment (Bocquier *et al.*, 1997), it took 23 days until the maximal difference in milk concentration occurred. During this period, fat and protein content of milk from ewes subjected to an increase in daylength (+3.30 h) decreased by 8 g l$^{-1}$ and 6 g l$^{-1}$ respectively compared with that from ewes subjected to a decrease (−3.30 h) in daylength.

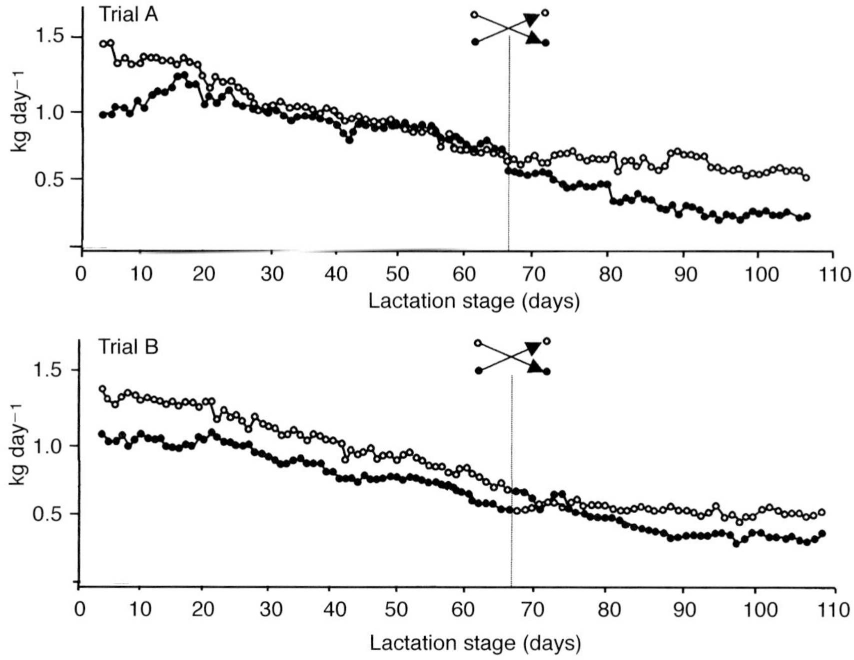


**Fig. 12.2.** Milk production of ewes subjected to either long (○) or short (●) photoperiod; in trial A (January lambing; $n = 24$ per group) and trial B (September lambing; $n = 22$ per group). Light treatments were established 42 days before lambing, and were reversed from the 9th week of lactation, by progressive changes within a fortnight (⋟⧗) (Bocquier, 1985).

This suggests that a single change from long or short days towards a 12L:12D constant photoperiod is sufficient to produce responses in milk yield and composition that could be as high as 50% of the response attained after an exposure of several weeks to a constant difference of 8 h in daily light exposure.

The negative relationship between the level of milk yield, and milk fat and protein concentrations may be due to a dilution effect since the increase in milk yield by long daylength (+20% in early lactation, i.e. between day 1 and day 30, and +67% in late lactation, i.e. between day 90 and day 110; Table 12.2) is accompanied by a decline in both fat and protein content of milk (−5% in early lactation and −19% in late lactation). This decline is, however, lower than the positive effect on milk yield, because both fat and protein yields are increased by long daylength in early (+14%) and late (+36%) lactation (Table 12.2). Surprisingly, the effects of photoperiod on milk lactose and mineral content have not been studied, despite the well-known contribution of these constituents to the osmotic regulation of water secretion in milk.

In *ad libitum*-fed ewes, voluntary intake was identical between groups during the first month of lactation and became higher (+16% after 150 days of lactation) in long-day exposed ewes (Bocquier *et al.*, 1997). This resulted in a lower calculated energy balance (−11% of energy requirements) during the first month of lactation for long-day

**Table 12.2.** Photoperiod and milk secretion by dairy ewes (Bocquier, 1985).

| | Early lactation[a] | | | Late lactation[b] | | |
|---|---|---|---|---|---|---|
| | Short[c] | Long[d] | Long % Short | Short[c] | Long[d] | Long % Short |
| Milk yield (g day$^{-1}$) | 980 | 1178 | 120 | 328 | 550 | 167 |
| Content (g kg$^{-1}$) | | | | | | |
| Fat | 73 | 69 | 95 | 99 | 80 | 81 |
| Protein | 50 | 47 | 94 | 71 | 57 | 80 |
| Yield (g day$^{-1}$) | | | | | | |
| Fat | 71 | 82 | 114 | 32 | 43 | 135 |
| Protein | 49 | 56 | 114 | 22 | 31 | 137 |

[a] From 1 to 30 days after lambing, data from trials A and B in Fig. 12.2.
[b] From 91 to 110 days after lambing.
[c] 8 h day$^{-1}$.
[d] 16 h day$^{-1}$.

exposed ewes which produced more milk energy, than for short-day exposed ewes (−5% of requirements), while the reverse was observed between days 30 and 150 of lactation (+22% and +16% of requirements, for long- and short-day exposed ewes, respectively). In order to isolate a direct effect of photoperiod on milk secretion Bocquier *et al.* (1986) fed ewes a limited and identical amount of food in both photoperiod treatments. Under these conditions, there was an increase in the apparent efficiency of milk yield at the expense of body energy deposition in ewes exposed to long days (Fig. 12.3). This partitioning of energy towards milk output occurred in ewes that were either in positive or equilibrated energy balance. Changes in body energy content were mainly due to changes in body lipids, suggesting that, in sheep, photoperiod acts on the partitioning of nutrients between mammary gland and adipose tissue independently of effects on food intake.

## Hormonal mediation of photoperiodic changes

### *Brain*

Secretion of plasma melatonin from the pineal gland, which is lowest during the light and greatest during the dark period, is considered as the hormonal signal related to the length of day. In sheep, pinealectomy greatly reduced the photoperiod-induced changes in food intake, growth rate and reproduction (Williams *et al.*, 1995; Malpaux *et al.*, 1996), and melatonin implants, which mimic short days, induced a decline in circulating prolactin (PRL) (Viguié *et al.*, 1997). Similarly, in hamsters, melatonin injections mimicked the effects of short days on body lipids, and these effects were initiated before any change in food intake. Pinealectomy, however, blocked the effects of short days on reproduction but not on body weight, suggesting that retinal photoreception could be involved in body weight regulation (Bartness and Wade, 1984). Considering the fact that sheep maintained in constant photoperiod eventually become cyclic,

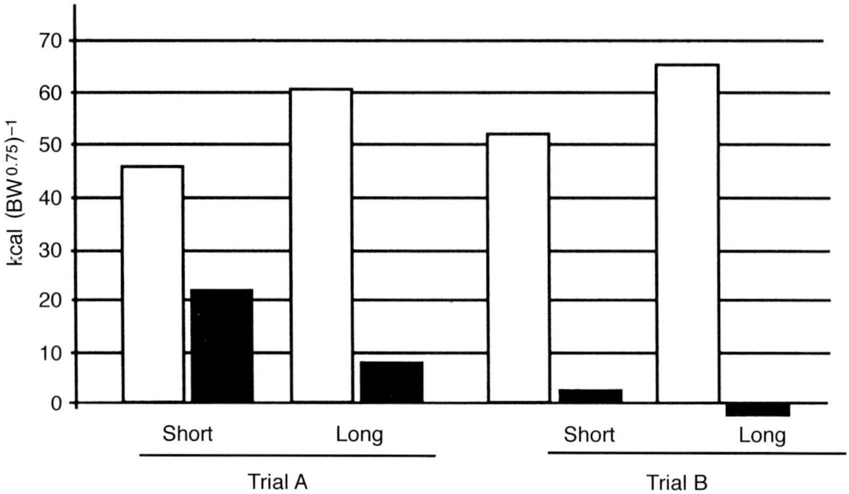


**Fig. 12.3.** Effect of photoperiod (long versus short) on energy partitioning between milk (open bars) and body reserves (shaded bars) in lactating (pair-fed) ewes (Bocquier *et al.*, 1986). Trials A and B are described in Fig. 12.2. Body energy was measured by $D_2O$ dilution technique (Bocquier *et al.*, 1999).

suggests that, in the absence of variation in melatonin production, both reproduction (Williams *et al.*, 1995) and metabolic heat production (Walker *et al.*, 1991) are driven by endogenous rhythms. Thus the natural photoperiod may serve to maintain the synchrony between the circannual cycle of physiological activities and the appropriate season (Williams *et al.*, 1995), although melatonin may not be the sole internal signal.

The complexity of the transfer of photoperiodic information is shown by the fact that hypothalamo-pituitary disconnection in rams suppresses the cyclical effects of photoperiod changes on α-melanocyte-stimulating hormone (α-MSH), β-endorphin, adrenocorticotrophic hormone (ACTH), luteinizing hormone (LH), follicle-stimulating hormone (FSH), feed intake and body weight, but does not suppress cyclical changes in PRL (Lincoln and Richardson, 1998). This means that photoperiod-driven signals act on pituitary hormones either via the hypothalamus by neurohormonal pathways (α-MSH, β-endorphin, ACTH, LH, FSH) or directly on pituitary secretion (PRL), and that PRL is not directly involved in body weight changes. In view of recent data indicating that leptin acutely stimulates *in vitro* prolactin release by the rat pituitary (Yu *et al.*, 1997), our results (Fig. 12.1) suggest that the photoperiod- and feeding-induced changes in PRL could be facilitated by leptin.

## *Peripheral hormones*

Prolactin is second only to melatonin in terms of responsiveness to photoperiod, and responds to an abrupt change of photoperiod within a week (Lincoln *et al.*, 1978). Prolactin is elevated both in intact (Kann, 1997) and ovariectomized (Bocquier *et al.*,

1998) ewes exposed to long daylength. Long photoperiod is also effective in the pregnant ewe (Perier *et al.*, 1986) and induces a 40% higher and earlier (20 h) pre-partum surge of plasma PRL (Bocquier, 1985). In lactating ewes exposed to long photoperiod, mean plasma PRL concentration was increased by 311% (Bocquier *et al.*, 1990). However, photoperiod-induced increments in serum concentration of PRL began to wane when sheep were maintained on 16L:8D for more than 16 weeks (Almeida and Lincoln, 1984). This indicates that PRL secretion becomes refractory to a prolonged constant photoperiodic stimulus.

Plasma growth hormone (GH) has been shown to be slightly elevated by long daylength in rams (Lincoln and Richardson, 1998) and in the pregnant ewe (Perier *et al.*, 1986), but this has never been observed in the lactating ewe (Bocquier, 1985; Bocquier *et al.*, 1990), nor in the dry non-pregnant ovariectomized ewe (Chilliard *et al.*, 1998b). It has been suggested recently that the effect of long photoperiod on milk yield in the cow could be mediated by circulating insulin-like growth factor-I (IGF-I). Cows exposed to long photoperiod produced more milk (+6%) without increased feed intake, and had an elevated concentration of IGF-I (+77%), without altered concentrations of GH (as in lactating sheep) or IGFBP-2 and -3 (Dahl *et al.*, 1997).

Other hormones were either unaffected by daylength (insulin; Bocquier *et al.*, 1998) or gave conflicting results (thyroid hormones and corticosteroids; Forbes 1982; Vernon *et al.*, 1986; Petitclerc and Zinn, 1991; Lincoln and Richardson, 1998). In *ad libitum*-fed sheep, there was an increase in plasma β-endorphin concentration at the beginning of a short-day period that could have stimulated insulin secretion and adipose tissue lipogenesis (Lincoln and Richardson, 1998). However, as this followed a period of high food intake at the end of a long-day period, it is not clear whether the changes in β-endorphin concentrations were due to photoperiod *per se*, or to the previous change in food intake.

## Adipose tissue

Peripheral effects of melatonin on adipose tissue are plausible, since specific binding sites in Siberian hamster brown adipose tissue have been described (Le Gouic *et al.*, 1997). However, the functionality of these receptors remains to be shown. Although plasma PRL concentration increases in response to increasing both daylength and feeding level (Bocquier *et al.*, 1998), this hormone is probably not directly involved in adipose tissue response to daylength because sheep adipose tissue appears to lack PRL receptors (Emane *et al.*, 1986; Knight and Flint, 1995). Furthermore, addition of PRL *in vitro* did not change lipolysis in bovine or ovine adipose tissue (Houseknecht *et al.*, 1996), nor was lipogenesis changed in ovine (Vernon and Finley, 1988) or bovine adipose tissue (Etherton *et al.*, 1987).

There could be, however, indirect effects of PRL *in vivo*. In the rat, PRL infusion was shown to stimulate the secretion of a mammotrophic and lactogenic hepatic factor (synlactin) (English *et al.*, 1990), the effects of which remain to be unravelled for adipose tissue. PRL injections increased mammary acetyl-CoA carboxylase activity and mRNA abundance in the lactating rat, and decreased it in adipose tissue (Barber *et al.*, 1992). The adipose tissue lipolytic response was increased by PRL injections in female rats during recovery from lactation (Vernon and Finley, 1986). Although these results suggest that the effects of PRL in rodents are antilipogenic and/or lipolytic (indirect), there was no *in vivo* effect of PRL on plasma non-esterified fatty acids in adult sheep (Luthman and Johnson, 1972), lactating dairy cows (Plaut *et al.*, 1987) or lactating goats (Jacquemet and Prigge, 1991). Furthermore, administration of bromocriptine (a dopamine agonist which inhibits PRL secretion) did not change adipose tissue lipogenic rate in post-partum beef cow (Mills *et al.*, 1989), and PRL administration did not affect adipose tissue lipogenic or lipolytic rates, although the weight of perirenal adipose tissue was increased in growing sheep (Eisemann *et al.*, 1984). Results from other studies with bromocriptine suggest that PRL has lipogenic (in synergy with insulin) rather than antilipogenic effects on the liver, and promotes body fat deposition in several rodent species and pig (Cincotta and Meier, 1989a; Cincotta *et al.*, 1989). Interestingly, the decrease in body fat after bromocriptine administration was observed without (or with very few) changes in food consumption, energy expenditure and lean body mass (Cincotta *et al.*, 1993a). However, the antilipogenic effects of bromocriptine could be due to dopaminergic effects in the hypothalamus (Luo *et al.*, 1997), rather than simply due to a decrease in PRL secretion.

In rodents, it appears that the lipogenic effects of photoperiod result from a synergy between insulin, corticosteroids and PRL which is dependent on the time of day, both *in vivo* (Cincotta *et al.*, 1993b) and *in vitro* (Cincotta and Meier, 1989b). These observations are related to a paradigm which was proposed more than 20 years ago by Meier and Burns (1976) from studies in several vertebrate species (including fishes, birds and mammals) that exhibit marked seasonal cycles in body fat stores:

> These cycles are controlled by the circadian rhythms of corticosteroids and prolactin. Prolactin stimulates increases or decreases in fat storage depending on whether it is present in larger quantities during daily intervals of lipogenic or lipolytic sensitivities. The intervals of sensitivity are entrained by the daily photoperiod and mediated by the adrenal corticosteroids. Thus the temporal synergism of the circadian rhythms of corticosteroid and prolactin hormones involves a relation between sensitivity rhythms of cells involved in lipid metabolism and rhythms of the stimulatory hormone, prolactin. The cumulative effects of various temporal hormonal patterns account for the seasonal changes in fat stores.

This results in annual cycles of metabolic activities that are the result of an interaction of the daily photoperiod with an endogenous seasonal timing mechanism, termed scotosensitivity. Alteration in PRL rhythm could reflect an altered phase of a central circadian pacemaker that entrains other circadian responses (Meier and Cincotta, 1996).

Alternating insulin–epinephrine infusions in the rat resulted in a complete inversion of the normal circadian distribution of sleeping and feeding patterns. This led to the hypothesis that the circadian light cycle might act on sleeping and feeding by influencing the tonus of secretion of metabolic hormones mediating the lipogenic vs. lipolytic ratio (Danguir and Nicolaidis, 1980). However, plasma insulin and GH are not changed by photoperiod in the non-pregnant ewe.

Meier's hypothesis can also be compared with the fact that the *in vitro* synergism between insulin and corticosteroids for lipogenic effects was more important in sheep than in cattle adipose tissue explants (Chilliard and Faulconnier, 1995; Faulconnier *et al.*, 1996), bearing in mind that the latter species is less photoperiod-sensitive. Nevertheless, the lack of a clear effect of season or photoperiod on plasma insulin and corticosteroid concentrations in sheep does not argue for a simple extrapolation to the ovine species of the PRL–glucocorticoid–insulin hypothesis, although it appears to be applicable to other species. There is a need for information on circadian variations of these three hormones in sheep, the sensitivity of adipose tissue to their effects, and on whether their secretion is altered by variations in daylength.

The mechanisms by which photoperiod influences leptin gene expression in adipose tissue remain unknown. It has been shown recently that intravenous injection of neuropeptide Y (NPY) increased leptin mRNA level in ovine adipose tissue (Dyer *et al.*, 1997), but it is not known whether daylength affects NPY or related peptides. It has also been shown that glucocorticoids can modulate leptin in monogastric and ruminant species (Houseknecht *et al.*, 1998; Chilliard *et al.*, 1999).

In the Siberian male hamster, exposure to short days activated the sympathetic nervous system and increased norepinephrine turnover (McElroy *et al.*, 1986). A similar catecholaminergic effect of short days, which remains to be demonstrated in sheep, could account for the increase in plasma non-esterified fatty acid concentrations, as well as for decreases in adipose tissue leptin and lipogenic activities (Fig. 12.1 and Table 12.1). Catecholamines are lipolytic and antilipogenic in sheep and it was recently shown that they lowered leptin concentrations in rodents (Trayhurn *et al.*, 1995) and cattle (Chilliard *et al.*, 1998a). Differences in catecholaminergic innervation between anatomical sites of adipose tissues could explain why short days deplete more internal than subcutaneous adipose tissue, a phenomenon which could serve to enhance insulation from the cold during winter (Youngstrom and Bartness, 1995). However, the effect of short days on the pattern of adipose tissue depletion, which was observed in males, was not apparent in female Siberian hamsters (Bartness, 1995). In ovariectomized ewes, short days decreased lipoprotein lipase activity to a greater extent in subcutaneous than in perirenal adipose tissue, and in skeletal than in cardiac muscle (Table 12.1). The noradrenergic sensitivity of adipocytes was not changed in Djungarian hamsters exposed to short days, suggesting that their rapid fat mobilization is determined centrally by the chronic activation of the sympathetic nervous systems, without desensitization of adipocyte β-adrenergic lipolysis (Mercer *et al.*, 1995).

## Mammary gland

The effects of photoperiod on lactation can be analysed either at the level of whole body metabolism and nutrient partitioning or at the level of mammogenesis (i.e. the formation of a functional mammary gland) and galactopoiesis (milk production).

The mammogenic effect of long photoperiod observed in ewes can be explained by the elevated and prolonged pre-partum PRL surge (Bocquier, 1985; Perier *et al.*, 1986); prolactin is necessary for the complete structural differentiation of epithelial cells and for the onset of copious milk production (Knight, 1993; Kann, 1997). A role for placental lactogen in photoperiod-induced differences in mammogenesis is, however, doubtful, since circulating concentrations were unchanged by daylength (Perier *et al.*, 1986). Artificial induction of lactation is an interesting model since it avoids the interference with numerous hormones that are prevalent during pregnancy. Prealpes maiden ewes, receiving an oestroprogestative treatment before being submitted to the machine-milking stimulus, produced more milk (+27%) under long-day treatment than under short-day treatment (Kann, 1997). In addition to the classical increment of circulating PRL, there was an increase in plasma concentrations of GH (+136%) and IGF-I (+23%), concurrently with an enhancement of GH hepatic receptor number.

Although galactopoeitic effects of GH have been clearly demonstrated in cows (Bauman and Vernon, 1993), its mechanism of action remains controversial. GH may not act directly on the mammary gland but via IGF-I. Although the plasma concentration of GH is generally not altered by photoperiod in the lactating sheep, IGF-I was elevated moderately in the ewe (Kann, 1997) and substantially in cattle (Dahl *et al.*, 1997) subjected to long daylength. These observations can be related to an increase in the number of GH receptors in the liver (Kann, 1997). It seems likely then that photoperiod-induced changes in IGF-I concentration may be partly responsible for effects on milk production.

Although systematically elevated by long daylength, the exact role of PRL on lactation is not clear, even if this hormone exerts some important galactopoietic effects in the ewe, a smaller effect in the goat (Knight and Flint, 1995) and is without effect in cattle (Plaut *et al.*, 1987). In the ewe it was demonstrated (Hooley *et al.*, 1978; Kann *et al.*, 1978; Gow *et al.*, 1983) that a pharmacological PRL depression, using bromocriptine, decreased milk yield by 20–30%. Furthermore, the negative effects of bromocriptine on milk yield were reversed by concurrent infusion of PRL (Hooley *et al.*, 1978). However, since milk production was not totally suppressed by the almost complete ($< 1$ ng ml$^{-1}$) depression of PRL, this hormone may not be absolutely necessary in the galactopoeitic hormonal complex of the ewe (Gow *et al.*, 1983). It is worth noting, however, that sheep, which have been less subjected to selection for milk production than dairy cattle, are also more sensitive to the effect of PRL and to the effects of photoperiod, while dairy cattle may have become exquisitely sensitive to small amounts of circulating PRL (Knight, 1993), and hence less dependent on photoperiod because PRL concentration would never be limiting. Separate consideration of the roles of GH or PRL is probably an oversimplification, since it has been shown in goats that GH response is higher when given together with prolactin than when given alone (Knight and Flint, 1995).

Although ewes exposed to long daylength had a higher milk-solid yield, their milk was more diluted than that of short-day exposed ewes. One hypothesis is that such a dilution effect may be attributed to high levels of PRL observed in ewes exposed to long daylength. In the case of dairy cattle, injection of exogenous PRL had no effect on milk concentration of fat, protein or lactose and had a small positive effect on α-lactalbumin (Plaut *et al.*, 1987). There is, however, evidence that PRL acts on water and solute transport across mammary epithelia, although no direct effect has been isolated (Sheenan, 1994).

The stimulatory effect of photoperiod on milk production is accompagnied by a readjustment of whole body energy metabolism, with a decrease in body fat (Fig. 12.3) before an increase in food intake. The effect of photoperiod may override and exacerbate the homeorhetic (from Greek meaning 'uniform flow'; Bauman and Currie, 1980) or teleophoretic (from Greek meaning 'a transport that is oriented towards a goal'; Chilliard, 1986, 1999) mechanisms whereby an animal re-adjusts its homeostatic setpoints to new physiological conditions. Besides the complex interactions linked to the circadian and seasonal rhythms of hormonal secretions and tissue sensitivity to hormones, the effects of PRL on adipose tissue are also dependent on the presence or the absence of a functional lactating mammary gland, as there are antilipogenic or lipolytic effects in lactating rats, but few effects in virgin or pregnant rats (Vernon and Flint, 1983). The mammary gland was, indeed, proposed to act as an endocrine gland (Maule Walker and Peaker, 1978) and thus could play its own role in teleophoretic adaptations which ensure a high priority of nutrient use for lactation.

## Conclusion

Wild animals have developed numerous non-exclusive adaptive strategies to survive in their environmental conditions. In non-equatorial regions, photoperiod is the most noise-free signal of the annual variations in food resources and general climatic conditions. Logically, the reproductive process is strongly linked to light duration, since it induces the succession of pregnancy and lactation – both of which lead to high nutritional requirements. As the feed resources available to meet this demand are generally maximal in spring and/or summer, the reproduction of the different species occurs at different periods of the year according to gestation length. Because of the wide variation in the length of pregnancy, there must be a great diversity of mechanisms whereby animal species translate the photoperiodic signals into biological adaptations, although there appears to be a strong communality of central neuroendocrine and peripheral signalling systems (melatonin and PRL responses) between species (Morgan and Mercer, 1994).

Although domesticated, sheep have not lost the ability to respond to photoperiod. It is of interest to note the differences in time-scale of the maximal response of physiological functions to changes in daylength, namely: milk composition (less than a week), milk production (a few weeks), adipose tissue metabolism (several weeks) and basal metabolic rate and food intake (several months). In parallel with this, hormonal changes are very rapid for prolactin (a few days), rapid for leptin and IGF-I (less than a month), or uncertain (GH, T3, T4, corticosteroids). These results together confirm the hypothesis formulated by Ortavant *et al.* (1988) that photoperiod may also be involved in metabolic adaptations that accompany the high nutritional requirements for pregnancy and lactation in the sheep.

Our proposed interpretation of the effect of photoperiod on metabolic regulation in sheep is illustrated in Fig. 12.4. Homeostasis in the dry non-productive ewe (Fig. 12.4a) reflects regulatory processes that ensure the constancy of internal conditions for vital processes (according to given set-points) in the face of changes in environmental conditions. We speculate that the set-point for body lipids is diminished by short days. Reproduction, which is favoured by short days, may be impaired if ewes are underfed. Hence, the increase in non-esterified fatty acids and decrease in leptin by short days can be seen as facilitating the general adaptation of ewes to winter undernutrition, and increasing the sensitivity of reproduction blockade in these adverse conditions.

When the reproductive process is successful, the organism is then subjected to teleophoresis, i.e. the orchestrated control of body tissue metabolism necessary to support physiological functions such as pregnancy or lactation. In the case of the ewe, lactation occurs naturally during long days due to the length of gestation. Here again, the physiological regulation is probably changed because milk yield is increased by daylength at the expense of body reserves (Fig. 12.4b), and later on food intake is also

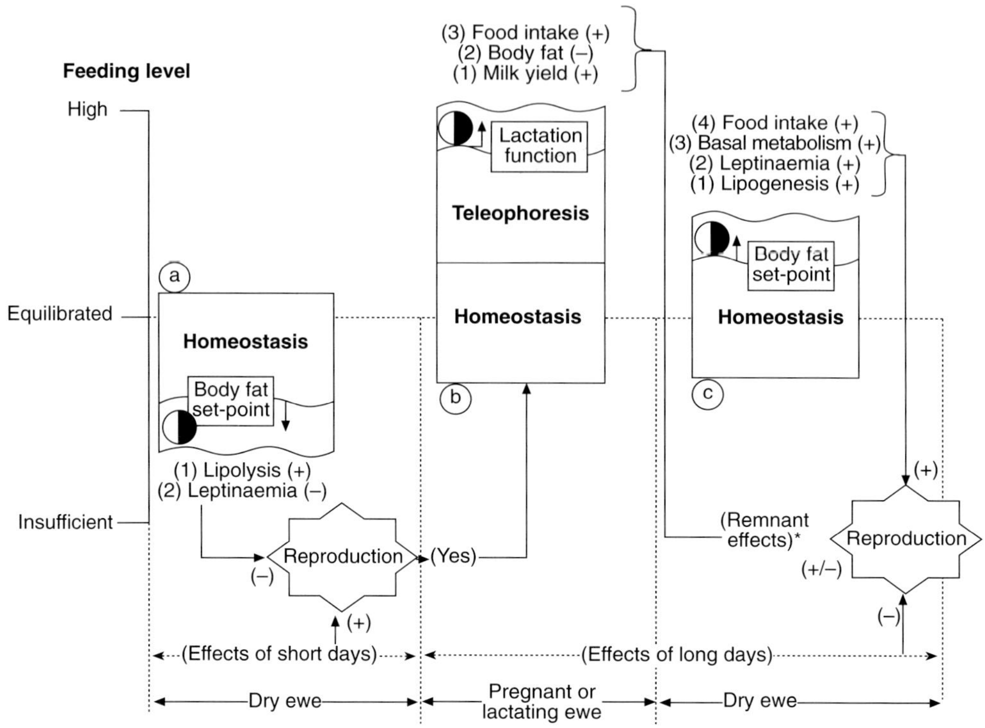


**Fig. 12.4.** Diagrammatic representation of the effects of photoperiod (◑) on energy metabolism regulation and body fat set-point during the reproductive cycle in the ewe (* remnant effects due to changes in body fatness during lactation).

increased. After drying off, ewes are generally still naturally subjected to long days (Fig. 12.4c), and body fat is restored towards a set-point that is higher than during the previous short-day season. This is in apparent contradiction with elevated leptin levels, which would decrease food intake and adipose tissue lipogenesis (Houseknecht *et al.*, 1998; Chilliard *et al.*, 1999). As this does not occur, central and peripheral resistance to leptin probably occurs. The sheep subjected to long days could thus furnish a new model to study leptin resistance, a key issue in obesity studies.

The biological significance of this ability to restore body reserves during long days is that it occurs at a period when long photoperiod postpones the occurrence of the next reproductive cycle and when an anticipatory body reserve replenishment, when food resource is abundant, is necessary to support future reproduction. Similarly, the increase in muscle lipoprotein lipase activity could reflect an adaptation for sparing body protein despite increasing physical activity and foraging, since the enzyme controls, in part, the entry of energy fuel, as fatty acids, into muscle cells. These observations can be viewed as annual rhythms regulating the expression of the thrifty genotype/phenotype, or fattening physiology, that enabled animals and human ancestors to adapt efficiently to large seasonal changes in food availability and in the nutrient content of food (Meier and Cincotta, 1996).

This concept of circannual adaptation through photoperiod re-adjusting set-points for both homeostasis and teleophoresis is in keeping with the views of Mercer (1998) for seasonal mammals, and helps to understand why a ewe placed in phase with optimal daylength is particularly efficient and well adapted to its environment. This set-point re-adjustment, which results from the photoperiodic rhythms, is anticipatory and preventive rather than corrective as is the homeostatic regulation (Mrosovsky, 1976). The independent effects of photoperiod on reproductive and non-reproductive functions such as cyclicity, milk yield, adipose tissue metabolism, energy expenditure and food intake indicates that sheep metabolism is temporally coordinated by different and complementary (or redundant) mechanisms towards a general adaptation to annual fluctuations in food resources. This is of practical interest when using these animals in extensive farming systems, in which their ability to cope with fluctuations of food resources may be of ecological and economic importance.

## Acknowledgements

We thank Yannick Faulconnier, Muriel Bonnet, Guy Kann and Daniel Sauvant for helpful discussions during the preparation of the manuscript, and Pascale Béraud for the artwork and secretarial assistance. This research was supported by the INRA 'Lipogenesis in farm animals' and 'Leptin and reproduction in farm animals' grants.

## References

Almeida, O.F.X. and Lincoln, G.A. (1984) Reproductive photorefractoriness in rams and accompanied changes in the pattern of melatonin and prolactin secretion. *Biology of Reproduction* 30, 143–153.

Barber, M.C., Travers, T., Finley, E., Flint, D.J. and Vernon, R. (1992) Growth-hormone–prolactin interactions in the regulation of mammary and adipose tissue-tissue acetyl-CoA carboxylase activity and gene expression in lactating rats. *Biochemical Journal* 285, 469–475.

Bartness, T.J. (1995) Short-day-induced depletion of lipid stores is fat pad- and gender-specific in Siberian hamsters. *Physiology and Behaviour* 58, 539–550.

Bartness, T.J. and Wade, G.N. (1984) Photoperiodic control of body weight and energy metabolism in Syrian hamsters (*Mesocricetus auratus*): Rôle of pineal gland, melatonin, gonads, and diet. *Endocrinology* 114, 492–498.

Bauman, D.E. and Currie, W.B. (1980) Partitioning of nutrients during pregnancy and lactation: a review of mechanism involving homeostasis and homeorhesis. *Journal of Dairy Science* 63, 1514–1529.

Bauman, D.E. and Vernon, R.G. (1993) Effects of exogenous bovine somatotropin on lactation. *Annual Review of Nutrition* 13, 347–461.

Blaxter, K.L. and Boyne, A.W. (1982) Fasting and maintenance metabolism of sheep. *Journal of Agricultural Sciences, Cambridge* 99, 611–620.

Bocquier, F. (1985) Influence de la photopériode et de la température sur certains équilibres hormonaux et sur les performances zootechniques de la brebis en gestation et en lactation. *Thèse Doc Ing Sci Agro, Institut National Agronomique Paris-Grignon*, 105 pp.

Bocquier, F., Thériez, M., Kann, G. and Delouis, C. (1986) Influence de la photopériode sur la partition de l'énergie nette entre la production laitière et les réserves corporelles chez la brebis laitière. *Reproduction Nutrition Development* 26, 389–390.

Bocquier, F., Kann. G. and Thériez, M. (1990) Relationships between secretory patterns of growth hormone, prolactin and body reserves and milk yield in dairy ewes under different photoperiod and feeding conditions. *Animal Production* 51, 115–125.

Bocquier, F., Ligios, S., Molle, G. and Casu, S. (1997) Effets de la photopériode sur la production, la composition du lait et sur les consommations volontaires chez la brebis laitière. *Annales de Zootechnie* 46, 427–438.

Bocquier, F., Bonnet, M., Faulconnier, Y., Guerre-Millo, M., Martin, P. and Chilliard, Y. (1998) Effects of photoperiod and feeding level on perirenal adipose tissue metabolic activity and leptin synthesis in the ovariectomized ewe. *Reproduction Nutrition Development* 38, 489–498.

Bocquier, F., Guilllouet, P., Barillet, F. and Chilliard, Y. (1999) Comparison of three methods for the *in vivo* estimation of body composition in dairy ewes. *Annales de Zootechnie* 48, 297–308.

Chilliard, Y. (1986) Revue bibliographique: Variations quantitatives et métabolisme des lipides dans les tissus adipeux et le foie au cours du cycle gestation-lactation. 1. Chez la ratte. *Reproduction Nutrition Development* 26, 1057–1103.

Chilliard, Y. (1987) Revue bibliographique: Variations quantitatives et métabolisme des lipides dans les tissus adipeux et le foie au cours du cycle gestation-lactation. 2. Chez la brebis et la vache. *Reproduction Nutrition Development* 27, 327–398.

Chilliard, Y. (1999) Metabolic adaptations and nutrient partitioning in the lactating animal. In: Martinet, J., Houdebine, L.M. and Head, H.H. (eds) *Biology of Lactation*. INRA Ed., Paris, France, pp. 503–552.

Chilliard, Y. and Faulconnier, Y. (1995) Insulin and(or) dexamethasone regulation of glucose and acetate utilization in ovine and bovine adipose tissue explants incubated for seven days. *Journal of Animal Science* 73, 2063–2070.

Chilliard, Y., Bocquier, F., Delavaud, C., Guerre-Millo, M., Bonnet, M., Martin, P., Faulconnier, Y. and Ferlay, A. (1998a) Leptin in ruminants: effects of species, breed, adiposity, photoperiod, beta-agonists and nutritional status. *Proceedings of the Cornell Nutrition Conference for Feed Manufacturers. October 1998, Cornell University, New York, USA,* pp. 65–74.

Chilliard, Y., Faulconnier, Y., Bonnet, M. and Bocquier, F. (1998b) Mid-term effect of photoperiod on adipose tissue lipogenesis, plasma metabolites, insulin, somatotropin and prolactin in underfed-refed ewes. In: McCracken, K., Unsworth, E.K. and Wylie, A.R.G. (eds) *Proceedings of the 14th Symposium on Energy Metabolism of Farm Animals*. CAB International, Wallingford, pp. 67–70.

Chilliard, Y., Bocquier, F., Delavaud, C., Faulconnier, Y., Bonnet, M., Guerre-Millo, M., Martin, P. and Ferlay, A. (1999) La leptine chez le ruminant. Facteurs de variation physiologiques et nutritionnels. *INRA Productions Animales* 12, 225–237.

Cincotta, A.H. and Meier, A.H. (1989a) Reduction of body fat stores and total plasma cholesterol and triglyceride concentrations in several species by bromocriptine treatment. *Life Sciences* 45, 2247–2254.

Cincotta, A.H. and Meier, A.H. (1989b) Prolactin influences the circadian rhythm of lipogenesis in primary cultured hepatocytes. *Hormone and Metabolic Research* 21, 64–68.

Cincotta, A.H., Meier, A.H. and Southern, L.L. (1989) Bromocriptine alters hormone rhythms and lipid metabolism in swine. *Annals of Nutrition and Metabolism* 33, 305–314.

Cincotta, A.H., MacEachern, T.A. and Meier, A.H. (1993a). Bromocriptine redirects metabolism and prevents seasonal onset of obese hyperinsulinemic state in Syrian hamsters. *American Journal of Physiology* 264, E285-E293.

Cincotta, A.H., Schiller, B.C., Landry, R.J., Herbert, S.J., Miers, W.R. and Meier, A.H. (1993b) Circadian neuroendocrine role in age-related changes in body fat stores and insulin sensitivity of the male Sprague-Drawley rat. *Chronobiology International* 10, 244–258.

Dahl, G.E., Elhasser, T.H., Capuco, A.V., Erdman, R.A. and Peters, R.R. (1997) Effects of a long daily photoperiod on milk yield and circulating concentrations of insulin-like growth factor-I. *Journal of Dairy Science* 80, 2784–2789.

Danguir, J. and Nicolaidis, S. (1980). Circadian sleep and feeding patterns in the rat: possible dependence on lipogenesis and lipolysis. *American Physiological Society* 238, E223–E230.

Dyer, C.J., Simmons, J.M., Matteri, R.L. and Keisler, D.H. (1997) Effects of an intravenous injection of NPY on the leptin and NPY-Y1 receptors mRNA expression in ovine adipose tissue. *Domestic Animal Endocrinology* 14, 325–333.

Eisemann, J.H., Bauman, D.E., Hogue, D.E. and Travis, H.F. (1984) Influence of photoperiod and prolactin on body composition and in vitro lipid metabolism in wether lambs. *Journal of Animal Science* 59, 95–104.

Emane, M.N., Delouis, C., Kelly, P.A. and Djiane, J. (1986) Evolution of prolactin and placental lactogen receptors in ewes during pregnancy and lactation. *Endocrinology* 118, 695–700.

English, D.E., Russel, S.M., Kartz, L.S. and Nicoll, C.S. (1990) Evidence for a role of the liver in the mammotrophic action of prolactin. *Endocrinology* 126, 2252–2256.

Etherton, T.D., Evock, C.M. and Kensinger, R.S. (1987) Native and recombinant bovine growth hormone antagonize insulin action in cultured bovine adipose tissue. *Endocrinology* 121, 699–703.

Faulconnier, Y., Guillon, L. and Chilliard, Y. (1996) Lipoprotein lipase and glucose-6-phosphate dehydrogenase activities in bovine and ovine adipose tissue incubated for 7 days : effects of insulin and/or dexamethasone. *Comparative Biochemistry and Physiology* 113B, 421–426.

Faulconnier, Y., Bonnet, M., Bocquier, F., Leroux, C., Hocquette, J.F., Martin, P. and Chilliard, Y. (1999) Régulation du métabolisme lipidique des tissus adipeux et musculaires chez le ruminant. Effets du niveau alimentaire et de la photopériode. *INRA Productions Animales* 12, 287–300.

Forbes, J.M. (1982) Effects of lighting patterns on growth, lactation and food intake in the sheep, cattle and deer. *Livestock Production Science* 9, 361–374.

Gordon, J.G. (1964) Effect of the time of the year on the roughage intake of housed sheep. *Nature* 204, 798–799.

Gow, C.B., McDowell, G.H. and Jenkin, G. (1983) The importance of prolactin for initiation of lactation in the pregnant ewe. *Australian Journal of Biological Science* 36, 357–367.

Hooley, R.D., Campbell, J.J. and Findlay, J.K. (1978) The importance of prolactin for lactation in the ewe. *Journal of Endocrinology* 79, 301–310.

Houseknecht, K.L., Bauman, D.E., Vernon, R.G., Byatt, J.C. and Collier, R.J. (1996) Insulin-like growth factors-I–II, somatotropin, prolactin, and placental lactogen are not acute effectors of lipolysis in ruminants. *Domestic Animal Endocrinology* 13, 239–249.

Houseknecht, K.L., Baile, C.A., Matteri, R.L. and Spurlock, M.E. (1998) The biology of leptin: a review. *Journal of Animal Science* 76, 1405–1420.

Jacquemet, N. and Prigge, E.C. (1991) Effect of increased postmilking prolactin concentrations on lactation, plasma metabolites, and pancreatic hormones in lactating goats. *Journal of Dairy Science* 74, 109–114.

Kann, G. (1997) Evidence for a mammogenic role of growth hormone in ewes: effects of growth hormone-releasing factor during artificial induction of lactation. *Journal of Animal Science* 75, 2541–2549.

Kann, G., Carpentier, M.C., Fevre, J., Martinet, J., Maubon, M., Meusnier, C., Paly, J. and Vermeyre, N. (1978) Lactation and prolactin in sheep, role of prolactin in initiation of lactation of milk secretion. In: Robyn, C. and Harter, M. (eds) *Progress in Prolactin Physiology and Pathology.* Elsevier North-Holland Biomedical Press, Amsterdam/New York, pp. 201–212.

Kay, R.N.B. (1985) Seasonal variation of appetite in ruminants. In: Haresign, W. and Cole, D.J.A. (eds) *Recent Advances in Animal Nutrition*. Butterworths, London, pp. 199–210.

Klingenspor, M., Dickopp, A., Heldmaier, G. and Klaus, S. (1996) Short photoperiod reduces leptin gene expression in white and brown adipose tissue of Djungarian hamsters. *FEBS Letters* 399, 290–294.

Knight, C.H. (1993) Prolactin revisited. *Hannah Research Institute Yearbook*, pp. 72–80.

Knight, C.H. and Flint, D.J. (1995) GH : Prolactin interactions in lactating rodents and ruminants. *Hannah Research Institute Yearbook*, pp. 73–79.

Larsen, T.S., Lagercrantz, H., Riemersma, R.A. and Blix, A.S. (1985a) Seasonal changes in blood lipids, adrenaline, noradrenaline, glucose and insulin in Norwegian reindeer. *Acta Physiologica Scandinavica* 124, 53–59.

Larsen, T.S., Nilsson, N.Ö. and Blix, A.S. (1985b) Seasonal changes in lipogenesis in isolated adipocytes from Svalbard and Norwegian reindeer. *Acta Physiologica Scandinavica* 123, 53–59.

Le Gouic, S., Atgié, C., Viguerie-Bascands, N., Hanoun, N., Larrouy, D., Ambid, L., Raimbault, S., Ricquier, D.P., Gardiola-Lemaitre, B., Pénicaud, L. and Casteilla, L. (1997) Characterisation of a melatonin binding site in Siberian hamster brown adipose tissue. *European Journal of Pharmacology* 339, 271–278.

Lincoln, G.A. and Richardson, M. (1998) Photo-endocrine control of seasonal cycles in body weight, pelage growth and reproduction: lessons from the HPD sheep model. *Comparative Biochemistry and Physiology* 119C, 283–294.

Lincoln, G.A., McNeilly, A.S. and Cameron, C.L. (1978) The effects of a sudden decrease or increase of daylength on prolactin secretion in the ram. *Journal of Reproduction and Fertility* 52, 305–311.

Linzell, J.L. (1973) Innate seasonal oscillations in the rate of milk secretion in goats. *Journal of Physiology* 230, 225–233.

Luo, S., Luo, J., Meier, A.H. and Cincotta, A.H. (1997) Dopaminergic neurotoxin administration to the area of the suprachiasmatic nuclei induces insulin resistance. *NeuroReport* 8, 3495–3499.

Luthman, J. and Johnson, G. (1972) Short term hormonal effects on blood glucose and non-esterified fatty acids in female sheep. *Acta Veterinaria Scandinavica* 13, 112–127.

McElroy, J., Mason, P., Hamilton, J. and Wade, G. (1986) Effects of diets and photoperiod on NE and GDP binding in Siberian hamster brown adipose tissue. *American Journal of Physiology* 250, R383–388.

Malpaux, B., Viguié, C., Thièry, J.C. and Chemineau, P. (1996) Contrôle photopériodique de la reproduction. *INRA Productions Animales* 9, 9–23.

Maule Walker, F.M. and Peaker, M. (1978) Production of oestradiol-17ß by the goat mammary gland during late pregnancy in relation to lactogenesis. *Journal of Physiology* 284, 71–72.

Meier, A.H. and Burns, J.T. (1976) Circadian hormone rhythms in lipid regulation. *American Zoology* 16, 649–659.

Meier, A.H. and Cincotta, A.H. (1996) Circadian rhythms regulate the expression of the thrifty genotype/phenotype. *Diabetes Reviews* 4, 464–487.

Mercer, J.G. (1998) Regulation of appetite and body weight in seasonal mammals. *Comparative Biochemistry and Physiology* 119C, 295–303.

Mercer, J.G., Bruce, C. and Morgan, P.J. (1995) Fat mobilisation in short days is not associated with altered noradrenergic sensitivity of adipocytes in Djungarian hamsters. *Physiology and Behaviour* 57, 249–253.

Mills, S.E., Lemenager, R.P. and Horstman, L.A. (1989) Effect of suppression of prolactin secretion on adipose tissue lipogenesis in the postpartum beef cow. *Journal of Animal Science* 67, 2904–2912.

Milne, J.A., MacRae, J.C., Spence, A.M. and Wilson, S. (1978) A comparison of the voluntary intake and digestion of a range of forages at different times of the year by sheep and the red deer (Cervus Elaphus). *British Journal of Nutrition* 40, 347–356.

Morgan, P.J. and Mercer, J.G. (1994) Control of seasonality by melatonin. *Proceedings of the Nutrition Society* 53, 483–493.

Mrosovsky, N. (1976) Lipid programmes and life strategies in hibernators. *American Zoology* 16, 685–697.

Ortavant, R., Bocquier, F., Pelletier, J.P., Ravault, J.P., Thimonier, J. and Volland-Nail, P. (1988) Seasonality of reproduction in sheep and its control by photoperiod. *Australian Journal of Biological Science* 41, 69–85.

Perier, A., Bocquier, F., Kann, G. and Martinet, J. (1986) Influence de la photopériode et des apports énergétiques pendant la gestation sur les taux plasmatiques de prolactine, de l'hormone de croissance et de l'hormone placentaire lactogène puis sur la production laitière des brebis traite. *Reproduction Nutrition Development* 26, 391–392.

Petitclerc, D. and Zinn, S.A. (1991) Effect of photoperiod on growth and development of ruminants. In: Pearson, A.M. and Dutson, T.R. (eds) *Growth Regulation in Farm Animals.* Elsevier, London, New York, pp. 427–453.

Plaut, K., Bauman, D.E., Agergaard, N. and Akers, R. (1987) Effect of exogenous prolactin administration on lactational performances of dairy cows. *Domestic Animal Endocrinology* 4, 279–290.

Reimers, E., Ringberg, T. and Sorumgaard, R. (1982) Body composition of Svalbard reindeer. *Canadian Journal of Zoology* 60, 1812–1821.

Sheenan, D.B. (1994) Regulation of water and solute transport across mammalian plasma cell membranes by prolactin. *Journal of Dairy Research* 61, 155–166.

Trayhurn, P., Duncan, J.S. and Rayner, D.V. (1995) Acute cold-induced suppression of ob (obese) gene expression in white adipose tissue of mice: mediation by the sympathetic system. *Biochemical Journal* 311, 729–733.

Vernon, R.G. and Finley, E. (1986) Lipolysis in rat adipocytes during recovery from lactation. *Biochemical Journal* 234, 229–231.

Vernon, R.G. and Finley, E. (1988) Roles of insulin and growth hormone in the adaptations of fatty acid synthesis in white adipose tissue during the lactation cycle in sheep. *Biochemical Journal* 256, 873–878.

Vernon, R.G. and Flint, D.J. (1983) Control of fatty acids synthesis in lactation. *Proceedings of the Nutrition Society* 42, 315–331.

Vernon, R.G., Clegg, R.A. and Flint, D.J. (1986) Adipose tissue metabolism in sheep: response to season and its modulation by reproductive state. *Hormone and Metabolic Research* 18, 308–312.

Viguié, C., Thibault, J., Thièry, J.C., Tillet, Y. and Malpaux, B. (1997) Characterization of the short day-induced decrease in median eminence tyrosine hydroxylase activity in the ewe: temporal relationship to the change in luteinizing hormone and prolactin secretion and short day-like effect of melatonin. *Endocrinology* 138, 499–506.

Walker, V.A., Young, B.A. and Walker, B. (1991) Does photoperiod directly influence energy metabolism? In: Wenk, C. and Boessinger, M. (eds) *Proceedings of the 12th Symposium on Energy Metabolism of Farm Animals* (EAAP Publication no 58), pp. 372–375.

Williams, L.M., Lincoln, G.A. and Morgan, P.J. (1995) Melatonin: Sites and mode of action in the sheep. In: Engelhardt, W. von, Leonhard-Marek, S., Breves, G. and Giesecke, D. (eds) *Proceedings of the Eighth International Symposium on Ruminant Physiology.* Ferdinand Enke Verlag, Stuttgart, pp. 541–557.

Youngstrom, T.G. and Bartness, T.J. (1995) Catecholaminergic innervation of white adipose tissue in Siberian hamsters. *American Physiological Society* 268, R744-R751.

Yu, W.H., Kimura, M., Walczewska, A., Karanth, S. and McCann, S.M. (1997) Role of leptin in hypothalamic-pituitary function. *Proceedings of the National Academy of Sciences USA* 94, 1023–1028.

# V Tissue Growth

# 13 Muscle Growth and Genetic Regulation

J.J. Bass, M. Sharma, J. Oldham and R. Kambadur

*AgResearch, Ruakura Agricultural Research Centre, Hamilton, New Zealand*

## Introduction

Growth of a whole animal, or its organs, or tissues, is clearly under a coordinated, integrated control system, of which there is limited understanding (Conlon and Raff, 1999). The size of an organ/tissue depends on the number and size of its cells and extracellular components. Therefore, the growth of a specific tissue, such as muscle, is dependent on the number of myogenic progenitor cells, proliferation of these committed myogenic cells (myoblasts) and the subsequent hypertrophic growth of these muscle fibres. An increase in total muscle mass is the result of a balance between cell proliferation, programmed cell death and cell/fibre growth. Growth of an individual muscle and the overall coordination of muscle growth throughout the body requires the integration of both local and systemic control systems. Research to date has largely focused on local control systems and how they interact with systemic systems, such as the growth hormone (GH) axis. Little work on the coordination of whole-body muscle growth has been undertaken, yet, in numerous selection trials, weight gains in cattle and sheep have shown that coordinated overall growth and associated muscle growth is highly heritable and therefore under genetic control (Notter, 1999).

This chapter will briefly review the stages and known control systems of muscle development and then will focus on some of the specific genes which have been shown to regulate muscle growth in ruminants. The continuous development of muscle can be divided into three basic phases for convenience: determination of progenitor cells, proliferation of myoblasts, differentiation and growth of muscle fibres (Fig. 13.1).

## Determination

Mesodermal cells are the source of all vertebrate skeletal muscle (Cossu *et al.*, 1996) and these originate from progenitor cells in the embryonic epiblast. The mesoderm, which initially consists of lateral plates, divides into somatic mesoderm, splanchnic

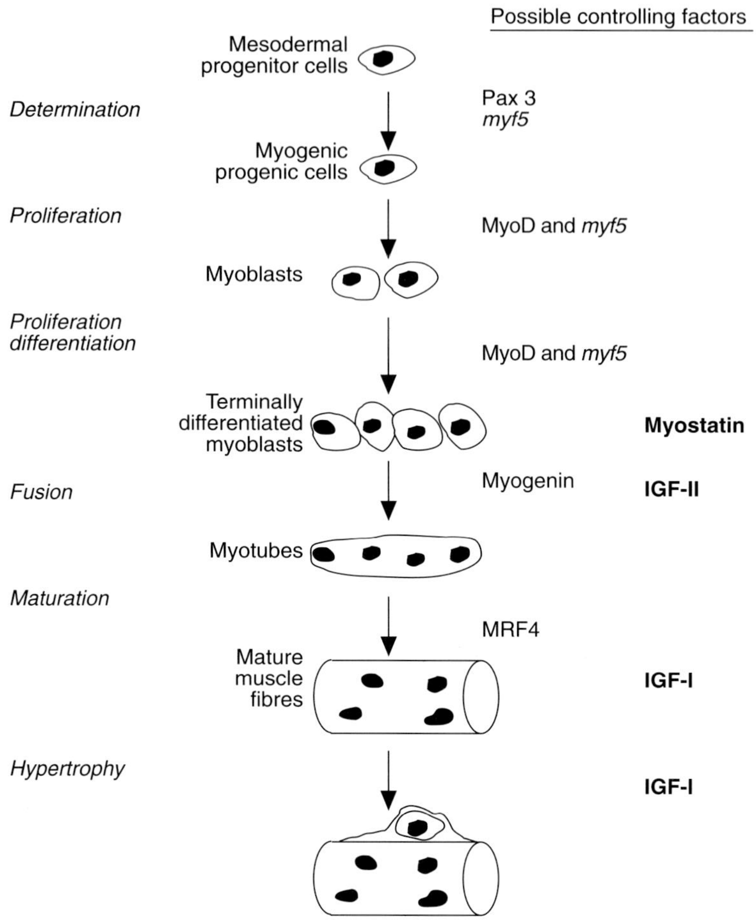


**Fig. 13.1.** Determination, proliferation, fusion and differentiation of myogenic cells and controlling factors.

mesoderm, intermediary mesoderm and paraxial mesoderm. The paraxial mesoderm then develops into somites (Hauschka, 1994). The ventral part of the somite mesoderm forms the sclerotome (Fig. 13.2) and the dorsal part becomes the dermomyotome that develops into dermatome and myotome. Both the axial and appendicular muscles are derived from myotome. The dorsal myotome forms the epiaxial back muscles, while cells from the lateral myotome migrate to the limbs to form limb muscles or hypoaxial muscles of the ventral body wall.

The fate of mesodermal cells appears to depend on external factors from surrounding tissue. The position of mesoderm relative to the neural tube, notochord, dorsal ectoderm and lateral plate mesoderm plays an important role in determining whether somatic cells will become myogenic precursor cells. A number of factors from these surrounding tissues have been identified and associated with the induction of

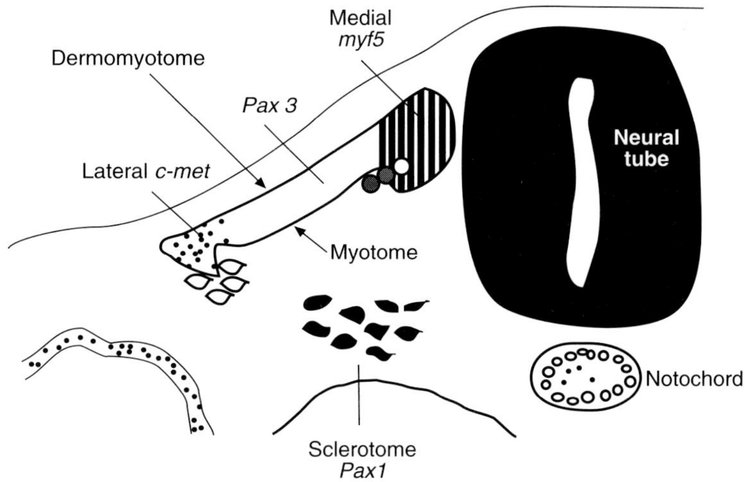


**Fig. 13.2.** Induction of myogenesis by factors from surrounding tissues.

myogenic cells (Cossu *et al.*, 1996; Capdevila *et al.*, 1998) (Fig. 13.1). Examples of inducing factors are: Pax 3, which has been shown to be essential for the determination of myogenic progenitor cells and their migration into limb muscles (Cossu *et al.*, 1996) and sonic hedgehog, which has been identified as a candidate for inducing the onset of myogenesis (Marcelle *et al.*, 1997).

The myogenic regulatory factors (MRFs), members of the helix-loop-helix group of transcription factors, are expressed in the medial dermomyotome and may have a role in determination of myogenic cells as well as late stages of muscle fibre development. Both Myf5 and MyoD are believed to be essential for the early development of muscle, as mice deficient in both lack normal myoblasts and skeletal muscle (Edmondson and Olson, 1993).

An understanding of the mechanisms that control the formation of myogenic progenitor cells will allow the future identification of critical gene pathways for ruminants as well as other species. At the moment, no observed variations in these early myogenic pathways have been associated with changes in muscle mass of ruminants.

## Proliferation

The second stage of muscle development is proliferation. During this stage, committed, mononucleated myogenic cells (myoblasts) first proliferate and then differentiate into fusion-competent myoblasts. These fusion-competent myoblasts divide and eventually fuse end-to-end, forming long, multinucleated myotubes. Myoblasts can be divided into three types on developmental age and morphology in culture (Franzini *et al.*, 1994). These are the embryonic or early myoblasts, isolated from embryonic muscles prior to and during primary myotube formation; the fetal or late myoblasts, isolated from fetal muscles throughout secondary fibre formation; and the satellite cells, isolated from adult muscles. The number of myoblast divisions may be important in regulating the skeletal musculature in the growing animal. Myoblasts have limited capacity to proliferate (Grounds and Yablonka-Reuveni, 1993), compared with other cells. Whether a myoblast continues dividing or differentiates appears to be dependent on extracellular cues, such as growth factors, extracellular matrix and cell-to-cell interactions. A large number of *in vitro* studies have indicated that myoblast proliferation can be regulated by growth factors (Florini and Magri, 1989).

To date, the major growth factors which affect myoblast proliferation are fibroblast growth factor, insulin, insulin-like growth factors (IGF), platelet-derived growth factor and transforming growth factor-β (TGF-β). Other mitogens such as epidermal growth factor and leukaemia inhibiting factor have also been shown to affect myoblast proliferation (Hauschka, 1994). Growth factors often have different biological effects which appear to depend on their concentration and the developmental stage of the myoblasts. This divergence of function is seen with IGF-I that can stimulate proliferation as well as maintain differentiation (reviewed by Florini *et al.*, 1996) and in TGF-β that inhibits both proliferation and differentiation of myogenesis (Massagué *et al.*, 1986).

An example of how a change in the control system of myoblast proliferation can affect muscling of cattle is to be found in the 'double-muscled' Belgian Blue breed. In certain breeds, intense selection over many generations has resulted in extreme muscle hypertrophy known variously as double-muscling, culard, etc. (Boccard, 1981). This type of muscle hypertrophy is mainly associated with an increase in the number of muscle fibres, especially secondary muscle fibres. It has been postulated that the increased fibre number results from increased numbers of late myoblasts, either because of increased myoblast proliferation or because of delayed differentiation into myotubes. The inheritance of double-muscling in Belgian Blue cattle has been identified as a monogenic autosomal segregation pattern (Hanset and Michaux, 1985; Charlier *et al.*, 1995). The *muscular hypertrophy* (*mh*) locus has been termed 'partially recessive', because, while a single copy of the allele may have some effect, both are required for full expression of the double-muscled phenotype. Gene mapping (Charlier *et al.*, 1995) of the Belgian Blue cattle localized the *mh* gene to the centromeric end of the bovine chromosome 2 (BTA2) linkage group. The *mh* locus, in more refined studies in the Piedmontese breed, localized the gene to a 3–5-cm interval near the centromere of BTA2 close to the position of the α-collagen type III (COL3A1) locus (Casas *et al.*, 1999).

Recently, growth differentiation factor-8 (GDF-8) (McPherron *et al.*, 1997), a member of the TGF-β superfamily, was disrupted in mice. These GDF-8 null mice were significantly larger than their wild type counterparts, and showed an increase in body weight resulting from a two- to threefold increase in muscle mass. This increase in muscle was due to an increase in the number of muscle fibres. As GDF-8 seemed to function as an inhibitor of muscle growth, it was renamed myostatin. The myostatin gene has since been mapped to the same interval as the *mh* locus in cattle by genetic linkage (Smith *et al.*, 1997). The finding that the myostatin null mice have a similar phenotype to double-muscled cattle and that the gene maps to a similar site, suggests that the myostatin and the muscular hypertrophy genes are one and the same. This hypothesis was further supported by finding an 11 bp mutation in the myostatin gene in Belgian Blue cattle and a point mutation in Piedmontese cattle, both double-muscled breeds (Kambadur *et al.*, 1997).

More recently, Grobet *et al.* (1997) have identified seven DNA sequence polymorphisms, of which five were predicted to disrupt the function of myostatin. These studies clearly demonstrate that the double-muscling phenotype in cattle is genetically heterogeneous and involves several mutations in the myostatin gene. The 11 bp deletion in the Belgian Blue cattle results in the loss of three amino acids, that causes a frame-shift after amino acid 273. This frame-shift leads to a stop codon after amino acid 286 that is predicted to produce a truncated, biologically inactive protein. In the Piedmontese breed, a mutation at position 941 bp results in the loss of a cysteine at amino acid 314 (Kambadur *et al.*, 1997). This cysteine has been shown to be essential in other TGF-β family members in order to form a cysteine knot, which stabilizes the TGF-β dimer. In general, these studies suggest that myostatin is probably the *mh* locus and that it acts as an inhibitor of muscle development by limiting muscle fibre number, and to some extent, muscle fibre size.

When the development of proliferating and fusing myoblasts was followed, it was found that myostatin mRNA was elevated during primary fibre formation through to the early stages of secondary fibre formation. This suggests that myostatin may participate in the proliferation and terminal differentiation of late myoblasts (Oldham *et al.*, 1998). This study also suggested that a link exists between MyoD and myostatin, as MyoD expression was increased during the fibre formation in muscle from double-muscled cattle, which lack myostatin. Interestingly, the different myostatin mutations found in cattle do not give similar increases in muscle mass. This suggests that the myostatin proprotein and the mutant proteins inhibit muscle growth differently. The possibility also exists that associated muscle-controlling genes have been differentially selected in the different breeds.

A further level of control of muscle development by myostatin has been identified in the compact hypermuscular mouse (Szabo *et al.*, 1998). The mutation that causes an increase in muscle mass in the compact mouse is a deletion in the pro-peptide region, which precedes the proteolytic processing site of myostatin. The pro-peptide region, by analogy with TGF-β (Miyazono *et al.*, 1991), may be involved in the folding and secretion of myostatin. Such changes possibly decrease the biological activity of myostatin, but do not completely remove it. Myostatin appears to be a tissue-specific inhibitor of myoblast proliferation that can have a direct effect on myoblast proliferation and so also on muscle fibre number.

Examples of factors that stimulate rather than inhibit myoblast proliferation include IGF-I and -II, as both have been shown to enhance myoblast proliferation in culture (Florini *et al.*, 1996). Whether both IGFs are local growth factors that are critical in controlling myoblast proliferation *in vivo* is unclear, because local expression of IGF-II increases at the end of secondary myofibre formation, whereas IGF-I does not show marked changes associated with stages of fetal muscle formation in cattle (Oldham *et al.*, 1998). In most species, serum concentrations of IGF-I, but not IGF-II, correlate positively with fetal weight (Gluckman and Brinsmead, 1976; Humbel, 1990). The IGFs have been related to fetal growth in general. When gene targeting is used to disrupt IGF-II, the offspring are only 30% of the weight of the wild type (Baker *et al.*, 1993), while IGF-I null mutants are growth retarded, depending on the magnitude of the reduction in IGF-I concentration (Liu *et al.*, 1998). While IGF-I appears to act locally on developing muscle (Tollefsen *et al.*, 1989), it dramatically increases IGF-II mRNA in regenerating adult muscle when satellite cells from healthy muscle divide and migrate into damaged areas to initiate regeneration (Kou and Rotwein, 1993; Kirk *et al.*, 1996). This indicates that IGF-II is important in myoblast proliferation and differentiation.

GH, which is the hormone controlling the expression of IGF-I, does not seem to be directly involved in myoblast proliferation, as mice overexpressing GH do not necessarily show increased muscle growth or increased IGF-I expression at birth (Liu *et al.*, 1998). However, the treatment of pregnant sows with GH early in gestation increases both the number of muscle fibres and the liveweight at birth of their offspring (Rehfeldt *et al.*, 1993). Such an effect of GH may be indirect via maternal changes rather than acting directly on the fetus.

## Differentiation

When myoblasts leave the cell cycle and become terminally differentiated, they fuse with one another to form myotubes (Fig. 13.1). Myotubes are immature muscle fibres which form in distinct phases. Primary myotubes are thought to arise from end-to-end fusion of early myoblasts. Late myoblasts then accumulate on the surface of the pre-existing primary myotubes and fuse to form secondary myofibres. In all muscles, there are many more secondary myotubes than primary myotubes. In large animals such as sheep and cattle, there is possibly a third phase of myogenesis, where the tertiary myofibres use the secondary fibres as a scaffolding, similar to the way developing secondary fibres use primary fibres for support. These tertiary fibres have a different myosin expression and are mainly distributed along the border of muscle fascicles (Franzini *et al.*, 1994). Any change in the number of myofibres which form could have a profound effect on the total muscle mass of the mature animal.

A number of the factors that regulate expression of muscle-specific genes following commitment to terminal differentiation have been established. The MRFs are the best characterized of these factors (Edmondson and Olson, 1993) and their sequential activation commits cells to induce genes required to establish terminally differentiated muscle cells.

Of the MRFs, MyoD and Myf5 are involved in the regulation of myoblasts and satellite cell proliferation, whereas myogenin induces differentiation (Olsen and Klein, 1994). MRF4 is expressed late in muscle differentiation and may share some of the functions of myogenin (Buckingham, 1994). MRFs, however, do not function alone, they require transcriptional co-activators such as the myocyte enhancer factor-2 (MEF2) before they can regulate myoblast and satellite cell proliferation (Olson *et al.*, 1995).

MRFs are also expressed in adult muscle while MyoD protein accumulates in satellite cells of regenerating myotubes and skeletal muscle fibres. This appears to be neurally regulated (Koishi *et al.*, 1995).

There are now indications that polymorphisms in the MRFs could be used as selection markers for improved growth rates in pigs. Polymorphic sites at the myogenin and Myf5 gene loci have been associated with increases in birth weight, postnatal growth and weight of lean meat in pigs (Te Pas *et al.*, 1998). However, Myf5 polymorphisms were not related to increased growth rates. Furthermore, expression levels of MyoD showed no relation to selection for liveweight gain (Te Pas *et al.*, 1998). Similar associations have not so far been reported for ruminants.

Terminal muscle cell differentiation eventually results in the sequential expression of muscle-specific genes, which include desmin, acetylcholine receptors and specific isoforms of actin, myosin heavy chain, troponin, tropomyosin and creatine kinase (Schiaffino and Reggiani, 1996). The expression of these and other genes leads to a differentiated muscle that is fully functional. Polymorphic variations have been reported for some of these genes and many, like myogenin, could be associated with improved meat production. An example is the myosin light chain-1-fast gene that modulates the interactions between actin and myosin during contraction. This gene undergoes differential splicing, which results in two products in chickens (Rushbrook and Somes, 1985) and in cattle (A. Clark, personal communication). However, as yet, no association has been found between these polymorphisms and any desirable muscle characteristics in cattle.

## Hypertrophy

The bulk of muscle growth results from satellite cell division and incorporation of daughter cells into the multinucleated fibre. This enables the nuclei/cytoplasmic ratios of the muscle fibre to be maintained, even though the nuclei of the fibre are unable to undergo division. During normal growth, and in response to stretch, muscle fibres lengthen by the addition of new sarcomeres at the ends of these fibres (Williams and Goldspink, 1973). Under conditions of normal growth, the longitudinal and radial growth of muscle fibres is highly correlated. However, in sheep heavier than 45 kg, there is little further increase in radial growth of muscle fibres, while in cattle, radial fibre growth is rapid at birth. This growth, as in sheep, does level off in older animals (Swatland, 1984).

Specific muscle hypertrophy can be work-induced, which results in the increased local expression of IGF-I and an increase in insulin sensitivity in association with increased glucose metabolism, amino acid transport, protein synthesis and protein degradation. More recently, the expression of the muscle growth inhibitor, myostatin, has been shown to decrease immediately after stretch-induced hypertrophy is initiated (J. Martyn *et al.*, personal communication). However, while muscle fibre hypertrophy is found in the myostatin knockout mouse (McPherron *et al.*, 1997), it has not been identified in double-muscled cattle that have a mutated, non-active myostatin (Boccard, 1981).

There are a number of animal models that show an abnormal increase in postnatal muscle size. The callipyge sheep was derived from a ram with extreme muscling, especially in the hindquarter (Cockell *et al.*, 1994). The fine mapping of the *CLPE* locus to ovine chromosome 18 has so far failed to identify a candidate gene for this trait. The compact mouse (Szabo *et al.*, 1998) also shows a postnatal increase in fibre size, resulting in enlarged hindquarter muscles. This muscle hypertrophy has been associated with a mutation in the myostatin pro-peptide region, which precedes the proteolytic processing site (Szabo *et al.*, 1998). This mutation may, as in the TGF-β proteins (Miyazono *et al.*, 1991), affect the folding of myostatin and hence its biological activity.

The lack of information concerning muscle hypertrophy compared with the number of publications on myoblast proliferation and differentiation results from a lack of suitable *in vitro* models, in which satellite cell division may be observed in association with increased muscle fibre growth. In future, the clever use of transgenic models and

Cre-Lox knockouts (Sauer, 1998), which may be both tissue- and time-based, will allow the identification of genes which may be controlling postnatal growth of muscle fibres, the period during which muscle size increases the most.

## Acknowledgements

Research into growth physiology by AgResearch is funded by the New Zealand Foundation of Research Science and Technology. We thank the organizers for inviting this review and the colleagues who have contributed.

## References

Baker, J., Liu, J-P., Robertson, E.J. and Efstratiadis, A. (1993) Role of insulin-like growth factors in embryonic and post-natal growth. *Cell* 75, 73–82.

Boccard, R. (1981) Facts and reflections on muscular hypertrophy in cattle: double muscling or culard. In: Ralson, L. (ed.) *Developments in Meat Science,* 2. Applied Science Publishers, pp. 1–28.

Buckingham, M. (1994) Which myogenic factors make muscle? *Current Biology* 4(1), 61–63.

Capdevila, J., Tobin, C. and Johnson, R.L. (1998). Control of dorsoventral somite patterning by Wnt-1 and β catenin. *Developmental Biology* 193, 182–194.

Casas, E., Shackelford, S.D., Keele, J.W., Stone, R.T., Kappes, S.M. and Koohmaraie, M. (2000) Quantitative trait loci affecting growth and carcass composition of cattle segregating alternate forms of myostatin. *Journal of Animal Science* 78(3), 560–569.

Charlier, C., Coppieters, W., Farnir, F., Grobet, L., Leroy, P.L., Michaux, C., Mni, M., Schwers, A., Vammanshoven, P., Hanset, R. and Georges, M. (1995) The mh gene causing double-muscling in cattle maps to bovine chromosome2. *Mammalian Genome* 6, 788–792.

Cockell, N.E., Jackson, S.P., Shay, T.L., Nielsen, D., Moore, S.S., Steele, M.R., Barendse, W., Green, R.D. and Georges, M. (1994) Chromosomal localization of the callipyge gene in sheep (Ovis aries) using bovine DNA markers. *Proceedings of the National Academy of Sciences USA* 91, 3019–3023.

Conlon, I. and Raff, M. (1999) Size control in animal development. *Cell* 96, 244–245.

Cossu, G., Tajbakhsh, S. and Buckingham, M. (1996) How is myogenesis initiated in the embryo. *Trends in Genetics* 12(6), 218–223.

Edmondson, D.G. and Olson, E.N. (1993) Helix-Loop-Helix proteins as regulators of muscle-specific transcription. *The Journal of Biological Chemistry* 268(2), 755–758.

Florini, J.R. and Magri, K.A. (1989). Effects of growth factors on myogenic differentiation. *American Journal of Physiology* 256 (*Cell Physiol* 25), C701–C711.

Florini, J.R., Ewton, D.Z. and Coolican, S.A. (1996) Growth hormone and the insulin-like growth factor system in myogenesis. *Endocrine Reviews* 17, 481–517.

Franzini, T., Armstrong, C. and Fischman, D.A. (1994) Morphogenesis of skeletal muscle fibres. In: Engel, A.G. and Franzini-Armstrong, C. (eds) *Myology* pp.74–96.

Gluckman, P.D. and Brinsmead, M.W. (1976) Somatomedin in cord blood: relationship to gestational age and birth size. *Journal of Clinical Endocrinology Metabolism* 43(6), 1378–1381.

Grobet, L., Royo Margin, L.J., Poncelet, D., Pirottini, D., Brouwers, B., Riquet, J., Schoeberlain, A., Punner, S., Menissier, F., Massabunda, J., Fries, R., Hanset, R. and George, M. (1997) A deletion in the myostatin gene causes double muscling in cattle. *Nature Genetics* 17, 71–73.

Grounds, M.D. and Yablonka-Reuveni, Z. (1993) Molecular and cell biology of skeletal muscle regeneration. In: Partridge, T.A. (ed.) *Molecular and Cell Biology of Muscular Dystrophy*, Chapman & Hall, New York, pp. 210–256.

Hanset, R. and Michaux, G. (1985) On the genetic determination of muscular hypertrophy in the Belgian White and Blue cattle breed. I. Experimental data. *Genetique, Selection, Evolution* 17, 354.

Hauschka, S. (1994) The embryonic origin of muscle. In: Engel, A.G. and Franzini-Armstrong, C. (eds) *Myology*, Vol. 1. pp. 3–73.

Humbel, R.E. (1990) Insulin-like growth factors I and II. *European Journal of Biochemistry* 190(3), 445–462.

Kambadur, R., Sharma, M., Smith, T.P.L. and Bass, J.J. (1997) Mutations in myostatin (GDF8) in double-muscled Belgian Blue and Piedmontese cattle. *Genome Research* 7, 910–916.

Kirk, S.P., Whittle, M.A., Oldham, J.M., Dobbie, P.M. and Bass, J.J. (1996) GH regulation of the type 2 IGF receptor in regenerating skeletal muscle of rats. *Journal of Endocrinology* 149, 81–91.

Koishi, K., Zhang, M., McLennan, I. and Harris, A.J. (1995) MyoD protein accumulates in satellite cells and is neurally regulated in regenerating myotubes and skeletal muscle fibers. *Developmental Dynamics* 202(3), 244–254.

Kou Kou, and Rotwein, P. (1993) Transcriptional activation of the insulin-like growth factor-II gene during myoblast differentiation. *Molecular Endocrinology* 7(2), 291–302.

Liu, J.I., Grinbery, A., Westphal, H., Sauer, B., Accili, D., Karas, M. and Le Roith, D. (1998) Insulin-like growth factor-I affect perinatal lethality and postnatal development in a gene dosage-dependent manner: manipulation and using the Cre/LoxP system in transgenic mice. *Molecular Endocrinology* 12(9), 1452–1462.

McPherron, A.C., Lawler, A.M. and Lee, S-J. (1997) Regulations of skeletal muscle mass in mice by a new TGFβ superfamily member. *Nature* 387, 87–90.

Marcelle, C., Stark, M.R. and Bronner-Fraser, M. (1997) Co-ordinate actions of BMPs, Wnts, shh and noggin mediate patterning of the dorsal somite. *Development* 124, 3955–3963.

Massagué, J., Cheifetz, S., Endo, T. and Nadal-Ginard, B. (1986). Type β transforming growth factor is an inhibitor of myogenic differentiation. *Proceedings of the National Academy of Sciences USA* 83, 8206–8210.

Miyazono, K., Olofosson, A., Colosetti, P. and Heldin, C.H. (1991) A role of the latent TGF-β1-binding protein in the assembly and secretion of TGF-β1. *European Molecular Biology Organisation Journal* 10, 1091–1101.

Notter, D.R. (1999) The importance of genetic diversity in livestock populations of the future. *Journal of Animal Science* 77, 61–69.

Oldham, J., Sharma, M., Martyn, J., Kambadur, R. and Bass, J. (1998) Myostatin and myogenic regulatory factors (MRFs), but not insulin-like growth factors (IGFs), differ between normal and double muscled cattle. *Proceedings of the 80th Conference of the Endocrine Society*, pp. 3–310.

Olson, E.N. and Klein, W.H. (1994) bHLH factors in muscle development. *Genes and Development* 8, 1–8.

Olson, E.N., Perry, M. and Schulz, R.A. (1995) Regulation of muscle differentiation by the MEF2 family of MADS box transcription factors. *Developmental Biology* 172, 2–14.

Rehfeldt, C., Fiedler, I., Weikard, R., Kanitz, E. and Ender, K. (1993) It is possible to increase skeletal muscle fibre number in utero. *Bioscience Reports* 13(4), 213–220.

Rushbrook, J.I. and Somes, R.G. (1985) Myosin light chain-1: Genetic analysis of three variants found in fast white chicken muscle and investigation of linkage with the muscular dystrophy gene. *Biochemical Genetics* 23(1/2), 17–27.

Sauer, B. (1998) Inducible gene targeting in mice using the Cre/Lox System. *Methods*. 14(4), 381–392.

Schiaffino, S. and Reggiani, C. (1996) Molecular diversity of myofibrillar proteins: Gene regulation and functional significance. *Physiological Reviews* 76(2), 371–423.

Smith, T.P.L., Lopez-Corrales, N.L., Kappes, S.M. and Sonstegaard, T.S. (1997) Myostatin maps to the interval containing the bovine mh locus. *Mammalian Genome* 8, 742–744.

Swatland, H.J. (1984) (ed.) *Structure and Development of Meat Animals.* Prentice-Hall, Englewood Cliffs, New Jersey, pp. 201–254.

Szabo, G., Dallmann, G., Muller, G., Patthy, L., Soller, M. and Varga, L. (1998) A deletion in the myostatin gene causes the compact (Cmpt) hypermuscular mutation in mice. *Mammalian Genome* 9, 671–672.

Te Pas, M.F.W., Harders, F.H., Soumillion, A., Verburg, F.J. and Meuwissen, T.H.E. (1998) The MyoD gene family influences birth weight, postnatal, growth rate and carcass meat content of pigs. *Proceedings of the XXVI International Conference on Animal Genetics*, E048.

Tollefsen, S.E., Lajara, R., McCusker, R.H., Clemmons, D.R. and Rotwein, P. (1989) Insulin-like growth factors (IGF) in muscle development. *The Journal of Biological Chemistry* 264(23), 13810–13817.

Williams, P. and Goldspink, G. (1973) The effect of immobilisation on the longitudinal growth of striated muscle fibres. *Journal of Anatomy* 116, 45–55.

# 14 Control and Manipulation of Hyperplasia and Hypertrophy in Muscle Tissue

P.J. Buttery, J.M. Brameld and J.M. Dawson

*Division of Nutritional Biochemistry, School of Biological Sciences, University of Nottingham, Loughborough, UK*

## Introduction – skeletal muscle growth

The growth of skeletal muscle involves increases in both cell number (hyperplasia) and cell size (hypertrophy). Hyperplasia involves the mononuclear skeletal muscle precursor cells (myoblasts and satellite cells), which subsequently become post-mitotic, align and fuse (differentiate) to form the multinuclear muscle fibres. Muscle fibres are then able to increase in cell size (hypertrophy) via the accretion of more protein. The total number of fibres in a muscle appears to be fixed at, or shortly after birth, with post-natal growth of muscle being entirely due to elongation and widening of the existing muscle fibres. However, this does not mean that the muscle precursor cells cease to grow at this stage, as the DNA content continues to increase until the animal approaches its mature size. Extra nuclei are recruited into muscle fibres from mono-nuclear myogenic cells known as satellite cells, which are enclosed beneath the basement membrane of the muscle fibres (see Harper and Buttery, 1992). Muscle cell differentiation therefore takes place during the mid–late gestation and neonatal periods, with differentiation in the neonatal period involving the fusion of satellite cells with the existing muscle fibres.

### *Myogenic cell determination and differentiation*

Whether a mesodermal cell becomes a skeletal muscle cell or some other cell type is determined at a very early stage of embryo development. A family of nuclear proteins has been identified which function as activators of muscle-specific gene transcription (i.e. transcription factors), and forced expression of any one of them can convert fibroblasts to muscle cells. They are called the MyoD family and are discussed in detail in Chapter 13.

## Control of hyperplasia and differentiation

### *Effects of growth factors, hormones and nutrients* in vitro

A wide variety of growth factors and metabolic hormones have been shown to affect skeletal muscle cell proliferation and/or differentiation *in vitro* (Table 14.1). Insulin-like growth factors (IGFs), epidermal growth factor (EGF) and insulin consistently stimulate both proliferation and differentiation, the fibroblast growth factors (FGFs) stimulate proliferation and inhibit differentiation, and the transforming growth factor-β (TGF-β) family tend to inhibit both proliferation and differentiation.

Differentiation of muscle cells in culture can be induced by the reduction of the serum content of the culture medium. The mechanism for this is thought to be via a reduction in the blood-borne inhibitors of differentiation, together with increases in locally produced activators, such as the IGFs. Differentiation has also been shown to be dependent upon other components of the cell culture medium, including the type of serum used (Doumit and Merkel, 1992), the type of medium and the substratum on which the cells are grown (Dodson *et al.*, 1990).

Studies investigating the manipulation of specific nutrients in the media are lacking. Published studies of direct nutritional effects on muscle cells in culture mainly involve micronutrients. A lack of zinc inhibits $C_2C_{12}$ myoblast differentiation and decreases expression of MyoD and myogenin mRNA (Petrie *et al.*, 1996). Similarly, a lack of calcium can also inhibit muscle cell fusion (Morris *et al.*, 1976). An increase in m-calpain, which requires calcium for activity, is seen during differentiation of fetal chicken myoblasts (Kwak *et al.*, 1993). The increase in m-calpain correlates with the elevated cleavage of filamin which occurs during the fusion process, suggesting that m-calpain plays an important role in the cytoskeletal reorganization that takes place during myoblast fusion. Addition of specific inhibitors for proteosome or calpains to the culture medium has also been shown to inhibit muscle differentiation (Ueda *et al.*, 1998), while addition of IGF-I, which stimulates differentiation, increases m-calpain mRNA concentrations (Hong and Forsberg, 1994) in rat L8 myotubes.

Retinoic acid induces myogenic differentiation and myogenin synthesis in a rat rhabdomyosarcoma cell line (Arnold *et al.*, 1992). The effects of retinoids are mediated through two receptor subtypes, namely the retinoic acid receptors (RARs) and the retinoid X receptors (RXRs). Expression of RAR mRNA has been shown to be repressed during $C_2C_{12}$ myoblast differentiation, while that of RXR mRNA was induced (Downes *et al.*, 1994). The unsaturated fatty acid, linoleic acid, has also been shown to stimulate myogenic differentiation in rat satellite cells (Allen *et al.*, 1985), but the mechanism for this effect is not clear. Dodson *et al.* (1990) suggested that differentiation is increased in low-glucose Dulbecco's modified Eagle's medium (DMEM), compared with high-glucose DMEM. However, we (Brameld *et al.*, 1999b) have found no effect of glucose concentration on differentiation of primary sheep muscle cells in culture, as measured by changes in intracellular creatine kinase activity, nor on the expression of IGF-I, IGF-II or growth hormone receptor (GHR) genes. Nutrients could affect myoblast differentiation via effects on locally produced growth factors or the hormones known to be influenced by diet (e.g. insulin and IGF-I).

**Table 14.1.** Effects of various growth factors and hormones on muscle precursor cell proliferation and differentiation (compiled from Allen *et al.*, 1985; Arnold *et al.*, 1992; Brameld *et al.*, 1998; Gal-Levi *et al.*, 1998).

| Factor | Insulin | IGF | GH | FGF | EGF/TGF-α | HGF | TGF-β | PDGF | Dex | T3 | Test | β-agonist | LIF | IL-6 | RA | LA | CGRP |
|---|---|---|---|---|---|---|---|---|---|---|---|---|---|---|---|---|---|
| Proliferation | ↑ | ↑ | → | ↑→ | ↑→ | ↑ | ↓→ | ↑→ | ↑→ | → | → | ↑→ | ↑ | ↑ | ND | ND | ND |
| Differentiation | ↑ | ↑ | ↑→ | ↓ | ↑ | ↓ | ↓ | ↓ | →↓ | ↑→ | ↓ | ↑→ | ND | ND | ↑ | ↑ | ↑ |

IGF, insulin-like growth factors -I and -II; GH, growth hormone; FGF, fibroblast growth factor; EGF/TGF-α, epidermal growth factor/transforming growth factor-α (same receptor); HGF, hepatocyte growth factor; TGF-β, transforming growth factor-β; PDGF, platelet-derived growth factor; Dex, Dexamethasone; T3, tri-iodothyronine; Test, testosterone; β-agonist, clenbuterol, ractopamine or isoproterenol; LIF, leukaemia inhibitory factor; IL-6, interleukin-6; RA, retinoic acid; LA, linoleic acid; CGRP, calcitonin gene related peptide; ↑, stimulates; ↓, inhibits; →, no effect; ND, no data found.

## *Genetic effects on hyperplasia and muscle fibre number*

The double-muscled (DM) syndrome, a hereditary condition in which cattle possess 40% more fibres than normal cattle at birth, is thought to be caused by an increase in muscle cell hyperplasia prior to differentiation. Increased growth factor activity during early fetal development has been detected in the serum of DM fetuses (Gerrard and Judge, 1993); myoblasts from DM fetuses proliferate more rapidly than those from normal fetuses (Quinn *et al.*, 1990). Gerrard and Grant (1994) suggested that maximal local IGF-II mRNA expression is delayed in DM fetuses. This may relate to the stimulatory effects of IGF-II on muscle differentiation (i.e. differentiation is also delayed in DM fetuses, enabling greater proliferation to occur). A similar conclusion was recently drawn from studies of the *high-growth* (*hg*) locus in mice (Summers and Medrano, 1997). The 40% increase in their adult body weight and the proportional increase in their skeletal muscle weight was associated with delayed fusion and prolonged proliferation of high-growth myoblasts. This resulted in muscle fibre hyperplasia at birth, with only moderate muscle fibre hypertrophy. However, other factors are probably also involved. Recent studies have demonstrated a number of mutations in the myostatin gene in DM cattle (see Bass *et al.*, Chapter 13, this volume). Myostatin (also called GDF-8) was originally identified in mice (McPherron *et al.*, 1997), and was shown to be a member of the TGF-β family and therefore an inhibitor of skeletal muscle cell hyperplasia and differentiation. Gene targeting to disrupt the GDF-8 gene in mice increased muscle mass by 2–3 times by a combination of hyperplasia and hypertrophy. The identification of other growth factors involved in double-muscling (be they DM-specific factors or altered concentrations of typical growth factors at critical stages of growth) may facilitate the increase of muscle fibre number in other species. For example, FGFs are expressed by muscle cells in culture (Moore *et al.*, 1991). They stimulate muscle cell proliferation (Table 14.1), but inhibit differentiation, myogenin gene expression (Brunetti and Goldfine, 1990) and also IGF-II gene expression (Rosenthal *et al.*, 1991), and could therefore play a role in controlling the timing of both proliferation and differentiation.

There is a strong positive correlation between muscle fibre number and postnatal growth. Fast-growing strains of animals have more muscle fibres than slower-growing strains (Ezekwe and Martin, 1975; Miller *et al.*, 1975). This correlation is apparent even within the same strain of animals; pigs between 50 and 80 kg body weight show a positive correlation between muscle fibre number and average daily gain (Dwyer *et al.*, 1993). In younger pigs (up to 25 kg body weight), however, growth rate correlates with birth weight and not muscle fibre number. Comparison of IGF-I gene expression in three breeds of pig with differing carcass characteristics and growth rates demonstrated significant differences in skeletal muscle (and adipose tissue) IGF-I expression (Brameld *et al.*, 1996a). IGF-I expression in longissimus dorsi (LD) muscle (and adipose tissue) was lower in the slowest growing Meishan × Landrace than in the other two breeds, with no difference in hepatic IGF-I expression. Thus, local muscle IGF-I expression and its subsequent effects on muscle growth may account for differences in growth rate.

### *Effects of prenatal nutrition/hormones (fetal programming)*

The supply of maternal nutrients can influence both the development of the fetus *in utero* and in the early neonatal period, when the young is dependent on mother's milk. Nutritional supply *in utero* is believed to induce programming of the fetal phenotype, with lifelong consequences. Variation in nutrient supply, particularly nutritional restriction, may have temporary effects on development from which the animal can fully recover when nutrient supply is restored. The effects may also be permanent and have long-term consequences for postnatal growth and body composition. In particular, fetal undernutrition at critical stages of development is believed to cause reduced or disproportionate fetal growth (see Desai and Hales, 1997).

Nutrition may influence fetal growth at all stages of development, from pre-implantation to parturition. The effects of nutrient modulation on fetal development depends upon the timing, duration and severity of the nutritional insult (see Robinson *et al.*, 1999), but the response may differ between species (Hegarty and Allen, 1978). *In utero* undernutrition may be caused either by maternal feed restriction or through maternal constraint in supplying nutrients to the developing fetus, such as impaired uterine blood supply or placental development, which in turn may be influenced by maternal nutrition (see Robinson *et al.*, 1999).

Nutrient restriction in early gestation can cause a permanent reduction in muscle fibre number. Muscle fibre formation begins in early development and is completed *in utero* in precocious species such as sheep, pigs and guinea pigs. In these species, fetal undernutrition can lead to runted animals with fewer muscle fibres (Powell and Aberle, 1981; Dwyer and Stickland, 1994). Fibre number is fixed soon after birth in all mammals and postnatal growth involves only hypertrophy of the existing fibres. Muscle fibre number is thus a critical determinant of muscle mass. Animals born with fewer muscle fibres remain smaller throughout postnatal life and never reach the same mature size as their well-nourished siblings, although fibre diameter may be larger in these animals (Powell and Aberle, 1981). Primary muscle fibres develop relatively early in gestation when the fetus is making little nutritional demand on the mother. Their development appears to be genetically determined and insensitive to environmental influences such as nutrition (Handel and Stickland, 1987). Primary muscle fibres generally develop into type I (slow-twitch) fibres with an oxidative metabolism. Muscles with a predominance of slow-twitch fibres tend to be less affected by undernutrition due to the lower proportion of secondary fibres in these muscles (Dwyer *et al.*,1995).

Secondary muscle fibres, which form around the primary fibres and constitute the major fibre population in developed muscle, are reduced in nutritionally-restricted animals (Dwyer *et al.*, 1995) as a result of reduced secondary myoblast proliferation. This may be due to reduced placental size and impaired nutrient transfer (Dwyer *et al.*, 1992). The reduction in secondary fibre number was prevented in maternal feed-restricted guinea pigs by replenishing either the protein or carbohydrate components of the diet, indicating that fibre number is reduced only when maternal diets are deficient in both protein and carbohydrate; fat supplementation had no effect (Dwyer and Stickland, 1994). This suggests that the effect is mediated by the carbohydrate, presumably glucose. Doubling feed intake of pregnant sows in the period immediately before muscle fibre hyperplasia (25–50 days of gestation), resulted in an increased mean secondary muscle fibre number in the litter (Dwyer *et al.*, 1994). An increase in both primary and secondary fibre size and in the secondary:primary fibre ratio is seen in muscles from large sheep fetuses obtained from *in vitro* embryo culture (Maxfield *et al.*, 1998). *In vitro* embryo culture in serum-supplemented media appears to significantly stimulate growth in some fetuses and this growth advantage is retained throughout gestation (Sinclair *et al.*, 1998). McCoard *et al.* (1997) demonstrated similar reductions in muscle fibre number in constrained ovine fetuses at 140 days gestation, with the constraints being the presence of twin or single fetuses, or fetuses due to be born in the spring or autumn. Both these constraints result in reduced fetal growth, presumably due to reduced availability of nutrients per fetus. The previously described relationships between muscle fibre number and both postnatal growth rates and gain:feed ratios (Dwyer *et al.*, 1993) demonstrate the importance of maternal nutrition during pregnancy on subsequent growth potential.

In species such as rats, a period of *in utero* undernutrition in early gestation may not result in permanent reduction in muscle fibre numbers if adequate nutritional realimentation occurs in the remaining gestation and early postnatal life (Hegarty and Allen, 1978; Wilson *et al.*, 1988); secondary fibre proliferation continues into early postnatal life in these species (see Dwyer *et al.*, 1995).

Administration of growth hormone (GH) to pregnant sows results in increased secondary muscle fibre number in the piglets at birth, but only in early (10–24 days) gestation when muscle cell proliferation is maximal (Rehfeldt *et al.*, 1993). Administration of GH in mid-gestation results in delayed muscle maturation, whilst administration in late-gestation results in an increased muscle fibre diameter, with no effect on muscle fibre number (i.e. hypertrophic effects – see later).

All of these increases in muscle fibre number are via increases in the ratio of secondary-to-primary fibres, due to increased secondary fibres. The number of primary fibres is thought to be determined by genotype, but the number of secondary fibres can be influenced by environmental factors. The mechanism for these effects of environmental factors on muscle fibre number is not known. However, a likely candidate is the fetal GH–IGF axis, which has been shown to be regulated by nutrition in exactly the same way as that of the mother (Bauer *et al.*, 1995). Our recent studies (J.M. Brameld *et al.*, unpublished data) indicate that nutrient restriction of pregnant ewes between 28 and 80 days' gestation, followed by feeding to control levels results in decreased expression of IGF-I in fetal liver at 80 days, but increased hepatic expression of IGF-II. The effects of nutritional and hormonal manipulations on postnatal growth, and in particular the GH–IGF axis, has recently been reviewed (e.g. Brameld, 1997). We have demonstrated local expression of the IGF-I gene in both skeletal muscle and adipose tissue in the pig, as well as in liver, the major site of production of circulating IGF-I. IGF-I gene expression is affected by diet in a tissue-specific manner (Weller *et al.*, 1994; Brameld *et al.*, 1996b). Evidence from studies in whole animals and cultured hepatocytes suggest that both the protein and energy constituents of a diet have direct effects on hepatic expression of growth-regulatory genes. In pigs, glucose directly increases hepatic expression of the GHR gene and therefore enhances the effects of GH on IGF-I expression, with protein (amino acids) controlling GH-stimulated expression of the IGF-I gene (Brameld *et al.*, 1999a). Interestingly, neither dietary energy nor protein alter IGF-I expression in skeletal muscle, despite having effects on muscle GHR expression opposite to those seen in liver (see Brameld, 1997). This is corroborated by our recent studies which show no effect of nutrient restriction on

IGF-I expression in fetal sheep skeletal muscle, although expression was higher at 80 days' than at 140 days' gestation (J.M. Brameld *et al.*, unpublished data). We hypothesized that a possible mechanism for nutrient restriction decreasing muscle fibre number may be via: (i) reduced fetal hepatic IGF-I expression, and therefore reduced mitogenic activity in the blood; and (ii) the reduced proliferation may induce an early peak of IGF-II expression in skeletal muscle, resulting in an early induction of differentiation and therefore reduced muscle fibre number. Indeed, our recent studies of nutrient-restricted ewes indicate an earlier peak of IGF-II expression in skeletal muscle of nutrient-restricted fetal sheep, although only two time points were investigated (J.M. Brameld *et al.*, unpublished data).

In some species, including man (Draeger *et al.*, 1987), sheep (Wilson *et al.*, 1992) and pigs (Mascarello *et al.*, 1992), a third generation of muscle fibres have been identified. These tertiary fibres occur during late fetal and very early neonatal development, after the completion of primary and secondary fibre production. Their origin is unclear, but it has been suggested that they are derived from satellite cells (Mascarello *et al.*, 1992) and their development may be affected by nutritional supply (see Dauncey and Gilmour, 1996), since they are present in maternal-fed piglets, but not in piglets fed sow-milk substitute. Whether this is due to differences in the composition of the milk or to the amount of food eaten is still unknown, since maternal-fed animals are likely to have ingested more than the sow-milk substitute-fed animals, which were only fed at 6-hourly intervals (M.J. Dauncey, Cambridge, 1999, personal communication). This suggests that nutrition in late pregnancy and very early neonatal life is important to maximize muscle fibre formation.

## Control and manipulation of hypertrophy

In postnatal life, increases in muscle size are due to hypertrophy, and not hyperplasia, as muscle fibre number does not increase significantly after birth. Muscle DNA content continues to increase throughout the growing phase due to satellite cell proliferation, differentiation and fusion with existing muscle fibres. This is accompanied by increased protein deposition within the cells. The cessation of DNA accretion occurs abruptly as the animal approaches its mature size and precedes the decline in protein accretion. Muscle protein accretion is the net balance between the relative rates of muscle protein synthesis (i.e. gene transcription and translation into protein) and degradation (proteolysis), and changes in either can result in increased muscle mass. Some of the factors shown to influence cell proliferation and differentiation in cultured cells (Table 14.1) can also influence protein metabolism within muscle cells (Table 14.2). *In vitro* studies have shown that EGF, insulin and the IGFs all increase protein synthesis and decrease protein breakdown in cultured muscle cells, whereas GH has no effect and dexamethasone increases protein breakdown (Harper and Buttery, 1992).

### *The use of exogenous hormones*

Increasing muscle protein accretion, via changes in either protein synthesis or protein degradation (or a combination of the two), forms the basis of several growth promoters

**Table 14.2.** Effects of various growth factors and hormones on differentiated muscle cell protein synthesis and breakdown (see Brameld *et al.*,1998).

| Factor | Insulin | IGF | GH | FGF | EGF/TGF-α | TGF-β | PDGF | Dex | T3 | Test | β-agonist | TNF-α | $PGE_2$ | $PGF_{2\alpha}$ |
|---|---|---|---|---|---|---|---|---|---|---|---|---|---|---|
| Protein synthesis | ↑ | ↑ | → | → | ↑ | ↓↑ | ↑→ | ↑↓ | → | → | ↑→ | ND | → | ↑ |
| Protein breakdown | ↓ | ↓ | → | ND | ↓ | ↓ | ND | ↑ | → | → | ↓ | ↑ | ↑ | → |

IGF, insulin-like growth factors -I and -II; GH, growth hormone; FGF, fibroblast growth factor; EGF/TGF-α, epidermal growth factor/transforming growth factor-α (same receptor); TGF-β, transforming growth factor-β; PDGF, platelet-derived growth factor; Dex, Dexamethasone; T3, tri-iodothyronine; Test, testosterone; β-agonist, clenbuterol, ractopamine or isoproterenol; TNF-α, tumour necrosis factor-α; $PGE_2$, prostaglandin $E_2$; $PGF_{2\alpha}$, prostaglandin $F_{2\alpha}$; ↑, stimulates; ↓, inhibits; →, no effect; ND, no data found.

which have been used to increase muscle mass in farm livestock. The anabolic growth promotors, although no longer available for commercial use in the UK, have been shown to effectively increase muscle mass through hypertrophy. The mode of action of these agents differs, depending on the compound. Longissimus DNA concentration is higher in bulls than in steers (Morgan *et al.*, 1993) and in animals treated with testosterone (Grigsby *et al.*, 1976), indicating an effect of testosterone on satellite cell proliferation. It has been suggested that trenbolone acetate, a testosterone analogue, increases skeletal muscle hypertrophy by enhancing the sensitivity of satellite cells to FGF and IGF-I (Thompson *et al.*, 1989). Testosterone increases both protein synthesis and degradation rates, with a greater effect on the former (Martinez *et al.*, 1984), while trenbolone acetate increases muscle mass predominantly by reducing protein degradation, with a lesser effect on protein synthesis (Vernon and Buttery, 1976). The mode of action of oestradiol is believed to involve enhancement of endogenous GH secretion (Gopinath and Kitts, 1984; Breier *et al.*, 1988), although measurable increases in protein synthetic rate are not always detected in treated animals (Dawson *et al.*, 1991). The response to anabolic agents, however, is clearly dependent on nutritional status (Gill *et al.*, 1987; Bauman *et al.*, 1994).

Treating animals with exogenous GH increases the fractional rates of protein synthesis and degradation in skeletal muscle, with the increase in synthesis exceeding the increase in degradation. This results in protein accretion (Pell and Bates, 1987; Eisemann *et al.*, 1989) and therefore hypertrophy. Total muscle RNA concentration is increased in GH-treated muscles suggesting increased protein synthetic capacity rather than increased protein synthesis efficiency (Pell and Bates, 1987). The mechanism for these effects is presumably via endocrine or autocrine/paracrine GH-stimulated IGF-I, since GH has no direct effect on protein metabolism in cultured muscle cells (see above). Although growth hormone administration to pigs increases fibre diameter (Beerman *et al.*, 1990; Rehfeldt and Ender, 1993), its effect on meat quality is variable. Shear force values for samples taken at slaughter were similar for control- and GH-treated pigs, but were higher in GH-treated samples from carcasses hung for 5 days in a cool room (Solomon *et al.*, 1994). This suggests that the post-mortem biochemical processes may be affected by GH, as has been demonstrated in animals treated with β-agonist (see later). The ambiguity as to whether GH affects meat quality may relate to the magnitude of the effect on proteolysis, since the rate of protein degradation before death affects the rate of protein degradation post mortem, and therefore the tenderization process.

Treatment of growing animals with other exogenous agents, such as β-adrenergic agonists, also results in increased muscle mass via muscle hypertrophy, with no proliferation of satellite cells. Indeed, muscle DNA concentration ($\mu g\ g^{-1}$ protein) is sometimes less in treated muscles than in control muscles (Kim *et al.*, 1987). The predominant mechanism of action of these agents is believed to be a reduction in protein degradation, although protein synthesis has also been shown to be stimulated in some studies (Dawson *et al.*, 1991). This is supported by the observation that translational efficiency (i.e. the amount of protein synthesized per unit RNA) is increased by β-agonists (Maltin *et al.*, 1992). Studies at Nottingham and in the USA have shown that, in sheep and cattle, β-adrenergic agonist administration over several weeks has a remarkable effect on the $Ca^{2+}$-dependent cysteine proteinase (calpain) system (Higgins *et al.*, 1988; Wang and Beermann, 1988; Parr *et al.*, 1992; Speck *et al.*, 1993). The calpain system comprises at least three components, two of which have proteolytic activity *in vitro* at micromolar and millimolar concentrations of calcium, namely μ-calpain and m-calpain respectively, with the third important factor being a specific endogenous inhibitor known as calpastatin. β-agonist treatment for several weeks results in significant increases in the level of activity of calpastatin and calpastatin mRNA in LD muscle (Higgins *et al.*, 1988; Wang and Beermann, 1988; Parr *et al.*, 1992; Speck *et al.*, 1993). However, not all muscles respond equally to treatment with β-agonists (Dawson *et al.*, 1991), probably related to differences in the content of fibre types of different muscles. The cross-sectional area of type II (fast contracting, mixed glycolytic-oxidative) fibres tends to be increased more consistently than type I (slow contracting, oxidative) fibres (see Yang and McElligott, 1989). There also appears to be a change in fibre type proportions in muscles of treated animals following long-term administration of these agents, with a shift towards fibres with increased anaerobic metabolism (Maltin *et al.*, 1986; Zeman *et al.*, 1988). These changes in fibre diameter and metabolism, along with the reduced protein degradation, are believed to contribute to the reduced tenderness of meat treated with β-agonists (Aalhus *et al.*, 1992).

### Effects of postnatal nutrition

Undernutrition has important influences on postnatal growth, as well as on prenatal growth, if it occurs during critical stages of development. Increasing postnatal growth rate via increased dietary protein intake is associated with increased muscle fibre diameter and decreased intramuscular lipid in porcine LD muscle (Essén-Gustavsson *et al.*, 1994). In general, increasing dietary protein above maintenance levels, when the supply of energy is adequate, results in increased protein synthesis and degradation (see Lobley, 1998). The rate of protein synthesis tends to be greater than the rate of degradation and therefore results in a net protein accretion and increased muscle fibre diameter. This increase in diameter has been associated with an increase in meat toughness (see later), although the increased rate of protein degradation would be expected to decrease meat toughness, since the rate of protein degradation before death affects the rate of post-mortem tenderization (see above). Indeed, it has been suggested that a period of *ad libitum* feeding prior to slaughter increases pork tenderness (Blanchard *et al.*, 1999). It should be stressed that all the effects of exogenous hormones described above are dependent upon an adequate supply of nutrients (see Bauman *et al.*, 1994).

### Genetic effects on hypertrophy

In sheep, an autosomal dominant gene (callipyge), which is associated with extreme muscling, has been identified on chromosome 18. The effect of this gene is not consistent among all skeletal muscles. The leg and loin muscles all exhibit hypertrophy (18–42%), but certain shoulder muscles (e.g. supraspinatus and infraspinatus) are unaffected (Koohmaraie *et al.*, 1995). The leg and loin muscle hypertrophy exhibited in sheep expressing the callipyge gene is similar to that seen in the double-muscle syndrome in cattle. However, the double-muscle syndrome in cattle is apparent at birth and often results in dystocia, whereas the sheep condition does not manifest itself until several weeks after birth. This suggests that the mechanisms that give rise to these conditions in cattle differ from those in sheep. The double-muscle syndrome is characterized by increased muscle fibre number, resulting from faster and more prolonged hyperplasia during prenatal growth and development (see above). The callipyge condition is associated with increased muscle DNA content, suggesting greater satellite cell proliferation and increased RNA content, which points to a greater capacity for protein synthesis (Koohmaraie *et al.*, 1995). This condition is thus due to increased hypertrophy rather than hyperplasia. Protein degradation is significantly reduced in these animals, contributing to the increased muscle mass but also reducing meat tenderness, and is due to extremely high levels of calpastatin (Koohmaraie *et al.*, 1995). Fibre type changes are also apparent, with an increase in both proportion and area of the type II fibres and a reduction (or no change) in the size of the red, type I fibres. Sheep with the callipyge gene therefore have 48–62% greater total muscle fibre areas compared with normal lambs (Koohmaraie *et al.*, 1995). These effects on muscle fibre area and fibre type proportions are similar to those often seen in β-agonist-treated muscles (Yang and McElligott, 1989), as are the changes in protein synthesis and degradation and meat tenderness.

## Meat quality

The final eating quality of meat depends on a number of organoleptic properties, including appearance (comprising colour and fat content), taste, texture and tenderness. Whilst colour and fat content are important in influencing meat purchase, consumer studies indicate that it is the degree to which muscle tenderizes after slaughter that is the most important factor contributing to overall meat quality in cattle, sheep and pigs (Koohmaraie, 1994). After slaughter, the loss of oxygen supply to tissues initiates anaerobic metabolism, resulting in the utilization of primary energy stores. In the case of skeletal muscle, this means that muscle glycogen is depleted, producing an increase in lactic acid in the muscle, thereby reducing muscle pH. The rate of postmortem glycogenolysis can be altered in ruminants by circumstances which produce or mimic stress. For example, adrenaline infusion in cattle reduces the rate of pH decrease post mortem by depleting glycogen stores prior to slaughter, giving rise to dark, firm, dry beef (Geesink *et al.*, 1992), with highly fragmented fibres and poor taste qualities. Similar observations have been made on pork carcasses following long-term stress (Warriss *et al.*, 1989), whilst short-term stress increases the rate of pH fall, giving rise to pale, soft, exudative meat. Stress also affects the biochemical processes involved in the tenderization process (see later).

The relationship between muscle fibre structure and meat-eating quality is complex, because a number of factors are involved. Attempts to manipulate growth by selective breeding, experimental diets or exogenous growth promoters may change muscle fibre type, mean fibre diameter and the physiological status of the muscle at the time of slaughter. Separating the effects of these factors on tenderness can be difficult, since, for example, glycolytic (white) fibres tend to have larger diameters than oxidative (red) ones. Some studies suggest that the toughness of uncooked meat, as judged from its shear value, may increase with overall fibre diameter. In a trial involving 120 crossbred steers, shear force was positively correlated with fibre diameter, and negatively correlated with the percentage of oxidative fibres (Seideman *et al.*, 1987). Pigs with more muscle fibres also tend to have less fat (Stickland and Goldspink, 1975) while other indices of quality (e.g. toughness) may have also improved. It would therefore seem that high fibre number, which correlates with smaller fibres, may be a relevant parameter that correlates with meat quality. Studies into the effects of dietary protein on tenderness have yielded variable results (see Soloman *et al.*, 1994). Essén-Gustavsson *et al.* (1994) found an increase in the toughness of the LD muscle in barrows and gilts fed a high protein diet. These changes were associated with a decrease in intramuscular lipid and an increase in muscle fibre diameter.

Among the most important reactions influencing toughness are those catalysed by certain endogenous proteolytic enzymes that act in a highly selective manner on a small number of key intracellular muscle proteins to initiate the tenderization process. Several candidate enzyme systems have been proposed over the last 10 years, including cathepsins and the multicatalytic protease (proteasome). However, in recent years a consensus has emerged that the most important proteolytic enzymes affecting tenderization belong to the calpain system (see above), with the most important substrates being myofibrillar, Z-line and costamere proteins in muscle fibres (Koohmaraie, 1994; Taylor *et al.*, 1995). The calpain system is highly sensitive to fluctuating levels of calcium ion, pH and temperature, all of which change rapidly in the immediate post-mortem period (Suzuki *et al.*, 1995). Unlike other proteolytic systems which may be active during the post-mortem period, the calpains are the only proteolytic enzymes in skeletal muscle which do not degrade the major myofibrillar proteins actin and myosin, both of which remain intact during the tenderization process (Goll *et al.*, 1991, 1992). Further evidence has shown that $CaCl_2$ infusion into the carcass increases the rate of tenderization in beef whilst infusion of $ZnCl_2$, a calpain inhibitor, reduces the rate of tenderization in both beef and lamb (Koohmaraie, 1990; Geesink *et al.*, 1994). Long-term β-agonist treatment is known to produce tough meat in cattle and sheep, presumably due to suppression of post-mortem proteolysis by the elevated activity of calpastatin (Kretchmar *et al.*, 1990; Wheeler and Koohmaraie, 1992). The rate of tenderization also varies between the principal meat-producing species, where (in order of greater tenderization) pork > lamb > beef. Significantly, calpastatin levels at slaughter are greatest in beef and lowest in pork, adding further evidence to the involvement of the calpain system in the tenderization process (Koohmaraie *et al.*, 1991). Callipyge lambs, characterized by their enhanced muscle growth and excessively tough meat, have extremely high levels of calpastatin (Koohmaraie *et al.*, 1995). Whilst the meat toughness is not further enhanced by β-agonist treatment, it can be reduced by infusion of $CaCl_2$ (Koohmaraie *et al.*, 1996; Clare *et al.*, 1997). Thus, a genetic condition has been shown to alter the calpain system in a manner consistent with its effects on tenderization. We have shown that the same system is involved in the conversion of pig muscle to pork and that pre-slaughter stress can alter the post-mortem responsiveness of the calpain system (Sensky *et al.*, 1996, 1999; Parr *et al.*, 1999a, b).

## Acknowledgements

The authors gratefully acknowledge the financial support of the BBSRC, Meat and Livestock Commission and the MAFF for much of their research.

## References

Aalhus, J.L., Schaefer, A.L., Murray, A.C. and Jones, S.D.M. (1992) The effect of ractopamine on myofiber distribution and morphology and their relation to meat quality in swine. *Meat Science* 31, 397–409.

Allen, R.E., Luiten, L.S. and Dodson, M.V. (1985) Effect of insulin and linoleic acid on satellite cell differentiation. *Journal of Animal Science* 60, 1571–1579.

Arnold, H.H., Gerharz, C.D., Gabbert, H.E. and Salminen, A. (1992) Retinoic acid induces myogenin synthesis and myogenic differentiation in the rat rhabdomyosarcoma cell line BA-Han-1C. *Journal of Cell Biology* 118, 877–887.

Bauer, M.K., Breier, B.H., Harding, J.E., Veldhuis, J.D. and Gluckman, P.D. (1995) The fetal somatotropic axis during long term maternal undernutrition in sheep: Evidence for nutritional regulation in utero. *Endocrinology* 136, 1250–1257.

Bauman, D.E., Beerman, D.H., Boyd, R.D., Buttery, P.J., Campbell, R.B., Chalupa, W.V., Etherton, T.D., Klasing, K., Schelling, G.T. and Steele, N.C. (1994) *Metabolic Modifiers: Effects on the Nutrient Requirements of Food-Producing Animals.* National Academy Press, Washington USA, 81 pp.

Beerman, D.H., Fishell, V.K., Roneker, K., Boyd, R.D., Armbruster, G. and Souza, L. (1990) Dose-response relationships between porcine somatotropin, muscle composition, muscle fiber characteristics and pork quality. *Journal of Animal Science* 68, 2690–2697.

Blanchard, P.J., Ellis, M., Warkup, C.C., Hardy, B., Chadwick, J.P. and Deans, G.A. (1999) The influence of rate of lean and fat tissue development on pork eating quality. *Animal Science* 68, 477–485.

Brameld, J.M. (1997) Molecular mechanisms involved in the nutritional and hormonal regulation of growth in pigs. *The Proceedings of the Nutrition Society* 56, 607–619.

Brameld, J.M., Atkinson, J.L., Budd, T.J., Saunders, J.C., Pell, J.M., Salter, A.M., Gilmour, R.S. and Buttery, P.J. (1996a) Expression of insulin-like growth factor-I (IGF-I) and growth hormone receptor (GHR) mRNA in liver, skeletal muscle and adipose tissue of different breeds of pig. *Animal Science* 62, 555–559.

Brameld, J.M., Atkinson, J.L., Saunders, J.C., Pell, J.M., Buttery, P.J. and Gilmour, R.S. (1996b) Effects of Growth Hormone administration and dietary protein intake on Insulin-like growth factor-I and Growth Hormone Receptor mRNA expression in porcine liver, skeletal muscle and adipose tissue. *Journal of Animal Science* 74, 1832–1841.

Brameld, J.M., Buttery, P.J., Dawson, J.M. and Harper, J.M.M. (1998) Nutritional and hormonal control of skeletal muscle cell growth and differentiation. *Proceedings of the Nutrition Society* 57, 207–217.

Brameld, J.M., Gilmour, R.S. and Buttery, P.J. (1999a) Glucose and amino acids interact with hormones to control expression of insulin-like growth factor-I (IGF-I) and growth hormone receptor (GHR) mRNA in cultured pig hepatocytes. *Journal of Nutrition* 129, 1298–1306.

Brameld, J.M., Smail, H., Imram, N., Millard, N. and Buttery, P.J. (1999b) Changes in expression of IGF-I, IGF-II and GH receptor (GHR) mRNA during differentiation of cultured muscle cells derived from adult and fetal sheep. *Proceedings of the IXth International Society for Ruminant Physiology*, Pretoria, South Africa (in press).

Breier, B.H., Gluckman, P.D. and Bass, J.J. (1988) Influence of nutritional status and oestradiol-17ß on plasma growth hormone, insulin-like growth factors-I and -II and the response to exogenous growth hormone in young steers. *Journal of Endocrinology* 118, 243–250.

Brunetti, A. and Goldfine, I.D. (1990) Role of myogenin in myoblast differentiation and its regulation by fibroblast growth factor. *Journal of Biological Chemistry* 265, 5960–5963.

Clare, T.L., Jackson, S.P., Miller, M.F., Elliott, C.T. and Ramsey, C.B. (1997) Improving tenderness of normal and callipyge lambs with calcium chloride. *Journal of Animal Science* 75, 377–385.

Dauncey, M.J. and Gilmour, R.S. (1996) Regulatory factors in the control of muscle development. *Proceedings of the Nutrition Society* 55, 543–559.

Dawson, J.M., Buttery, P.J., Lammiman, M.J., Soar, J.B., Essex, C.P., Gill, M. and Beever, D.E. (1991) Nutritional and endocrinological manipulation of lean deposition in forage-fed steers. *British Journal of Nutrition* 66, 171–185.

Desai, M. and Hales, C.N. (1997) Role of fetal and infant growth in programming metabolism in later life. *Biological Reviews* 72, 329–348.

Dodson, M.V., Mathison, B.A. and Mathison, B.D. (1990) Effects of medium and substratum on ovine satellite cell attachment, proliferation and differentiation in vitro. *Cell Differentiation and Development* 29, 59–66.

Doumit, M.E. and Merkel, R.A. (1992) Conditions for isolation and culture of porcine myogenic satellite cells. *Tissue and Cell* 24, 253–262.

Downes, M., Mynett-Johnson, L. and Muscat, G.E.O. (1994) The retinoic acid and retinoid X receptors are differentially expressed during myoblast differentiation. *Endocrinology* 134, 2658–2661.

Draeger, A., Weeds, A.G. and Fitzsimons, R.B. (1987) Primary, secondary and tertiary myotubes in developing skeletal muscle: a new approach to the analysis of human myogenesis. *Journal of Neurological Science* 81, 19–43.

Dwyer, C.M. and Stickland, N.C. (1994) Supplementation of a restricted maternal diet with protein or carbohydrate alone prevents a reduction in fetal muscle fibre number in the guinea-pig. *British Journal of Nutrition* 72, 173–180.

Dwyer, C.M., Madgwick, A.J.A., Crook, A.R. and Stickland, N.C. (1992) The effect of maternal undernutrition on the growth and development of the guinea-pig placenta. *Journal of Developmental Physiology* 18, 295–302.

Dwyer, C.M., Fletcher, J.M. and Stickland, N.C. (1993) Muscle cellularity and postnatal growth in the pig. *Journal of Animal Science* 71, 3339–3343.

Dwyer, C.M., Stickland, N.C. and Fletcher, J.M. (1994) The influence of maternal nutrition on muscle fiber number development in the porcine fetus and on subsequent postnatal growth. *Journal of Animal Science* 72, 911–917.

Dwyer, C.M., Madgwick, A.J.A., Ward, S.S. and Stickland, N.C. (1995) Effect of maternal undernutrition in early gestation on the development of fetal myofibres in the guinea-pig. *Reproduction, Fertility and Development* 7, 1285–1292.

Eisemann, J.H., Hammond, A.C. and Rumsey, T.S. (1989) Tissue protein synthesis and nucleic acid concentrations in steers treated with somatotropin. *British Journal of Nutrition* 62, 657–671.

Essén-Gustavsson, B., Karlsson, K., Lundström, K. and Enfält, A-C. (1994) Intramuscular fat and muscle fiber lipid contents in halothane-gene-free pigs fed high or low protein diets and its relation to meat quality. *Meat Science* 38, 269–277.

Ezekwe, M.O. and Martin, R.J. (1975) Cellular characteristics of skeletal muscle in selected strains of pigs and mice and the unselected controls. *Growth* 39, 95–106.

Gal-Levi, R., Leshem, Y., Aoki, S., Nakamura, T. and Halevy, O. (1998) Hepatocyte growth factor plays a dual role in regulating skeletal muscle satellite cell proliferation and differentiation. *Biochimica et Biophysica Acta* 1402, 39–51.

Geesink, G.H., Ouali, A., Smulders, F.J.M., Talmant, A., Tassy, C., Guignot, F. and van Laack, H.L.J.M. (1992) The role of ultimate pH in proteolysis and calpain calpastatin activity in bovine muscle. *Biochimie* 74, 283–289.

Geesink, G.H., Smulders, F.J.M. and Van Laack, R.L.J.M. (1994) The effects of calcium-chlorides, sodium-chlorides and zinc-chlorides treatment on the quality of beef. *Sciences des Aliments* 14, 485–502.

Gerrard, D.E. and Grant, A.L. (1994) Insulin-like growth factor-II expression in developing skeletal muscle of double muscled and normal cattle. *Domestic Animal Endocrinology* 11, 339–347.

Gerrard, D.E. and Judge, M.D. (1993) Induction of myoblast proliferation in L6 myoblast cultures by fetal serum of double muscled and normal cattle. *Journal of Animal Science* 71, 1464–1470.

Gill, M., Beever, D.E., Buttery, P.J., England, P., Gibb, M.J. and Baker, R.D. (1987) The effect of oestradiol-17β implantation on the response in voluntary intake, live-weight gain and body composition, to fishmeal supplementation of silage offered to growing calves. *Journal of Agricultural Science, Cambridge* 108, 9–16.

Goll, D.E., Dayton, W.R., Singh, I. and Robson, R.M. (1991) Studies of the alpha-actinin actin interaction in the z-disk by using calpain. *Journal of Biological Chemistry* 266, 8501–8510.

Goll, D.E., Thompson, V.F., Taylor, R.G. and Christiansen, J.A. (1992) Role of the calpain system in muscle growth. *Biochimie* 74, 225–237.

Gopinath, R. and Kitts, W.D. (1984) Growth hormone secretion and clearance rate in growing beef steers implanted with oestrogenic anabolic compounds. *Growth* 48, 499–514.

Grigsby, J.S., Bergen, W.G. and Merkel, R.A. (1976) The effect of testosterone on skeletal muscle development and protein synthesis in rabbits. *Growth* 40, 303–316.

Handel, S.E. and Stickland, N.C. (1987) Muscle cellularity and birth weight. *Animal Production* 44, 311–317.

Harper, J.M.M. and Buttery, P.J. (1992) Muscle cell growth. In: Boorman, K.N., Buttery, P.J. and Lindsay, D.B. (eds) *The Control of Fat and Lean Deposition.* Butterworths, London, pp. 27–58.

Hegarty, P.V.J. and Allen, C.E. (1978) Effect of pre-natal runting on the post-natal development of skeletal muscle in swine and rats. *Journal of Animal Science* 46, 1634–1640.

Higgins, J.A., Lasslett, Y.V., Bardsley, R.G. and Buttery, P.J. (1988) The relation between dietary restriction or clenbuterol (a selective β-2 agonist) treatment on muscle growth and calpain proteinase (ec3.4.22.17) and calpastatin activities in lambs. *British Journal of Nutrition* 60, 645–652.

Hong, D.H. and Forsberg, N.E. (1994) Effects of serum and insulin-like growth factor I on protein degradation and protease gene expression in rat L8 myotubes. *Journal of Animal Science* 72, 2279–2288.

Kim, Y.S., Lee, Y.B. and Dalymple, R.H. (1987) Effect of the repartitioning agent, cimaterol on growth, carcass and skeletal muscle characteristics in lambs. *Journal of Animal Science* 65, 1392–1399.

Koohmaraie, M. (1990) Inhibition of postmortem tenderization in ovine carcasses through infusion of zinc. *Journal of Animal Science* 68, 1476–1483.

Koohmaraie, M. (1994) Muscle proteinases and meat aging. *Meat Science* 36, 93–104.

Koohmaraie, M., Whipple, G., Kretchmar, D.H., Crouse, J.D. and Mersmann, H.J. (1991) Postmortem proteolysis in longissimus muscle from beef, lamb and pork carcasses. *Journal of Animal Science* 69, 617–624.

Koohmaraie, M., Shackelford, S.D., Wheeler, T.L., Longergan, S.M. and Doumit, M.E. (1995) A muscle hypertrophy condition in lamb (callipyge): characterization of effects on muscle growth and meat quality traits. *Journal of Animal Science* 73, 3596–3607.

Koohmaraie, M., Shackelford, S.D. and Wheeler, T.L. (1996) Effects of a beta-adrenergic agonist (L-644,969) and male sex condition on muscle growth and meat quality of callipyge lambs. *Journal of Animal Science* 74, 70–79.

Kretchmar, D.H., Hathaway, M.R., Epley, R.J. and Dayton, W.R. (1990) Alterations in postmortem degradation of myofibrillar proteins in muscle of lambs fed a β-adrenergic agonist. *Journal of Animal Science* 68, 1760–1772.

Kwak, K.B., Chung, S.S., Kim, O.M., Kang, M.S., Ha, D.B. and Chung, C.H. (1993) Increase in the level of m-calpain correlates with the elevated cleavage of filamin during myogenic differentiation of embryonic muscle cells. *Biochimica et Biophysica Acta* 1175, 243–249.

Lobley, G.E. (1998) Nutritional and hormonal control of muscle and peripheral tissue metabolism in farm species. *Livestock Production Science* 56, 91–114.

McCoard, S.A., Peterson, S.W., McNabb, W.C., Harris, P.M. and McCutcheon, S.N. (1997) Maternal constraint influences muscle fibre development in fetal lambs. *Reproduction, Fertility and Development* 9, 675–681.

McPherron, A.C., Lawler, A.M. and Lee, S.J. (1997) Regulation of skeletal muscle mass in mice by a new TGF-β superfamily member. *Nature* 387, 83–90.

Maltin, C.A., Delday, M.I. and Reeds, P.J. (1986) The effect of a growth promoting drug, clenbuterol, on fibre frequency and area in hind limb muscles from young rats. *Bioscience Reports* 6, 293–299.

Maltin, C.A., Hay, S.M., McMillan, D.N. and Delday, M.I. (1992) Tissue specific responses to clenbuterol: temporal changes in protein metabolism of striated muscle and visceral tissues from rats. *Growth Regulation* 2, 161–166.

Martinez, J.A., Buttery, P.J. and Pearson, J.T. (1984) The mode of action of anabolic agents, the effects of testosterone on growth rates and muscle protein metabolism in female rats. *British Journal of Nutrition* 54, 515–521.

Mascarello, F., Stecchini, M.L., Rowlerson, A. and Ballocchi, E. (1992) Tertiary myotubes in postnatal growing pig muscle detected by their myosin isoform composition. *Journal of Animal Science* 70, 1806–1813.

Maxfield, E.K., Sinclair, K.D., Broadbent, P.J., McEvoy, T.G., Robinson, J.J. and Maltin, C.A. (1998) Short-term culture of ovine embryos modifies fetal myogenesis. *American Journal of Physiology* 274, E1121-E1123.

Miller, L.R., Garwood, V.A. and Judge, M.D. (1975) Factors affecting porcine muscle fiber type, diameter and number. *Journal of Animal Science* 41, 66–77.

Moore, J.W., Dionne, C., Jaye, M. and Swain, J.L. (1991) The mRNAs encoding acidic FGF, basic FGF and FGF receptor are coordinately downregulated during myogenic differentiation. *Development* 111, 741–748.

Morgan, J.B., Wheeler, T.L., Koohmaraie, M., Crouse, J.D. and Savell, J.W. (1993) Effect of castration on myofibrillar protein turnover, endogenous proteinase activities and muscle growth in bovine skeletal muscle. *Journal of Animal Science* 71, 408–414.

Morris, G.E., Piper, M. and Cole, R. (1976) Differential effects of calcium ion concentration on cell fusion, cell division and creatine kinase activity in muscle cell cultures. *Experimental Cell Research* 99, 106–114.

Parr, T., Bardsley, R.G., Gilmour, R.S. and Buttery, P.J. (1992) Changes in calpain and calpastatin messenger-RNA induced by β-adrenergic stimulation of bovine skeletal-muscle. *European Journal of Biochemistry* 208, 333–339.

Parr, T., Sensky, P.L., Scothern, G., Bardsley, R.G., Buttery, P.J., Wood, J.D. and Warkup, C. (1999a) Immunochemical study of the calpain system in porcine longissimus muscle with high and low shear force values. *Journal of Animal Science* 77(suppl. 1), 164.

Parr, T., Sensky, P.L., Warkup, C., Bardsley, R.G. and Buttery, P.J. (1999b) Changes in porcine skeletal muscle calpastatin expression induced by a single dose of a $\beta_2$-agonist. *Journal of Animal Science* 77(suppl. 1), 164–165.

Pell, J.M. and Bates, P.C. (1987) Collagen and non-collagen protein turnover in skeletal muscle of growth hormone-treated lambs. *Journal of Endocrinology* 115, R1-R4.

Petrie, L., Buskin, J.N. and Chesters, J.K. (1996) Zinc and the initiation of myoblast differentiation. *Journal of Nutritional Biochemistry* 7, 670–676.

Powell, S.E. and Aberle, E.D. (1981) Skeletal-muscle and adipose-tissue cellularity in runt and normal birth-weight swine. *Journal of Animal Science* 52, 748–756.

Quinn, L.S., Ong, L.D. and Roeder, R.A. (1990) Paracrine control of myoblast proliferation and differentiation by fibroblasts. *Developmental Biology* 140, 8–19.

Rehfeldt, C. and Ender, K. (1993) Skeletal-muscle cellularity and histochemistry in response to porcine somatotropin in finishing pigs. *Meat Science* 34, 107–118.

Rehfeldt, C., Fiedler, I., Weikard, R., Kanitz, E. and Ender, K. (1993) It is possible to increase skeletal-muscle fiber number *in utero*. *Bioscience Reports* 13, 213–220.

Robinson, J.J., Sinclair, K.D. and McEvoy, T.G. (1999) Nutritional effects on foetal growth. *Animal Science* 68, 315–331.

Rosenthal, S.M., Brown, E.J., Brunetti, A. and Goldfine, I.D. (1991) Fibroblast growth factor inhibits insulin-like growth factor-II (IGF-II) gene expression and increases IGF-I receptor abundance in BC3H-1 muscle cells. *Molecular Endocrinology* 5, 678–684.

Seideman, S.C., Koohmaraie, M. and Crouse, J.D. (1987) Factors associated with tenderness in young beef. *Meat Science* 20, 281–291.

Sensky, P.L., Parr, T., Bardsley, R.G. and Buttery, P.J. (1996) The relationship between plasma epinephrine concentration and the activity of the calpain enzyme-system in porcine longissimus muscle. *Journal of Animal Science* 74, 380–387.

Sensky, P.L., Parr, T., Lockley, A.K., Bardsley, R.G., Buttery, P.J., Wood, J.D. and Warkup, C. (1999) Altered calpain levels in longissimus muscle from normal pigs and heterzygotes with the ryanodine receptor mutation. *Journal of Animal Science* 77, 2956–2964.

Sinclair, K.D., McEvoy, T.G., Carolan, C., Maxfield, E.K., Maltin, C.A., Young, L.E., Wilmut, I., Robinson, J.J. and Broadbent, P.J. (1998) Conceptus growth and development following *in vitro* culture of ovine embryos in media supplemented with bovine serum. *Theriogenology* 49, 218.

Solomon, M.B., Caperna, T.J., Mroz, R.J. and Steele, N.C. (1994) Influence of dietary protein and recombinant porcine somatotropin administration in young pigs, III. Muscle fiber morphology and shear force. *Journal of Animal Science* 72, 615–621.

Speck, P.A., Collingwood, K.M., Bardsley, R.G., Tucker, G.A., Gilmour, R.S. and Buttery, P.J. (1993) Transient changes in growth and in calpain and calpastatin expression in ovine skeletal-muscle after short-term dietary inclusion of cimaterol. *Biochimie* 75, 917–923.

Stickland, N.C. and Goldspink, G. (1975) A note on porcine skeletal muscle parameters and their possible use in early progeny testing. *Animal Production* 21, 93–96.

Summers, P.J. and Medrano, J.F. (1997) Delayed myogenesis associated with muscle fiber hyperplasia in High-Growth mice. *Proceedings of the Society for Experimental Biology and Medicine* 214, 380–385.

Suzuki, K., Sorimachi, H., Yoshizawa, T., Kinbara, K. and Ishiura, S. (1995) Calpain, novel family members, activation, and physiological function. *Biological Chemistry Hoppe-Seyler* 376, 523–529

Taylor, R.G., Geesink, G.H., Thompson, V.F., Koohmaraie, M. and Goll, D.E. (1995) Is z-disk degradation responsible for postmortem tenderization. *Journal of Animal Science* 73, 1351–1367.

Thompson, S.H., Boxhorn, L.K., Wong, W. and Allen, R.E. (1989) Trenbolone alters the responsiveness of skeletal muscle satellite cells to fibroblast growth factor and insulin-like growth factor I. *Endocrinology* 124, 2110–2117.

Ueda, Y., Wang, M.-C., Ou, B.R., Huang, J., Elce, J., Tanaka, K., Ichihara, A. and Forsberg, N.E. (1998) Evidence for the participation of the proteasome and calpain in early phases of muscle cell differentiation. *International Journal of Biochemistry and Cell Biology* 30, 679–694.

Vernon, B.G. and Buttery, P.J. (1976). Protein turnover in rats treated with trenbolone acetate. *British Journal of Nutrition* 36, 575–579.

Wang, S.Y. and Beermann, D.H. (1988) Reduced calcium-dependent proteinase activity in cimaterol-induced muscle hypertrophy in lambs. *Journal of Animal Science* 66, 2245–2550.

Warriss, P.D., Bevis, E.A. and Ekins, P.J. (1989) The relationships between glycogen stores and muscle ultimate pH in commercially slaughtered pigs. *British Veterinary Journal* 145, 378–383.

Weller, P.A., Dauncey, M.J., Bates, P.C., Brameld, J.M., Buttery, P.J. and Gilmour, R.S. (1994) Regulation of porcine insulin-like growth factor-I and growth hormone receptor mRNA expression by energy status. *American Journal of Physiology* 266, E776–E785.

Wheeler, T.L. and Koohmaraie, M. (1992) Effects of the ß-adrenergic agonist $L_{644,969}$ on muscle protein-turnover, endogenous proteinase activities, and meat tenderness in steers. *Journal of Animal Science* 70, 3035–3043.

Wilson, S.J., Ross, J.J. and Harris, A.J. (1988) A critical period for formation of secondary myotubes defined by prenatal undernourishment in rats. *Development* 102, 815–821.

Wilson, S.J., McEwan, J.C., Sheard, P.W. and Harris, A.J. (1992) Early stages of myogenesis in a large mammal: formation of successive generations of myotubes in sheep tibialis cranialis muscle. *Journal of Muscle Research and Cell Motility* 13, 534–550.

Yang, Y.T. and McElligott, M.A. (1989) Multiple actions of β-adrenergic agonists on skeletal muscle and adipose tissue. *Biochemical Journal* 261, 1–10.

Zeman, R.J., Ludemann, R., Easton, T.G. and Etlinger, J.D. (1988) Slow to fast alterations in skeletal muscle fibers caused by clenbuterol, a β2-receptor agonist. *American Journal of Physiology* 254, E726–E732.

# 15 Regulation of Protein Synthesis for Wool Growth

N.R. Adams, S. Liu and D.G. Masters

*CSIRO Division of Animal Production and CRC for Premium Quality Wool, Wembley, Western Australia*

## Introduction

Despite numerous studies on wool growth, there is not yet an integrated biological understanding of how wool growth rate is controlled. Genetic and nutritional effects on wool growth have been studied intensively, but the biological mechanisms underlying the substantial interactions between these factors remain ill-defined (Jackson and Roberts, 1970; Woolaston, 1987). This chapter aims to develop an integrative framework that will facilitate collaboration between nutritionists, physiologists, and cell and molecular biologists to develop practical systems to improve wool growth. Such a framework is needed, for example, to assess the likely value of individual genes for wool growth, as new information becomes available. Subsequent work is planned to use the framework to develop more quantitative applied models relating wool growth rate to staple strength, meat production and reproduction.

We have approached this task by considering follicles as scattered units of a single organ that produces wool, and treating the organ as a whole (much as milk production is normally considered as a function of the mammary gland, rather than of individual mammary alveoli). This contrasts with the more traditional approach of considering wool growth as an aggregation of the characteristics of individual follicles, and describing wool growth by detailed analysis of follicle biology. Although our approach glosses over much of the detailed information about function within the follicle, it enables the integration of follicle function with the overall protein synthetic activity in the skin, and in turn, in the body. This approach is possible because wool growth represents a relatively constant proportion of protein synthesis in skin (Table 15.1).

Figure 15.1 presents the framework developed in this chapter. Wool growth rate is considered as a function of the mass of follicular tissue, the rate at which this tissue synthesizes protein, and the proportion of that protein extruded as wool. In this schema, genotype may affect the capacity for wool growth at a number of sites, including skin mass, the fraction of skin made up of follicular tissue, the capacity of follicles to use amino acids, follicle efficiency, uptake of amino acids by other tissues,

 *Ruminant Physiology: Digestion, Metabolism, Growth and Reproduction* (ed. P.B. Cronjé)

**Table 15.1.** The proportion of skin protein synthesis deposited into wool (skin efficiency) in sheep with different fractional synthesis rates (FSR) of protein in the skin.

| Sheep | Dietary level (M) | FSR (% day$^{-1}$) | Skin efficiency for wool | Source |
|---|---|---|---|---|
| Romney ewes | 1.4 | 15.3 | 0.15 | Harris *et al.* |
| | 0.6 | 5.6 | 0.14 | (1994b) |
| Merino lambs | 1 | 17.3 | 0.18 | Liu *et al.* |
| | 0.6 | 15.3 | 0.20 | (1998) |
| Merino wethers, | | | | |
| Fleece+ | 1 | 18.2 | 0.19 | D.G. Masters *et al.* |
| Fleece− | 1 | 14.4 | 0.16 | (unpublished) |
| Merino ewes, dry | 1 | 11.7 | 0.16–0.21[a] | S. Liu *et al.* |
| Pregnancy | 1 | 10.1 | 0.10[a] | (unpublished) |
| Lactation | 1 | 11.9 | 0.16[a] | |

[a] Ewes fed to maintain empty body weight. Skin efficiency was calculated from wool growth (Masters and Mata, 1996) and estimated skin protein content.

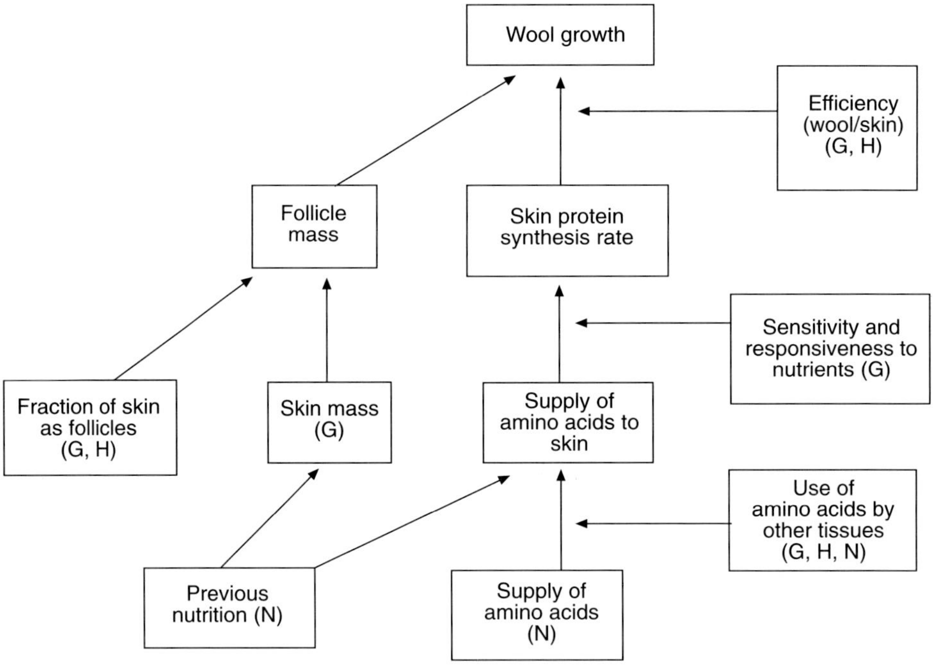


**Fig. 15.1.** Diagrammatic model of the control of wool growth, indicating points at which genetic (G), hormonal (H) and nutritional (N) control may be exerted.

responsiveness of the body to nutrient supply, and feed intake. Nutrition may affect wool growth in one of two ways. Firstly, it may increase total mass of follicular tissue. As with other body tissues (e.g. skin or muscle mass), such a change will depend primarily on the level of feed intake. Secondly, the rate of protein synthesis in follicles depends primarily on the supply of amino acids. Hormonal and physiological state may modulate and mediate these effects at several sites.

## Models of wool growth

The growth of wool can be conceptualized in three main ways. Efforts to integrate wool growth rate with wool fibre qualities have focused on the characteristics of individual wool follicles. Prediction of wool growth under practical husbandry conditions has utilized nutrient supply models. In this review, protein synthesis rate is used to provide a basis to integrate both nutritional and genetic mechanisms affecting wool growth.

### *Follicle models*

There has been considerable work relating cell dynamics in the wool follicle to wool growth. The pioneering model in this area used the supply of amino acids to the follicle to drive the rate of mitosis in the follicle bulb, which had a genetic limit to its rate of response (Black and Reis, 1979). This concept has been developed further (e.g. Scobie and Woods, 1992; Bowman *et al.*, 1993).

Subsequent work showed that enhanced nutrient supply not only increased mitotic rate (or decreased cell turnover time), but also increased the number of cells in the bulb germinal epithelium, and increased the volume of fibre cortical cells (Table 15.2). The relative importance of these three components of wool growth varied, presumably depending on the diet and duration or rate of feeding, but this cannot be determined from these studies. Although the role of cell apoptosis in determining the availability of cells to the fibre is unknown, the total number of cells generated (i.e. the product of bulb volume and mitotic rate) correlated well with changes in wool fibre production, while cortical cell size played a less important role (Table 15.2).

Genetic differences in fleece weight appear to be independent of cell turnover time (Williams and Winston, 1987; Hynd, 1989), but are also reflected in the amount of germinal tissue in follicle bulbs, which in turn is a product of bulb volume and follicle density. Seasonal changes in wool growth in Romney rams were also independent of changes in cell turnover time (Holle *et al.*, 1994). Thus, despite the conclusion of Wilson and Short (1979), cell turnover time alone does not provide a reliable means for modelling wool growth.

Total cell production rate by the bulb correlates closely with wool growth rate, but this does not prove causation. As with many other tissues, the rate of cell production by the follicle is inversely related to the rate of cell differentiation, so that cells labelled with [$^{3}$H]thymidine take longer to differentiate and mature as the cell production rate increases (Chapman *et al.*, 1980). When wool production rate was doubled as a result of increased feed intake, the time cells took to pass through the keratinization zone of

**Table 15.2.** Percentage changes in fibre production and follicle characteristics associated with increased feeding in Merino sheep.

| Increase in wool growth | Duration of study (days) | Increase in cell turnover rate | Increase in bulb volume | Increase in cortical cell volume | Reference |
|---|---|---|---|---|---|
| 250 | 28 | 47 | 12 | 19 | Short *et al.* (1968) |
| 164 | 365 | 20 | 22 | 33 | Wilson and Short (1979) |
| 130 | 106 | 0 | 3 | 0 | Williams and Winston (1987) |
| 133 | 56 | 6 | 30 | 5 | Hynd (1989) |
| 186 | 49 | 25 | 94 | 76 | Hynd (1994b) |

the follicle increased from 19 to 25 h, whilst the time taken for incorporation of [$^{35}$S]cystine into insoluble keratin doubled from 7 to 14.5 h (Chapman *et al.*, 1980). It is therefore equally sustainable to hypothesize that wool growth rate is limited by the capacity for cell maturation in the developing fibre shaft.

Robust and complex control mechanisms ensure close coordination between cell production, cell differentiation and cell maturation in the wool follicle. This makes it impossible to nominate any one of these functions as the ultimate controller of wool growth, because they all correlate closely with each other and also with wool growth. Nevertheless, the simplest hypothesis is that nutritional control of wool growth is exerted by affecting the capacity of the follicle to support protein synthesis for cell maturation, rather than by direct control of cell production in the bulb.

## Nutritional models

The earliest nutritional models were simply equations derived to show a linear relationship between feed intake and wool growth. These were later adjusted to linear relationships between wool growth and digestible dry matter intake, to take account of variation in feed quality between experiments (Allden, 1979). However, wool growth is even more closely related to the amount of protein absorbed from the small intestine (Black *et al.*, 1973). The amount of protein arriving in the small intestine is a function of the fermentable energy available for microbial protein synthesis, the protein content of the diet and the extent to which that protein is degraded in the rumen. The most successful model to date (GrazFeed; Freer *et al.*, 1997) estimates 'Digestible Protein Leaving the Stomach' from a range of feed and intake variables, including those described above.

This model achieves accurate prediction in many circumstances and has been applied widely in practical animal husbandry to calculate appropriate levels of pasture and supplement for sheep. However, differences between genotypes are dealt with by classifying sheep broadly by age, sex and breed type, and therefore the model does not directly account for interactions between genotype and nutrient supply. In addition, GrazFeed predicts wool growth as a constant proportion of protein absorbed, regardless of the composition of that protein. However, the amino acid composition of the absorbed protein can have a profound effect on the efficiency of conversion of protein to wool. For example, the low level of sulphur amino acids in microbial protein usually limits wool growth, and provision of additional methionine may increase efficiency of conversion of absorbed protein to wool by 30% (Mata and Masters, 1999). It is likely that nutritional models will develop further to include absorption and use of specific amino acids for wool.

The rate of blood flow to the skin regulates the nutrient supply to the follicle, but it is almost certainly not the primary control mechanism (Hales and Fawcett, 1993). This does not mean it is not important. Changes in wool growth rate are usually accompanied by changes in blood flow to the skin, and Harris *et al.* (1994b) suggested that changes in blood flow were as important in determining the rate of protein synthesis in the skin as they are in mammary gland.

### Protein synthesis

Most nutritional models focus on protein deposition, which is the difference between protein synthesis rate and protein degradation rate. Thus, they model mass balance, rather than rate constants. However, wool protein is excreted by the wool follicle so it does not undergo 'turnover'. Wool growth therefore depends only on protein synthesis rate. Models of protein turnover rates are in their infancy (Sainz and Wolff, 1990; Knap and Schrama, 1996), but offer a basis for specialized wool models of the future.

If we consider wool growth as part of the function of skin as whole, a simple model to relate wool growth to protein synthesis in skin can be described as follows:

$$\begin{aligned}\text{Wool growth} &= \text{Protein content (Pc)} \times \text{Synthesis rate (Ps)} \times \text{Skin efficiency (r)}\\ &= [\text{Pc}_0 + \text{Pc}_0 \times (\text{Ps} - \text{Pd})] \times \text{Ps} \times \text{r}\end{aligned}$$

$Pc_0$ may be defined as Pc at a maintenance state, Pd is protein degradation rate in skin and r is the proportion of skin protein synthesis committed into wool. We can replace Pc with $[Pc_0 + Pc_0 \times (Ps - Pd)]$ to allow for the change in skin protein mass, which is determined by both synthesis and degradation rates. Based on this model, wool growth will reflect three main components: skin (protein) mass and its change, synthesis rate and skin efficiency. The following discussion explores each of these components in greater detail, as a basis for developing such a model.

## Control of skin and follicular mass

The mass of follicular tissue depends on both the fraction of the skin made up by follicles, and on the mass of skin. During normal growth, changes in skin mass generally follow changes in the mass of the whole body. For example, as Suffolk-Finn Dorset lambs grew from 25 kg to 40 kg and then 55 kg, the proportion of nitrogen in skin to the total nitrogen in the whole body was almost constant at 0.091, 0.11 and 0.106

(MacRae *et al.*, 1993, recalculated to exclude wool N). However, skin mass can be altered differentially by the level of nutrition.

## Nutrition

As shown in Table 15.2, changes in the amount of germinal epithelium in skin contribute to the effect of nutrition on wool growth. With increased levels of feed intake, both the weight of skin per unit area and the amount of non-collagenous protein per unit mass increase (Williams and Morley, 1994). However, there are no reliable data on the effects of nutrition on the proportion of skin mass made up by follicles.

While skin and muscle mass generally move in the same direction when feed intake changes, the overall rates and amplitudes of change can be quite different. For example, Murray and Slezacek (1994) reported a 35% reduction in skin weight, compared with a 13% reduction in muscle and a 53% reduction in alimentary tract, associated with an 18% body weight loss in young sheep. Similarly, when Merino hoggets were fed at 0.44 M, 1.15 M or 1.8 M, the carcass weight increased by 29% from the lowest to highest intake, while the skin weight increased by 65% (N.R. Adams, unpublished observations). The protein concentration in skin did not change with intake, so the weight change indicated similar changes in protein mass in the skin.

Although changes in skin mass are relatively large, the rate of change is slower than changes in fractional protein synthesis rate (Liu *et al.*, 1998) or in the mitotic rate of bulb cells (A.C. Schlink, unpublished observations). The slow rate of change in mass would contribute to the delayed response of wool growth to changes in nutrition summarized by Nagorka (1977). In addition to slow changes in follicle mass, the substantial delays in wool growth response to re-feeding in sheep that have been subjected to restricted nutrition (Butler-Hogg, 1984) are likely to be affected also by diversion of amino acids to tissues with a greater demand (see p. 264).

## Genetics

Genetic factors affect the mass of follicles mainly through effects on the proportion of skin mass made up of follicles. Jackson *et al.* (1975) calculated that 83% of genetic difference between sheep in fleece weight was correlated with four descriptors of follicle morphology: follicle length, follicle curvature, the number of follicles per unit area of skin, and the ratio of primary to secondary follicles. Subsequent work supports these observations. Hocking Edwards and Hynd (1992) found that the relative volume of germinative tissue in skin of sheep is correlated with their genetic capacity for wool growth, and Williams and Winston (1987) found that the major difference between sheep that differed genetically for fleece weight was the length of follicles, rather than their diameter.

Genetic effects on wool production through total skin mass are more limited. Low and variable genetic correlations have been reported between skin wrinkles, skin thickness, body size and wool production, summarized by Williams (1987). Gregory (1982) reported a positive genetic correlation (0.39) between skin thickness and clean fleece weight, but Hynd *et al.* (1996) observed the opposite, with thinner-skinned sheep producing more wool than thick-skinned animals. Williams and Morley (1994) reported that sheep that were genetically low wool producers had a higher skin collagen content per unit skin mass than high wool producers.

### Hormonal control

Hormones affect the mass of active follicular tissue in two different ways. Firstly, hormones influence the follicular cycle of anagen (development), telogen (constant secretion) and catagen (degeneration), which occurs as part of an annual moulting cycle. Secondly, hormones are involved in follicular 'shut-down' from nutritional or other stresses. Such shut-down can be distinguished from normal follicle cycles (Schlink and Dollin, 1995).

#### Follicle cycles

The impact of annual follicular cycles, coordinated by photoperiod, varies among breeds. The Wiltshire Horn breed may shed the entire fleece, the Romney undergoes seasonal fluctuations of up to 40% in wool growth, while changes in the Merino are relatively limited. Seasonal cycles also appear to affect follicle efficiency (see p. 267). Prolactin mediates photoperiodic effects by synchronizing endogenous follicular cycles, but it does not drive the follicular events. Thus, pharmacological manipulation of prolactin does not affect wool growth (Wallace, 1979), although prolactin receptors are widely distributed in dermal papilla, inner and outer root sheath, germinal matrix, and the sebaceous and sweat glands (Choy *et al.*, 1997). An increased concentration of receptors for insulin-like growth factor-I (IGF-I) has been associated with catagen (Nixon *et al.*, 1997). The effects of the follicular cycle on protein synthesis in skin have not been measured.

#### Follicle shut-down

Increased cortisol causes loss of bulb and root sheath cells, resulting in fibre shedding (Chapman and Bassett, 1970). Prolonged treatment with cortisol causes the skin to become thinner, with loss of collagen from the dermis, reduction in the size of sebaceous glands and regression of follicles. Epidermal growth factor can also cause follicle shut-down. Other parts of the glucocorticoid axis, including corticotrophin releasing factor, pro-opiomelanocorticotropin and adrenocorticotropic hormone (ACTH), are found in skin (Slominski *et al.*, 1998), but their involvement in fibre production is unknown. However, local treatment with ACTH can stimulate growth of fibre during the anagen stage in mink (Rose, 1998).

## The rate of protein synthesis in skin and follicles

The contribution of variation in the rate of protein synthesis in skin to changes in wool growth can be substantial (Table 15.1). However, the proportion of skin protein synthesis that occurs within the follicle is unknown, because the current methods of measurement of protein synthesis rate with isotope tracers treat skin (including follicles) as an integrated organ. Approximately 20% of protein synthesis in skin is secreted as wool fibre (Table 15.1), but the proportion of skin protein synthesis accounted for by the follicle must be greater than this. The fibre represents only 10–20% of bulb matrix cells (see p. 266), although these undergo a 13-fold increase in protein mass during keratinization (Short *et al.*, 1968). Additional protein synthesis is associated with the generation, maintenance and maturation of the remaining 80–90% of cells generated in the bulb. Even if these only double in total protein synthesized during maturation and function as inner root sheath, their number is such that they would account for more protein synthesis than that involved in fibre formation. Further protein synthesis occurs in the generation, maturation and maintenance of cells of the outer root sheath, not associated with the bulb (Chapman *et al.*, 1980). Thus, up to half the protein synthesis in sheep skin is likely to be associated with the wool follicle. This value would explain the close relationship between protein synthesis in skin and wool fibre production recorded in Table 15.1.

## Dietary supply of amino acids

The fractional rate of protein synthesis in the skin, and consequently the rate of wool growth, is highly responsive to changes in feed intake and type of dietary protein (Table 15.3). However, the extent of the increase, and its relativity with increases in other tissues, varies substantially between experimental models (Table 15.3). These differences must be related at least in part to the relative demands of skin and other tissues (see below).

Increasing the supply of sulphur amino acids increases the rate of wool growth and the concentration of sulphur amino acids in wool. The wool growth response tends to be higher in sheep with a low overall feed intake (Harris *et al.*, 1994a). Few studies have measured directly the rate of protein synthesis in skin or whole body in response to feeding or infusing sulphur amino acids. Harris *et al.* (1997) measured an increase in the uptake of cysteine for protein synthesis following cysteine infusion into an isolated skin patch. However, it was not possible to determine whether total protein synthesis increased in skin, or just the ultra-high sulphur protein component. S. Liu (unpublished observations) measured an increase in fractional synthesis rate (FSR) of protein in the skin of 11%, and a significant increase in keratin mRNAs, following methionine infusion into the jugular vein. In this experiment wool growth increased by 37%, indicating that some of the observed increase in wool growth resulted from increased efficiency of incorporation of synthesized protein into wool.

Amino acids available for protein synthesis are derived from the diet and also potentially from protein degradation. Protein synthesis for the wool fibre is essentially non-degraded (Harris and Lobley, 1991), but the contribution of other protein degradation in the skin to subsequent protein synthesis is unknown. For example, skin protein content of 46 kg sheep can be estimated as 590 g from MacRae *et al.* (1993) and at maintenance the amount of protein degradation is 59 g day$^{-1}$ assuming a fractional degradation rate of 10% day$^{-1}$, similar to skin FSR (Rocha *et al.*, 1993). In terms of amino acids, this amount of degraded protein could provide 0.6 g day$^{-1}$ methionine, calculated from amino acid concentrations in skin protein (MacRae *et al.*, 1993). On the other hand, amino acids available to the skin from the circulation can be estimated to be considerably less than this. Using a cardiac output of 4.9 l min$^{-1}$ for a 46 kg sheep,

**Table 15.3.** Fractional synthesis rate (FSR, % day$^{-1}$) of protein in midside skin and skeletal muscle of sheep in response to different diets.

| Sheep | Diet or treatment | Muscle FSR | Skin FSR | Whole-body FSR | Source |
|---|---|---|---|---|---|
| Merino wethers 35–47 kg | 0.44 M<br>1.15 M<br>1.8 M | 1.41<br>2.27<br>2.93 | 12.8<br>18.2<br>20.6 | 2.20<br>2.52 | N.R. Adams (unpublished observations) |
| Merino lambs 24–26 kg, fed 0.83 M | Canola meal<br>Lupin seeds | 2.33<br>2.21 | 17.3<br>15.8 | | D.G. Masters (unpublished observations)[a] |
| Merino lambs 25–33 kg | 0.6 M<br>1 M | 1.64<br>2.28 | 15.4<br>16.9 | | Liu *et al.* 1998 |
| Suffolk cross wether, fed 1.7 M 35–41 kg | – Cimaterol<br>+ Cimaterol | 1.45<br>2.76–3.01 | 11.6<br>6.3–10.9 | | Nash *et al.* (1994) |
| Suffolk cross wether 26–35 kg | 0.6 M<br>1.2 M<br>1.8 M | 1.16<br>1.67<br>3.30 | | 2.83<br>3.76<br>4.73 | Harris *et al.* (1992)[b] |
| Romney sheep 42–44 kg | 1.5 M<br>+ 2 g cysteine | | 12.8<br>22.8 | | Harris *et al.* (1994b)[c] |
| Suffolk cross wether 26–35 kg | 0.6 M<br>1.8 M | 1.91<br>2.84 | 5.6<br>9.5 | | Lobley *et al.* (1992) |

[a]Source of protein was lupin seed or canolameal.
[b]FSR in muscle was measured using arterio-venous technique. All the others were measured using flooding-dose techniques. Isotope-labelled phenylalanine was used as the tracer.
[c]FSR in skin was measured from the uptake of [$^{35}$S]cysteine by skin using arterio-venous technique.

a proportion of 0.06 to the whole skin, and concentrations in blood of 20 μM for methionine (Harris *et al.*, 1989, 1997), calculations indicate a supply of 1.3 g day$^{-1}$ methionine. However, only 10–35% is taken up from the blood (Harris *et al.*, 1997), so the effective amino acid supply to the skin from the blood is considerably less than the total amount available to skin from protein degradation *in situ*. Despite these calculations, evidence that protein degradation makes a significant contribution to wool growth is still weak. For example, lines of sheep that differ in weaning weight differed primarily in protein degradation rate (Oddy *et al.*, 1995), yet wool growth in these lines was similar (Pattie and Williams, 1967).

## Competition for amino acids

### Growth

At high nutrient intake, Graham and Searle (1972) found a constant relationship between wool production and feed intake from birth, indicating that wool was similar to other body tissues in the age-associated reduction in the efficiency of incorporation of amino acids into protein. However, at lower feed intakes Doyle and Egan (1983) found that body tissues competed more strongly than wool for nutrients in young sheep, while in older sheep wool growth was maintained at the expense of body tissues. Further evidence of competition for nutrients between wool growth and other body tissues was obtained by Graham and Searle (1982) and Cronje and Smuts (1994).

While the FSR of protein responds to changes in dietary intake in a number of tissues, including skin, muscle and gut (Lobley *et al.*, 1992, 1994), the magnitude of the response is not always consistent between tissues. As shown in Table 15.3, increasing feed intake from 0.6 M to 1.8 M in the Suffolk-cross lambs resulted in large increases in skin FSR (68%) but less in muscle (29%), while the opposite was observed in the Merino, with a change in feeding from 0.6 M to 1.6 M increasing skin FSR by 10% and muscle FSR by 60% (Liu *et al.*, 1998). Furthermore, skin FSR was reported to be three times that of muscle in Suffolk-cross lambs (Lobley *et al.*, 1992) and six times that of muscle in Merino lambs (Liu *et al.*, 1998). These differences may be related to the high rates of wool growth and less muscle growth in Merino sheep, but other factors such as the growth demand by the animal and the length of re-feeding period must also be important.

### Pregnancy and lactation

During pregnancy there is usually a specific reduction in the proportion of whole body protein deposited in wool, relative to maternal and fetal tissue. Merino ewes fed to maintain maternal liveweight increased non-wool body protein in the ewe and fetus by 6% between days 127 and 144 of pregnancy. During the same period, wool growth decreased by 6% (Masters and Mata, 1996). In lactation, the situation is reversed with maintenance or even an increase in wool protein production at the same time as a reduction in maternal body protein (Masters and Stewart, 1990; Masters and Mata, 1996). These findings are consistent with the hormonal milieu increasing the priority of fetal over maternal tissues during pregnancy and lactation.

Measurements of protein synthesis in skin and muscle during the final month of pregnancy indicated that reduced wool growth resulted from reduced skin mass (skin thickness) and a slight but variable reduction in FSR per unit of skin relative to non-pregnant sheep of the same maternal weight (Table 15.1; S. Liu, D.G. Masters and H. O'Donohoe, unpublished observations). The situation during lactation remains to be explored.

### Hormonal effects

Effects of hormones on protein synthesis have been well reviewed (MacRae and Lobley, 1991; Bell *et al.*, 1998) and are summarized in Table 15.4. In contrast to most other areas of animal production, there are no hormones with a homeorrhoetic function to direct nutrients to the skin. However, hormones do affect wool growth through their effects on whole body protein synthesis, or by affecting protein metabolism in other

**Table 15.4.** Summary of increases (↑), decreases (↓) or no significant change (→) in whole body protein metabolism and in wool growth, reported to result from treatment with hormones.

| Hormone | Protein synthesis | Protein degradation | Specific follicle effects | Wool effect |
|---|---|---|---|---|
| Thyroxine | ↑ | ↑ | Follicle efficiency | ↑or → |
| Androgens | ↑? | ↓ | Nil | ↑ |
| β-adrenergic agonist | ↑ | ↓ | Nil | ↓ |
| Growth hormone | ↑ | → | Nil | ↑ or ↓ |
| Insulin | → | ↓ | Nil | → |
| IGF-I | →, ↑? | ↓ | Nil | → or ↑ |
| Cortisol | → | ↑ | Follicle shut-down | ↓ |
| Prolactin | → | → | Follicle efficiency | ↓ or → |

organs. In addition to these effects, specific effects of hormones on the mass or efficiency of wool follicles are dealt with on pp. 261 and 264.

Most of the effects of hormones on wool growth are consistent with their effects on general protein metabolism. For example, the small increase in wool growth reported in wethers treated with androgens (Southcott and Royal, 1971; Hynd and James, 1987) probably resulted from an increase in feed intake. Insulin and cortisol, which affect protein degradation rather than synthesis, have no effect on wool growth under normal physiological concentrations. High concentrations of cortisol may have specific effects on follicular function (p. 261). Growth hormone may increase (Johnsson *et al.*, 1985) or decrease (Wynn *et al.*, 1988) wool growth, depending on the relative balance between an increase in overall body protein synthesis resulting from increased feed intake, and the diversion of nutrients to tissues more responsive to growth hormone, such as gut and muscle. Diversion of nutrients is even more marked with the adrenergic β-agonist cimaterol, which decreased wool growth by 16% and shorn skin weight by 9%, while increasing carcass protein and decreasing lipid deposition (Fennessy *et al.*, 1990). Nash *et al.* (1994) found that the reduction in wool growth was directly counter-balanced by the increase in protein deposition in muscle of treated sheep.

Recent work has explored the effects of IGF-I on wool growth. Expression of IGF-I with a keratin promoter in transgenic sheep increased wool growth during spring and summer, but not in winter, so that staple strength was decreased (Damak *et al.*, 1996). More extensive studies (Su *et al.*, 1998) on the subsequent generation of these transgenic animals were unable to detect a significant increase in wool growth or in plasma IGF-I. Sheep treated with IGF-I had a transient increase in skin protein synthesis which disappeared by 24 h (Lobley *et al.*, 1998). The extensive buffering of IGF-I through specific binding proteins makes assessment of its role difficult. It mediates the stimulation of protein synthesis by growth hormone (Bell *et al.*, 1998) and may affect the interaction between cortisol and low nutritional status (Chapman and Bassett, 1970), because treatment with IGF-I can prevent the deleterious effects of glucocorticoids on nitrogen metabolism (Tomas, 1998).

### Responsiveness to nutrient supply

Merino sheep with a high genetic capacity for wool growth respond more to nutrition, although the magnitude of the response varies among genotypes (Jackson and Roberts, 1970; Woolaston, 1987). Differences in responsiveness probably reflect, in part, differences in overall protein metabolism. For example, the responsiveness of protein synthesis rate to nutrition is reduced in both skin and muscle of sheep selected for high staple strength (Fig. 15.2). These sheep also differed in nutrient partitioning between muscle and wool, such that the sheep with a less volatile wool growth response to nutrition had greater changes in liveweight (Adams *et al.*, 1997). A similar apparent competition for protein between wool and muscle was reported by Cronje and Smuts (1994).

Follicle-specific mechanisms may also affect the responsiveness to nutrient supply. For example, Harris *et al.* (1994b) suggested that the rate of uptake of cyst(e)ine by follicle cells may be genetically determined. Fibre length and diameter can respond differently to nutrition in different flocks (Jackson and Roberts, 1970), depending on factors such as cortical cell length and the dimensions of the dermal papilla (Hynd, 1994a). The relative importance of specific follicle effects and whole body mechanisms, in determining the effects of nutrition on wool growth, remains unknown.

## The proportion of skin protein synthesis in wool

The remaining factor that may affect wool growth is the proportion of skin protein synthesis that becomes fibre. Although this is similar across genotypes (Table 15.1), unpublished work indicates that efficiency differences of 24% between high and low fleece weight genotypes contributed to a total difference of 53% in wool growth rate (S. Liu, D.G. Masters and M.J. Nancarrow, unpublished observations). At a different level,

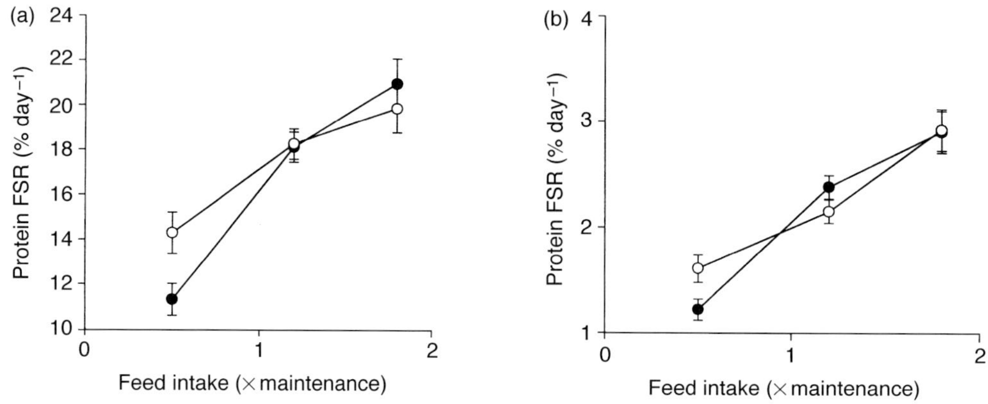


**Fig. 15.2.** Fractional synthesis rate (FSR) of protein in skin (a) and muscle (b) from ewes selected for low (●) and high (○) staple strength. Significant ($P < 0.05$) interaction between genotype and feed intake for both tissues.

the phenomenon has been studied as the proportion of cells from the bulb matrix that enter the fibre (as opposed to the root sheath). Normally, between 9 and 40% of cells formed in the bulb enter the fibre, the remaining cells becoming root sheath and subsequently lost (Wilson and Short, 1979; Hynd, 1989). This factor has been called the 'follicle efficiency'. Its exact relationship to protein synthesis is not defined.

### Nutrition and genotype

Short *et al.* (1968) found that a dramatic increase in nutrition, which increased wool production by 250%, only increased the average proportion of cells entering the fibre from 14.6% to 18.3%. Hynd (1989) concluded that follicle efficiency was independent of nutrition. Both Williams and Winston (1987) and Hynd (1989) concluded that follicle efficiency contributes about 10% to genetic differences between sheep in wool production. There was a general trend to higher follicle efficiency in sheep with a higher fibre diameter that produced more wool (Hocking Edwards and Hynd, 1992).

### Hormones

In addition to its role in mediating the effect of photoperiod on the follicular cycle described on p. 261, prolactin may also mediate seasonal changes in the efficiency of wool growth. Holle *et al.* (1994) reported that seasonal changes in wool growth were accompanied by changes to the relative distribution of cells between the fibre and inner root sheath. This was estimated by the 'production ratio', the ratio of area of fibre to root sheath in horizontal sections of skin (Butler and Wilkinson, 1979). However, Hynd (1989) found a poor correlation between the production ratio and follicle efficiency. Measures of seasonal changes in the ratio of protein synthesis to wool production might offer valuable insights.

Hynd (1994b) showed that abnormally low thyroxine concentrations reduced follicle efficiency by 40%, and also reduced the production ratio. Cell division was only reduced by 16%, although the proportion of inactive follicles was increased. Thus, it appears that thyroxine affects follicle efficiency more than bulb activity. However, it is likely that such low levels of thyroxine do not occur in normal husbandry.

## Conclusions

The impact of genotype, nutrition and physiological state on wool growth can be integrated with other bodily functions by considering it as part of the overall protein synthesis in the body. Indeed, the rate of protein synthesis for wool growth broadly reflects that in other tissues, so wool growth offers a simple index of changes in protein synthesis rate in the body. Furthermore, protein synthesis and protein degradation rates together provide a sounder basis for understanding of nutrient partitioning between wool and other tissues than do concepts based on mass flow.

The rate of wool growth is affected both by specific follicle mechanisms and by factors controlling protein metabolism in the whole body, but their relative importance is unclear in most situations. Although specific morphological characteristics of follicles correlate closely with wool growth (e.g. Jackson *et al.*, 1975), some of these may simply reflect systemic effects, such as nutrient supply. The relative importance of local and systemic factors remains a major challenge to wool scientists.

Much of the information needed to estimate wool growth as a part of protein synthesis is still sketchy. There are limited data on protein synthesis rates in the skin, and none that distinguish the follicle from the remaining skin. Even simple characteristics, such as the total fraction of skin volume made up by germinative epithelium, are poorly documented. However, as the missing data become available and models of protein synthesis become more widely used, sheep farmers and animal scientists will be able to apply a greater mechanistic understanding to enable more accurate and robust prediction of the outcomes of breeding and feeding.

## Acknowledgements

Research by CSIRO Animal Production and the Co-operative Research Centre for Premium Quality Wool is supported by Australian wool growers through the Australian Wool Research and Promotion Organization.

## References

Adams, N.R., Briegel, J.R. and Ritchie, A.J.M. (1997) Wool and liveweight responses to nutrition by Merino sheep genetically selected for high or low staple strength. *Australian Journal of Agricultural Research* 48, 1129–1137.

Allden, W.G. (1979) Feed intake, diet composition and wool growth. In: Black, J.L. and Reis, P.J. (eds) *Physiological and Environmental Limitations to Wool Growth.* The University of New England Publishing Unit, Armidale, New South Wales, pp. 61–78.

Bell, A.W., Bauman, D.E., Beerman, D.H. and Harrell, R.J. (1998) Nutrition, development and efficiency of growth modifiers in livestock species. *Journal of Nutrition* 128, 360S–363S.

Black, J.L. and Reis, P.J. (1979) Speculation on the control of nutrient partition between wool growth and other body functions. In: Black, J.L. and Reis, P.J. (eds) *Physiological and Environmental Limitations to Wool Growth.* The University of New England Publishing Unit, Armidale, New South Wales, pp. 269–293.

Black, J.L., Robards, G.E. and Thomas, R. (1973) Effects of protein and energy intake on the wool growth of merino wethers. *Australian Journal of Agricultural Research* 24, 399–412.

Bowman, P.J., Cottle, D.J., White, D.H. and Bywater, A.C. (1993) Simulation of wool growth rate and fleece characteristics of merino sheep in southern Australia. Part 1. Model description. *Agricultural Systems* 43, 287–299.

Butler-Hogg, B.W. (1984) Growth patterns in sheep: wool growth during weight loss and subsequent compensatory growth. *Journal of Agricultural Science, Cambridge* 102, 105–109.

Butler, L.G. and Wilkinson, B.R. (1979) Efficiency of wool production and the mechanics of wool follicle production. *New Zealand Journal of Agricultural Research* 22, 543–545.

Chapman, R.E. and Bassett, J.M. (1970) The effects of prolonged administration of cortisol on the skin of sheep on different planes of nutrition. *Journal of Endocrinology* 48, 649–663.

Chapman, R.E., Downes, A.M. and Wilson, P.A. (1980) Migration and keratinization of cells in wool follicles. *Australian Journal of Biological Science* 33, 587–603.

Choy, V.J., Nixon, A.J. and Pearson, A.J. (1997) Distribution of prolactin receptor immunoreactivity in ovine skin and changes during the wool follicle growth cycle. *Journal of Endocrinology* 155, 265–275.

Cronjé, P.B. and Smuts, M. (1994) Nutrient partitioning in Merino rams with different wool growth rates. *Animal Production* 59, 55–60.

Damak, S., Su, H.-Y., Jay, N.P. and Bullock, D.W. (1996) Improved wool production in transgenic sheep expressing insulin-like growth factor 1. *Biotechnology* 14, 185–188.

Doyle, P.T. and Egan, J.K. (1983) The utilization of nitrogen and sulfur by weaner and mature sheep. *Australian Journal of Agricultural Research* 34, 433–439.

Fennessy, P.F., McEwan, J.C., Lord, E.A., Greer, G.J., Bain, W.E., Johnstone, P.D., Knowler, M.A., Dalrymple, R.H. and Ingle, D.L. (1990) Effect of Cimaterol implants on lamb growth and carcass traits. *New Zealand Journal of Agricultural Research* 33, 413–427.

Freer, M., Moore, A.D. and Donnelly, J.R. (1997) Decision support systems for Australian grazing enterprises. II The animal biology model for feed intake, production and reproduction and the Grazfeed DSS. *Agricultural Systems* 53, 77–126.

Graham, N.M. and Searle, T.W. (1972) Growth in sheep. II Efficiency of energy and nitrogen utilization from birth to 2 years. *Journal of Agricultural Science, Cambridge* 79, 383–389.

Graham, N.M. and Searle, T.W. (1982) Energy and nitrogen utilisation for body growth in young sheep from two breeds with differing capacities for wool growth. *Australian Journal of Agricultural Research* 33, 607–615.

Gregory, I.P. (1982) Genetic studies of South Australian Merino sheep. IV genetic, phenotypic and environmental correlations between various wool and body traits. *Australian Journal of Agricultural Research* 33, 363–373.

Hales, J.R.S. and Fawcett, A.A. (1993) Wool production and blood supply to skin and other tissues in sheep. *Journal of Animal Science* 71, 492–498.

Harris, P.M. and Lobley, G.E. (1991) Amino acid and energy metabolism in the peripheral tissues of ruminants. In: Tsude, T., Sasaki, Y. and Kawashima, R. (eds) *Physiological Aspects of Digestion and Metabolism in Ruminants.* Academic Press, San Diego, California, pp. 201–230.

Harris, P.M., Dellow, D.W. and Sinclair, B.R. (1989) Preliminary 'in vivo' measurements of protein and energy metabolism in the skin of sheep. *Australian Journal of Agricultural Research* 40, 879–888.

Harris, P.M., Skene, P.A., Buchan, V., Milne, E., Calder, A.G., Anderson, S.E., Connell, A. and Lobley, G.E. (1992) Effect of food intake on hind-limb and whole-body protein metabolism in young growing sheep: chronic studies based on arterio-venous techniques. Species comparisons of tissue protein metabolism: effects of age and hormonal metabolism. *British Journal of Nutrition* 68, 389–407.

Harris, P.M., Lee, J., Sinclair, B.R. and Treloar, B.P. (1994a) The effect of whole body cysteine supplementation on cysteine utilization by the skin of a well-fed sheep. *Proceedings of the New Zealand Society of Animal Production* 54, 139–142.

Harris, P.M., Lee, J., Sinclair, B.R., Treloar, B.P. and Gurnsey, M.P. (1994b) Effect of food intake on energy and protein metabolism in the skin of Romney sheep. *British Journal of Nutrition* 71, 647–660.

Harris, P.M., Sinclair, B.R., Treloar, B.P. and Lee, G.J. (1997) Short-term changes in whole body and skin sulfur amino acid metabolism of sheep in response to supplementary cysteine. *Australian Journal of Agricultural Research* 48, 137–146.

Hocking Edwards, J.E. and Hynd, P.I. (1992) Cellular characteristics of wool follicles and fibres in finewool and strongwool Merinos. *Australian Journal of Agricultural Research* 43, 355–365 (Abstract).

Holle, S.A., Harris, P.M., Davies, A.S. and Birtles, M.J. (1994) Wool follicle morphology and cell proliferation in New Zealand Romney sheep; a seasonal comparison of fleeceweight-selected and control rams. *Australian Journal of Agricultural Research* 45, 769–782.

Hynd, P.I. (1989) Effects of nutrition on wool follicle cell kinetics in sheep differing in efficiency of wool production. *Australian Journal of Agricultural Research* 40, 409–417.

Hynd, P.I. (1994a) Follicular determinants of the length and diameter of wool fibres. I. Comparison of sheep differing in fibre length/diameter ratio at two levels of nutrition. *Australian Journal of Agricultural Research* 45, 1137–1147.

Hynd, P.I. (1994b) Follicular determinants of the length and diameter of wool fibres. II. Comparison of sheep differing in thyroid hormone status. *Australian Journal of Agricultural Research* 45, 1149–1157.

Hynd, P.I. and James, R.E. (1987) The effect of trenbolone acetate and trenbolone acetate plus oestradiol-17β on wool growth. *Journal of Agricultural Science, Cambridge* 108, 501–503.

Hynd, P.I., Ponzoni, R.W., Grimson, R.J., Jaensch, K.S., Smith, D. and Kenyon, R.V. (1996) Wool follicle and skin characters – their potential to improve wool production and quality in Merino sheep. *Wool Technology and Sheep Breeding* 44, 167–177.

Jackson, N. and Roberts, E.M. (1970) Comparison of three Australian Merino strains for wool and body traits. I. Genetic means of studs and strains and their interactions with years and sexes. *Australian Journal of Agricultural Research* 21, 815–835.

Jackson, N., Nay, T. and Turner, H.N. (1975) Response to selection in Australian Merino sheep. VII Phenotypic and genetic parameters for some wool follicle characteristics and their correlation with wool and body traits. *Australian Journal of Agricultural Research* 26, 937–957.

Johnsson, I.D., Hart, I.C. and Butler-Hogg, B.W. (1985) The effects of exogenous bovine growth hormone and bromocryptine on growth, body development, fleece weight and plasma concentrations of growth hormone, insulin and prolactin in female lambs. *Animal Production* 41, 207–217.

Knap, P.W. and Schrama, J.W. (1996) Simulation of growth in pigs: approximation of protein turnover characteristics. *Animal Science* 63, 533–547.

Liu, S., Mata, G., O'Donoghue, H. and Masters, D.G. (1998) The influence of live weight, live-weight change and diet on protein synthesis in the skin and skeletal muscle in young Merino sheep. *British Journal of Nutrition* 79, 267–274.

Lobley, G.E., Harris, P.M., Skene, P.A., Brown, D., Milne, E., Calder, A.G., Anderson, S.E., Garlick, P.J., Nevison, I. and Connell, A. (1992) Response in tissue protein synthesis to sub- and supra-maintenance intake in young growing sheep: comparison of large-dose and continuous-infusion techniques. *British Journal of Nutrition* 68, 373–388.

Lobley, G.E., Connell, A., Milne, E., Newman, A. and Ewing, T.A. (1994) Protein synthesis in splanchnic tissues of sheep offered two levels of intake. *British Journal of Nutrition* 71, 3–12.

Lobley, G.E., Lee, J., Hocking Edwards, J.E. and Harris, P.M. (1998) A comparison of changes in whole body and skin amino acid metabolism of sheep in response to 24 h continuous infusions of variants of insulin-like growth factor 1. *Canadian Journal of Animal Science* 77, 695–706.

MacRae, J.C. and Lobley, G.E. (1991) Physiological and metabolic implications of conventional and novel methods for the manipulation of growth and production. *Livestock Production Science* 27, 43–59.

MacRae, J.C., Walker, A., Brown, D. and Lobley, G.E. (1993) Accretion of total protein and individual amino acids by organs and tissues of growing lambs and the ability of nitrogen balance techniques to quantitate protein retention. *Animal Production* 57, 237–245.

Masters, D.G. and Mata, G. (1996) Responses to feeding canola meal or lupin seed to pregnant, lactating, and dry ewes. *Australian Journal of Agricultural Research* 47, 1291–1303.

Masters, D.G. and Stewart, C.A. (1990) Wool growth and reproduction. In: Oldham, C.M., Martin, G.B. and Purvis, I.W. (eds) *Reproductive Physiology of Merino Sheep. Concepts and Consequences.* School of Agriculture, The University of Western Australia, Perth, pp. 265–274.

Mata, G. and Masters, D.G. (1999) Rumen protected methionine improves the efficiency of use of metabolisable protein for wool growth. In: Stone, G., Forbes, T.D.A., Stith, J.W. and Byers, F.M. (eds) *Proceedings of the 5th International Symposium on the Nutrition of Herbivores, San Antonio Texas.* Online poster http://cnrit.tamu.edu/conf/isnh/post-online/post 0045.

Murray, D.M. and Slezacek, O. (1994) Effects of a 50-day period of slow liveweight loss on the body composition of crossbred wethers. *Proceedings of the Australian Society for Animal Production* 20, 397.

Nagorka, B.N. (1977) The description and analysis of wool growth. *Australian Journal of Agricultural Research* 28, 737–746.

Nash, J.E., Rocha, H.J.G., Buchan, V., Calder, G.A., Milne, E., Quirke, J.F. and Lobley, G.E. (1994) The effect of acute and chronic administration of the β-agonist, cimaterol, on protein synthesis in ovine skin and muscle. *British Journal of Nutrition* 71, 501–513.

Nixon, A.J., Ford, C.A., Oldham, J.M. and Pearson, A.J. (1997) Localisation of insulin-like growth factor receptors in skin follicles of sheep (Ovis aries) and changes during an induced growth cycle. *Comparative Biochemistry and Physiology* 118A, 1247–1257.

Oddy, V.H., Speck, P.A., Warren, H.M. and Wynn, P.C. (1995) Protein metabolism in lambs from lines divergently selected for weaning weight. *Journal of Agricultural Science, Cambridge* 124, 129–137.

Pattie, W.A. and Williams, A.J. (1967) Selection for weaning weight in Merino sheep. 3. Maintenance requirements and the efficiency of conversion of feed to wool in mature ewes. *Australian Journal of Experimental Agriculture and Animal Husbandry* 7, 117–125.

Rocha, H.J.G., Nash, J.E., Connell, A. and Lobley, G.E. (1993) Protein synthesis in ovine muscle and skin: sequential measurements with three different amino acids based on the large-dose procedure. *Comparative Biochemistry and Physiology* 105B, 301–307.

Rose, J. (1998) Adrenocorticotropic hormone (ACTH) but not alpha-melanocyte stimulating hormone (a-MSH) as a mediator of adrenalectomy induced hair growth in mink. *Journal of Investigative Dermatology* 110, 456–457.

Sainz, R.D. and Wolff, J.E. (1990) Development of a dynamic, mechanistic model of lamb metabolism and growth. *Animal Production* 51, 535–549.

Schlink, A.C. and Dollin, A.E. (1995) Abnormal shedding contributes to the reduced staple strength of tender wool in Western Australian Merinos. *Wool Technology and Sheep Breeding* 43, 268–284.

Scobie, D.R. and Woods, J.L. (1992) A review implicating a two compartment model for the processes of cell division and differentiation in the wool follicle. *Proceedings of the New Zealand Society of Animal Production* 52, 265–271.

Short, B.F., Wilson, P.A. and Schinckel, P.G. (1968) Proliferation of follicle matrix cells in relation to wool growth. In: Lyne, H.G. and Short, B.F. (eds) *Biology of Skin and Hair Growth.* Angus and Robertson, Sydney, pp. 409–426.

Slominski, A., Ermak, G., Mazurkiewicz, J.E., Baker, J. and Wortsman, J. (1998) Characterization of corticotropin-releasing hormone (CRH) in human skin. *Journal of Clinical Endocrinology and Metabolism* 83, 1020–1024.

Southcott, W.H. and Royal, W.M. (1971) Effect of implanted testosterone propionate on the growth and wool production of merino wethers. *Australian Journal of Agricultural Research* 22, 271–282.

Su, H.Y., Jay, N.P., Gourley, T.S., Kay, G.W. and Damak, S. (1998) Wool production in transgenic sheep – results from first-generation adults and second-generation lambs. *Animal Biotechnology* 9, 135–147.

Tomas, F.M. (1998) The anti-catabolic efficiency of insulin-like growth factor-1 is enhanced by its early administration to rats receiving dexamethasone. *Journal of Endocrinology* 157, 89–97.

Wallace, A.L.C. (1979) The effect of hormones on wool growth. In: Black, J.L. and Reis, P.J. (eds) *Physiological and Environmental Limitations to Wool Growth.* University of New England Publishing Unit, Armidale, New South Wales, pp. 257–268.

Williams, A.J. (1987) Physiological consequences of selection for increased fleece weight. In: McGuirk, B.J. (ed.) *Merino Improvement Programs in Australia.* Australian Wool Corporation, Melbourne, pp. 481–494.

Williams, A.J. and Morley, F.C. (1994) Influence of dietary intake and genetic capacity for wool growth on the composition of mid-trunk skin of Merino sheep. *Australian Journal of Agricultural Research* 45, 1715–1729.

Williams, A.J. and Winston, R.J. (1987) A study of the characteristics of wool follicle and fibre in merino sheep genetically different in wool production. *Australian Journal of Agricultural Research* 38, 743–755.

Wilson, P.A. and Short, B.F. (1979) Cell proliferation and cortical cell production in relation to wool growth. *Australian Journal of Biological Science* 32, 317–327.

Woolaston, R.R. (1987) Genotype × environment interactions and their possible impact on breeding programs. In: McGuirk, B.J. (ed.) *Merino Improvement Programs in Australia.* Australian Wool Corporation, Melbourne, pp. 421–435.

Wynn, P.C., Wallace, A.L.C., Kirby, A.C. and Annison, E.F. (1988) Effects of growth hormone administration on wool growth in merino sheep. *Australian Journal of Biological Science* 41, 177–187.

# VI Reproduction, Pregnancy and Lactation

# 16 Regulation of Macronutrient Partitioning between Maternal and Conceptus Tissues in the Pregnant Ruminant

A.W. Bell and R.A. Ehrhardt

*Department of Animal Science, Cornell University, Ithaca, New York, USA*

## Introduction

The concept that pregnant animals partition available nutrients to favour their developing offspring was formally postulated by Hammond (1947) who considered that tissues compete for circulating nutrients on the basis of their relative metabolic rates. This idea was reinforced by early demonstrations of high metabolic rates in tissues of the conceptus relative to those of the dam (Meschia *et al.*, 1980). However, more recent thinking on the extracellular regulation of nutrient partitioning has focused on endocrine coordination rather than tissue competition as a general explanatory mechanism, building on the concept of homeorhesis elaborated by Bauman and Currie (1980). This concept more satisfactorily accommodates the coordinated metabolic compromises implicit in the constraint of fetal growth during late pregnancy, to levels that optimize opportunity for neonatal survival and postnatal development, yet minimize excessive depletion of maternal energy and protein reserves during late pregnancy.

This chapter will describe patterns of partitioning of macronutrients between maternal and conceptus tissues, as a background to more detailed consideration of their mechanistic explanations involving placental functions and adaptive responses of maternal (non-uterine) tissues to pregnancy.

## Patterns of nutrient utilization and partitioning during pregnancy

### *Fetal nutrient requirements*

The conceptus (fetus(es), placenta, associated fetal membranes, and supporting uterine tissues) makes extensive, direct demands upon maternal supplies of glucose and amino acids, especially during late pregnancy. Numerous *in vivo* studies of conceptus metabolism in sheep have focused mostly on fetal metabolism in terms of umbilical exchanges of oxygen, nutrients and metabolites (see Battaglia and Meschia, 1988; Bell, 1993). These and similar data from pregnant cows (Comline and Silver, 1976; Reynolds *et al.*, 1986; Ferrell, 1991) have been used to construct metabolic balance sheets for fetal sheep (Bell, 1993) and cattle (Bell, 1995). In both species during late gestation, 35–40% of fetal energy is supplied as glucose and its fetal-placental metabolite, lactate, and a further 55% is taken up as amino acids. Most of the residual 5–10% is contributed by acetate, placental transfer of which is meagre relative to its abundance and energetic importance in the maternal system. Placental capacity for maternal–fetal transport of long-chain, non-esterified fatty acids (NEFAs) and ketoacids is even more limited (see Bell, 1993), denying fetal access to substrates derived from maternal fat mobilization and thus constraining fetal ability to grow at the direct expense of maternal energy reserves. Almost all of the nitrogen acquired by the fetus is in the form of amino acids, although a small net umbilical uptake of ammonia, derived from placental deamination of amino acids, is detectable in fetal sheep during mid (Bell *et al.*, 1989a) and late gestation (Holzman *et al.*, 1977).

Patterns of utilization of glucose and amino acids for oxidative metabolism and growth in fetal ruminants are reviewed elsewhere (Battaglia and Meschia, 1988; Bell, 1993, 1995; Hay, 1996, 1998).

### Metabolism of non-fetal conceptus tissues

The major contribution of the non-fetal components of the gravid uterus, especially the placenta, to oxygen and nutrient requirements of the conceptus is sometimes ignored. However, these requirements greatly affect the partitioning of nutrients within the gravid uterus and add substantially to the nutrient demands upon the dam. In late-pregnant ewes and cows, the aggregate weight of placentomes, consisting of fetal (cotyledonary) and maternal (caruncular) tissues, is less than 15% of that of the attached fetus. However, the weight-specific metabolic rate of the placenta is so great that the uteroplacental tissues (placentomes, endometrium, myometrium) consume 35–50% of the oxygen and 60–70% of the glucose taken up by the uterus in ewes (Meschia *et al.*, 1980) and cows (Reynolds *et al.*, 1986). Net uteroplacental consumption of amino acids is lower relative to fetal uptake, because growth of the placenta and uterine tissues is negligible in sheep (Ehrhardt and Bell, 1995) and small in cattle (Bell *et al.*, 1995) during late pregnancy. Nevertheless, net removal by the uteroplacental tissues has been estimated to account for 24% of uterine uptake of amino acid nitrogen in well-fed ewes during late pregnancy (Chung *et al.*, 1998).

### Partitioning of glucose and amino acids between conceptus and maternal tissues

In well-fed, monotocous ewes during late pregnancy, uterine uptake of glucose consumes 30–50% of maternal glucose supply (Prior and Christenson, 1978; Hay *et al.*, 1983b; Oddy *et al.*, 1985; Leury *et al.*, 1990), which accounts for all of the pregnancy-induced increment in whole-body production and utilization of glucose (Petterson *et al.*, 1993; Freetly and Ferrell, 1998). About 10% of this glucose carbon is returned to the maternal circulation as lactate (Meschia *et al.*, 1980; Faichney *et al.*, 1981), presumably resulting from glycolysis in uteroplacental tissues of maternal origin, and contributing to the increased cycling between glucose and lactate that is characteristic of ruminants during late pregnancy (van der Walt *et al.*, 1983).

No studies have directly addressed the partitioning of amino acids between the gravid uterus and maternal non-uterine tissues. However, we have used traditional nitrogen balance and comparative slaughter techniques to estimate the partition of apparently digested crude protein in ditocous ewes that were fed to predicted energy and protein requirements between days 110 and 140 of pregnancy (McNeill *et al.*, 1997). Calculations suggest that as much as 80% of apparently digested crude protein was partitioned to the gravid uterus, the remainder being used to support increased metabolism and net deposition of amino acids in the developing mammary glands and visceral organs (Bell and Ehrhardt, 1998). However, it is notable that even though the ewes were fed to predicted requirement for dietary protein (> 90 g day$^{-1}$ digestible crude protein), the available pool of circulating amino acids was augmented by the net mobilization of protein from maternal carcass tissues (mostly skeletal muscle), amounting to almost 10% of the digestible crude protein intake.

## Modifying effects of maternal nutrition and placental size

### Direct effects of maternal nutrition

Partitioning of glucose between conceptus and maternal tissues during late pregnancy is affected by maternal plane of nutrition (see Bell, 1993; Bell and Ehrhardt, 1998). Previously well-conditioned, ditocous ewes fed at 60% of predicted energy requirement for 3 weeks were able to maintain absolute rates of uterine uptake of glucose despite a 25% reduction in maternal glucose entry rate. Fetal growth was unimpaired but maternal loss of body weight and condition was appreciable. In contrast, more severe undernutrition (30–40% energy requirement for 2–3 weeks) or fasting (5 days) caused a marked decrease in uterine uptake of glucose that was proportional to that in maternal glucose production. Possible mechanisms for these phenomena, involving adaptations in the placenta and maternal tissues, are discussed in later sections.

Effects of maternal nutrition on partitioning of amino acids to the conceptus have been little studied. In ewes fasted for 5 days, umbilical net uptake of amino acids was relatively unaffected despite reduced maternal blood concentrations of most amino acids (Lemons and Schreiner, 1983; Liechty *et al.*, 1991). This implies that potentially negative effects of severe, short-term nutrient deprivation on fetal access to maternal amino acids were offset by maternal mobilization of tissue protein reserves and active transport of amino acids by the placenta. Even if these putative compensatory mechanisms are successful in sustaining fetal amino acid supply, a decrease in fetal tissue protein synthesis and growth is to be expected because of increased fetal catabolism of amino acids to substitute for lack of glucose as an energy source (see Bell, 1993).

Specific effects of protein deprivation on partitioning of amino acids between maternal tissues and the conceptus have not been examined in pregnant ruminants. However, in ditocous ewes fed a protein-deficient (80 g crude protein kg$^{-1}$ dry matter) but energy-sufficient diet for 30 days in late pregnancy, uterine (mostly fetal) nitrogen accretion was reduced by 20% compared with that in ewes fed to predicted requirement (120 g $kg^{-1}$ crude protein) (McNeill *et al.*, 1997). This occurred despite a substantial net loss of protein from maternal carcass tissues and reduced net accretion of protein in maternal visceral and mammary tissues, indicating a major shift in partitioning of maternal amino acids to partially offset the effects of dietary deficiency on conceptus growth (Fig. 16.1).

### *Indirect effects of maternal nutrition*

Effects of maternal nutrition in early and mid pregnancy may have later effects on nutrient partitioning to the conceptus via influences on growth and highly correlated indices of functional capacity of the placenta. Moderate undernutrition of ewes during early to mid pregnancy, when placental growth is rapid (Ehrhardt and Bell, 1995), has caused conflicting positive (Faichney and White, 1987; McCrabb *et al.*, 1992) and negative (McCrabb *et al.*, 1992; Clarke *et al.*, 1998) effects on placental size. Variation in body condition during early pregnancy appears to at least partly explain this confusion, in that fatter ewes responded to underfeeding with an increase in placental size, whereas the opposite occurred in lean ewes (McCrabb *et al.*, 1992). This suggests that if maternal energy stores are available, the dam will mobilize them for her own use, permitting a compensatory response in placental growth.

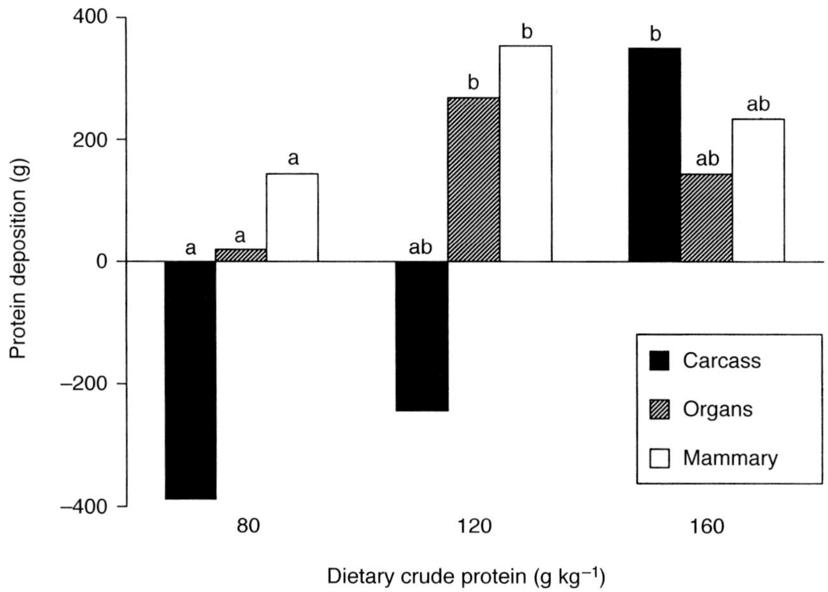


**Fig. 16.1.** Protein deposition between days 110 and 140 of pregnancy in maternal tissue components of ditocous ewes fed diets containing different levels of dietary crude protein. Histograms are means for eight ewes. Pooled standard errors were 214 g for carcass, 84 g for organs, and 44 g for mammary gland. Within tissue components, means with different letters are significantly different ($P < 0.05$). Adapted from the data of McNeill *et al.* (1997).

In distinct contrast, recent novel studies have shown that overfeeding and rapid growth of primiparous ewes during early–mid pregnancy causes profound reductions in placental growth, followed by severe fetal growth retardation with all the hallmarks of placental insufficiency (Wallace *et al.*, 1996, 1999). Intriguingly, the negative effects of overfeeding before day 50 of pregnancy were ameliorated by feed restriction after day 50; conversely, overfeeding after day 50 partially induced placental and fetal growth retardation in ewes fed at a moderate level up to that time (Table 16.1; Wallace *et al.*, 1999).

Maternal body condition may also affect the partitioning of nutrients during late pregnancy by mechanisms not directly related to placental size or function. For example, McNeill *et al.* (1998) observed that when given *ad libitum* access to feed during late pregnancy, lean ewes ate more than fatter ewes but partitioned the extra nutrients to maternal rather than conceptus tissues. Also, fatter ewes were better able than lean ewes to sustain conceptus growth during moderate undernutrition in late pregnancy (McNeill *et al.*, 1999). Placental size did not differ appreciably between lean and fatter ewes in either of these studies, and was not considered to be an important determinant of differences in nutrient partitioning between maternal and conceptus tissues.

### *Effects of non-nutritional influences on placental size*

Environmental heat stress or pre-mating carunclectomy can cause dramatic reductions in placental size and functional capacity (see Bell, 1987; Robinson *et al.*, 1995), and might, therefore, be expected to influence the partitioning of nutrients between maternal and conceptus tissues. Plasma glucose concentrations were elevated in chronically heat-treated (Bell *et al.*, 1989b) and carunclectomized ewes (Falconer *et al.*, 1985) fed at the same levels as controls during late pregnancy, suggesting that placental restriction may have caused a repartitioning of maternal glucose away from the conceptus. Unfortunately, the only experiment designed to test this hypothesis on carunclectomized ewes was inconclusive (Owens *et al.*, 1989). Ewes were allowed *ad libitum* access to feed and intakes were not reported. This presumably contributed to the highly variable rates of maternal glucose production in this study, and a confounding trend for these rates to be lower in carunclectomized than in control ewes.

**Table 16.1.** Effect of switching maternal nutrient intake at day 50 of gestation on placental development and fetal growth in monotocous adolescent ewes. Values are means for 11 or 12 ewes. Data from Wallace et *al.* (1999).

| | Plane of nutrition[1] | | | | |
|---|---|---|---|---|---|
| Variable | MM | HH | HM | MH | PSE[2] |
| Placental weight (g) | 457[a] | 258[b] | 381[ac] | 312[bc] | 44 |
| Placentome number | 103[a] | 79[b] | 96[a] | 109[a] | 6.3 |
| Birth weight (kg) | 4.94[a] | 3.03[b] | 4.45[a] | 3.11[b] | 0.39 |

[1]M-, moderate intake, growth rate 57 g day$^{-1}$, days 0–50; -M, moderate intake, days 50–100; H-, high intake, growth rate 280 g day$^{-1}$, days 0–50; -H, high intake, days 50–100. All ewes were fed to maintain body condition from day 100 to term.
[2] PSE, pooled standard error.
Row means with different letters are significantly different ($P < 0.05$).

## Placental transport and metabolism of nutrients

### *Mechanisms of placental transport of glucose and amino acids*

#### Glucose

Glucose, which accounts for about 60% of the net uptake of carbon by the gravid uterus in the well-nourished, late-pregnant ewe (Carver and Hay, 1995; Chung *et al.*, 1998), is placentally transported from the maternal to the fetal circulations by carrier-mediated, facilitated diffusion (see Hay, 1995). The predominant glucose transporter protein isoforms in sheep placenta are GLUT-1 and GLUT-3 (Ehrhardt and Bell, 1997; Das *et al.*, 1998), mRNA and protein abundance of which increase with gestational development, especially in GLUT-3 (Ehrhardt and Bell, 1997; Currie *et al.*, 1997). This appears to account for much of the fivefold increase in glucose transport capacity of the sheep placenta *in vivo* between mid and late gestation (Molina *et al.*, 1991). The relative importance of GLUT-3 may also be inferred from its immunolocalization at the apical surface of the trophoblastic cell layer that forms the fetomaternal tissue barrier in the ovine placenta (U.G. Das, R.A. Ehrhardt, W.W. Hay, Jr and S.U. Devaskar, unpublished observations). In contrast, GLUT-1 was localized on the basolateral surface of this cell layer, and in the plasma membrane of chorionic binucleate cells within the maternal caruncular epithelium. This differentially polarized localization of the two isoforms is consistent with the notion that GLUT-3, with a lower $K_m$, efficiently transfers maternal glucose into trophoblastic cells surrounding the fetal vasculature. Exit of accumulated glucose from these cells into the umbilical microcirculation may then be mediated by GLUT-1, which has a higher $K_m$ and basolateral distribution.

Negligible abundance of the insulin-responsive transporter protein, GLUT-4 (Ehrhardt and Bell, 1997), together with minimal expression of its mRNA in the sheep placenta (R.A. Ehrhardt, unpublished observations), is consistent with the lack of a direct effect of maternal or fetal insulinaemia on uteroplacental uptake and placental transport of glucose *in vivo* in the pregnant ewe (see Bell, 1993; Hay, 1995).

#### Amino acids

Most amino acids taken up by the sheep placenta are transported against a fetal–maternal concentration gradient, implying the use of energy-dependent, active transport processes (Carter *et al.*, 1991). The specific protein transporters involved have yet to be identified and characterized in ruminant placentae. However, it is assumed that, as in human placental microvesicles, there are at least ten sodium-dependent and sodium-independent transporter systems which have varying levels of activity at different placental membrane surfaces (see Hay, 1998).

Recent investigations suggest that rapid maternal–fetal transport of neutral amino acids may require not only sodium-dependent transport at the maternal surface of the placental epithelial surface, but affinity for highly reversible, sodium-independent transporters located at the fetal surface (Jozwik *et al.*, 1998). This interpretation refutes the long-held assumption that passive diffusion is responsible for transporting neutral amino acids from placenta to fetus down an intracellular–extracellular concentration gradient created by sodium-dependent transporters (see Bell, 1993).

Fractional exchange of all amino acids across the uterine circulation is small ($\leq$ 10%) in well-fed, late-pregnant ewes, implying that, under normal conditions, maternal supply is not a limiting factor for uterine uptake of amino acids (Chung *et al.*, 1998). Negligible uterine uptake of glycine, which is most abundant in maternal blood, and glutamate suggest that the sodium-dependent transport systems for these acids, observed in other tissues, are not expressed in the ovine placenta.

## Impact of placental metabolism on maternal–fetal nutrient transfer

### Glucose metabolism

Glucose entry into the gravid uterus and its component tissues is determined by maternal arterial glucose concentration (Hay and Meznarich, 1988; Leury *et al.*, 1990), while glucose transport to the fetus is determined by the transplacental (maternal–fetal) concentration gradient (Hay *et al.*, 1984). In turn, the transplacental gradient is directly related to both placental and fetal glucose consumption, which are dependent on fetal arterial glucose concentration (Hay *et al.*, 1990). Thus, as fetal glucose concentration changes relative to that of the mother, thereby changing the transplacental gradient, placental transfer of glucose to the fetus varies reciprocally with placental glucose consumption.

In addition to its quantitative impact on placental transfer of glucose, placental glucose metabolism has a major qualitative influence on the pattern of carbohydrate metabolites delivered to the fetus. Rapid metabolism to lactate (~35%), fructose (~4%), and $CO_2$ (~17%) accounted for about 56% of uteroplacental glucose consumption in late-pregnant ewes, and was directly related to placental glucose supply (Aldoretta *et al.*, 1994). The fate of the remaining 44% of glucose metabolized by the placenta must include synthesis of alanine and other non-essential amino acids (Timmerman *et al.*, 1998), directly or via lactate (Carter *et al.*, 1995). Umbilical uptake and fetal oxidation of placentally derived lactate (Sparks *et al.*, 1982; Hay *et al.*, 1983a) and fructose (Meznarich *et al.*, 1987) are estimated to contribute approximately 20% and 5%, respectively, to fetal $CO_2$ production, in addition to the 30% contributed by the rapid oxidation of glucose (Hay *et al.*, 1983a).

### Amino acid metabolism

Placental metabolism substantially affects both the quantity and composition of amino acids delivered to the fetus. The significant net consumption by uteroplacental tissues of glutamate, serine, and the branched chain amino acids (Liechty *et al.*, 1991; Chung *et al.*, 1998) implies catabolism or transamination of these acids. An additional, small fraction of this net loss of amino acids will be in the form of secreted peptides.

The ovine placenta has very little enzymatic capacity for urea synthesis but produces considerable amounts of ammonia, much of which is released into maternal and, to a lesser extent, fetal circulations (Holzman *et al.*, 1977; Bell *et al.*, 1989a). This is consistent with extensive placental deamination of branched chain amino acids to their respective keto acids, which are released into fetal and maternal bloodstreams (Smeaton *et al.*, 1989; Loy *et al.*, 1990), and with rapid rates of glutamate oxidation in the placenta (Moores *et al.*, 1994). Transamination of branched chain amino acids accounts for some of the net glutamate acquisition by the placenta, the remainder of which is taken up from the umbilical circulation (Moores *et al.*, 1994). That which is not quickly oxidized combines with ammonia to synthesize glutamine, which is then released back into the umbilical bloodstream (Chung *et al.*, 1998). Some of this glutamine is converted back to glutamate by the fetal liver, which produces most of the glutamate consumed by the placenta (Vaughn *et al.*, 1995). This establishes a glutamate–glutamine shuttle which promotes placental oxidation of glutamate and fetal hepatic utilization of the amide group of glutamine.

Similarly, the placenta almost quantitatively converts serine, mostly taken up from maternal blood, to glycine (Chung *et al.*, 1998), reconciling the discrepancy between the negligible net uptake of glycine by the uterus and substantial net release of this amino acid into the umbilical circulation (see Hay, 1998).

The complexity of interrelations among placental uptake, metabolism, and transport of amino acids was further illustrated by a recent study of alanine metabolism in ewes during late pregnancy (Timmerman *et al.*, 1998). Application of tracer methodology showed that negligible net placental consumption or production of alanine masks an appreciable metabolism of maternal alanine entering the placenta which exchanges with endogenously produced alanine. Thus, most of the alanine delivered to the fetus is of placental origin, derived from placental protein turnover and transamination.

## Factors affecting placental capacity for nutrient transport

### Placental size

Placental capacity for glucose transport was substantially reduced, as were uteroplacental glucose consumption rate and fetal glycaemia, in carunclectomized (Owens *et al.*, 1987a) and heat-treated ewes (Bell, 1987; Thureen *et al.*, 1992). At least part of the absolute reduction in glucose transport capacity is presumed to be due to reduction in exchange surface area of the trophoblastic membrane, as shown in carunclectomized ewes (Robinson *et al.*, 1995). In previously heat-treated ewes (Thureen *et al.*, 1992), placental weight-specific glucose transport capacity was also reduced. This implies that chronic heat stress, which reduces average weight but not total number of placentomes, additionally reduces number and/or activity of specific glucose transport proteins at maternal and/or fetal exchange surfaces. In contrast, carunclectomy, which reduces placentome number but may stimulate a compensatory increase in average weight of individual placentomes, caused a modest increase in the placental weight-specific clearance of the non-metabolizable glucose analogue, 3-*O*-methyl glucose (Owens *et al.*, 1987b). This implies that glucose transporter expression was preserved or increased in the remaining placentomes.

Placental insufficiency in heat-treated ewes also extends to impaired capacity for amino acid transport, including major reductions in placental uptake and fetal transfer of leucine (Ross *et al.*, 1996) and threonine (Anderson *et al.*, 1997). The normally extensive placental catabolism of leucine was also greatly reduced (Ross *et al.*, 1996).

### Maternal nutrition

The placenta appears to play an active role in ameliorating the effects of maternal undernutrition on glucose partitioning to the conceptus. We recently found that restriction of the energy intake of ditocous ewes to 60% of predicted requirements for 2 weeks caused moderate maternal and fetal hypoglycaemia and a 26% reduction in the maternal–fetal glucose concentration gradient (Table 16.2). This was offset by a 50%

**Table 16.2.** Maternal weight change, fetal weight, and indices of placental glucose transport at day 135 of pregnancy in ditocous ewes fed 100% or 60% of predicted energy requirements for the preceding 14 days (R.A. Ehrhardt and A.W. Bell, unpublished observations).

| | Energy intake, % requirement[a] | | | |
|---|---|---|---|---|
| Variable | 100 | 60 | PSE[b] | *P*[c] |
| Δ maternal weight (kg) | 5.3 | −2.7 | 1.0 | < 0.001 |
| Fetal weight (kg) | 3.58 | 3.46 | 0.16 | NS[d] |
| Plasma glucose (mM) | | | | |
| Maternal | 3.72 | 2.84 | 0.09 | < 0.001 |
| Fetal | 0.57 | 0.49 | 0.03 | < 0.05 |
| Δ maternal–fetal | 3.15 | 2.33 | 0.03 | < 0.001 |
| Placental 3MG[e] clearance (ml $min^{-1}$ $kg^{-1}$ placenta) | 117 | 176 | 7 | < 0.001 |
| CB[f] sites (pmol $mg^{-1}$ protein) | 105 | 126 | 3 | < 0.01 |
| GLUT-3 protein (arbitrary units[g]) | 1.00 | 1.19 | 0.04 | < 0.05 |

[a] Values are means for five ewes and ten conceptuses.
[b] PSE, pooled standard error.
[c] *P*, significance of difference between means.
[d] NS, not significant.
[e] 3MG, 3-*O*-methyl glucose.
[f] CB, cytochalasin B binding sites.
[g] Expressed relative to control (100% energy intake).

increase in maternal–fetal glucose transport capacity *in vivo*, at least partly explained by a 20% increase in placental glucose transporter abundance *in vitro*, assessed by binding of cytochalasin B and GLUT-3 protein abundance (Table 16.2). Possible adaptations in cellular distribution, in addition to total abundance, of placental GLUT-3 remain to be investigated. The lack of impairment of fetal growth attests to the success of adaptive responses to maintain placental glucose transfer (Table 16.2).

During more severe maternal undernutrition or starvation for several days, the development of profound fetal hypoglycaemia helps to sustain the maternal–fetal gradient in glucose concentration by restricting the reverse transfer of glucose to the placenta, and reducing placental glucose consumption (Hay, 1995). More specific manipulation of maternal and fetal glycaemia by prolonged maternal infusion with insulin has shown that the decline in fetal glucose concentration is less than that of the mother. This tends to decrease the maternal–fetal glucose concentration gradient, protecting placental glucose consumption at the expense of the fetus. In response, fetal glucose needs are diminished by a reduction in fetal growth rate (Carver and Hay, 1995), associated with decreased placental transfer and fetal accretion of leucine, and increased release of leucine from proteolysis in fetal tissues (Carver *et al.*, 1997).

## Metabolic adaptations in maternal tissues

### *Pregnancy-specific effects on glucose and amino acid metabolism*

Maternal strategies for accommodating the substantial requirements for glucose and amino acids of the conceptus include adaptations in carbohydrate and protein metabolism, briefly discussed below. They also involve changes in lipid metabolism which result in increased mobilization and maternal metabolism of NEFAs (see Bell and Bauman, 1994). Although maternal NEFAs are not readily used by the conceptus, their increased utilization in maternal tissues serves to spare maternal glucose and, perhaps, amino acids to meet requirements of the conceptus.

#### *Glucose metabolism*

Hepatic gluconeogenesis increases in ewes during late pregnancy even when feed intake is not increased above non-pregnant levels (Freetly and Ferrell, 1998), consistent with earlier observations of pregnancy-specific effects on whole-body glucose kinetics (see Bell, 1993). This was concomitant with increased hepatic uptake of lactate (Freetly and Ferrell, 1998), apparently derived from uteroplacental metabolism (Meschia *et al.*, 1980; Faichney *et al.*, 1981) and increased glycolysis in maternal peripheral tissues (Hough *et al.*, 1985). Part of the moderate net mobilization of amino acids from carcass tissues of late-pregnant ewes fed to predicted nutrient requirements (Fig. 16.1; McNeill *et al.*, 1997) may also be used to sustain increased hepatic gluconeogenesis as term approaches.

Glucose uptake by maternal peripheral tissues such as hindlimb muscle and adipose tissue tends to decline during late pregnancy, although the evidence for some of these responses in 'well-fed' animals may have been confounded by variations in voluntary feed intake (see Bell and Bauman, 1997).

#### *Amino acid metabolism*

Effects of pregnancy on the quantitative metabolism of amino acids have not been systematically studied in ruminants. However, in agreement with data from rats (Ling *et al.*, 1987), the fractional rate of hepatic protein synthesis increases 45% during late pregnancy in dairy cows (Bell, 1995). This is consistent with the moderate increase in hepatic protein accretion (Campbell and Fell, 1970), and an apparent decrease in hepatic amino acid catabolism (Freetly and Ferrell, 1998) in late-pregnant ewes. These changes in hepatic metabolism occurred despite a decreased (dairy cows) or unchanged (ewes) protein intake and hepatic uptake of amino acids, implying endogenous regulation of hepatic disposal of amino acids as in late-pregnant rats (Casado *et al.*, 1987).

### *Modulation by maternal nutrition*

#### *Energy*

Most of the relatively modest metabolic adaptations that occur in maternal tissues in well-fed pregnant ruminants are readily exacerbated or modified by maternal energy intake, especially during late pregnancy. Thus, the tendency for maternal insulin-responsive peripheral tissues to reduce glucose utilization during late pregnancy is exaggerated by moderate underfeeding, consistent with reductions in abundance of GLUT-4 protein in skeletal muscle and perirenal adipose tissue of ewes fed at 60% of predicted requirements for 2 weeks in late pregnancy (Fig. 16.2; Ehrhardt *et al.*, 1998).

On the other hand, the decline in maternal glucose production in moderately underfed ewes was not as great as predicted by their reduction in energy intake (Petterson *et al.*, 1993). This, together with the fact that glucose production in energy-restricted pregnant ewes was 40% greater than that in non-pregnant ewes consuming 13% more energy (Petterson *et al.*, 1993), implies that hepatic gluconeogenesis in underfed pregnant ewes was supported by much greater reliance on endogenous glucogenic substrates such as amino acids, glycerol and lactate. Mobilization of amino acids from skeletal muscle to meet this need is consistent with the shrinkage of muscle fibres in dairy cows underfed during late pregnancy (Reid *et al.*, 1980).

### *Protein*

As discussed earlier, protein deprivation of ditocous ewes during late pregnancy increased the net loss of protein from carcass tissues and abolished the normal, pregnancy-related increase in protein accretion of visceral organs. Conversely, feeding protein at levels above predicted requirements (160 g $kg^{-1}$ versus 120 g $kg^{-1}$ dry matter), caused an appreciable reversal of the net flux of nitrogen from maternal carcass tissues and allowed significant net accretion of tissue protein during late pregnancy (Fig. 16.1; McNeill *et al.*, 1997).

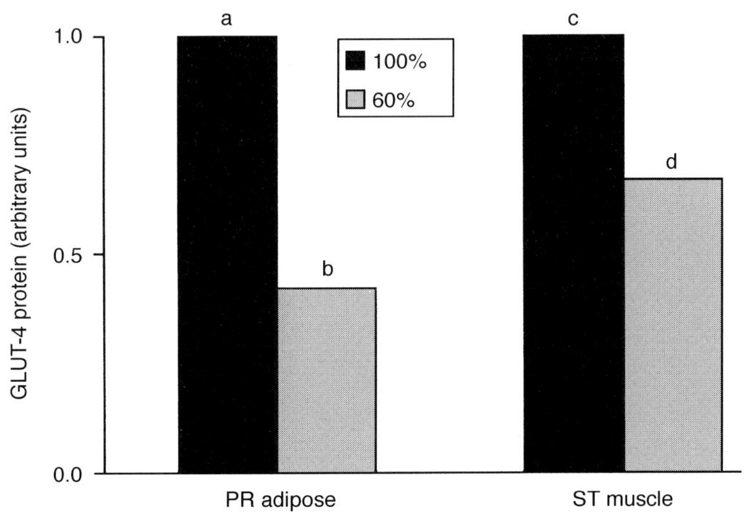


**Fig. 16.2.** Relative abundance of GLUT-4 protein in perirenal (PR) adipose tissue and semitendinosus (ST) muscle at day 135 of pregnancy in ditocous ewes that had been fed at 100% or 60% of predicted energy requirements for the preceding 14 days. Histograms are means for five ewes, expressed relative to control values (100% energy intake) for each tissue. Pooled standard errors were 0.07 units for PR adipose, and 0.06 units for ST muscle. Within tissues, means with different letters are significantly different ($P < 0.05$). From the data of Ehrhardt *et al.* (1998).

## Homeorhetic regulation

### General concept

The concept of homeorhesis as applied to regulation of nutrient partitioning (Bauman and Currie, 1980) is discussed elsewhere in this volume (Chapter 18). Key postulates of this concept include its simultaneous influence on multiple tissues and functional systems, implying extracellular mediation, and its operation through altered tissue responses to homeostatic effectors such as insulin, at various levels of extracellular and intracellular signalling.

### Altered tissue responses to insulin

Application of the euglycaemic, hyperinsulinaemic clamp technique has demonstrated that in sheep, as in humans and laboratory animals, late pregnancy is associated with moderate insulin resistance, and that this condition is exacerbated by undernutrition. This was manifested as diminished sensitivity to insulin of several variables of whole-body glucose utilization (Petterson *et al.*, 1993) and decreased insulin responsiveness of lipolysis and NEFA mobilization (Petterson *et al.*, 1994). Tissue specificity of these whole-body phenomena was not assessed in these studies, but it is likely that adaptations in the major insulin-responsive peripheral tissues, skeletal muscle and adipose tissue, were mostly responsible (see Bell and Bauman, 1997).

As pregnancy advances, ovine adipose tissue *in vitro* becomes refractory to the stimulation of lipogenesis by insulin (Vernon *et al.*, 1985; Guesnet *et al.*, 1991). Although glucose is not an important carbon precursor for lipogenesis in ruminants, it is required for synthesis of glyceride glycerol, and for part of the NADPH necessary for *de novo* fatty acid synthesis (see Bauman and Davis, 1975). This is consistent with our recent observation of progressive development of insulin resistance in terms of whole-body glucose disposal during pregnancy in fat-tailed Karakul ewes (Slepetis *et al.*, 1999).

The role of skeletal muscle in the evolution of insulin resistance during pregnancy has not been addressed in ruminants. However, it seems likely that the ability of insulin to promote glucose uptake by muscle is diminished in late pregnancy as in early lactation (Vernon *et al.*, 1990) because of the similar degrees of whole-body insulin resistance observed in ewes in these two states (Slepetis *et al.*, 1999). The reduction in muscle abundance of the insulin-responsive GLUT-4 protein in underfed versus well-fed pregnant ewes (Fig. 16.2; Ehrhardt *et al.*, 1998) provides a mechanistic clue.

### Possible homeorhetic agents

Various hormones, including progesterone, oestradiol and placental lactogen (PL), may act as homeorhetic regulators of observed changes in tissue responses to insulin and associated metabolic adaptations to the state of pregnancy in ruminants (Bell and Bauman, 1994, 1997). The argument for PL is especially hard to dismiss, despite a continuing lack of direct experimental evidence. This uniquely placental peptide cross-reacts with both growth hormone (GH) and prolactin receptors in ruminant tissues (Byatt *et al.*, 1992), and its specific binding in adipose tissue increases with advancing pregnancy in sheep (N'Guema *et al.*, 1986). However, it remains unclear whether high-affinity binding of PL in adipose tissue or liver is mediated through structurally unique receptors (Anthony *et al.*, 1995). Cross-reactivity with the GH receptor would be consistent with the development of insulin resistance in adipose tissue during late pregnancy because GH is a potent homeorhetic effector of this response in ruminant adipose tissue (see Bell and Bauman, 1997; Chapter 18). Other indirect evidence for a homeorhetic role for PL includes enhanced placental gene expression and secretion of ovine placental lactogen in moderately undernourished, late-pregnant ewes, coincident with decreased expression of GLUT-4 in maternal insulin-responsive tissues (Fig. 16.2; Ehrhardt *et al.*, 1998) and exaggeration of indices of whole-body insulin resistance discussed earlier in this section.

It is also possible that leptin, the *ob* gene product expressed almost exclusively in adipose tissue (see Houseknecht *et al.*, 1998), plays a role in the homeorhetic coordination of conceptus nutrient demands and metabolic adaptations in maternal tissues of pregnant ruminants. In addition to its commonly postulated role as a signal of peripheral energy status to the central nervous system, leptin may have pleiotropic effects on peripheral tissues, including mediation of insulin resistance in adipocytes (Muller *et al.*, 1997). Coincident with the evolution of insulin resistance, we have recently observed a consistent, threefold increase in expression of leptin mRNA in tail adipose tissue from fat-tailed Karakul ewes during mid (50–60 days) and late (125–135 days) pregnancy compared with levels in the same animals when non-pregnant (Fig. 16.3). A similar, pregnancy-induced increase in leptin mRNA expression in white adipose tissue was associated with a marked increase in serum concentrations of leptin in mice (Tomimatsu *et al.*, 1997).

Leptin may also act on the placenta, which, in sheep, strongly expresses the leptin receptor splice variant OB-Rb (Ehrhardt *et al.*, 1999) that is considered essential for intracellular signal transduction after binding leptin. This raises the possibility that, in

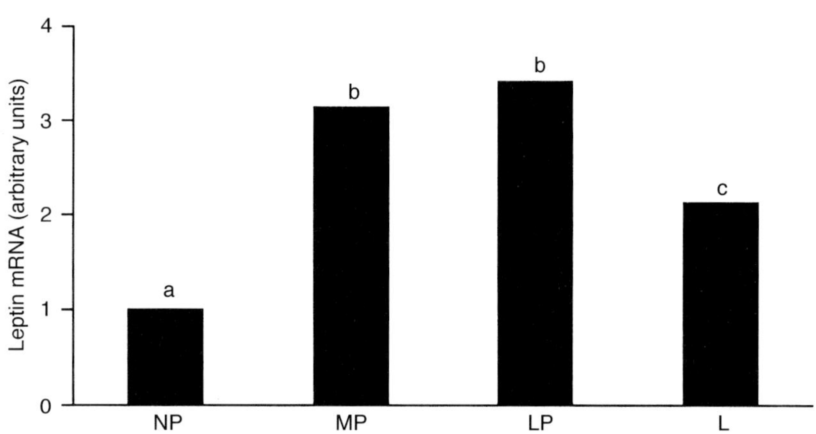


**Fig. 16.3.** Relative abundance of leptin mRNA in tail adipose tissue of Karakul ewes ($n = 8$) sequentially sampled while non-pregnant and non-lactating (NP), in mid pregnancy (MP, day 50–60), in late pregnancy (LP, day 125–135), and in early lactation (L, day 15–22 post-partum). Pooled standard error was 0.46 units. Means with different letters are significantly different ($P < 0.05$). From the unpublished data of R.A. Ehrhardt, R.M. Slepetis, Y.R. Boisclair and A.W. Bell.

addition to its putative influences on adipose and other maternal tissues, leptin may directly or indirectly mediate observed adaptations in placental capacity for glucose transport during advancing gestation and moderate maternal undernutrition.

## Conclusions

We propose that the placenta plays a pivotal, multi-faceted role in the regulation of nutrient partitioning between maternal tissues and the conceptus, especially during late pregnancy when most of fetal growth occurs. First, it is a highly regulated nutritional and excretory conduit that provides a means of coordinating fetal nutrient demands with maternal capacity to supply these nutrients, especially glucose and amino acids. Under optimal conditions, this will result in appropriate constraint of late-gestation fetal growth such that birth weight will be sufficient to permit a good chance of neonatal survival, but insufficient to cause dystocia or an inappropriate depletion of maternal nutrient reserves prior to the onset of lactation. The mechanistic bases for this general role of the placenta are beginning to be understood, although much of the details about amino acid transport and its regulation have yet to be elucidated.

Second, through its capacity to synthesize and secrete a plethora of bioactive molecules, including proteins, steroids, and eicosanoids, into maternal and fetal circulations, the placenta offers unique opportunities for direct and indirect communication between the conceptus and its maternal host. Although lacking direct evidence, the hypothesis that PL and/or other placental hormones are responsible for effecting insulin resistance and other metabolic adaptations in maternal tissues remains attractive. We also speculate that communication of maternal energy status to the placenta may be mediated through the leptin system, adding a further layer of complexity to homeorhetic coordination of maternal nutrition and energy stores with nutrient requirements of the conceptus.

## References

Aldoretta, P.W., Anderson, S. and Hay, W.W., Jr (1994) Ovine uteroplacental glucose and oxygen metabolism in relation to chronic changes in maternal and fetal glucose concentrations. *Placenta* 15, 753–764.

Anderson, A.H., Fennessey, P.V., Meschia, G., Wilkening, R.B. and Battaglia, F.C. (1997) Placental transport of threonine and its utilization in the normal and growth-restricted fetus. *American Journal of Physiology* 272, E892–E900.

Anthony, R.V., Pratt, S.L., Liang, R. and Holland, M.D. (1995) Placental-fetal hormonal interactions: impact on fetal growth. *Journal of Animal Science* 73, 1861–1871.

Battaglia, F.C. and Meschia, G. (1988) Fetal nutrition. *Annual Review of Nutrition* 8, 43–61.

Bauman, D.E. and Currie, W.B. (1980) Partitioning of nutrients during pregnancy and lactation: a review of mechanisms involving homeostasis and homeorhesis. *Journal of Dairy Science* 63, 1514–1529.

Bauman, D.E. and Davis, C.L. (1975) Regulation of lipid metabolism. In: McDonald, I.W. and Warner, A.C.I. (eds) *Digestion and Metabolism in the Ruminant.* The University of New England Publishing Unit, Armidale, pp. 496–509.

Bell, A.W. (1987) Consequences of severe heat stress for fetal development. In: Hales, J.R.S. and

Richards, D.A.B. (eds) *Heat Stress: Physical Exertion and Environment.* Elsevier, Amsterdam, pp. 313–333.

Bell, A.W. (1993) Pregnancy and fetal metabolism. In: Forbes, J.M. and France, J. (eds) *Quantitative Aspects of Ruminant Digestion and Metabolism.* CAB International, Wallingford, pp. 405–431.

Bell, A.W. (1995) Regulation of organic nutrient metabolism during transition from late pregnancy to early lactation. *Journal of Animal Science* 73, 2804–2819.

Bell, A.W and Bauman, D.E. (1994) Animal models for the study of adipose regulation in pregnancy and lactation. In: Allen, L., King, J. and Lonnerdal, B. (eds) *Nutrient Regulation during Pregnancy, Lactation, and Infant Growth.* Plenum Press, New York, pp. 71–84.

Bell, A.W. and Bauman, D.E. (1997) Adaptations of glucose metabolism during pregnancy and lactation. *Journal of Mammary Gland Biology and Neoplasia* 2, 265–278.

Bell, A.W. and Ehrhardt, R.A. (1998) Placental regulation of nutrient partitioning during pregnancy. In: Hansel, W., Bray, G.A. and Ryan, D.H. (eds) *Nutrition and Reproduction.* Louisiana State University Press, Baton Rouge, pp. 229–254.

Bell, A.W., Kennaugh, J.M., Battaglia, F.C. and Meschia, G. (1989a) Uptake of amino acids and ammonia at mid-gestation by the fetal lamb. *Quarterly Journal of Experimental Physiology* 74, 635–643.

Bell, A.W., McBride, B.W., Slepetis, R., Early, R.J. and Currie, W.B. (1989b) Chronic heat stress and prenatal development in sheep: 1. Conceptus growth and maternal plasma hormones and metabolites. *Journal of Animal Science* 67, 3289–3299.

Bell, A.W., Slepetis, R. and Ehrhardt, R.A. (1995) Growth and accretion of energy and protein in the gravid uterus during late pregnancy in Holstein cows. *Journal of Dairy Science* 78, 1954–1961.

Byatt, J.C., Warren, W.C., Eppard, P.J., Staten, N.R., Krivi, G.G. and Collier, R.J. (1992) Ruminant placental lactogens: structure and biology. *Journal of Animal Science* 70, 2911–2923.

Campbell, R.M. and Fell, B.F. (1970) Observations on hypertrophy of the liver in breeding ewes. *Research in Veterinary Science* 11, 540–547.

Carter, B.S., Moores, R.R. and Battaglia, F.C. (1991) Placental transport and fetal and placental metabolism of amino acids. *Journal of Nutritional Biochemistry* 2, 4–13.

Carter, B.S., Moores, R.R., Jr, Teng, C., Meschia, G. and Battaglia, F.C. (1995) Main routes of plasma lactate carbon disposal in the midgestation fetal lamb. *Biology of the Neonate* 67, 295–300.

Carver, T.D. and Hay, W.W., Jr (1995) Uteroplacental carbon substrate metabolism and $O_2$ consumption after long-term hypoglycemia in pregnant sheep. *American Journal of Physiology* 269, E299–E308.

Carver, T.D., Quick, A.A., Teng, C.C., Pike A.W., Fennessey, P.V. and Hay, W.W., Jr (1997) Leucine metabolism in chronically hypoglycemic hypoinsulinemic growth-restricted fetal sheep. *American Journal of Physiology* 272, E107–E117.

Casado, J., Remesar, X. and Pastor-Anglada, M. (1987) Hepatic uptake of amino acids in late-pregnant rats. *Biochemical Journal* 248, 117–122.

Chung, M., Teng, C., Timmerman, M., Meschia, G. and Battaglia, F.C. (1998) Production and utilization of amino acids by ovine placenta in vivo. *American Journal of Physiology* 274, E13–E22.

Clarke, L., Heasman, L., Juniper, D.T. and Symonds, M.E. (1998) Maternal nutrition in early-mid gestation and placental size in sheep. *British Journal of Nutrition* 79, 359–364.

Comline, R.S. and Silver, M. (1976) Some aspects of foetal and utero-placental metabolism in cows with indwelling umbilical and uterine vascular catheters. *Journal of Physiology* 260, 571–586.

Currie, M.J., Bassett, N.S. and Gluckman, P.D. (1997) Ovine glucose transporter-1 and -3: cDNA partial sequences and developmental gene expression in the placenta. *Placenta* 18, 393–401.

Das, U.G., Sadiq, H.F., Soares, M.J., Hay, W.W., Jr and Devaskar, S.U. (1998) Time-dependent physiological regulation of rodent and ovine placental glucose transporter (GLUT-1) protein. *American Journal of Physiology* 274, R339–R347.

Ehrhardt, R.A. and Bell, A.W. (1995) Growth and metabolism of the ovine placenta during mid-gestation. *Placenta* 16, 727–741.

Ehrhardt, R.A. and Bell, A.W. (1997) Developmental increases in glucose transporter concentration in the sheep placenta. *American Journal of Physiology* 273, R1132–R1141.

Ehrhardt, R.A., Slepetis, R.M. and Bell, A.W. (1998) Moderate undernutrition alters glucose transporter levels in maternal insulin-responsive tissues and in the placenta. *Journal of Animal Science* 76 (Suppl. 1), 130.

Ehrhardt, R.A., Bell, A.W. and Bosclair, Y.R. (1999) Analysis of leptin and leptin RNA expression in the sheep placenta. *Journal of Animal Science* 77 (suppl. 1), 159.

Faichney, G.J. and White, G.A. (1987) Effects of maternal nutritional status on fetal and placental growth and on fetal urea synthesis in sheep. *Australian Journal of Biological Sciences* 40, 365–377.

Faichney, G.J., Barker, P.J., Setchell, B.P. and Lindsay, D.B. (1981) The utilization of lactic acid by sheep in late pregnancy. *Quarterly Journal of Experimental Physiology* 66, 195–201.

Falconer, J., Owens, J.A., Allotta, E. and Robinson, J.S. (1985) Effect of restriction of placental growth on the concentrations of insulin, glucose and placental lactogen in the plasma of sheep. *Journal of Endocrinology* 106, 7–11.

Ferrell, C.L. (1991) Maternal and fetal influences on uterine and conceptus development in the cow: II. Blood flow and nutrient flux. *Journal of Animal Science* 69, 1954–1965.

Freetly, H.C. and Ferrell, C.L. (1998) Net flux of glucose, lactate, volatile fatty acids, and nitrogen metabolites across the portal-drained viscera and liver of pregnant ewes. *Journal of Animal Science* 76, 3133–3145.

Guesnet, P.M., Massoud, M.J. and Demarne, Y. (1991) Regulation of adipose tissue metabolism during pregnancy and lactation in the ewe: the role of insulin. *Journal of Animal Science* 69, 2057–2065.

Hammond, J. (1947) Animal breeding in relation to nutrition and environmental conditions. *Biological Reviews* 22, 195–213.

Hay, W.W., Jr (1995) Regulation of placental metabolism by glucose supply. *Reproduction, Fertility and Development* 7, 365–375.

Hay, W.W., Jr (1996) Nutrition and development of the fetus: carbohydrate and lipid metabolism. In: Walker, W.A. and Watkins, J.B. (eds) *Nutrition in Pediatrics*. B.C. Decker, Hamilton, pp. 364–378.

Hay, W.W., Jr (1998) Fetal requirements and placental transfer of nitrogenous compounds. In: Polin, R.A. and Fox, W.W. (eds) *Fetal and Neonatal Physiology*, 2nd edn. Saunders, Philadelphia, pp. 619–634.

Hay, W.W., Jr and Meznarich, H.K. (1988) Effect of maternal glucose concentration on utero-placental glucose consumption and transfer in pregnant sheep. *Proceedings of the Society for Experimental Biology and Medicine* 190, 63–69.

Hay, W.W., Jr, Myers, S.A., Sparks, J.W., Wilkening, R.B., Meschia, G. and Battaglia, F.C. (1983a) Glucose and lactate oxidation in the fetal lamb. *Proceedings of the Society for Experimental Biology and Medicine* 173, 553–563.

Hay, W.W., Jr, Sparks, J.W., Wilkening, R.B., Battaglia, F.C. and Meschia, G. (1983b) Partition of maternal glucose production between conceptus and maternal tissues in sheep. *American Journal of Physiology* 245, E347–E350.

Hay, W.W., Jr, Sparks, J.W., Wilkening, R.B., Battaglia, F.C. and Meschia, G. (1984) Fetal glucose uptake and utilization as functions of maternal glucose concentration. *American Journal of Physiology* 246, E237–E242.

Hay, W.W., Jr, Molina, R.D., DiGiacamo, J.E. and Meschia, G. (1990) Model of placental glucose consumption and transfer. *American Journal of Physiology* 258, R569–R577.

Holzman, I.R., Lemons, J.A., Meschia, G. and Battaglia, F.C. (1977) Ammonia production by the pregnant uterus. *Proceedings of the Society for Experimental Biology and Medicine* 156, 27–30.

Hough, G.M., McDowell, G.H., Annison, E.F. and Williams, A.J. (1985) Glucose metabolism in hindlimb muscle of pregnant and lactating ewes. *Proceedings of the Nutrition Society of Australia* 10, 97.

Houseknecht, K.L., Baile, C.A., Matteri, R.L. and Spurlock, M.E. (1998) The biology of leptin: a review. *Journal of Animal Science* 76, 1405–1420.

Jozwik, M., Teng, C., Timmerman, M., Chung, M., Meschia, G. and Battaglia, F.C. (1998) Uptake and transport by the ovine placenta of neutral nonmetabolizable amino acids with different transport system affinities. *Placenta* 19, 531–538.

Lemons, J.A. and Schreiner, R.L. (1983) Amino acid metabolism in the ovine fetus. *American Journal of Physiology* 244, E459–E466.

Leury, B.J., Bird, A.R., Chandler, K.D. and Bell, A.W. (1990) Glucose partitioning in the pregnant ewe: effects of undernutrition and exercise. *British Journal of Nutrition* 64, 449–462.

Liechty, E.A., Kelley, J. and Lemons, J.A. (1991) Effect of fasting on uteroplacental amino acid metabolism in the pregnant sheep. *Biology of the Neonate* 60, 207–214.

Ling, P.R., Bistrian, B.R., Blackburn, G.L. and Istfan, N. (1987) Effect of fetal growth on maternal metabolism in postabsorptive rat. *American Journal of Physiology* 252, E380–E390.

Loy, G.L., Quick, A.N., Jr, Hay, W.W., Jr, Meschia, G., Battaglia, F.C. and Fennessey, P.V. (1990) Fetoplacental deamination and decarboxylation of leucine. *American Journal of Physiology* 259, E492–E497.

McCrabb, G.J., Egan, A.R. and Hosking, B.J. (1992) Maternal undernutrition during mid-pregnancy in sheep: variable effects on placental growth. *Journal of Agricultural Science, Cambridge* 118, 127–132.

McNeill, D.M., Slepetis, R., Ehrhardt, R.A., Smith, D.M. and Bell, A.W. (1997) Protein requirements of sheep in late pregnancy: partitioning of nitrogen between gravid uterus and maternal tissues. *Journal of Animal Science* 75, 809–816.

McNeill, D.M., Kelly, R.W. and Williams, I.H. (1998) Partition of nutrients in moderately fat ewes compared with lean ewes given ad libitum access to feed in late pregnancy. *Australian Journal of Agricultural Research* 49, 575–580.

McNeill, D.M., Kelly, R.W. and Williams, I.H. (1999) Maternal fatness influences fetal size in ewes underfed in late pregnancy. *Australian Journal of Agricultural Research* 50, 1171–1177.

Meschia, G., Battaglia, F.C., Hay, W.W., Jr and Sparks, J.W. (1980) Utilization of substrates by the ovine placenta in vivo. *Federation Proceedings* 39, 245–249.

Meznarich, H.K., Hay, W.W., Jr, Sparks, J.W., Meschia, G. and Battaglia, F.C. (1987) Fructose disposal and oxidation rates in the ovine fetus. *Quarterly Journal of Experimental Physiology* 72, 617–625.

Molina, R.D., Meschia, G., Battaglia, F.C. and Hay, W.W., Jr (1991) Gestational maturation of placental glucose transfer capacity in sheep. *American Journal of Physiology* 261, R697–R704.

Moores, R.R., Jr, Vaughn, P.R., Battaglia, F.C., Fennessey, P.V., Wilkening, R.B. and Meschia, G. (1994) Glutamate metabolism in the fetus and placenta of late-gestation sheep. *American Journal of Physiology* 267, R89–R96.

Muller, G., Ertl, J., Gerl, M. and Preibisch, G. (1997) Leptin impairs metabolic actions of insulin in isolated rat adipocytes. *Journal of Biological Chemistry* 272, 10585–10593.

N'Guema, M., Delouis, C., Kelly, P.A. and Djiane, J. (1986) Evolution of prolactin and placental lactogen receptors in ewes during pregnancy and lactation. *Endocrinology* 118, 695–700.

Oddy, V.H., Gooden, J.M., Hough, G.M., Teleni, E. and Annison, E.F. (1985) Partitioning of nutrients in Merino ewes. II Glucose utilization by skeletal muscle, the pregnant uterus and the lactating mammary gland in relation to whole body glucose utilization. *Australian Journal of Biological Sciences* 38, 95–108.

Owens, J.A., Falconer, J. and Robinson, J.S. (1987a) Effect of restriction of placental growth on fetal and utero-placental metabolism. *Journal of Developmental Physiology* 9, 225–238.

Owens, J.A., Falconer, J. and Robinson, J.S. (1987b) Restriction of placental size in sheep enhances efficiency of placental transfer of antipyrine, 3-O-methyl-D-glucose but not urea. *Journal of Developmental Physiology* 9, 457–464.

Owens, J.A., Falconer, J. and Robinson, J.S. (1989) Glucose metabolism in pregnant sheep when placental growth is restricted. *American Journal of Physiology* 257, R350–R357.

Petterson, J.A., Dunshea, F.R., Ehrhardt, R.A. and Bell, A.W. (1993) Pregnancy and undernutrition alter glucose metabolic responses to insulin in sheep. *Journal of Nutrition* 123, 1286–1295.

Petterson, J.A., Slepetis, R., Ehrhardt, R.A., Dunshea, F.R. and Bell, A.W. (1994) Pregnancy but not moderate undernutrition attenuates insulin suppression of fat mobilization in sheep. *Journal of Nutrition* 124, 2431–2436.

Prior, R.L. and Christenson, R.K. (1978) Insulin and glucose effects on glucose metabolism in pregnant and nonpregnant ewes. *Journal of Animal Science* 46, 201–210.

Reid, I.M., Roberts, C.J. and Baird, G.D. (1980) The effects of underfeeding during pregnancy and lactation on structure and chemistry of bovine liver and muscle. *Journal of Agricultural Science, Cambridge* 94, 239–245.

Reynolds, L.P., Ferrell, C.L., Robertson, D.A. and Ford, S.P. (1986) Metabolism of the gravid uterus, foetus and utero-placenta at several stages of gestation in cows. *Journal of Agricultural Science, Cambridge* 106, 437–444.

Robinson, J., Chidzanja, S., Kind, K., Lok, F., Owens, P. and Owens, J. (1995) Placental control of fetal growth. *Reproduction, Fertility and Development* 7, 333–344.

Ross, J.C., Fennessey, P.V., Wilkening, R.B., Battaglia, F.C. and Meschia, G. (1996) Placental transport and fetal utilization of leucine in a model of fetal growth retardation. *American Journal of Physiology* 270, E491–E503.

Slepetis, R.M., Ehrhardt, R.A., Block, S.S., Boisclair, Y.R. and Bell, A.W. (1999) Alterations in in vivo whole-body insulin resistance according to physiological state in Karakul ewes. *Journal of Animal Science* 77 (Suppl. 1), 163.

Smeaton, T.C., Owens, J.A., Kind, K.L. and Robinson, J.S. (1989) The placenta releases branched-chain amino acids into the umbilical and uterine circulations in the pregnant sheep. *Journal of Developmental Physiology* 12, 95–99.

Sparks, J.W., Hay, W.W., Jr, Bonds, D., Meschia, G. and Battaglia, F.C. (1982) Simultaneous measurements of lactate turnover rate and umbilical lactate uptake in the fetal lamb. *Journal of Clinical Investigation* 70, 179–192.

Thureen, P., Trembler, K.A., Meschia, G., Makowski, E.L. and Wilkening, R.B. (1992) Placental glucose transport in heat-induced fetal growth retardation. *American Journal of Physiology* 263, R578–R585.

Timmerman, M., Chung, M., Wilkening, R.B., Fennessey, P.V., Battaglia, F.C. and Meschia, G. (1998) Relationship of fetal alanine uptake and placental alanine metabolism to maternal alanine concentration. *American Journal of Physiology* 275, E942–E950.

Tomimatsu, T., Yamaguchi, M., Murakami, T., Ogura, K., Sakata, M., Mitsuda, N., Kanzaki, T., Kurachi, H., Irahara, M., Miyake, A., Shima, K., Aono, T. and Murata, Y. (1997) Increase of mouse leptin production by adipose tissue after midpregnancy: gestational profile of serum leptin concentration. *Biochemical and Biophysical Research Communications* 240, 213–215.

van der Walt, J.G., Baird, G.D. and Bergman, E.N. (1983) Tissue glucose and lactate metabolism and interconversions in pregnant and lactating sheep. *British Journal of Nutrition* 50, 267–280.

Vaughn, P.R., Lobo, C., Battaglia, F.C., Fennessey, P.V., Wilkening, R.B. and Meschia, G. (1995)

Glutamine-glutamate exchange between placenta and fetal liver. *American Journal of Physiology* 268, E705–E711.

Vernon, R.G., Clegg, R.A. and Flint, D.J. (1985) Adaptation of adipose tissue metabolism and number of insulin receptors in pregnant sheep. *Comparative Biochemistry and Physiology* 81B, 909–913.

Vernon, R.G., Faulkner, A., Hay, W.W., Jr, Calvert, D.T. and Flint, D.J. (1990) Insulin resistance of hind-limb tissues *in vivo* in lactating sheep. *Biochemical Journal* 270, 783–786.

Wallace, J.M., Aitken, R.P. and Cheyne, M.A. (1996) Nutrient partitioning and fetal growth in rapidly growing adolescent ewes. *Journal of Reproduction and Fertility* 107, 183–190.

Wallace, J.M., Bourke, D.A., Aitken, R.P. and Cruickshank, M.A. (1999) Switching maternal dietary intake at the end of the first trimester has profound effects on placental development and fetal growth in adolescent ewes carrying singleton fetuses. *Biology of Reproduction* 61, 101–110.

# 17 The Thermal Physiology of the Ruminant Fetus

H. LABURN, A. FAURIE AND D. MITCHELL

*Department of Physiology and Brain Function Research Unit, University of the Witwatersrand, Parktown, Johannesburg, South Africa*

## Introduction

We have come to know much about the thermal physiology of the ruminant fetus not because the thermal physiology of ruminants is peculiar, but because so much of the research on the physiology of fetal mammals has been carried out on sheep. That includes our own research. The experimental evidence, however, leads us to believe that much of what is known about the thermal physiology of the sheep fetus also applies to other ruminant species, and some non-ruminant species, including our own. It would be a reasonable expectation that the thermal relationship the fetal lamb has with the pregnant ewe is representative of the fetal–maternal relationship in other ruminants, where fetuses typically number one or two per pregnancy, and where total fetal mass constitutes approximately 10% maternal body mass at term.

## Fetal body temperature homeostasis

### *Fetal heat production and heat loss*

Fetal growth and development generates considerable metabolic heat. Oxygen consumption per kg of fetal body mass increases progressively during gestation with a peak prior to term (Bell *et al.*, 1987), but even at term, the metabolic rate (about 3 W $kg^{-1}$; Gilbert *et al.*, 1985) of the sheep fetus is approximately twice that per kg body mass of the pregnant ewe or adult non-pregnant sheep (Abrams *et al.*, 1970). Most fetal metabolic heat derives from the developing 'vital organs', including the brain, which although constituting only about 6% of body mass, contribute some 40% of total metabolic heat in late gestation (Bell *et al.*, 1987). To avoid potentially dangerous heat storage, the fetus requires an efficient route for heat loss to dissipate the heat produced, and, as for many other metabolic by-products, it is the placenta which provides that route. Based on measurement of fetal lamb and pregnant ewe body temperatures,

 *Ruminant Physiology: Digestion, Metabolism, Growth and Reproduction* (ed. P.B. Cronjé)

and model predictions of the feto–maternal thermal relationship, Gilbert *et al.* (1985) estimated that approximately 85% of fetal heat is lost via the utero-placental circulation (Fig. 17.1). That heat is delivered to the placenta by the umbilical vessels, and leaves the placenta in the uterine circulation, so increases or decreases in umbilical and/or uterine blood flow will lead to increased or decreased heat transfer respectively, between the fetus and the mother animal (Schröder *et al.*, 1988). The remainder of the heat produced is dissipated via convection and conduction through the amniotic fluid and subsequently the uterine wall. Ultimately, the heat is lost to the environment, via typical heat loss strategies employed by the pregnant animal, such as vasodilation in peripheral blood vessels, and panting.

## *Fetal body temperature*

The equilibrium between fetal heat production and heat loss via the placenta or uterine wall occurs when fetal body temperature is approximately 0.5°C higher than that of its mother's body core; a situation which prevails in all species of mammal tested to date, including rabbits (Hart and Faber, 1965), baboons (Morishima *et al.*, 1975), dogs (Assali and Westin, 1962), sheep (Gunn and Gluckman, 1983; Laburn *et al.*, 1992)

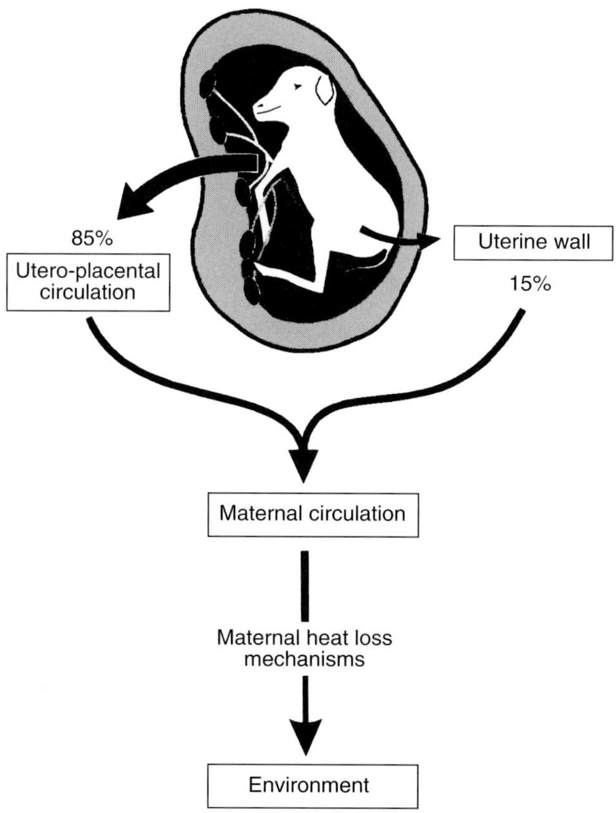


**Fig. 17.1.** Diagrammatic representation of the routes of heat loss from the fetal lamb.

and humans (Wood and Beard, 1964; Adamsons and Towell, 1965). Our measurements reveal typical values for body temperatures of the sheep fetus to be 39.6°C, and of the pregnant ewe to be 39.2°C. We recently have measured these temperatures in pregnant nanny goats and their fetuses, and have found almost identical values to those in sheep for fetal and maternal body temperatures in this ruminant species too. Why the equilibrium feto-maternal temperature difference is the same in species of different mass remains unknown.

## The feto-maternal thermal relationship

### *Body temperatures in laboratory and field conditions*

Not only is the feto-maternal (F-M) thermal gradient remarkably similar in different species, but it also appears to be remarkably constant, at least in sheep, in the face of potential thermal perturbations during the latter stages of gestation. Figure 17.2 shows measurements made using temperature-sensitive radiotelemeters, which had been implanted into the abdomens of a group of pregnant ewes and into the abdomens of their fetuses, over the last five weeks of gestation, when the animals were kept in thermally stable indoor conditions. On a day-to-day basis, the feto-maternal temperature relationship is very stable, despite large gains in fetal body mass, and metabolic heat produced (Bell *et al.*, 1987). Moreover, Fig. 17.3 shows that, despite circadian fluctuations in the ewe's body temperature, the F-M temperature difference does not show significant deviations on an hour-to-hour basis.

In their natural environments or in field conditions, pregnant animals are subject to greater fluctuations in ambient conditions than are laboratory-housed animals, and one may ask whether the F-M thermal relationship is as constant then. Using temperature-sensitive data loggers, rather than radiotelemeters, implanted chronically into the fetal sheep and pregnant ewes, we compared the F-M relationship in laboratory conditions with that in field conditions. Figure 17.4 shows typical data obtained from one pregnant ewe. Ewe body core temperature varied much more in the field, with day- and night-time environmental temperature changes (Fig. 17.4b). Fetal body temperature changed too, but not nearly to the same extent as occurred in the mother animal. These data are the first recordings of which we are aware, of continuous body temperature measurements in a fetus under field conditions. Our measurements were made while the animals were sedentary and exposed to mild winter conditions – day-time dry-bulb temperature varied between 10° and 20°C, with relative humidity approximately 40%. Night-time temperatures were generally in the range 0–3°C. Measurements of fetal body temperature need to be made in animals living in harsher environments, to establish the robustness of the F-M thermal relationship, especially as many ruminants are bred and/or housed or live naturally in conditions considered to be extreme (Alexander and Williams, 1971; Barlow *et al.*, 1987).

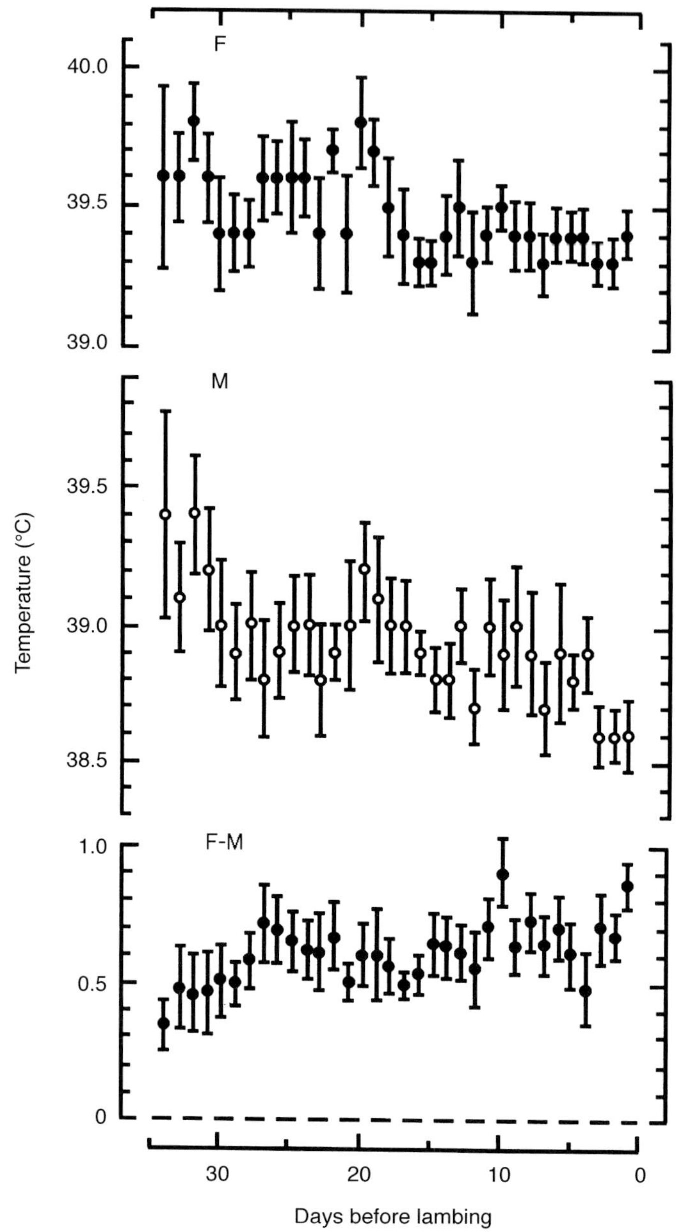


**Fig. 17.2.** Fetal lamb (F), and pregnant ewe (M) body temperatures as measured by radiotelemetry in a group of seven animals over the last 35 days of pregnancy (lambing at day 0). Lowest panel shows the feto-maternal (F–M) body temperature gradient. All points are means ± standard error of a mean. Reprinted with permission from the American Physiological Society, from Laburn *et al.* (1992).

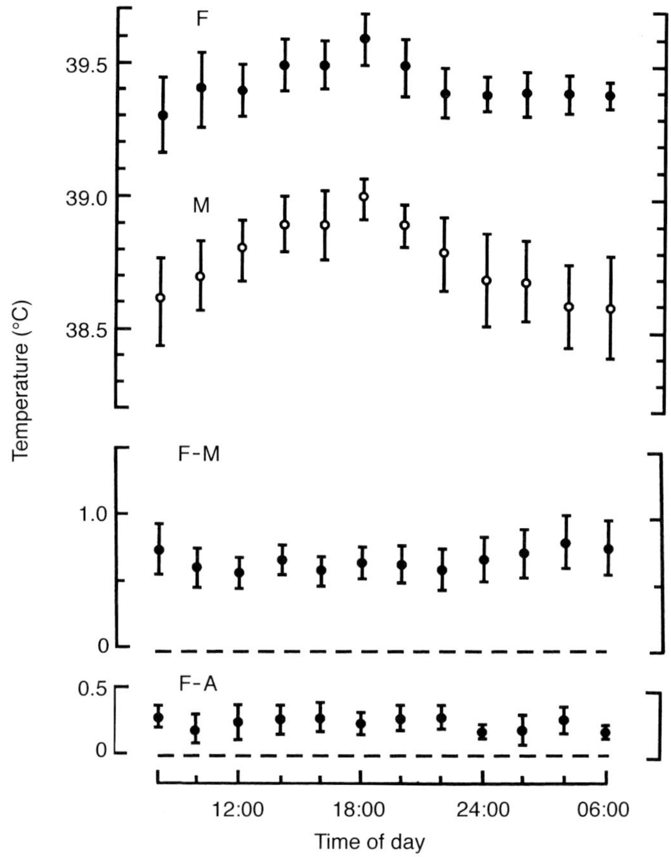


**Fig. 17.3.** Top panel shows circadian variations in fetal lamb and pregnant ewe body temperatures over a 24-h period, and lower panels show changes in the feto-maternal (F-M) temperature gradient, and feto-amniotic (F-A) temperature gradient. The F-A gradient is approximately midway between the F-M gradient, confirming the amnion as a conduit of heat between fetus and maternal tissues. Other details as in Fig. 17.1. Reprinted with permission from the American Physiological Society, from Laburn *et al.* (1992).

## Fetal hyperthermia

A potentially serious consequence of the fetus having a higher body temperature than that of its mother, and of being thermally clamped to its mother, is that the fetus experiences greater hyperthermia than the mother does, when maternal body temperature rises. Heat is damaging, especially to developing tissues, and the developing central nervous system is particularly susceptible. A rise in temperature of 1.5°C arrests fetal brain cell division, and a 3°C rise kills dividing cells (Gericke *et al.*, 1989). As a result, maternal hyperthermia in early pregnancy leads to congenital abnormalities (Edwards *et al.*, 1995; Chambers *et al.*, 1998). In later pregnancy, raised intrauterine temperature is

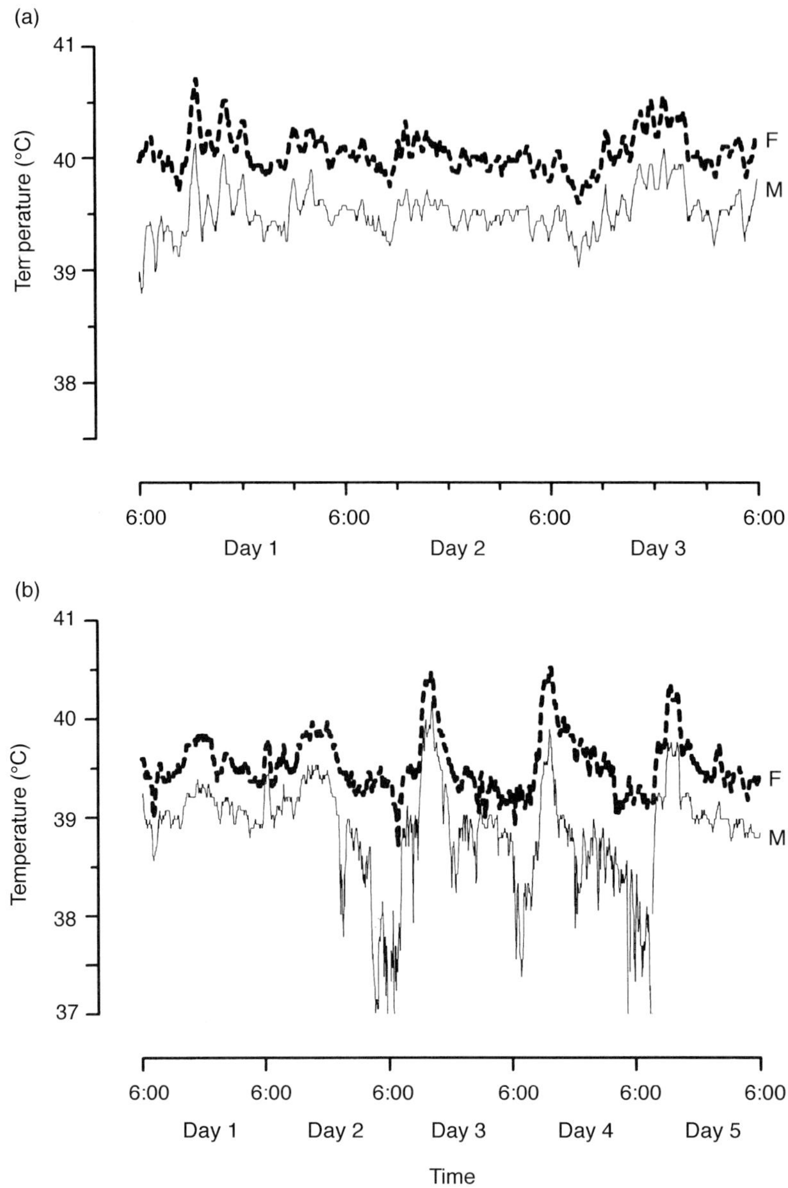


**Fig. 17.4.** Body temperatures of a pregnant ewe (M) and her fetal lamb (F). Measurements were made while the ewe was housed in laboratory conditions (a), and in field conditions (see text, b). Data for a 3-day (a) or a 5-day period (b) are shown.

associated with fetal growth retardation, such that the lambs of animals kept at abnormally high ambient temperatures during their pregnancy have significantly lower body mass than their counterparts whose mothers were kept in more moderate conditions (Alexander and Williams, 1971; Dreiling *et al.*, 1991).

### Fetal body temperature during maternal hyperthermia and hypothermia

The fetus apparently has no way of controlling its body temperature independently of that of its mother, nor can the fetus escape the thermal influence of its immediate environment, that is, maternal tissue. Thus, irrespective of the prevailing levels of fetal heat production and heat loss, fetal body temperature rises whenever maternal body temperature does. Rises in maternal body temperature occur in ruminants, for example, during exposure to hot environments and when animals exercise. We have raised pregnant ewe body temperature in three different experimental protocols, and have examined fetal lamb body temperature in those circumstances (Laburn *et al.*, 1992). Figure 17.5 summarizes our results. Exposure of the ewes to a hot environment for 2 h (Fig. 17.5a) or to 30 min of treadmill exercise (Fig. 17.5b) caused their body temperature to rise, and fetal lamb temperature rose too. However, we observed that the fetal lamb temperature rise did not exactly parallel the ewe's temperature. Instead, the difference between fetal and maternal temperature fell significantly. The result was that, in these situations of raised ewe body temperature, fetal body temperature was prevented from rising to the same extent by a reduction in the F-M gradient, and a consequent reduction in the risk of hyperthermia for the fetus.

In a single experiment on a pregnant baboon, external heat exposure reduced the F-M gradient in that species too (Morishima *et al.*, 1975), and we recently have found a similar response in goats (unpublished oberservations). Given the apparent ubiquity of a reduction in the F-M gradient during maternal ambient heat stress and during exercise in sheep (we also have observed the phenomenon during labour, see below), we have asked ourselves whether there could be a common mechanism for reducing the risk of fetal hyperthermia. Increased blood flow in umbilical or uterine blood vessels could account for the significant fall in the F-M gradient. Intuitively, one might expect a decrease in perfusion of visceral vascular beds in heat-stressed animals, to divert blood flow to the periphery. Indeed there are several reports claiming that uterine blood flow does decrease in pregnant ewes exposed to heat (Alexander *et al.*, 1987) and exercise (Bell *et al.*, 1984), although the effects do not occur in pregnant women (Vähä-Eskeli *et al.*, 1991). If uterine blood flow fell, then the F-M thermal gradient should have increased, but the researchers who measured uterine flow did not measure fetal temperature. Evidence in support of our hypothesis is that from Cefalo and Hellegers (1978), who showed that uterine blood flow rose during ewe hyperthermia, diminishing only after maternal body temperature had risen by more than 2.0°C, and that umbilical blood flow also increased as ewe body temperature rose, falling only after ewe temperature had risen by 2.5°C. Walker *et al.* (1995) demonstrated a rise in blood flow to many tissues in fetal lambs during maternal heat exposure, and a consequent rise in vascular conductance, which would increase heat loss from the fetus via placental and non-placental routes. Thus, at moderate levels of maternal hyperthermia, both mother and fetal animal may activate mechanisms to reduce the risk of fetal hyperthermia. Under more severe heat stress, the mother may attend to her own thermoregulatory needs, even though that means compromising her fetus.

Fever is the specific form of hyperthermia which accompanies inflammation and infection, and in our third hyperthermic protocol, we examined the consequences for fetal lamb body temperature when the pregnant ewe is febrile (Laburn *et al.*, 1992). Fig. 17.5c shows the typical fever response when ewes were injected with the purified

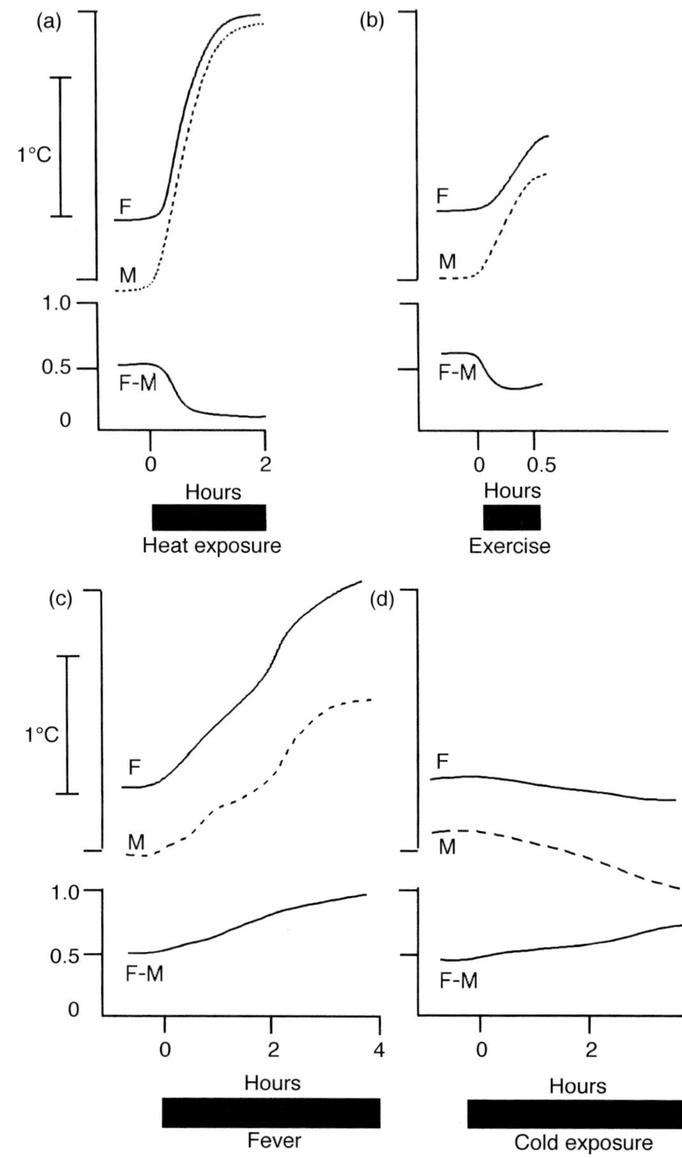


**Fig. 17.5.** Summary of changes in fetal lamb (F) and pregnant ewe (M) body temperatures and effects on the feto-maternal (F-M) gradient of exposure of the ewes to various conditions of thermal stress. (a) exposure at rest to 40°C and 60% relative humidity. (b) Treadmill exercise (2 km h$^{-1}$, 5° gradient) for 30 min at room temperature. (c) Experimental fever following intravenous injection of bacterial lipopolysaccharide into the ewe at time zero. (d) Exposure to 4°C. Reprinted with permission from the American Physiological Society, from Laburn (1996).

lipopolysaccharide of a Gram-negative bacterium. Ewe body core temperature rose after the pyrogen injection and, as with the other forms of hyperthermia, body temperature of the fetus rose also, but during maternal fever, contrary to what occurred in the other hyperthermias, the F-M gradient increased significantly. The rise in temperature therefore was significantly greater for the fetus than it was for the mother animal, during maternal fever.

For hyperthermias associated with increased thermal loads, of external (hot environments) or internal (exercise) origin, the mother's physiological response was to increase her heat loss, using peripheral vasodilation and/or evaporation to dissipate heat. During fever, the body reacts as if it were exposed to cold to achieve the elevated temperature. Typically, animals vasoconstrict and shiver. We have hypothesized that during the febrile episode, vasoconstriction in the uterine circulation, as well as in other peripheral vascular beds, compromised fetal heat loss. The thermal consequences of fever in pregnant ruminants, of whatever cause, therefore may be a significant risk to fetal well-being.

Peripheral vasoconstriction also occurs in response to a lowering of body temperature, and we suspect that, in a way similar to that which we postulate occurred during the febrile episode, pregnant ewes exposed to cold ambient conditions reduce blood flow in the uterine circulation, and fetal heat loss is decreased. In the case of maternal hypothermia, however, the resulting rise in F-M gradient, attenuates the fall in fetal body temperature, a response which benefits the fetus. The effect is shown in Fig. 17.5d, and also has been observed by Morishima *et al.* (1975) in a baboon. Thus, apart from the specific case of maternal fever, maternal or fetal mechanisms involving changes in blood flow conspire to reduce changes in fetal body temperature when the mother is subjected to thermal insult.

## Birth-related changes in body temperatures

### *Changes in body temperature of the ewe and fetal lamb*

There are three phases of body temperature change accompanying parturition. The first is that which occurs prior to lambing, in which we have observed that ewe body temperature falls by a few tenths of a degree over the few days leading up to labour (Fig. 17.2). This phenomenon has been observed in horses (Shaw *et al.*, 1988; Haluska and Wilkins, 1989), cattle (Ewbank, 1963), dogs (Concannon *et al.*, 1977) and sheep (Ewbank, 1969; Laburn *et al.*, 1992), and would be a useful predictor of parturition if daily body temperature measurements were made routinely in ruminants. It may have to do with the fall in progesterone plasma concentration as labour becomes imminent (Janowski *et al.*, 1995).

The second phase is that of the exertion of labour, and myometrial contractions, which produce sufficient heat in the active tissues (Gemzell *et al.*, 1957; Marx and Loew, 1975) to cause body temperature to rise. Fetal temperature inevitably rises too (Fig. 17.6), but is accompanied by a reduced F-M thermal gradient (Laburn *et al.*, 1994), as it is in maternal locomotor exercise. The effect of attenuating the F-M gradient during labour may have to do with reducing any increase in body temperature in the fetal/neonatal animal. The lower the body temperature during the process of birth,

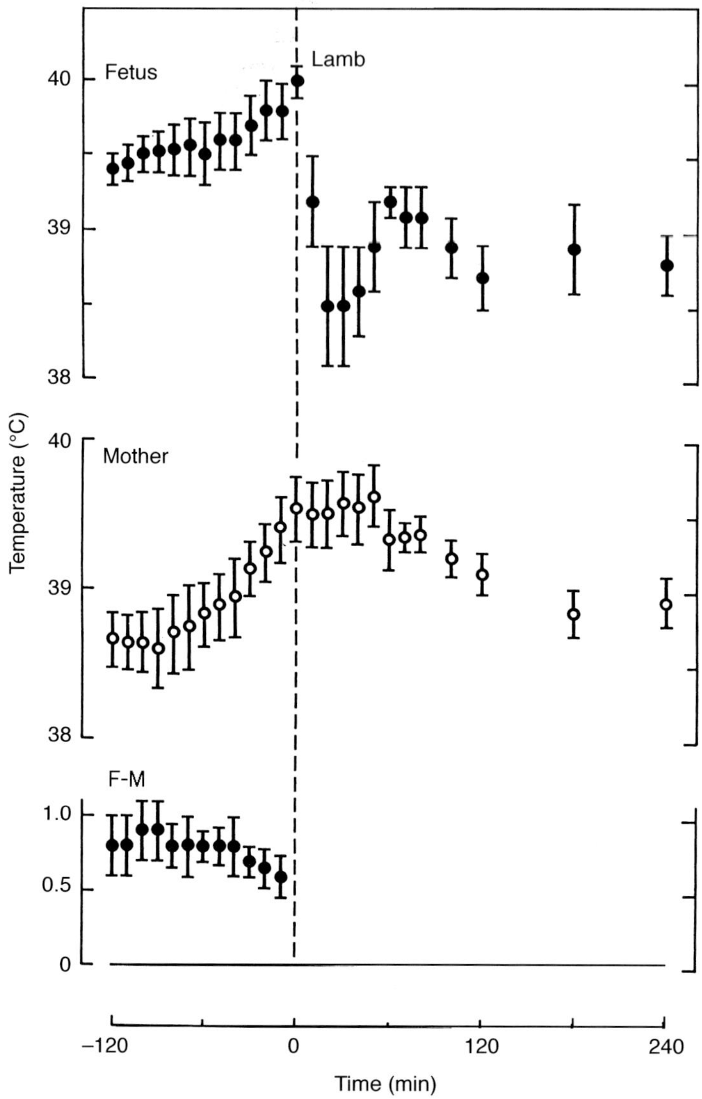


**Fig. 17.6.** Changes in body temperatures of fetus or lamb, and mother animal, and the feto-maternal (F-M) temperature gradient, measured for 2 h before and 4 h after lambing (at time zero). Each point is the mean ± standard error of a mean (SEM) of seven ewes and eight fetuses/lambs. Reprinted with permission from Birkhäuser Verlag AG, from Laburn *et al.* (1994).

the lower the risk of ischaemic brain damage, should neonatal circulation or respiration be compromised in the birth process (Kuroiwa *et al.*, 1990).

The third phase of the thermal response to parturition is that more dramatic phase, for the newborn ruminant animal at least, immediately after birth. The measurements shown in Fig. 17.6 were taken in a group of ewes giving birth in indoor conditions. The neonatal lamb, delivered soaked in amniotic fluid, is confronted by an environment which, compared with the intrauterine environment, is cold and dry. Evaporation of fluid from the lamb's skin, and its predisposition to heat loss by virtue of its high surface area to mass ratio, results in a precipitous fall in body temperature, by about 1.5°C within minutes of birth. Lambs in the field are delivered into much colder conditions (Barlow *et al.*, 1987) so one should expect that the fall in neonatal body temperature would be more precipitous than it was in our experiments. Nevertheless, within an hour, in the indoor environment at least, thermoregulatory effector mechanisms come into play, not only to halt the plummeting body temperature, but to reverse it. By 3 h post-partum, lamb and ewe body temperatures are not significantly different (Fig. 17.6).

## Development of thermoregulatory mechanisms

The remarkable ability of neonatal animals to cope with the environment into which they are born is the result, in large part, of their ability, very soon after birth in ruminant neonates, to activate non-shivering thermogenesis (NST) in the brown adipose tissue with which they are well endowed at birth (Nedergaard and Cannon, 1992). NST appears when an uncoupling protein, in brown adipose cells, causes short-circuiting of protons in the process of oxidative phosphorylation (Cannon and Nedergaard, 1985) such that extra heat, rather than adenosine triphosphate, is produced. In sheep, the concentration of mRNA for uncoupling protein peaks just before birth (Casteilla *et al.*, 1989). Cold exposure stimulates beta-adrenergic activity and in turn receptors on the brown adipose cells, and synthesis of the uncoupling protein. Because of the large blood supply to brown adipose tissue, heat produced is transported from its source to other body sites, and body temperature of the neonate rises, unless the ambient thermal conditions and heat loss from the lamb overcome the capacity for NST (Barlow *et al.*, 1987).

The ability to activate NST within minutes of birth is crucial to the survival of the neonatal lamb. How does the fetal lamb acquire the mechanisms for activation of NST, when the necessary stimuli for inducing the development of the mechanisms are absent *in utero*? The late Tania Gunn and her colleagues (Gunn and Gluckman, 1983; Gunn *et al.*, 1991) carried out elegant experiments in which they were able to simulate the exposure to cold, and other features that the neonate encounters at birth, but in fetal lambs, *in utero*. They measured plasma glycerol concentration and the temperature in brown adipose tissue as indicators of the activation of NST (Fig. 17.7). Neither oxygenation nor sympathetic stimulation alone, nor jointly, could activate NST. What was necessary was stimulation of cold receptors in the fetal lamb, and a simulation of cutting the umbilical cord. Thus, NST is possible in the fetal lamb, but the intrauterine thermal environment, and an inhibitor of the response emanating from the placenta, possibly prostaglandins (Andrianakis *et al.*, 1989; Gunn *et al.*, 1993), appear to suppress NST *in utero*.

Shivering too can be induced experimentally in certain circumstances, in fetal lambs (Dawes, 1968) and so can vasoconstriction in response to a cold stimulus (Gunn

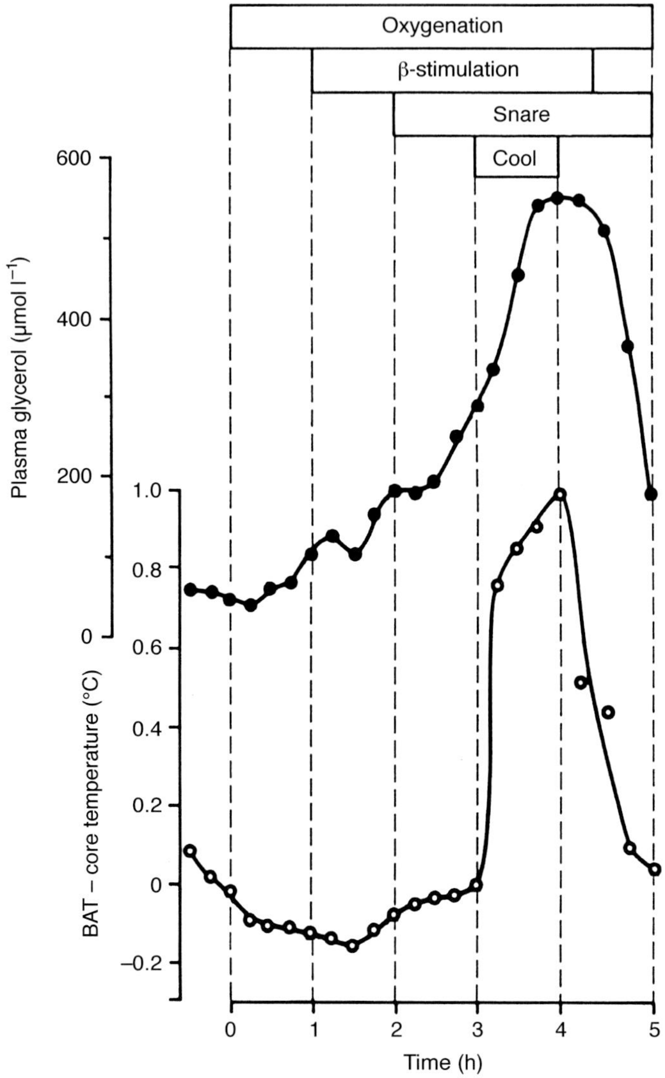


**Fig. 17.7.** Data from experiments performed by Gunn *et al.* (1991) showing the activation of non-shivering thermogenesis (NST) in a group of fetal lambs during 5 h of simulated post-birth conditions. Top curve, changes in plasma glycerol concentrations. Bottom curve, increase in brown adipose tissue (BAT) temperature above that of the lamb's core body temperature. Lambs were ventilated, and primed with a β-stimulant. NST was stimulated by cooling the fetus *in utero,* but only after the umbilical cord was snared. Reprinted with permission from the American Physiological Society, from Laburn (1996).

and Gluckman, 1983). Normally, however, fetal lambs probably are peripherally vasodilated (Schröder *et al.*, 1988) and presumably shivering, like NST, is suppressed *in utero*. Other thermoregulatory mechanisms, including the ability to pant in response to high ambient conditions, also probably are mature in the late-gestation fetal lamb, but presumably are suppressed.

### *Post-partum body temperatures*

The body temperatures of both neonatal lamb and its mother remain unexpectedly high in the days and weeks following parturition. Figure 17.8 shows that at constant ambient temperature, neonatal body temperaure peaks at about 7 days post-partum, and then declines towards adult levels by a month of age. In the post-partum ewe, lactation appears to promote a sustained elevation of body temperature. A comparison between Figs 17.6 and 17.8 shows that the body temperature of the lactating ewe is

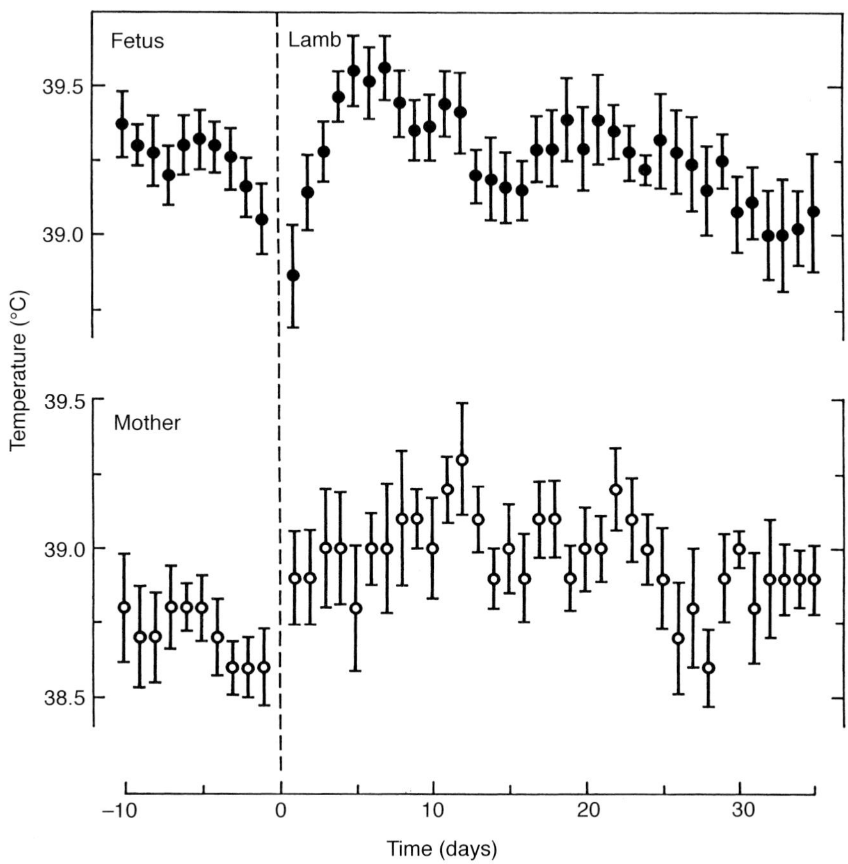


**Fig. 17.8.** Changes in fetal and lamb (upper panel) and ewe (bottom panel) body temperatures measured for 10 days prior to lambing and for 7 weeks thereafter. Each point is the mean ± standard error of a mean for seven ewes and eight fetuses/lambs. Lambing is indicated by a vertical dashed line.

higher even than that towards the end of pregnancy, at least for 3 weeks. What the mechanism is of the sustained rise in body temperature is unknown, but it may be a defended elevation in temperature; lactating ewes have attenuated fever responses in response to administration of a Gram-negative pyrogen (S. Glassom *et al.*, South Africa, unpublished observations).

## Preterm labour and abortion

There is anecdotal evidence that acute ambient heat or cold stress may precipitate preterm labour in ruminants. Infection, too, can cause premature delivery, and in this instance the cause appears to be fetal death. We simulated maternal, fetal and intrauterine infection by injecting Gram-positive and Gram-negative pyrogens directly into the pregnant ewe, its fetus and amniotic cavity. Even though the dose of pyrogen administered to adult or fetal animal was not sufficient to lead to severe fetal hyperthermia, the pyrogens had potentially serious effects on fetal well-being. Table 17.1 shows the percentage of abortions we observed after administering Gram-positive or Gram-negative pyrogens to ewe, fetus, or into the amniotic fluid. The presence of bacterial material (even killed organisms or purified extracts) induced abortion in a proportion of sheep, with a greater risk of abortion when the pyrogen was present in the fetus than in the mother animal, and the highest risk being associated with the presence of pyrogenic material in the amniotic cavity. It is likely that cytokines, released from fetal and/or placental tissue (Mitchell *et al.*, 1991; Gibbs *et al.*, 1992) as a result of the pyrogen presence, induced inflammatory changes in the placenta, which led to severely compromised fetal function, and ultimately fetal death.

## Acknowledgements

The work reported in this article was supported by the Council of the University of the Witwatersrand, Johannesburg, and the South African Medical Research Council.

**Table 17.1.** Abortion following pyrogen injection. Percentage of abortions, and time to abortion (in days, mean ± standard error of a mean (SEM), after either Gram-positive or Gram-negative pyrogens were injected directly into the circulation of pregnant ewes (*n* = 12), or fetal animals (*n* = 14), or into the amniotic fluid (*n* = 4).

| | Abortion (%) | Days to abortion ± SEM |
|---|---|---|
| Pyrogen to ewe | 41 | 10.5 ± 2.6 |
| Pyrogen to fetus | 64 | 4.6 ± 0.9 |
| Pyrogen to amnion | 75 | 1.6 ± 0.7 |

## References

Abrams, R., Caton, D., Clapp, J. and Barron, D.H. (1970) Thermal and metabolic features of life in utero. *Clinical Obstetrics and Gynecology* 13, 549–564.

Adamsons, K. and Towell, M.E. (1965) Thermal homeostasis in the fetus and newborn. *Anesthesiology* 26, 531–548.

Alexander, G. and Williams, D. (1971) Heat stress and development of the conceptus in domestic sheep. *Journal of Agricultural Science, Cambridge* 76, 53–72.

Alexander, G., Hales, J.R.S., Stevens, D. and Donnelly, J.B. (1987) Effects of acute and prolonged exposure to heat on regional blood flows in pregnant sheep. *Journal of Developmental Physiology* 9, 1–15.

Andrianakis, P., Walker, D.W., Ralph, M.M. and Thorburn, G.D. (1989) Effect of inhibiting prostaglandin synthesis in pregnant sheep with 4-aminoantipyrine under normothermic and hyperthermic conditions. *American Journal of Obstetrics and Gynecology* 161, 241–247.

Assali, N.S. and Westin, B. (1962) Effects of hypothermia on uterine circulation and on the fetus. *Proceedings of the Society for Experimental Biology and Medicine* 109, 485–488.

Barlow, R.M., Gardiner, A.C., Angus, K.W., Gilmour, J.S., Mellor, D.J., Cuthbertson, J.K., Newlands, G. and Thompson, R. (1987) Clinical, biochemical and pathological study of perinatal lambs in a commercial flock. *Veterinary Record* 120, 357–362.

Bell, A.W., Hales, J.R.S., Fawcett, A.A. and King, R.B. (1984) Contribution of the uterus to circulatory requirements of exercise and heat stress in pregnant sheep. In: Hales, J.R.S. (ed.) *Thermal Physiology*. Raven Press, New York, pp. 377–380.

Bell, A.W., Battaglia, F.C. and Meschia, G. (1987) Relation between metabolic rate and body size in the ovine fetus. *Journal of Nutrition* 117, 1181–1186.

Cannon, B. and Nedergaard, J. (1985) The biochemistry of an inefficient tissue: Brown adipose tissue. *Essays in Biochemistry* 20, 110–164.

Casteilla, L., Champigny, O., Bouilland, F., Robelin, J. and Ricquier, D. (1989) Sequential changes in the expression of mitochondrial protein mRNA during the development of brown adipose tissue in bovine and ovine species. *Biochemical Journal* 257, 665–671.

Cefalo, R.C. and Hellegers, A.E. (1978) The effects of maternal hyperthermia on maternal and fetal cardiovascular and respiratory function. *American Journal of Obstetrics and Gynecology* 131, 687–694.

Chambers, C.D., Johnson, K.A., Dick, L.M., Felix, R.J. and Jones, K.L. (1998) Maternal fever and birth outcome: a prospective study. *Teratology* 58, 251–257.

Concannon, P.W., Powers, M.E., Holder, W. and Hansel, W. (1977) Pregnancy and parturition in the bitch. *Biology of Reproduction* 16, 517–526.

Dawes, G.S. (1968) *Fetal and Neonatal Physiology*. Year Book Medical Publishers Inc., Chicago, pp. 191–209.

Dreiling, C.E., Carman, F.S. and Brown, D.E. (1991) Maternal endocrine and fetal metabolic responses to heat stress. *Journal of Dairy Science* 74, 312–327.

Edwards, M.J., Shiota, K., Smith, M.S.R. and Walsh, D.A. (1995) Hyperthermia and birth defects. *Reproductive Toxicology* 9, 411–425.

Ewbank, R. (1963) Predicting the time of parturition in the normal cow. *The Veterinary Record* 75, 367–371.

Ewbank, R. (1969) The fall in rectal temperature seen before parturition in sheep. *Journal of Reproduction and Fertility* 19, 569–571.

Gemzell, C.A., Robbe, H., Stern, B. and Ström, G. (1957) Observations on circulatory changes and muscular work in normal labour. *Acta Obstetrica et Gynecologica Scandinavica* 36, 75–92

Gericke, G.S., Hofmeyr, G.J., Laburn, H.P. and Isaacs, H. (1989) Does heat damage fetuses? *Medical Hypotheses* 29, 275–278.

Gibbs, R.S., Romera, R., Hillier, S.L., Eschenbach, D.A. and Sweet, R.L. (1992) A review of premature birth and subclinical infection. *American Journal of Obstetrics and Gynecology* 166, 1515–1528.

Gilbert, R.D., Schröder, H., Kawamura, T., Dale, P.S. and Power, G.G. (1985) Heat transfer pathways between fetal lamb and ewe. *Journal of Applied Physiology* 59(2), 634–638.

Gunn, T.R. and Gluckman, P.D. (1983) Development of temperature regulation in the fetal sheep. *Journal of Developmental Physiology* 5, 167–179.

Gunn, T.R., Ball, K.T., Power, G.G. and Gluckman, P.D. (1991) Factors influencing the initiation of nonshivering thermogenesis. *American Journal of Obstetrics and Gynecology* 164, 210–217.

Gunn, T.R., Ball, K.T. and Gluckman, P.D. (1993) Withdrawal of placental prostaglandins permits thermogenic responses in fetal sheep brown adipose tissue. *Journal of Applied Physiology* 74(3), 998–1004.

Haluska, G.J. and Wilkins, K. (1989) Predictive utility of pre-partum temperature changes in the mare. *Equine Veterinary Journal* 21, 116–118.

Hart, F.M. and Faber, J.J. (1965) Fetal and maternal temperatures in rabbits. *Journal of Applied Physiology* 20, 737–741.

Janowski, T., Zdunczyk, S., Podhalicz-Dziegielewska, M., Ras, A. and Chmielewski, A. (1995) Course of pregnancy, steroid hormone levels and maturation process of placentomes after oestrone infusion in cows near term. *Zentralbl Veterinarmed A* 42(5), 345–350.

Kuroiwa, T., Bonnekoh, P. and Hossmann, K.-A. (1990) Prevention of postischemic hyperthermia prevents ischemic injury of $CA_1$ neurons in gerbils. *Journal of Cerebral Blood Flow and Metabolism* 10, 550–556.

Laburn, H.P. (1996) How does the fetus cope with thermal challenges? *News in Physiological Sciences* 11, 96–100.

Laburn, H.P., Mitchell, D. and Goelst, K. (1992) Fetal and maternal body temperatures measured by radiotelemetry in near-term sheep during thermal stress. *Journal of Applied Physiology* 72(3), 894–900.

Laburn, H.P., Goelst, K. and Mitchell, D. (1994) Body temperatures of lambs and their mothers measured by radio-telemetry during parturition. *Experientia* 50, 708–711.

Marx, G.F. and Loew, D.A.Y. (1975) Tympanic temperature during labour and parturition. *British Journal of Anaesthesia* 47, 600–602.

Mitchell, M.D., Branch, D.W., Lundin-Schiller, S., Romero, R.J., Daynes, R.A. and Dudley, D.J. (1991) Immunologic aspects of preterm labor. *Seminars in Perinatology* 15, 210–224.

Morishima, H.O., Glaser, B., Niemann, W.H. and James, L.S. (1975) Increased uterine activity and fetal deterioration during maternal hyperthermia. *American Journal of Obstetrics and Gynecology* 121, 531–538.

Nedergaard, J. and Cannon, B. (1992) Brown adipose tissue: development and function. In: Polin, R.A. and Fox, W.W. (eds) *Fetal and Neonatal Physiology*. Saunders, Philadelphia.

Schröder, H., Gilbert, R.D. and Power, G.G. (1988) Computer model of fetal-maternal heat exchange in sheep. *Journal of Applied Physiology* 65(1), 460–468.

Shaw, E.B., Houpt, K.A. and Holmes, D.F. (1988) Body temperature and behaviour of mares during the last two weeks of pregnancy. *Equine Veterinary Journal* 20, 199–202.

Vähä-Eskeli, K., Pirhonen, J., Seppänen, A. and Erkkola, R. (1991) Doppler flow measurement of uterine and umbilical arteries in heat stress during late pregnancy. *American Journal of Perinatology* 8, 385–389.

Walker, D.W., Hales, J.R.S., Fawcett, A.A. and Pratt, N.M. (1995) Cardiovascular response to heat stress in late gestation fetal sheep. *Experimental Physiology* 80, 755–766.

Wood, C. and Beard, R.W. (1964) Temperature of the human foetus. *Journal of Obstetrics and Gynaecology of the British Commonwealth* 71, 768–769.

# 18 Regulation of Nutrient Partitioning During Lactation: Homeostasis and Homeorhesis Revisited

D.E. BAUMAN

*Department of Animal Science, Cornell University, Ithaca, New York, USA*

## Introduction

Nature has accorded a high priority to lactation, and the ability of mammals to synthesize milk is essential for survival of the newborn. The mammary glands have a high metabolic rate during lactation, yet this organ is unique because its biosynthetic products represent no direct benefit to the mother. Rather, the extensive rate of nutrient use to make milk imposes a substantial demand on the mother and mandates that the metabolism of the mammary glands and other maternal tissues be coordinated. Thus, during lactation many physiological changes occur and there are profound alterations in the metabolism of many tissues. The extensive physiological adaptations which occur during lactation have the overall effect of providing the proper quantity and pattern of nutrients for milk synthesis. Thus, lactation provides an excellent opportunity to elucidate the broad concepts of metabolic regulation and identify specific mechanisms involved in the partitioning of nutrients. The regulation of nutrient use is not only important for a successful lactation, but it also represents the physiological basis for differences in productive efficiency, and this is of special importance in agricultural species. Furthermore, if the coordination of nutrient use is inadequate then animal well-being is compromised which may result in stress, subclinical conditions and metabolic disorders.

In 1980, Currie and I reviewed the regulation of nutrient partitioning during pregnancy and lactation. In discussing the concepts of regulation, we crystallized the concept of homeorhesis, proposed mechanisms, and addressed the interrelations between homeostasis and homeorhesis (Bauman and Currie, 1980). This review will elaborate and trace the development of the concepts of homeostasis and homeorhesis, focus on the regulation of nutrient use during lactation, and review current understanding of the mechanisms and applications of these concepts. Because of its breadth, reviews will be cited frequently for various aspects of the biology.

## Concepts of regulation

At 12 minutes past eight on 3 February 1783, Lavoisier initiated experiments with a guinea pig in what is now recognized as the first animal calorimeter (Blaxter, 1989; Welch, 1991). In collaboration with Laplace, he carried out two experiments in a 24 h period, which for the first time firmly linked the evolution of heat by animals with the consumption of oxygen and the formation of carbon dioxide. These simple, but elegant, experiments established that life is a chemical process and represented an initial important step in the search to understand bioenergetics and the regulation of metabolism. It was almost 100 years later that another milestone occurred when Bernard, a physiologist, recognized the ability of living organisms to maintain their own constancy. He cogently observed that 'all the vital mechanisms, however varied they may be, have only one object, that of preserving constant the conditions of life in the internal environment' (Bernard, 1878).

Using Bernard's concept of a stable '*milieu interieur*' as the cornerstone, Cannon concluded that the coordinated physiological reactions which maintain most of the steady states in a living organism were complex, involving many different tissues and organs 'all working cooperatively' (Cannon, 1932). He recognized that higher organisms had more elaborate and effective systems to maintain steady-state conditions, and suggested a special designation for these states – homeostasis. Cannon (1929) chose the term homeostasis with a great deal of thought and foresight. He concluded that the prefix 'homeo' was preferred because it meant 'like' or 'similar', thereby indicating some variation. He rejected the prefix 'homo' because it meant 'same' and implied a rigid constancy. Likewise Cannon (1929) chose 'stasis' because it represented a 'condition'. By choice of terminology and examples to elaborate the concept, Cannon (1929, 1932) emphasized that dynamic regulation and coordination were key features of homeostasis.

Today the concept of homeostasis is well known to biologists, and there are many systems where the positive and negative feedback controls to preserve steady state are well established. Glucose was an example used by Bernard (1878) in developing the concept of *milieu interieur* and by Cannon (1932) in crystallizing the concept of homeostasis. The homeostatic controls to maintain steady-state conditions for glucose are also of special significance during lactation. Glucose is critical during lactation because its uptake by the mammary gland is essential for the synthesis of milk lactose, the major osmotic regulator of milk volume. The pancreatic hormones, insulin and glucagon, are key controls of glucose homeostasis. Thus, acute regulation of plasma glucose concentration by the reciprocal actions of insulin and glucagon ensure the proper balance in glucose supply and utilization by body tissues and organs during lactation.

While the concept of homeostasis is universally accepted today, it resulted in substantial debate over several decades following its introduction (Waddington, 1942, 1953; Lerner, 1954; Lewontin, 1956). In part, this debate focused on the fact that the concept was not adequate to address regulation over a wide range of physiological and developmental situations. Waddington (1957) reasoned that there are at least three types of temporal change in biological adaptations, all occurring simultaneously. The first, on the longest time scale, is evolution. The second, of medium time scale relates to development throughout the life cycle and the third, on the shortest time scale, applies to the day-to-day events of physiology. Thus, physiological changes do more than preserve a day-to-day constant state of animal well-being, they are also altered as time passes which produces the longer-term changes associated with ontogeny. To illustrate this, Waddington (1957) compared biological processes in an animal with a musical composition. The short-scale physiology of the animal would be similar to the vibrations of individual notes. The medium-scale life-history would be analogous to the melodic phrases into which the notes build themselves; and the animal's long-scale evolution would represent the structure of the whole musical composition, in which the melodies are repeated and varied.

In the three decades following Cannon's crystallization of homeostasis, several terms were introduced to accommodate the developmental aspect of regulation. Waddington played a central role in this discussion, and introduced the terms 'developmental homeostasis' and 'canalization' (Waddington, 1942, 1953). Later, he suggested the term 'homeorhesis' by analogy to homeostasis, and related it to describe 'flow' to a new equilibrium or pathway of developmental change over the life cycle (Waddington, 1957). In reviewing this early debate, one cannot help but marvel at the creativeness and originality of Waddington and others, bearing in mind that this was a period prior to discovery of DNA and the genetic code, prior to the identification of second message cascades and signal transduction systems, and prior to the development of analytical techniques and molecular probes for measuring proteins and gene transcripts. Obviously, some aspects of this early discussion are incorrect as shown by subsequent discoveries, and it is often difficult to relate some of the discussion to current knowledge of biology. Nevertheless, the reasoning and insight of these early pioneers in biological regulation are impressive.

In 1980, Currie and I crystallized the concept of homeorhesis in the regulation of nutrient partitioning and physiological processes. We defined homeorhesis as 'the orchestrated or coordinated changes in metabolism of body tissues necessary to support a physiological state' (Bauman and Currie, 1980). The Greek derivation of homeorhesis is 'uniform-flow'. Thus, homeorhetic regulation involves a coordination of metabolism to ensure a uniform flow of nutrients in support of a physiological state. Of particular importance, we characterized key features of homeorhetic control: its chronic nature, i.e. hours or days versus the seconds or minutes required for most examples of homeostatic regulation, its simultaneous influence on multiple tissues and systems with apparently unrelated functions, and its mediation through altered responses to homeostatic signals (Bauman and Currie, 1980; Bauman, 1984).

The physiological states of lactation and pregnancy were used as the initial examples to illustrate the concept of homeorhesis, and we later extended it to regulation during growth (Bauman *et al.*, 1982). Kennedy (1967) also applied the concept of homeorhesis to growth in an earlier review which detailed the changes in the homeostatic regulation of food intake that occur at different stages of development. Others have extended this concept to a broad array of physiological processes, and in some cases other terms have been introduced which, for all practical purposes, are synonymous and/or consistent with the concept of homeorhesis. These have included homeorheusis (Nicoläidis, 1983), teleophoresis (Chilliard, 1986), rheostasis (Mrosovsky, 1990) and poikilostasis (Kuenzel *et al.*, 1999). For example, Mrosovsky (1990) introduced the term rheostasis (rheo = change) and defined it as 'a condition in which, at any one instant, homeostatic defenses are present but over a span of time, there is a change in the level that is defended'. Likewise, Kuenzel *et al.* (1999) introduced the term poikilostasis (poikilo = various) to describe the dynamic shifts in homeostatic regulation of metabolism and food intake that occur in birds over different physiological states such as growth, migration and reproduction.

In recognition of the original work of Waddington (1957) and the earlier use by Kennedy (1967), we chose to use the term homeorhesis (Bauman and Currie, 1980). Nevertheless, in all cases the concept relates to the ability of the animal to adjust biological processes in a manner to support a dominant physiological state for animal well-being and survival of the species. As illustrated in Table 18.1, the general concept represented by homeorhesis has been extended to an impressive range of biological situations encompassing many different physiological, nutritional and even pathological states. In particular, Mrosovsky (1990) provided an extensive list of examples of this regulation in terms of both physiological states and biological processes. A key physiological state in the survival of mammals is lactation, and the remainder of this review will relate the concepts of regulation to lactation.

## Adaptations to lactation

Lactation represents an impressive example of homeorhesis, and at peak lactation the proportion of nutrients used by the mammary gland can be extraordinary. For example, in high-producing dairy cows nutrient utilization by the mammary glands exceeds that of the rest of the body causing Brown (1969) to suggest that the cow should be envisioned as an appendage to the udder rather than vice versa. Nutrient supply and use during lactation are similar for all species, although there are qualitative and quantitative differences (Linzell, 1967; Williamson *et al.*, 1995). In all species, the extent of nutrient use for milk synthesis requires integrated regulation of the metabolism of the mammary glands and other body tissues.

Using the dairy cow as an example, a partial list of the physiological adaptations which occur during lactation is presented in Table 18.2. The importance accorded to lactation in mammals is demonstrated by the fact that the physiological adaptations involve many, perhaps most, of the body tissues and relate to the metabolism of all nutrient classes. The net effect is that the increase in mammary gland metabolic rate and nutrient use which occurs during lactation coincides with alterations in the metab-

**Table 18.1.** Partial list of physiological situations where the general concept representing homeorhetic regulation has been applied.[a]

| | |
|---|---|
| Lactation | Hibernation |
| Pregnancy | Premigration/migration |
| Growth | Egg laying |
| Puberty | Incubation anorexia |
| Ageing | Seasonal cycles |
| Chronic undernutrition | Exercise |
| Chronic illness | |

[a] References include Kennedy (1967), Bauman and Currie (1980), Dilman (1982), Mrosovsky (1990), Wade and Schneider (1992), Vernon (1998), and Kuenzel *et al.* (1999).

**Table 18.2.** Partial list of physiological adaptations which occur in lactating dairy cows.

| Process or tissue | Response |
|---|---|
| Mammary tissue | Increased number of secretory cells<br>Increased nutrient use<br>Increased supply of blood |
| Food intake | Increased quantity |
| Digestive tract | Increased size<br>Increased absorptive capacity<br>Increased rates of nutrient absorption |
| Liver | Increased size<br>Increased rates of gluconeogenesis<br>Increased glycogen mobilization<br>Increased protein synthesis |
| Adipose tissue | Decreased *de novo* fat synthesis<br>Decreased uptake of preformed fatty acids<br>Decreased re-esterification of fatty acids<br>Increased lipolysis |
| Skeletal muscle | Decreased glucose utilization<br>Decreased protein synthesis<br>Increased protein degradation |
| Bone | Increased mobilization of Ca and P |
| Heart | Increased cardiac output with a larger percentage going to the mammary glands |
| Plasma hormones | Decreased insulin<br>Increased somatotropin<br>Increased prolactin<br>Increased glucocorticoids<br>Decreased thyroid hormones<br>Decreased IGF-I |

olism of other body tissues so that an adequate quantity and pattern of nutrients to support milk synthesis is ensured. Several reviews have detailed the physiological adaptations which occur during lactation and readers are referred to these for quantitative information (Williamson, 1980; Bauman and Elliot, 1983; Chilliard, 1986, 1987; Vernon, 1989; Williamson and Lund, 1994; McNamara, 1995). Nevertheless, some of these adaptations merit mention as examples which illustrate the concepts of homeostasis and homeorhesis.

During lactation, food intake is increased in many species. Corresponding adaptations also occur in the size and absorptive capacity of the gastrointestinal tract, thereby allowing for an increased absorption of nutrients (Bauman and Elliot, 1983; Vernon, 1989). In rodents where the milk demand of the nursing pups gradually increases, the increase in food intake is also gradual and can amount to a 300–400% increase compared to the non-lactating state (Cripps and Williams, 1975). Furthermore, the magnitude of the increase in intake is related to the number of nursing pups (Millar, 1979), providing convincing evidence of the coordinated nature of these changes. Increases in feed intake and digestive tract size also occur in dairy cows, with the magnitude of the increase in voluntary intake being related to milk yield (Bauman and Elliot, 1983). However, for this species the increase in feed intake occurs over a longer interval so that high-producing dairy cows may not achieve a positive energy balance until 8–12 weeks post-partum (Bauman and Currie, 1980).

Despite adjustments in food intake, many species rely on body reserves during early lactation. Vernon (1998) pointed out that this results in many body tissues having metabolic characteristics that are typical of the adaptations which occur with chronic undernutrition. However, the extent to which reserves are needed in early lactation to meet nutrient requirements, especially energy requirements, varies among species. In some species, the use of energy reserves is minimal to modest because either the amount of milk required by the nursing young is limited (e.g. human and guinea pig), or the increase in voluntary intake is nearly adequate to meet requirements (e.g. rat and mouse). In other species such as the cow, goat and pig, the use of body reserves is more extensive during the lactation cycle. For example, lactational yields in dairy cows are related to the magnitude of body reserve utilization (Bauman *et al.*, 1985), and in high-producing cows the mobilization of body fat during the first month of lactation can be energetically equivalent to over one-third of the milk produced (Bauman and Currie, 1980).

The use of body reserves is extraordinary in several species and some examples of these merit special mention because they dramatically illustrate the concepts of metabolic regulation. Seals make spectacular use of body reserves to meet their nutrient requirements during lactation (Riedman, 1990; Oftedal, 1993). This is illustrated by the pregnant elephant seal which gives birth to a single pup and sustains a 28 day lactation. The nursing pup averages a rate of body weight gain of approximately 10% day$^{-1}$. In contrast, the mother loses weight, as she neither eats nor drinks throughout the 4 week lactation. Maternal use of body reserves over the lactation interval results in a 42% loss of body weight, and this represents a 58% reduction in body fat content and a 14% reduction in body lean weight (Costa *et al.*, 1986). Whales, especially baleen whales, also make extraordinary use of body reserves during lactation and the blue whale provides an example (Lockyer, 1981; Oftedal, 1993). Female blue whales have a body weight of about 80,000 kg and during pregnancy their 40,000–55,000 kg gain in body weight primarily represents the accretion of body reserves. The blue whale calf weighs about 2500 kg at birth and gains at the rate of 80–100 kg day$^{-1}$ during the 7 month lactation. During this interval the blue whale mother produces about 90 kg of milk day$^{-1}$. Even more impressive, she relies almost exclusively on body reserves to support the nutrient needs for milk synthesis and her own sustenance, as she eats little, if at all, throughout the 7 month lactation.

In seals and whales, as well as other lactating mammals, the physiological adaptations to support lactation are extensive. Based on work with laboratory and farm animals, we can envision how body fat reserves may serve to meet the energy requirements of the lactating elephant seal and blue whale. However, body reserves in these species must also account for the protein, carbohydrate, mineral and vitamin components of milk. To a large extent the qualitative and quantitative regulation of maternal metabolism of these reserves is unknown. Nevertheless, the regulation of nutrient utilization involves homeostatic controls operating on a minute-by-minute basis to maintain constant conditions every day of the mother's life, and at the same time homeorhetic controls must function during lactation to orchestrate body processes in such a manner to partition the appropriate pattern and quantity of nutrients to support milk synthesis. Furthermore, the success of these regulatory processes is essential to ensure well-being of the lactating mother and survival of the nursing young.

## Mechanisms

The metabolic adaptations occurring with the onset of lactation are undoubtedly related to the plethora of hormonal changes occurring throughout this period. Some of the hormones undergoing major changes are listed in Table 18.2. Obviously, there must be synergisms and redundancy in the hormonal signalling systems, but our approaches to date to investigate these have been relatively simple. Nevertheless, somatotropin, prolactin and glucocorticoids have the clearest identified effects which are consistent with homeorhetic controls (Bauman and Elliot, 1983; Bell and Bauman, 1997; Vernon, 1998; Chilliard, 1999).

The overall mechanisms for shifting nutrient partitioning and metabolism involve alterations in the set-points for physiological responses to homeostatic controls. These alterations can be reflected by changes in the sensitivity or the magnitude of the biological response. The former is reflected by a change in the effective dose to obtain a 50% response ($ED_{50}$) for the homeostatic signal whereas the latter is reflected by a change in the maximum response ($R_{max}$) to the homeostatic signal (Kahn, 1978). Specific mechanisms for the alterations of the response to homeostatic signals can include alterations in tissue receptors and binding kinetics, changes in the intracellular signal transduction systems and effects on the expression and activity of key enzymes in the biochemical pathways. The net effect is that homeorhetic controls have tissue-specific effects on both the amounts and activity of critical metabolic enzymes and the signalling proteins that regulate them. Therefore, homeorhetic adaptations allow for chronic alterations or even redirection of physiological processes while still allowing homeostatic systems to preserve constant conditions. These same general mechanisms involving alterations in the set-points for responses to homeostatic controls have been demonstrated for many biological processes in the different physiological situations listed in Table 18.1.

Several of the tissues and processes in which alterations to homeostatic controls occur during lactation are listed in Table 18.3. Feed intake provides a general example, and a number of homeostatic controls for regulating feed intake have been identified (Forbes, 1996; Langhans, 1999). These homeostatic controls are obviously functioning in non-lactating and lactating animals, but in species such as the rat and cow the set points are altered during lactation. Thus, homeostatic controls of feed intake still occur but the altered set-points allow for a greater voluntary intake so that nutrient supply more adequately meets the nutrient requirement.

Insulin is an especially powerful mediator of many different physiological effects, most of which serve to acutely maintain metabolic equilibrium in the face of short-term variations in nutrient supply and demand. Thus, this acute regulatory signal is a pivotal target for chronic metabolic adaptations. As illustrated by examples in Table 18.3, many tissues have specific responses to insulin which are attenuated with the onset of lactation. This includes liver (gluconeogenesis inhibition), adipose tissue (fat synthesis), skeletal muscle (glucose uptake) and whole body (glucose oxidation) (Vernon and Sasaki, 1991; Williamson and Lund, 1994; Bell and Bauman, 1997; Vernon, 1998). These adaptations during early lactation are frequently referred to as reflecting an 'insulin resistance'. However, judicious use of this term is needed. An attenuated response is not observed for all acute regulatory functions of insulin. For example, the antilipolytic effect of insulin is greater in lactating sheep compared with non-lactating sheep (Vernon *et al.*, 1990), and the inhibition of whole-body rates of protein degradation is enhanced during early lactation (Tesseraud *et al.*, 1993). Thus, the changes in response to insulin are specific for certain tissues and certain biochemical processes within those tissues, rather than representing any generalized phenomenon. Overall, the physiological adaptations in the response of various processes to insulin have the net effect of enhancing hepatic production of glucose, and sparing glucose use by non-mammary tissues, consistent with the increased glucose requirement of the mammary gland.

An example of the attenuated response to insulin which occurs during lactation is presented in Fig. 18.1. In this example of glucose uptake by the hindlimb, the responsiveness to insulin is substantially reduced during lactation but the sensitivity to insulin is relatively unaltered. Similar studies on glucose and acetate uptake by adipose tissue indicates the response to insulin is virtually abolished in early lactation (Burnol *et al.*, 1986; Vernon, 1989; Vernon and Sasaki, 1991; Chilliard, 1999). A second example is

**Table 18.3.** A partial list of adaptations in metabolic regulation which occur during lactogenesis and early lactation in ruminants.

| Tissue/processes | Homeostatic control | Response to altered set-points |
|---|---|---|
| Feed intake | Multiple controls | ↑Appetite and satiety set-point |
| Adipose tissue | Insulin | ↓Lipogenesis<br>↓Uptake of preformed fatty acids |
| | Catecholamines | ↑Stimulation of lipolysis |
| | Adenosine | ↑Inhibition of lipolysis |
| Skeletal muscle | Insulin | ↓Glucose uptake |
| | Insulin(?) | ↓Protein synthesis<br>↓Amino acid uptake<br>↑Protein degradation |
| Liver | Insulin | ↑Gluconeogenesis |
| Pancreas | Insulinotropic agents | ↓Insulin release |
| Whole animal | Insulin | ↓Glucose oxidation<br>↓Glucose utilization by non-mammary tissues |

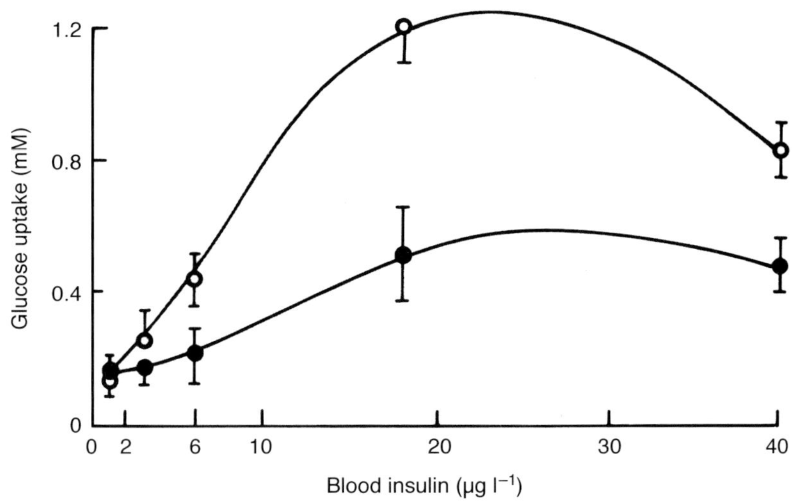


**Fig. 18.1.** Effect of insulin on the arterio-venous difference for glucose across the hindlimb of non-lactating (○) and lactating (●) sheep. Results represent mean ± standard error of the mean. From Vernon (1986).

the effect of β-adrenergics on adipose tissue rates of lipolysis (Bauman and Elliot, 1983; Vernon and Sasaki, 1991; Vernon, 1996; Chilliard, 1999). As illustrated by the elegant studies of Guesnet *et al.* (1987), the ability of isoproterenol, a specific β-receptor agonist, to stimulate lipolysis is markedly altered by stage of pregnancy and early lactation (Fig. 18.2).

Somatotropin (ST) is the homeorhetic control for which mechanisms have been most extensively investigated. Much of these data come from investigations of the response to exogenous ST in lactating or growing farm animals (Bauman and Vernon, 1993; Burton *et al.*, 1994; Etherton and Bauman, 1998). In the case of dairy cows, the studies convincingly demonstrate that exogenous bovine somatotropin (bST) results in an increase in milk yield of the treated animal and a series of coordinated adaptations in body tissues to support the greater use of nutrients for milk synthesis. Many of the coordinated adaptations are manifested by alterations in tissue responses to homeostatic signals. In fact, the changes which occur with bST treatment are similar to those illustrated in Figs 18.1 and 18.2 for the onset of lactation. While our understanding of the signal transduction systems regulating enzyme activity and gene expression is still incomplete, the specific effects have been reviewed recently for both the onset of lactation (Vernon, 1998; Chilliard, 1999) and treatment with exogenous ST (Bauman and Vernon, 1993; Etherton and Bauman, 1998).

Overall, the changes which occur with the onset of lactation or the initiation of bST treatment allow for a chronic alteration of nutrient utilization. This is illustrated by the above examples. When a meal is consumed and circulatory insulin increases, less nutrients are directed to body fat reserves and other non-mammary tissues because of their altered response to insulin, and more nutrients are taken up by the mammary gland consistent with the increased milk synthesis. Likewise, if nutrient supply is

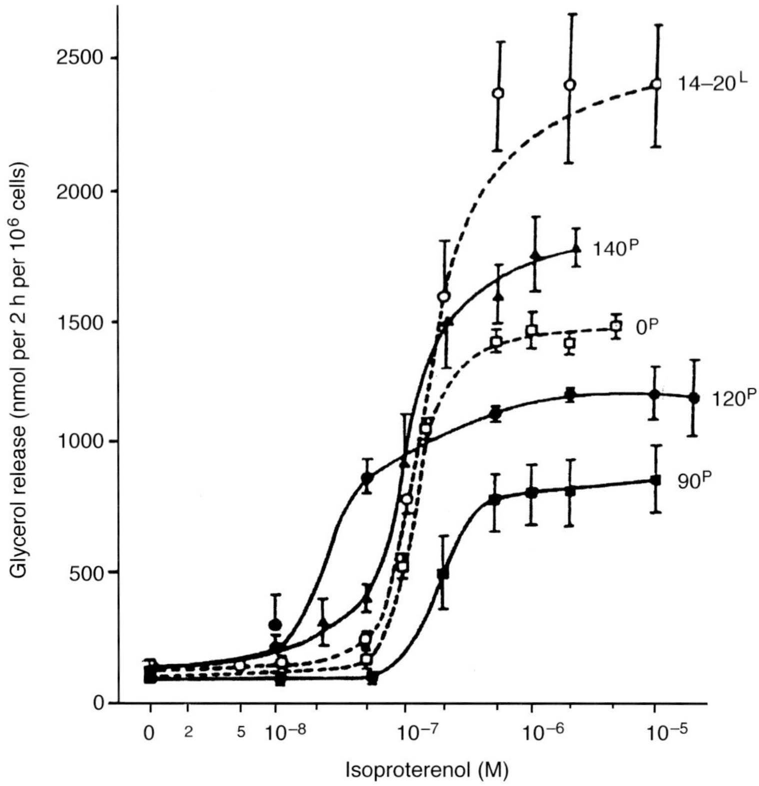


**Fig. 18.2.** Effect of β-adrenergic stimulation on rates of lipolysis in sheep adipocytes during pregnancy and lactation. Isoproterenol is a specific β-receptor agonist and results represent mean ± standard error of a mean for different days of pregnancy (P) and lactation (L). From Guesnet *et al.* (1987).

inadequate the coordinated responses require a greater mobilization of energy reserves to meet the requirements associated with the increased milk synthesis, and this is accommodated by a greater response to signals which stimulate lipolysis. Thus, adaptations in the response to homeostatic signals affect metabolic processes in an orchestrated manner to match the mammary gland need for nutrients for milk synthesis.

## Coordination of biological processes

Coordination represents a key feature of biological regulation. It was emphasized in the original ideas which led to the concept of homeostasis (Bernard, 1878; Cannon, 1929, 1932) and was central in the more recent crystallization of homeorhesis (Bauman and Currie, 1980) and analogous concepts (Chilliard, 1986; Mrosovsky, 1990; Kuenzel *et al.*, 1999). However, coordination of tissue processes has not been universally accepted, particularly in the early years. An alternative concept articulated by Hammond (1944) was that regulation of nutrient use represented a 'competition' between organs. Applying this to the onset of lactation, Hammond (1952) concluded that regulation increased the metabolic rate of the mammary gland which allowed it to 'compete more successfully with other tissues of the body for nutriments in the bloodstream'. The concept that regulation involved competition between organs also dominated many of the early experimental approaches and the interpretation of results. One example is the early studies with bST; based on the proposed mechanism of action it was anticipated that bST treatment would create a competition between organs which would cause metabolic problems and burnout (see review by Bauman, 1999). Instead, the research demonstrated that bST treatment of dairy cows resulted in a series of coordinated responses as discussed previously, and today it is used commercially in many countries (Bauman, 1999). Recent analysis of US dairy herds ($n > 200,000$ lactations) demonstrated use of bST improved lactational yield and persistency consistently over the 4-year period of commercial use and animal well-being was maintained as indicated by performance, milk quality, stayability and herd-life (Bauman *et al.*, 1999).

Tepperman and Tepperman (1970) were among the first to elaborate specific examples of coordination relative to metabolic processes. They referred to this as the 'Sherrington metaphor' based on Sir Charles Sherrington's (Sherrington, 1947) famous 'principle of reciprocal inhibition of antagonistic muscles' which stated:

> where two muscles would antagonize each other's action the reflex arc, instead of merely activating one of the two, when it activates the one causes depression of the activity of the other. The latter is an inhibitory effect.

Examples of the Sherrington metaphor provided by Tepperman and Tepperman (1970) included the coordination between metabolic pathways within a cell (e.g. lipogenesis and gluconeogenesis in the rat hepatocyte) and between organs in different nutritional states (e.g. nutrient flow during fasting and re-feeding).

During lactation, the exquisite nature of the coordination of biological processes is illustrated by many examples of the Sherrington metaphor. Coordination occurs among biochemical pathways within a cell as indicated by the reduction in the pathways of fatty acid synthesis and the increased importance of the pathways of lipolysis which occurs in adipose tissue with the onset of lactation (Table 18.2). The orchestrated changes are also tissue specific. For example, lipoprotein lipase activity and the uptake of preformed fatty acids as well as *de novo* lipogenesis and activities of key enzymes in fatty acid synthesis are reduced in adipose tissue with the onset of lactation, while these exact same enzymes and biochemical pathways are dramatically increased in the mammary gland during the same time interval. Furthermore, the coordinated responses frequently involve many different organs and physiological processes as illustrated by the aforementioned example of glucose. The substantial mammary demand for glucose which occurs with the onset of lactation is provided for by a series of orchestrated adaptations which include increased hepatic synthesis of glucose, reduced glucose uptake by several non-mammary tissues and an overall reduction in whole-body oxidation of glucose (Table 18.2). Of special importance, if the coordinated regulation of these processes is inadequate so that glucose supply fails to meet the overall needs during lactation, then a chain of metabolic events occurs leading to a metabolic disease, ketosis, and a compromise in animal well-being.

The fact that biological regulation involves a series of orchestrated responses is frequently overlooked in articles relating to animal welfare. At regular intervals over the last 50 years some have expressed concern that practices to improve the productive efficiency of dairy cows may be pushing them too far, thereby compromising animal health and shortening the lifespan. Hammond (1952) and Bauman *et al.* (1985) reviewed the physiological limits to production, and they did not share that concern. Indeed, milk yield and productive efficiency have continued to increase throughout this century and today we have herds with annual milk yields over 15,000 kg per cow and individual cows that have produced over 27,000 kg year$^{-1}$ (Bauman, 1999). Nevertheless, some continue to assume that genetic selection and improved management practices which lead to more efficient, high-producing animals are at variance with the physiological controls for animal well-being (Rauw *et al.*, 1998; Broom, 1999). For example, a recent review on welfare of dairy cattle suggested that it 'may well be necessary to stop using genetic selection and some feeding methods to increase milk yield' because these practices have resulted in stressed cows in which 'their normal biological functioning controls are overtaxed' (Broom, 1999). This viewpoint clearly fails to appreciate that genetic selection and management improvements are successful because they have altered the biological controls in a coordinated manner. Rather than the biological controls being at discord with increased performance, it is the improvements in the biological control systems which are responsible for the increases in milk yield and the gains in productive efficiency.

The role of coordinated responses in metabolic regulation is also important in the development of dynamic models of metabolism. Recent studies on the regulation of milk protein synthesis provide an example of this coordination. Most investigations of milk protein production in dairy cows have examined adequacy of amino acid (AA) supply. In general, results indicate that supplemental amino acids increase milk protein content and yield when the AA supply from microbial and bypass protein is inadequate, but little or no response occurs in well-fed cows (Sutton, 1989; Rulquin *et al.*, 1995). We were interested in the endocrine regulation of milk protein synthesis, and focused on the chronic effects of insulin because of data summaries showing a high correlation between dietary energy intake and milk protein content (Sporndly, 1989). By using the hyperinsulinaemic-euglycaemic clamp technique we were able to examine the role of insulin without the confounding effects of hypoglycaemia. This technique involves intravenous infusion of insulin to achieve a constant elevated concentration and simultaneous infusion of sufficient glucose to maintain normal blood concentrations. Results indicated that a 4-day insulin clamp resulted in a dramatic increase in milk protein content and yield; in the well-fed cow milk protein yield increased by over 25% (Griinari *et al.*, 1997a; Mackle *et al.*, 1999). Plasma urea nitrogen concentrations were markedly reduced during the insulin clamp, demonstrating that one component of the coordinated response was a reduction in whole-body oxidation of AA (McGuire *et al.*, 1995; Griinari *et al.*, 1997a; Mackle *et al.*, 1999). The means by which the mammary gland obtained sufficient AA to support the increased output of milk protein was of special interest because plasma concentrations of essential AA were reduced by 30–50% during the insulin clamp. We found that during the insulin clamp both blood flow (ml ml$^{-1}$ of milk) and extraction of essential AA (per cent) were increased (Mackle *et al.*, 2000). Thus, the mammary gland component of the coordinated responses included alterations in the local control of blood flow and transport mechanisms for essential AA as well as an increased ability to synthesize milk proteins. Overall, the orchestrated series of changes which occurred during the chronic insulin clamp involved adaptations in whole-body processes to allow a greater proportion of AA to be used for milk synthesis and adaptations by the mammary gland to increase AA uptake and rates of milk protein synthesis.

The coordination of metabolic regulation also needs to be considered in evaluating research results. An example is research relating to the low-fat milk syndrome in dairy cows (Davis and Brown, 1970). This syndrome is characterized by a marked reduction in both yield and percentage of milk fat, and diets which cause milk fat depression (MFD) typically result in a more positive energy balance, an increase in circulating insulin, and an increase in body fat accretion. Variations in circulating insulin have no acute effects on mammary lipid metabolism of dairy cows, but as discussed earlier they can affect adipose tissue rates of lipogenesis and lipolysis. One theory for the cause of MFD is the glucogenic-insulin theory which postulates that when diets cause an increase in circulating insulin, the mammary gland is deprived of milk fat precursors due to vigorous competition by adipose tissue (see discussion by Griinari *et al.*, 1997b). We have evaluated this theory using a 4-day hyperinsulinaemic-euglycaemic clamp described previously. Despite the substantial challenge to the mammary gland supply of lipogenic precursors imposed by the fourfold increase in circulating insulin, body metabolism was coordinated so that the rate of milk fat synthesis was relatively constant during the insulin clamp (McGuire *et al.*, 1995; Griinari *et al.*, 1997b; Mackle *et al.*, 1999). Thus, our work demonstrated the coordinated regulation of nutrient partitioning, but it provided no support for the glucogenic-insulin theory.

What then is the cause of dietary-induced MFD? Another theory proposes that MFD is caused by a direct inhibition of mammary fatty acid synthesis by products from incomplete or unusual biohydrogenation of polyunsaturated fatty acids in the rumen (Davis and Brown, 1970). Trans fatty acids have received special attention as the cause (Davis and Brown, 1970; Erdman, 1996), and more recently we have expanded this to include trans fatty acids and related metabolites (Griinari *et al.*, 1998). Consistent with the trans fatty acid theory, MFD is observed when partially hydrogenated vegetable oils are abomasally infused and there is a close relationship between the decrease in milk fat percentage and the increase in milk fat content of *trans*-$C_{18:1}$ over a wide range of diets (Erdman, 1996; Griinari *et al.*, 1998). The typical ruminal biohydrogenation of linoleic acid to stearic acid produces *cis*-9, *trans*-11 conjugated linoleic acid (CLA) and *trans*-11 $C_{18:1}$ as intermediates. We found that diets which cause MFD resulted in a shift in ruminal biohydrogenation with the reduction in milk fat percentage corresponding to increases in the milk fat content of *trans*-10 $C_{18:1}$ and *trans*-10, *cis*-12 CLA (Griinari *et al.*, 1998, 1999). To directly examine the role of the CLA on milk fat synthesis in dairy cows we abomasally infused CLA isomers. The *cis*-9, *trans*-11 CLA isomer had no impact on milk fat whereas a 4-day abomasal infusion of less than 10 g $day^{-1}$ of *trans*-10, *cis*-12 CLA resulted in over a 40% reduction in milk fat content and yield (Fig. 18.3; Baumgard *et al.*, 2000). Furthermore, the shifts in milk fatty acid composition found with *trans*-10, *cis*-12 CLA paralleled those observed with classical MFD. Thus, our results are consistent with diet-induced MFD involving a direct inhibition of milk fat synthesis by intermediates formed in rumen biohydrogenation of polyunsaturated fatty acids. In this scenario the increase in body fat accretion which is particularly evident with MFD induced by high grain–low fibre

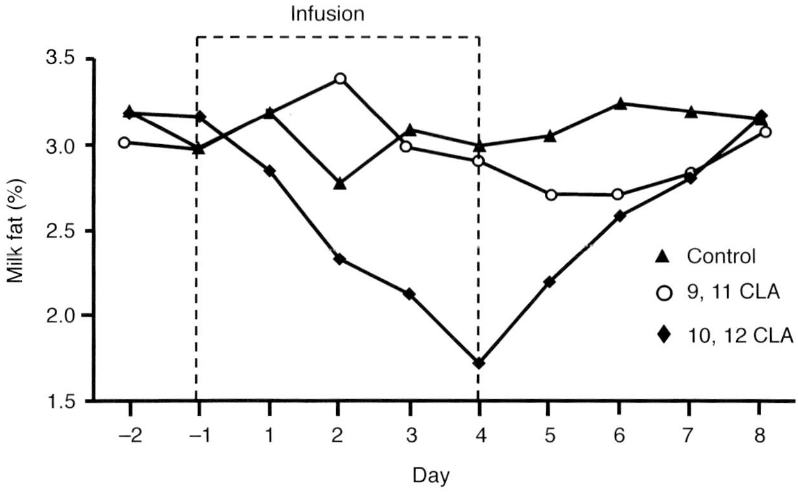


**Fig. 18.3.** Temporal pattern of milk fat content during abomasal infusion of conjugated linoleic acid (CLA) isomers. Infusions represented 10 g day$^{-1}$ of *cis*-9, *trans*-11 CLA or *trans*-10, *cis*-12 CLA. From Baumgard *et al.* (2000).

diets would represent a shift in nutrient partitioning which occurs as a consequence rather than a cause of the reduced mammary fat synthesis. Dietary addition of CLA has also been shown to inhibit body fat accretion in several species, but the required dose appears to be 10–20-fold greater than the dietary level which achieves MFD (Baumgard *et al.*, 2000).

## Conclusions

As our knowledge of biology increases we appreciate even more the remarkable system for the regulation of metabolic processes which occurs in different physiological states. This review has re-examined the concepts of homeostasis and homeorhesis in the regulation of metabolism. Lactation provides an impressive example of homeostasis and homeorhesis in action, and examples were used to provide an overview of the integrated mechanisms. Overall, the homeostatic and homeorhetic mechanisms provide a coordinated regulation of the metabolism of different organs and tissues to ensure the proper nutrient supply to the mammary gland.

In crystallizing the concept of homeorhesis and the interrelationships between homeostasis and homeorhesis, we originally concluded with a quotation (Bauman and Currie, 1980). This quotation from Duclaux, an eminent French scientist, also provides an appropriate perspective for this review. Duclaux (1920) reviewed the scientific contributions of Louis Pasteur and noted that many of his ideas were incorrect. However, he pointed out ' … we see clearly how much a matter of indifference it is whether a theory or a doctrine is right, provided, it incites to work, and results in the discovery of new facts'.

## References


Bauman, D.E. (1984) Regulation of nutrient partitioning. In: Gilchrist, F.M.C. and Mackie, R.I. (eds) *Herbivore Nutrition in the Subtropics and Tropics.* The Science Press, Craighall, South Africa, pp. 505–524.

Bauman, D.E. (1999) Bovine somatotropin: from basic science to commercial application. *Domestic Animal Endocrinology* 17, 101–116.

Bauman, D.E. and Currie, W.B. (1980) Partitioning of nutrients during pregnancy and lactation: a review of mechanisms involving homeostasis and homeorhesis. *Journal of Dairy Science* 63, 1514–1529.

Bauman, D.E. and Elliot, J.M. (1983) Control of nutrient partitioning in lactating ruminants. In: Mepham, T.B. (ed.) *Biochemistry of Lactation.* Elsevier Science Publishers B.V., Amsterdam, The Netherlands, pp. 437–468.

Bauman, D.E. and Vernon, R.G. (1993) Effects of exogenous bovine somatotropin on lactation. *Annual Review of Nutrition* 13, 437–461.

Bauman, D.E., Eisemann, J.H. and Currie, W.B. (1982) Hormonal effects on partitioning of nutrients for tissue growth: role of growth hormone and prolactin. *Federation Proceedings* 41, 2538–2544.

Bauman, D.E., McCutcheon, S.N., Steinhour, W.D., Eppard, P.J. and Sechen, S.J. (1985) Sources of variation and prospects for improvement of productive efficiency in the dairy cow: a review. *Journal of Animal Science* 60, 583–592.

Bauman, D.E., Everett, R.W., Weiland, W.H. and Collier, R.J. (1999) Production responses to bovine somatotropin in northeast dairy herds. *Journal of Dairy Science* 82, 2564–2573.

Baumgard, L.H., Corl, B.A., Dwyer, D.A., Sæbø, A. and Bauman, D.E. (2000) Identification of the conjugated linoleic acid isomer which inhibits milk fat synthesis. *American Journal of Physiology* 278, R179–R184.

Bell, A.W. and Bauman, D.E. (1997) Adaptations of glucose metabolism during pregnancy and lactation. *Journal of Mammary Gland Biology and Neoplasia* 2, 265–278.

Bernard, C. (1878) *Leçons sur les phénomènes de la vie communs aux animaux et aux végétaux.* Librairie J-B. Baillière et fils, Paris, France.

Blaxter, K. (1989) *Energy Metabolism in Animals and Man.* Cambridge University Press, Cambridge, UK.

Broom, D.M. (1999) The welfare of dairy cattle. In: Aagaard, K. (ed.) *Future Milk Farming. FIL-IDF 25th International Dairy Congress.* Danish Natl. Com. IDF, Aarhus, Denmark, pp. 32–39.

Brown, R.E. (1969) The conversion of nutrients into milk. In: Swan, H. and Lewis, D. (eds) *The University of Nottingham Third Nutrition Conference for Feed Manufacturers.* J&A Churchill Ltd., London, UK, pp. 23–29.

Burnol, A.-F., Guerre-Millo, M., Lavau, M. and Girard, J. (1986) Effect of lactation on insulin sensitivity of glucose metabolism in rat adipocytes. *Federation of European Biochemical Societies* 194, 292–296.

Burton, J.L., McBride, B.W., Block, E., Glimm, D.R. and Kennelly, J.J. (1994) A review of bovine growth hormone. *Canadian Journal of Animal Science* 74, 167–201.

Cannon, W.B. (1929) Organization for physiological homeostasis. *Physiological Reviews* IX, 399–431.

Cannon, W.B. (1932) *The Wisdom of the Body.* W.W. Norton & Company, Inc., New York.

Chilliard, Y. (1986) Revue bibliographique: Variations quantitatives et métabolisme des lipides dans les tissus adipeux et le foie au cours du cycle gestation-lactation. 1re partie: chez la ratte. *Reproduction, Nutrition, Développement* 26, 1057–1103.

Chilliard, Y. (1987) Revue bibliographique: Variations quantitatives et métabolisme des lipides dans les tissus adipeux et le foie au cours du cycle gestation-lactation. 2e partie: chez la brebis et la vache. *Reproduction, Nutrition, Développement* 27, 327–398.

Chilliard, Y. (1999) Metabolic adaptations and nutrient partitioning in the lactating animal. In: Martinet, J., Houdebine, L.M. and Head, H.H. (eds) *Biology of Lactation.* INRA Editions, Paris, France, pp. 503–552.

Costa, D.P., Le Boeuf, B.J., Huntley, A.C. and Ortiz, C.L. (1986) The energetics of lactation in the Northern elephant seal, *Mirounga angustirostris. Journal of The Zoological Society of London* 209, 21–33.

Cripps, A.W. and Williams, V.J. (1975) The effect of pregnancy and lactation on food intake, gastrointestinal anatomy and the absorptive capacity of the small intestine in the albino rat. *British Journal of Nutrition* 33, 17–32.

Davis, C.L. and Brown, R.E. (1970) Low-fat Milk Syndrome. In: Phillipson, A.T. (ed.) *Physiology of Digestion and Metabolism in the Ruminant.* Oriel Press Limited, Newcastle upon Tyne, UK, pp. 545–565.

Dilman, V.M. (1982) Endocrinology and diseases of aging. In: Essman, W.B. (ed.) *Regulatory Processes in Clinical Endocrinology.* Spectrum Publications, Inc., New York, pp. 123–161.

Duclaux, E. (1920) *Pasteur: The History of a Mind.* (Translated by Smith, E.F. and Hedtes, F.) W.B. Saunders Co., Philadelphia.

Erdman, R. (1996) Milk fat depression: some new insights. Tri-State Dairy Nutrition Conference, Fort Wayne, Indiana, pp. 1–16.

Etherton, T.D. and Bauman, D.E. (1998) Biology of somatotropin in growth and lactation of domestic animals. *Physiology Reviews* 78, 745–761.

Forbes, J.M. (1996) Integration of regulatory signals controlling forage intake in ruminants. *Journal of Animal Science* 74, 3029–3035.

Griinari, J.M., McGuire, M.A., Dwyer, D.A., Bauman, D.E., Barbano, D.M. and House, W.A. (1997a) The role of insulin in the regulation of milk protein synthesis in dairy cows. *Journal of Dairy Science* 80, 2361–2371.

Griinari, J.M., McGuire, M.A., Dwyer, D.A., Bauman, D.E. and Palmquist, D.L. (1997b) Role of insulin in the regulation of milk fat synthesis in dairy cows. *Journal of Dairy Science* 80, 1076–1084.

Griinari, J.M., Dwyer, D.A., McGuire, M.A., Bauman, D.E., Palmquist, D.L. and Nurmela, K.V.V. (1998) *Trans*-octadecenoic acids and milk fat depression in lactating dairy cows. *Journal of Dairy Science* 81, 1251–1261.

Griinari, J.M., Nurmela, K., Dwyer, D.A., Barbano, D.M. and Bauman, D.E. (1999) Variation of milk fat concentration of conjugated linoleic acid and milk fat percentage is associated with a change in ruminal biohydrogenation. *Journal of Animal Science* 77(Suppl. 1), 117–118 (abstract).

Guesnet, P., Massoud, M. and Demarne, Y. (1987) Effects of pregnancy and lactation on lipolysis of ewe adipocytes induced by β-adrenergic stimulation. *Molecular and Cellular Endocrinology* 50, 177–181.

Hammond, J. (1944) Physiological factors affecting birth weight. *Proceedings of the Nutrition Society* 2, 8–12.

Hammond, J. (1952) Physiological limits to intensive production in animals. *British Agricultural Bulletin* 4, 222–224.

Kahn, C.R. (1978) Insulin resistance, insulin insensitivity, and insulin unresponsiveness: A necessary distinction. *Metabolism* 27, 1893–1902.

Kennedy, G.C. (1967) Ontogeny of mechanisms controlling food and water intake. In: Code, C.F. (ed.) *Handbook of Physiology. Section 6: Alimentary Canal . Control of Food and Water Intake*, Vol. I. American Physiological Society, Washington DC, pp. 337–351.

Kuenzel, W.J., Beck, M.M. and Teruyama, R. (1999) Neural sites and pathways regulating food intake in birds: A comparative analysis to mammalian systems. *Journal of Experimental Zoology* 283, 348–364.

Langhans, W. (1999) Appetite regulation. In: Lobley, G.E., White, A. and MacRae, J.C. (eds)

*Protein Metabolism and Nutrition.* EEAP Publication No. 96, Wageningen Pers, Wageningen The Netherlands, pp. 225–251.

Lerner, I.M. (1954) *Genetic Homeostais.* Oliver and Boyd, London, UK.

Lewontin, R.C. (1956) Studies on homeostasis and heterozygosity I. General considerations. Abdominal bristle number in second chromosome homozygotes of *Drosophila melanogaster. The American Naturalist* 90, 237–255.

Linzell, J.L. (1967) The magnitude and mechanisms of the uptake of milk precursors by the mammary gland. *Proceedings of the Nutrition Society* 27, 44–52.

Lockyer, C. (1981) Growth and energy budgets of large baleen whales from the southern hemisphere. In: *Mammals in the Sea.* Fisheries Series. No. 5, Vol. 3, Food and Agriculture Organization. United Nations, Rome, Italy, pp. 379–487.

McGuire, M.A., Griinari, J.M., Dwyer, D.A. and Bauman, D.E. (1995) Role of insulin in the regulation of mammary synthesis of fat and protein. *Journal of Dairy Science* 78, 816–824.

Mackle, T.R., Dwyer, D.A., Ingvartsen, K.L., Chouinard, P.Y., Lynch, J.M., Barbano, D.M. and Bauman, D.E. (1999) Effects of insulin and amino acids on milk protein concentration and yield from dairy cows. *Journal of Dairy Science* 82, 1512–1524.

Mackle, T.R., Dwyer, D.A., Ingvartsen, K.L., Chouinard, P.Y., Ross, D.A. and Bauman, D.E. (2000) Effects of insulin and post ruminal supply of protein on use of amino acids by the mammary gland for milk protein synthesis. *Journal of Dairy Science* 83, 93–105.

McNamara, J.P. (1995) Role and regulation of metabolism in adipose tissue during lactation. *Nutritional Biochemistry* 6, 120–129.

Millar, J.S. (1979) Energetics of lacation in *Peromyscus maniculatus. Canadian Journal of Zoology* 57, 1015–1019.

Mrosovsky, N. (1990) *Rheostasis: The Physiology of Change.* Oxford University Press, Inc., New York.

Nicolaïdis, S. (1983) Physiologie du comportement alimentaire. In: Meyer, P. (ed.) *Physiologie Humaine.* Flammarion Médecine-Sciences, Paris, France, pp. 1059–1076.

Oftedal, O.T. (1993) The adaptation of milk secretion to the constraints of fasting in bears, seals, and baleen whales. *Journal of Dairy Science* 76, 3234–3246.

Rauw, W.M., Kanis, E., Noordhuizen-Stassen, E.N. and Groomers, F.J. (1998) Undesirable side effects of selection for high production efficiency in farm animals: a review. *Livestock Production Science* 56, 15–33.

Riedman, M. (1990) *The Pinnipeds: Seals, Sea Lions and Walruses.* University of California Press, Berkeley, California.

Rulquin, H., Verite, R., Guinard, G. and Pisulewski, P.M. (1995) Dairy cows' requirement for amino acids. In: Ivan, M. (ed.) *Animal Science Research and Development.* Canadian Society of Animal Science, Ottawa, Ontario, Canada, pp. 143–160.

Sherrington, C. (1947) *The Integrative Action of the Nervous System,* 2nd edn. Yale University Press, New Haven, Connecticut.

Sporndly, E. (1989) Effects of diet on milk composition and yield of dairy cows with special emphasis on milk protein content. *Swedish Journal of Agricultural Research* 19, 99–106.

Sutton, J.D. (1989) Altering milk composition by feeding. *Journal of Dairy Science* 72, 2801–2814.

Tepperman, J. and Tepperman, H.M. (1970) Gluconeogenesis, lipogenesis and the Sherringtonian metaphor. *Federation Proceedings* 29, 1284–1293.

Tesseraud, S., Grizard, J., Debras, E., Papet, I., Bonnet, Y., Bayle, G. and Champredon, C. (1993) Leucine metabolism in lactating and dry goats: effect of insulin and substrate availability. *American Journal of Physiology* 265, E402–E413.

Vernon, R.G. (1986) The response of tissues to hormones and the partition of nutrients during lactation. In: *Research Reviews, Hannah Yearbook.* Hannah Research Institute, Ayr, UK, pp. 115–121.

Vernon, R.G. (1989) Endocrine control of metabolic adaptation during lactation. *Proceeedings of the Nutrition Society* 48, 23–32.

Vernon, R.G. (1996) Control of lipid mobilisation in ruminant adipose tissue. *Journal of Reproduction and Development* 42, 118–122.

Vernon, R.G. (1998) Homeorhesis. In: *Research Reviews, Hannah Yearbook.* Hannah Research Institute, Ayr, UK, pp. 64–73.

Vernon, R.G. and Sasaki, S. (1991) Control of responsiveness of tissues to hormones. In: Tsuda, T., Sasaki, Y. and Kawashima, R. (eds) *Physiological Aspects of Digestion and Metabolism in Ruminants.* Proceedings of the 7th International Symposium on Ruminant Physiology. Academic Press, San Diego, California, pp. 155–182.

Vernon, R.G., Faulkner, A., Hay, W.W., Jr, Calvert, D.T. and Flint, D.J. (1990) Insulin resistance of hind-limb tissues *in vivo* in lactating sheep. *Biochemistry Journal* 270, 783–786.

Waddington, C.H. (1942) Canalization of development and the inheritance of acquired characters. *Nature* 150, 563–565.

Waddington, C.H. (1953) The 'Baldwin effect,' 'genetic assimilation' and 'homeostasis.' *Evolution* 7, 386–387.

Waddington, C.H. (1957) *The Strategy of the Genes.* The MacMillan Company, New York.

Wade, G.N. and Schneider, J.E. (1992) Metabolic fuels and reproduction in female mammals. *Neuroscience and Biobehavioral Reviews* 16, 235–272.

Welch, G.R. (1991) Thermodynamics and living systems: Problems and paradigms. *Journal of Nutrition* 121, 1902–1906.

Williamson, D.H. (1980) Integration of metabolism in tissues of the lactating rat. *FEBS Letters* 117, K93–K105.

Williamson, D.H. and Lund, P. (1994) Cellular mechanisms for the regulation of adipose tissue lipid metabolism in pregnancy and lactation. In: Allen, L., King, J. and Lönnerdal, B. (eds) *Nutrient Regulation During Pregnancy, Lactation and Infant Growth.* Plenum Press, New York, pp. 45–70.

Williamson, D.H., Lund, P. and Evans, R.D. (1995) Substrate selection and oxygen uptake by the lactating mammary gland. *Proceedings of the Nutrition Society* 54, 165–175.

# 19 The Insulin-like Growth Factor (IGF) System in the Mammary Gland: Role of IGFBP-3 Binding Protein

C.R. Baumrucker

*Department of Dairy and Animal Science, Penn State University, Pennsylvania, USA*

## Introduction to the insulin-like growth factor (IGF) system

The IGF system (Fig. 19.1) consists of the three ligands, insulin, IGF-I and IGF-II, three corresponding high affinity receptors, and six IGF binding proteins (IGFBP) that associate with the IGFs. The IGF ligands are known to be ubiquitous and act as mediators of growth, development and differentiation (Daughaday and Rotwein, 1989; LeRoith *et al.* 1995). The three peptides have around 50% amino acid homology (LeRoith, 1991). Insulin, synthesized in pancreatic β-cells as proinsulin, is degraded to insulin and the C-peptide, both of which are found in the circulation. IGFs, which are synthesized in many tissues, retain the C-peptide and have an extended carboxy terminus sometimes called the D-peptide (Daughaday and Rotwein, 1989). While insulin circulates in the blood freely and has a short half-life, IGFs circulate at higher concentrations, but are largely bound to one of six IGFBPs. The potential for the IGFs to bind to any one of the six known high affinity IGFBPs imparts potential for signal inhibition, signal potentiation, extended ligand half-life and specific tissue targeting (Zapf, 1995).

### *IGF receptors*

The biological effects of the IGFs are mediated through specific cell surface protein receptors. These include the IGF-I receptor (IGF-IR), the insulin receptor, and the IGF-II (IGF-IIR)/mannose-6-phosphate receptor (m-6-pR). The IGF-IR is structurally similar to the insulin receptor. Both couple tyrosine kinase activity to a series of intracellular signalling pathways (LeRoith *et al.*, 1993). The similarity between these ligands and receptors explains much of the ambiguity of research findings when high levels of either ligand are applied to an experimental system. Under such conditions, IGF-I and insulin are known to cross-react with the other species' receptor and trigger intracellular signal cascades associated with both receptors. In addition, both IGF ligands

 *Ruminant Physiology: Digestion, Metabolism, Growth and Reproduction* (ed. P.B. Cronjé)

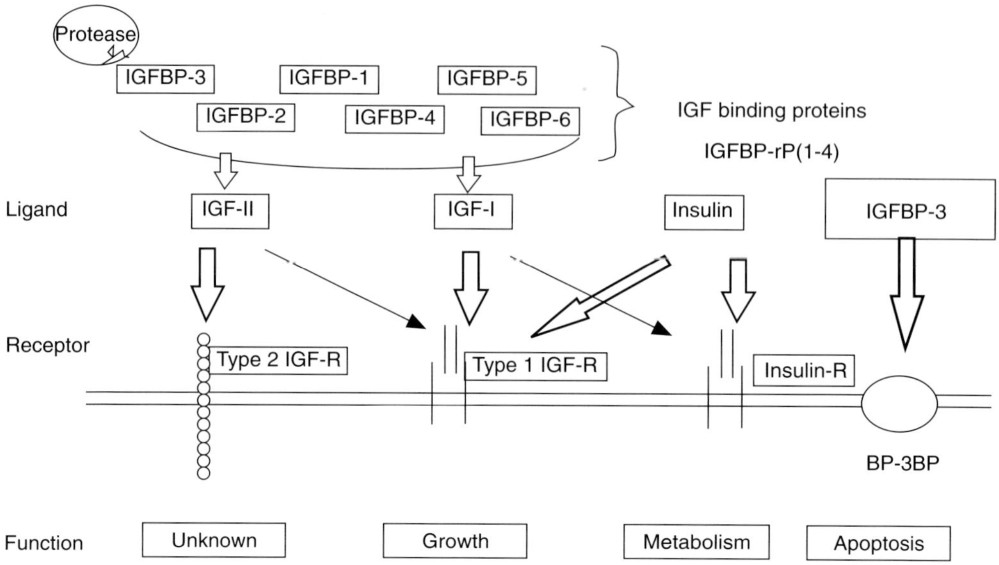


**Fig. 19.1.** Figure showing the components of the insulin-like growth factor (IGF) system. Arrows indicate the binding and interactions that occur between the components.

bind with high affinity with the IGF-IR and IGF-I has a reduced affinity with the IGF-IIR.

The IGF-IIR is not linked to tyrosine kinase activity, but rather appears to be coupled to a G protein (Nishimoto, 1993). The receptor expression is developmentally regulated, with high expression in fetal and neonatal tissues (Nissley *et al.*, 1993). The IGF-IIR is also the cation-independent m-6-pR, although the ligand binding sites are separate. The latter is known to function as a lysosomal enzyme targeting protein, the activation of transforming growth factor-β (TGF-β) (Dennis and Rifkin, 1991), and the degradation of IGF-II (Oka *et al.*, 1985). Most recently, it has been reported that the m-6-p/IGF-IIR is also a receptor for retinoic acid (Kang *et al.*, 1997).

## *IGFBPs*

The majority of IGF ligands are bound to IGFBPs *in vivo*. Unlike the transmembrane IGF receptors, the IGFBPs are secreted. They are present in blood serum, all biological fluids, and conditioned media from all *in vitro* cell cultures. While IGF-I and IGF-II bind to all IGFBPs with high affinity ($\sim 10^{-10}$ M), insulin does not bind to IGFBPs. Six distinct human IGFBPs have been identified and have been termed IGFBP-1 through to IGFBP-6 (Anonymous, 1992). Additionally, four IGFBP-related proteins that exhibit reduced binding affinity (four to tenfold less) for IGFs have been recently identified (Baxter *et al.*, 1998). All of these proteins exhibit the presence of the specific IGFBP motif (GCGCCXXC) (Rosenfeld, 1998). Table 19.1 shows the proposed nomenclature for the superfamily of IGFBPs.

**Table 19.1.** Insulin-like growth factor binding protein (IGFBP) superfamily.

| Name | IGF affinity | Other/previous names |
|---|---|---|
| IGFBP-1 | High | PP12, BP-28 |
| IGFBP-2 | High | |
| IGFBP-3 | High | BP-53 |
| IGFBP-4 | High | |
| IGFBP-5 | High | |
| IGFBP-6 | High | |
| *IGFBP-related proteins (CNN family)* | | |
| IGFBP-rP1 | Low | Mac25, TAF, PSF, IGFBP-7 |
| IGFBP-rP2 | Low | CTGF, fisp-12, IGFBP-8 |
| IGFBP-rP3 | Low | Nov, IGFBP-9 |
| IGFBP-rP4 | Low | Cry-61, cyr10, IGFBP-10 |

The role of IGFBPs in the regulation of the IGF system has been the source of many reviews (Jones and Clemmons, 1995; Kelley *et al.*, 1996; Murphy, 1998) and concepts relating to cancer have emerged (Werner and LeRoith, 1996). Generally, IGFBP bio-action is thought to be: (i) modulation action of IGF-I or IGF-II by competition with the IGF-IR or IGF-IIR; (ii) alteration of the action of IGF-I or IGF-II; (iii) transport and/or extended half-life of IGFs; (iv) interactions with other growth factor systems or (v) IGF-independent action via (a) nuclear localization sequence, (b) IGFBP binding proteins, or (c) yet unknown mechanisms (Murphy, 1998). The most important insight into these suggested bio actions is that they mainly emerge from *in vitro* cell culture experiments.

The IGFBP related proteins (IGFBPrP1–4) are a family of proteins (CCN family; connective tissue growth factor; cef10/cry61 and nov) (Bork, 1998) that have been recently identified as low-affinity IGFBPs. These proteins are expressed in cells within minutes (called immediate-early genes) of cellular stimulation by growth factors or transforming oncogenes (Williams *et al.*, 1992). Because of the high level of sequence similarity among the members of the CCN family, they probably have common molecular functions. Because of their low IGFBP affinity and rather new entry into the IGF arena, these proteins will not be considered in the remainder of this chapter.

The differential nature of responses to IGF ligands in various experiments in different tissues has amplified the interest of the IGFBPs in these interactions. Because IGFBP bind to IGFs with high affinity that effectively competes with the IGF-IR (Clemmons, 1997), they may be viewed as agents that prevent IGFs action. Apparently to counteract this blocking action, proteases have been shown to attack IGFBPs and release free IGF resulting from the altered IGFBP affinity (Fowlkes, 1997). For example, IGFBP-5 that is released by cells in culture is rapidly cleaved into 23 and 16 kDa fragments by a cation-dependent serine protease (Imai *et al.*, 1997). It is not known if this protease is present in mammary cell cultures. Other means of altering IGFBP action with IGFs is by protein phosphorylation (IGFBP-1) (Jones *et al.*, 1991) and cell surface binding that lowers IGFBP-1 and IGFBP-3 affinity (Mccusker *et al.*, 1990).

While IGFBP-1 binds to integrin via an arginine–glycine–aspartate (RGD) sequence, the binding of IGFBP-3 to the cell surface has only recently been shown to be due to specific binding to a number of low molecular weight proteins (Oh and Rosenfeld, 1998).

Although the ability of IGFBPs to bind IGFs is thought to be an inhibitory action, IGF action has been potentiated with low concentrations of IGFBP-3 and is thought to occur by concentrating the IGFs on the cell surface in the region of the IGF receptors (Rechler, 1993). Because the affinity of IGFBP-3 is reduced tenfold when bound to the cell surface, the current concept for IGF potentiation suggests that endogenous IGF-I interaction with the bound IGFBP-3 favours a slow release and possibly avoiding down-regulation or the type-1 receptors.

### *IGFBP-3*

Because IGFBP-3 is the major IGFBP and a reliable recombinant source became available some time ago, innumerable studies have focused upon its biological action. Although the binding and inhibition of IGF action can explain the inhibition of cellular growth by the application of exogenous IGFBP-3, several lines of evidence have emerged to suggest that other mechanisms may also be involved in the actions of the IGF system. First, IGFBP-3 has been shown to negatively regulate cell proliferation through an IGF receptor-independent pathway. Stable transfection of BALB/c 3T3 mouse fibroblasts with human IGFBP-3 cDNA decreased the rate of cell proliferation (Lamson *et al.*, 1993). Studies using a fibroblast cell line developed from mice with targeted disruption of the IGF-I receptor established that the growth inhibitory effects of IGFBP-3 do not involve IGF binding to the IGF-IR (Valentinis *et al.*, 1996). IGF receptor-independent actions of IGFBP-3 have also been observed in breast cancer cells (Oh *et al.*, 1993b) and are believed to be responsible for the predisposition of breast cancer cells to apoptosis (Gill *et al.*, 1997). Secondly, both IGF-I and IGFBP-3 have been shown to be present in the cell nucleus (Radulescu, 1994; Li *et al.*, 1998; Peralta Soler *et al.*, 1990; Radulescu and Wendtner, 1993) including that of breast cancer cells (Schedlich *et al.*, 1998) and primary cultures of mammary cells (Baumrucker *et al.*, 1999). In one of these reports (Li *et al.*, 1998), IGF-I and IGFBP-3 have been observed to be co-localized in the nucleus of opossum kidney cells. Based on the presence of a putative nuclear localization sequence in the structure of IGFBP-3, it was proposed that IGFBP-3 might act as a carrier for IGF-I (Li *et al.*, 1998).

A recent report indicates that IGFBP-3 competes for binding with the TGF-β receptor (Leal *et al.*, 1997) and the retinoic acid receptor (Kang *et al.*, 1997). Utilizing a yeast two-hybrid system, Murphy (1998) suggests that a cDNA encoding latent TGF-β-binding protein 1 interacts with IGFBP-3. In support of the IGFBP-3 role associated with negative growth regulation action, the expression of IGFBP-3 has also been shown to be regulated by other growth-inhibitory (and apoptosis-inducing) agents such as TGF-β (Gucev *et al.*, 1996; Huynh *et al.*, 1996; Rajah *et al.*, 1997) anti-oestrogens (Buckbinder *et al.*, 1995) and tumour necrosis factor-α (TNF-α) (Yateman *et al.*, 1993). In normal human mammary epithelial cells, growth inhibition by atRA is independent of p53 expression (Seewaldt *et al.*, 1999).

### *IGFBP proteases*

Degradation of IGFBP-3 by specific proteases has been reported in the milk and serum of humans and rats (Lamson *et al.*, 1991). Generally, protease modified binding proteins have decreased affinity for IGFs. Plasmin, a protease found in bovine milk (Politis *et al.*, 1988) that has been projected to be involved in mammary involution, is an IGFBP protease (Politis *et al.*, 1995). Since the initial discovery of the IGFBP-3 protease, other proteases have been described that cleave IGFBP-2, -4, -5 and -6 (Collett-Solberg and Cohen, 1996). These proteases are known as prostate-specific antigen (Cohen *et al.*, 1994), cathepsins (Conover and De Leon, 1994), matrix metalloproteinases and others (Rajah *et al.*, 1995). While the proteases have been suggested as a local tissue capacity to free bound IGF from the IGFBP and thereby provide free IGF to the IGF-IR, independent actions of IGFBP fragments remain a possibility since Yamanaka *et al.* (1997) demonstrated that an IGFBP-3 fragment derived from protelolytic attack binds to insulin.

## IGF system change and mammary physiology

### *Blood versus milk changes*

The initial focus of animal scientists towards the IGF system was linked to the availability of tools to conduct temporal survey analyses and the drive to explain the galactopoietic effect of bovine growth hormone (Bauman and Vernon, 1993). Both IGF-I and -II have been characterized in the ruminant blood relative to the stage of lactation. Ronge *et al.* (1988) showed that blood IGF-I concentration was inversely related to milk production with a drastic drop after parturition followed by a gradual increase as lactation persisted. IGF-II does not appear to be affected by lactation (Vicini *et al.*, 1991). During the dry period of a pregnant animal, serum IGF-I is high in concentration, but declines rapidly after parturition while IGF-II is largely unchanged (Vega *et al.*, 1991). While exogenous injections of bovine growth hormone increase IGF-I concentrations in the blood, the increase occurs in a period when normal IGF-I concentrations are very low and the low concentration was partially attributed to lower liver mRNA levels suggesting decreased synthesis (Sharma *et al.*, 1994). However, blood clearance (tissue utilization) was not examined.

IGFBP concentration in the serum of ruminant species has been reported to not change for IGFBP-3 during lactation, but is higher during the pre-partum period (Vicini *et al.*, 1991; Sharma *et al.*, 1994). We reported prior to these reports that blood IGFBP activity was very high during the pre-partum colostral phase and then declined to lower levels essentially paralleling the IGF-I pattern (Vega *et al.*, 1991). Recently, we have shown that the bovine blood and milk IGFBP, identifiable by Western ligand blots, are IGFBP-2, -3, -4 and -5 (Gibson *et al.*, 1999). The blood identification confirms that established by others (Funston *et al.*, 1995; Roberts *et al.*, 1997). Figure 19.2 shows a survey of circulating serum milk IGFBP occurring during the full course of lactation. As has been previously demonstrated, Fig. 19.2c shows that IGFBP-3 is the predominant IGFBP in bovine circulation (Roberts *et al.*, 1997). The comparison between milk and blood from the same animals indicates that although IGFBP-3 and

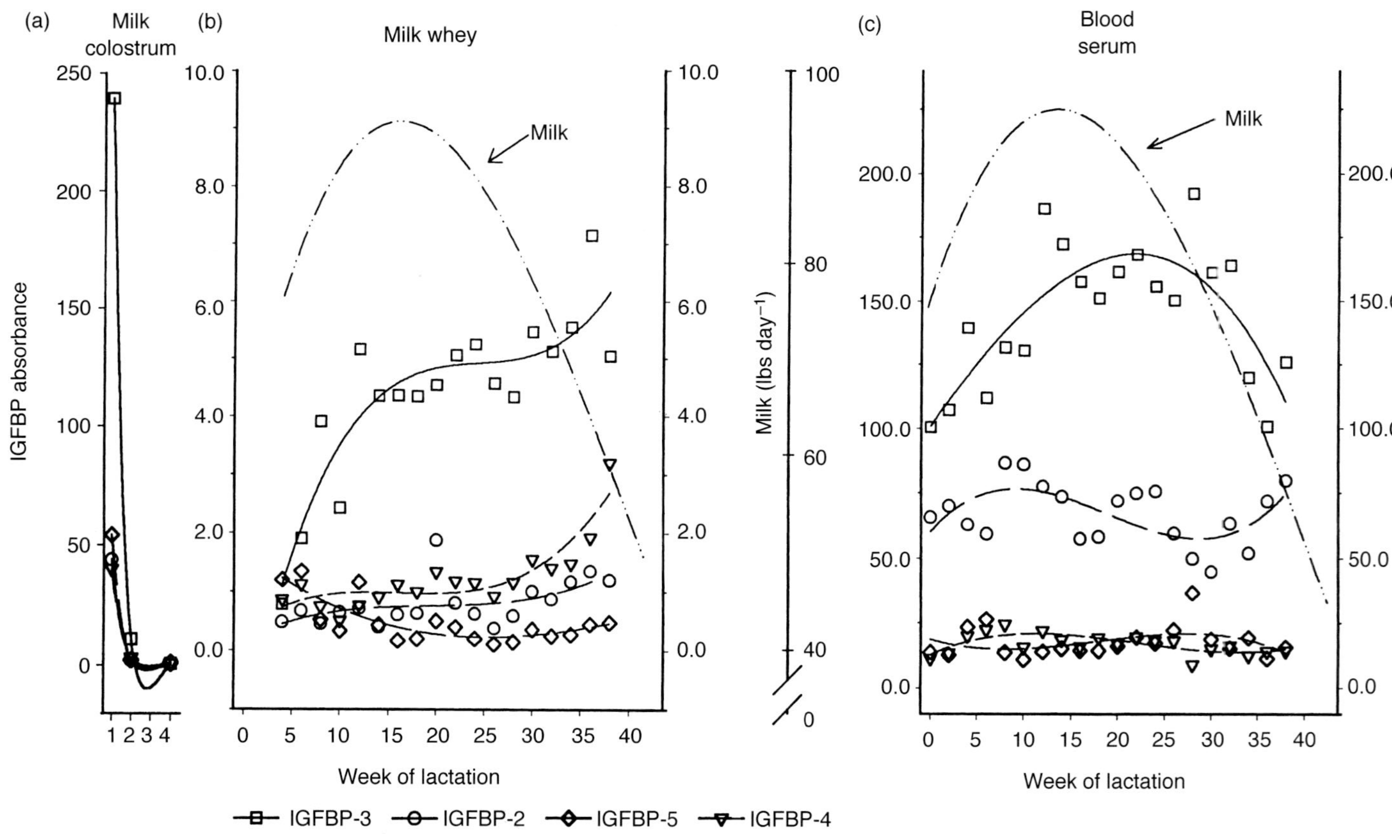


**Fig. 19.2.** Comparison of the changes in insulin-like growth factor binding proteins (IGFBP) in (a) milk colostrum, (b) milk whey, and (c) blood serum during lactation. Cows were sampled for blood and milk at 2-week intervals. Western blots utilizing [$^{125}$I]IGF-II as a ligand. Data was standardized by the use of an internal blood serum standard. Lines are regressions for each IGFBP. $n = 33$ cows.

IGFBP-2 change in both blood and milk during the course of lactation, their patterns of change are significantly different (Fig. 19.2b and c). While milk is approximately tenfold less in concentration of IGFBPs than that of blood (Fig. 19.2b versus 2c), the pattern clearly shows that blood IGFBP changes (Fig. 19.2c) are not reflected in milk changes (Fig. 19.2b). Milk IGFBPs all drop precipitously after the colostrum phase and the onset of copious milk secretion and IGFBP-3 shows an increase towards the end of the milk production period. The inverse relationship with milk production is evident.

Milk IGFBP-2 is relatively lower in concentration in both fluids and may exhibit some changes, but the variation between animals precludes interpretation. Milk and blood IGFBP-4 and -5 change little during the course of lactation. While nutrition has an impact upon the IGF axis (Thissen *et al.*, 1994), feed restriction experiments show little influence over IGFBP-3 serum concentrations (McGuire *et al.*, 1995). Because IGF-I is low when serum bovine growth hormone concentrations are high, the growth homone/IGF axis is said to be uncoupled. The 33 animals used for the data shown in Fig. 19.2 were fed adequately relative to NRC requirements (National Research Council, 1988).

Statistical analysis of the milk changes for all of the IGFBPs showed that there were two main effects upon IGFBP changes. The first was time of lactation and the second was an effect of pregnancy. The repetitive sample study with 33 cows from the Pennsylvania State University dairy herd, shown in Fig. 19.2, had eight cows that did not become pregnant. Figure 19.3a–c shows that there are large differences between the appearance of IGFBPs in milk of cows that do not become pregnant (Fig. 19.3a) compared with those that become pregnant (Fig. 19.3b). Most notable is that the cows that do not become pregnant show high concentrations and higher variances of IGFBP-3 and -2 in their milk. This is not reflected in the IGFBP concentrations occurring in blood (Fig. 19.3c and d). These findings support three major concepts. First, some event related to pregnancy is communicating to the mammary gland to alter milk IGFBP profiles; second, that milk changes occur independent of blood changes clearly suggests that the mammary gland regulates IGFBPs occurring in milk; and lastly, the blood changes, being different from that of milk, indicate that blood IGFBPs must be reflections of more than mammary gland tissue and therefore changes in blood IGFBP components will not reflect impacts upon the mammary gland.

Finally, many investigations with humans and rodents have shown that IGFBPs may be affected by the presence of proteases (Giudice, 1995). These enzymes have been shown to specifically modify IGFBPs so that they exhibit lowered binding of IGFs such that Western blotting may fail to detect them. Since our recently published characterization of IGFBPs in blood and milk utilized Western blotting, we needed to know if the milk IGFBP patterns were influenced by the presence of proteases. Extensive investigations for the presence of the pregnancy-related IGFBP-3 protease reported for the rodent and human (Staley, 1998) has been negative for bovine blood and milk. Thus, the changes observed in milk IGFBP-3 are not due to protease activity. The presence of other IGFBP proteases is currently unknown.

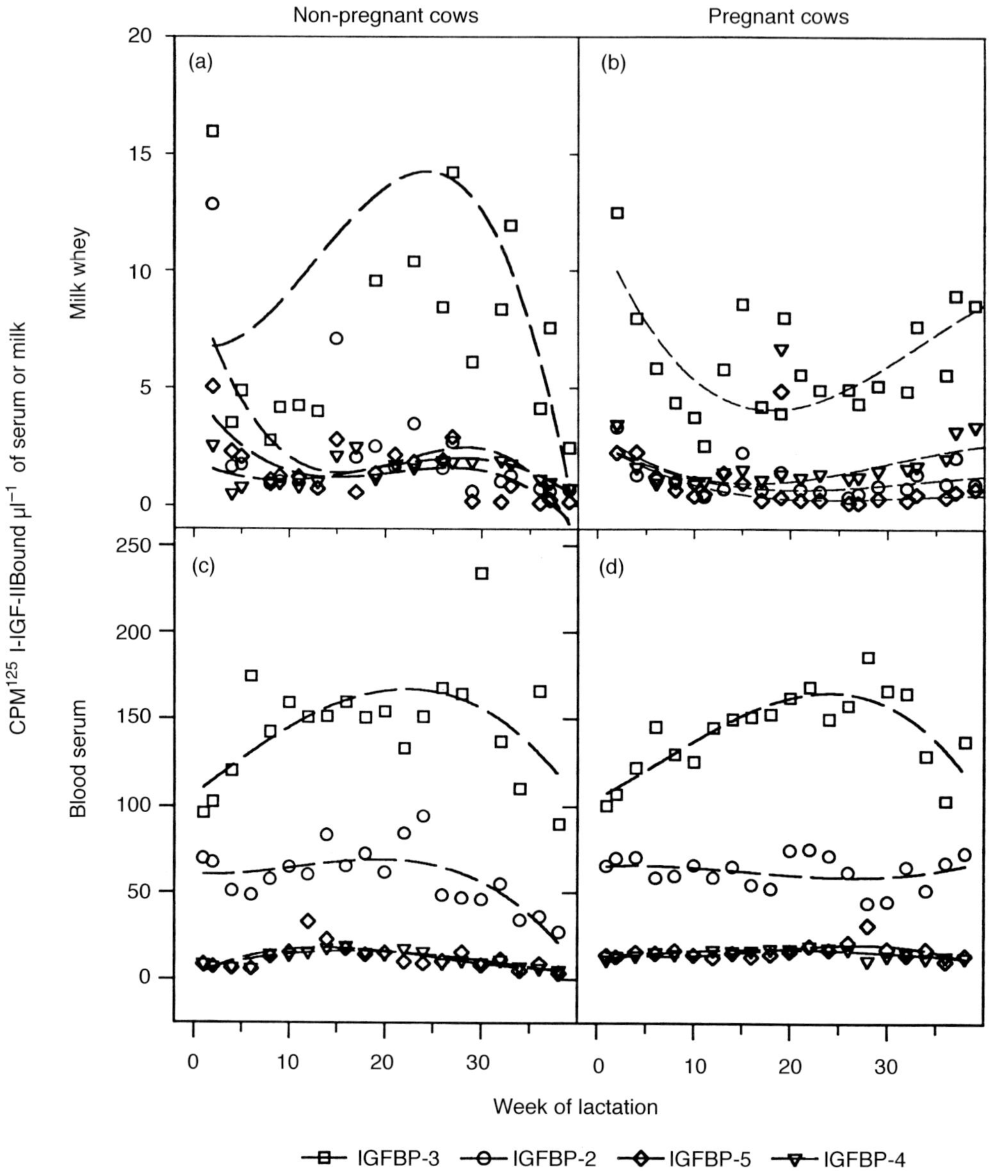


**Fig. 19.3.** Comparison of insulin-like growth factor binding proteins (IGFBP) in milk whey and blood serum of pregnant and non-pregnant cow changes during lactation. (a) Cow milk whey from non-pregnant cows and (c) serum ($n = 8$). (b) Cow milk whey from pregnant cows and (d) serum ($n = 25$). Samples are the same as those shown in Fig. 19.2. Data was standardized by the use of a internal blood serum standard. Lines are regressions for each IGFBP.

### *Local IGF ligands*

The initial interest in circulating IGF system components soon expanded when information began to accumulate indicating that all tissues and cells have the capacity to synthesize and secrete a number of IGF system components. Thus, the autocrine or paracrine concept of IGF system component regulation upon mammary gland physiology was expanded. To conduct these studies, tissue, organ or cell cultures are favoured. Mammary tissue of pregnant and lactating cows shows the presence of IGF-I mRNA (Glimm *et al.*, 1992; Sharma *et al.*, 1994). Subsequent studies with tissue indicated that stroma cells (Yee *et al.*, 1989), not secretory epithelial cells, were responsible for the tissue production of IGF-I. In contrast, IGF-II appears to be synthesized by rodent mammary cells (Campana *et al.*, 1994) and normal bovine mammary cells *in vitro* (Baumrucker *et al.*, 1993).

### *IGF receptors*

High-affinity receptors for IGF-I, IGF-II and insulin have been demonstrated for bovine mammary tissue and mammary epithelial cells (Oscar *et al.*, 1986; Hadsell *et al.*, 1990). Studies have shown that while type-1 IGF-IR binding increases with lactogenesis (Dehoff *et al.*, 1988), this is an apparent increase that results from a decrease that occurs during the pre-partum period (Hadsell *et al.*, 1990). Although the type-2 IGF-II receptor shows approximately fivefold greater binding (number of receptors), no microsomal binding changes were observed during lactation (Hadsell *et al.*, 1990). Since the mammary type-2 receptor does not change during the course of lactation, its capacity to invoke changes in the mammary gland becomes less significant, yet it probably contributes to the disappearance of IGFs from the cellular environment. Perhaps this is the mechanism of IGF appearance in milk (Prosser and Fleet, 1992; Donovan *et al.*, 1995). The significance of the type-1 receptor change (1.6-fold increase) is perhaps debatable. This is apparent when considering insulin receptor action. Characterization of the insulin receptor has shown that less than 5% of receptor binding leads to maximal insulin stimulated events (White and Kahn, 1994; Taylor *et al.*, 1996) leading to the concept of spare receptors. Is this also true for the type-1 receptor? Studies by Neuenschwander *et al.* (1995) with breast cancer cells (MCF-7) indicated that there appears to be no spare receptors for the type-1 receptors, at least in MCF-7 cells, since the antisense suppression of the expression of the type-1 receptor translated directly into a linear decrease in cellular growth in the presence of a constant supply of IGF-I. Thus, the decline observed for the type-1 receptor in mammary tissue during the late pre-partum period may be a signal for less growth and perhaps provide more opportunity for cellular differentiation.

This raises the question, what is the IGF signal to the mammary gland: growth, differentiation, or apoptosis? Three transgenic studies have shed some light upon the explanation of this question. When milk protein promoters were utilized to express excess IGF-I (and analogues) during lactation, no significant difference was observed in the mammary gland milk production (pup weight gain) or structural morphology (Brem *et al.*, 1994; Benito *et al.*, 1996; Hadsell *et al.*, 1996) Some reports of smaller alveoli were reported. The main effect observed is the delay in cellular decline during involution providing strong support for a role of IGF-I in the resistance of apoptotic mechanisms, at least during involution. In light of the low IGF-I in milk and blood during early lactation, perhaps the increase in IGF-I induced by bovine somatotropin (bST) treatments are decreasing lactational apoptosis and thereby increasing milk production.

In the cow, although circulating levels of IGF are increased by bovine growth hormone, the overall pattern of IGF-I concentration is inversely related to levels of milk production (Ronge *et al.*, 1988). This correlative data may indicate that the IGFs are low during peak lactation in order to allow maximal differentiation and continued milk production from those cells established into lactation. However, systemic circulation of IGFs is not the only source of IGF for mammary tissue. Mammary stromal fibroblasts are a source of IGF-I while epithelial cells have not been shown to have this capacity (Glimm *et al.*, 1988; Yee *et al.*, 1989). This suggests that although IGF-I is synthesized in mammary tissue, the level of synthesis is low and discourages the concept that local IGF-I production accounts for the galactopoietic potential during lactation. On the other hand, we have demonstrated that IGF-II is synthesized and regulated in mouse mammary epithelial cells (COMMA-D1) (Campana and Baumrucker, 1994) and primary cultures of bovine mammary epithelial cells (Fig. 19.4).

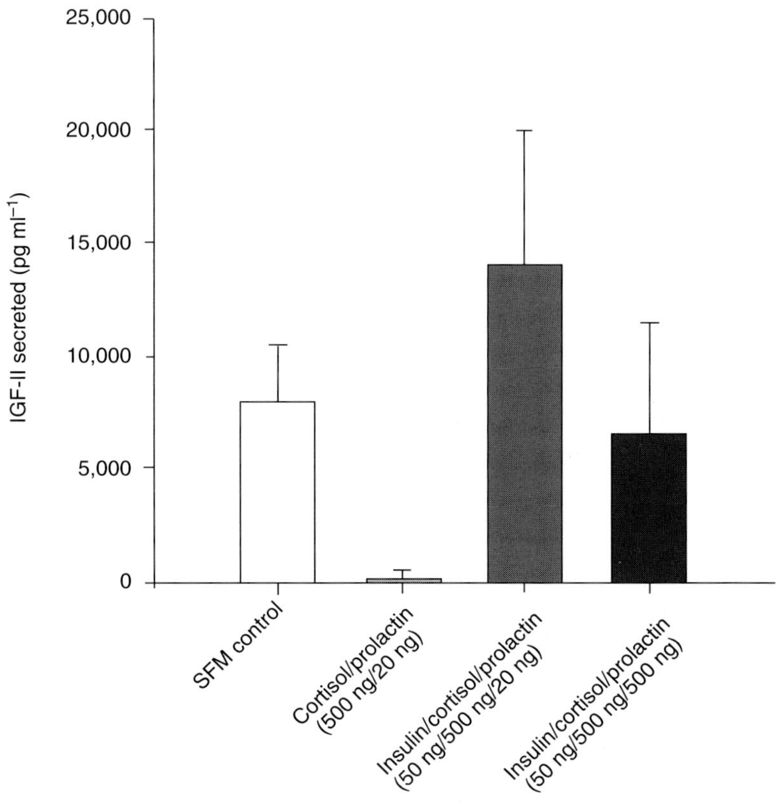


**Fig. 19.4.** Secretion of insulin-like growth factor II (IGF-II) into conditioned media (CM) from primary cultures of bovine mammary epithelial cells. CM was changed each day and was analysed for IGF-II by RIA after acetic acid column separation. Values are the mean ± standard error of a mean ($n$ = 3).

How can receptor numbers and ligand changes be reconciled with changes in milk production? We believe that changes in IGFs and their receptors need to be considered in the whole IGF system (Fig. 19.1). It has been demonstrated that the application of IGFs and other endocrine factors to most tissues and mammary cells (Skaar and Baumrucker, 1993; Cohick, 1998) stimulates a change in the appearance of IGFBPs.

## Mammary cell IGFBP synthesis

Numerous sources show that IGFBPs are synthesized and secreted by mammary epithelial cells (Clemmons *et al.*, 1990; Figueroa and Yee, 1992; Werner and LeRoith, 1996). This is not surprising since almost all mammalian cells synthesize and secrete IGFBPs (Jones and Clemmons, 1995). IGFBP-3 and IGFBP-2 have been identified by Western ligand blot analysis in bovine mammary pre-partum secretions (McGrath *et al.*, 1991). Recently, we have demonstrated that primary cultures of bovine mammary epithelial cells secrete IGFBP-2, -3, -4 and -5, which by molecular weight analysis, appear to be identical to those shown in bovine milk (Gibson *et al.*, 1999). Several hormones and growth factors regulate the synthesis and secretion of IGFBPs by mammary epithelial cells. These factors that affect mammary IGFBPs (IGFs, epidermal growth factor, prolactin, cortisol, retinol) are also known to be primary effectors of mammary growth and differentiation (Fielder *et al.*, 1992; Skaar and Baumrucker, 1993).

We have utilized frozen mammary tissue and mammary epithelial cells from bovine (MEBo) frozen recovered cells for subsequent experiments described below. These cells were prepared from bovine mammary tissue by first making mammary acini (Baumrucker *et al.*, 1988) and then freezing the cells for subsequent culture and passages. Figure 19.4 shows that while IGF-II is secreted by bovine mammary cells, lactogenic hormones (cortisol or prolactin) appear to inhibit production, while insulin is a stimulator as shown for the rodent (Campana et al. 1994). Low levels of insulin (50 ng $ml^{-1}$) do not displace IGF-I from the type-1 receptor in mammary tissue (Hadsell *et al.*, 1990) suggesting that this change is attributable to the insulin receptor.

Figure 19.5 shows that primary cultures of bovine mammary epithelial cells secrete the four IGFBPs into conditioned media over time and that IGFBP-3 remains the dominant IGFBP in serum-free media (Fig. 19.5b). Furthermore, we show that retinoic acid, one of the known IGFBP-3 regulatory agents, stimulates IGFBP-3 appearance with time in culture. Cohick and Turner (1998) recently utilized a SV-40 large T antigen immortalized bovine mammary cell line (MAC-T) (Huynh *et al.*, 1991) to demonstrate some of the regulatory factors in IGFBP expression in these cells. Most interesting is that the pattern of IGFBP released by these cells in serum-free media was IGFBP-2, -4 and -6. No IGFBP-5 and little IGFBP-3 is detectable unless the latter was stimulated by IGF-I. The lack of IGFBP-3 is similar to other immortalized mammary cells such as MCF-7 cells where IGFBP-3 is also low or undetectable. However, MCF-7 cells (Shang *et al.*, 1998, 1999) and primary bovine mammary cells (Fig. 19.5) are stimulated to produce IGFBP-3 with retinoic acid while the MAC-T cells are stimulated by IGF-I (Cohick and Turner, 1998). Thus, the regulation of IGFBP in cell cultures shows differential effects, either due to immortalization or to culture conditions.

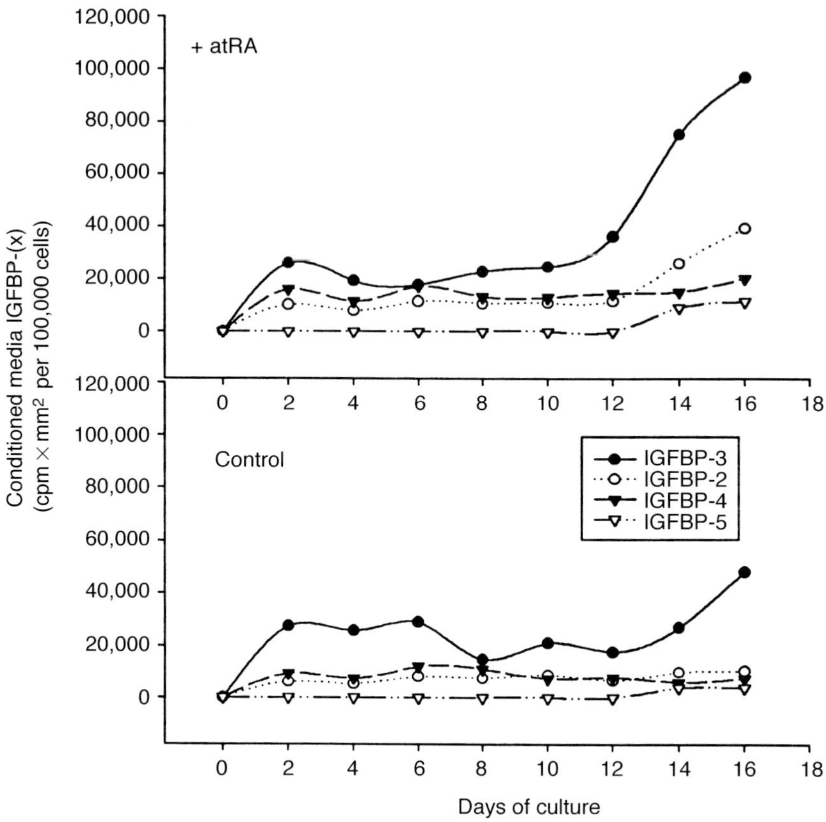


**Fig. 19.5.** Primary cultures of bovine mammary epithelial cells secrete insulin-like growth factor binding proteins (IGFBP) into conditioned media. Cells were plated on plastic in serum-free media and media changed every 2 days. IGFBPs were detected by Western blots utilizing [$^{125}$I]IGF-II as a ligand. All trans retinoic acid (1 μM; atRA) stimulates the appearance of IGFBP-3 in the conditioned media after 12 days of treatment.

## *IGFBP-3 binding to membrane proteins*

It is now established that IGFBP-3 binds to membrane-bound proteins (Oh *et al.*, 1993; Hodgkinson *et al.*, 1995) Recombinant human (rh) IGFBP-3 binds to membrane proteins of 50, 26, and 20 kDa from breast cancer cells (Hwa *et al.*, 1999). A personal communication with Youngman Oh indicated that although these proteins are detectable, the expression is low. In addition to these proteins, TGF-β receptor-V is also reported to be an IGFBP-3 receptor (Leal *et al.*, 1997). A recent study (Schedlich *et al.*, 1998) in which mutant IGFBPs were produced indicated that mutants of rhIGFBP-3 exhibit differential binding to cell surfaces and concluded that residues 228–232 were essential for cell association. Schedlich *et al.* (1998) also produced a mutant RGD IGFBP-3 protein (not found in the wild type) and demonstrated that not all binding was due to the introduced RGD sequence that delineates the integrin binding site (Ruoslahti and Pierschbacher, 1987).

It has been speculated that the mechanism of action of IGFBP-3 may be attributed to its ability to bind the serum and autocrine IGF-II and prevent action at the type-1 IGF receptor as shown with a mouse mammary cell line (Campana *et al.*, 1994) and the primary bovine cells shown in Fig. 19.5. However, the presence of a type-1 receptor-independent action of IGFBP-3 cannot be excluded and we now have descriptive evidence for such a mechanism as reported in preliminary data described below.

## IGFBP-3: IGF-I dependent versus independent action

Because IGF-I acts as a survival factor through stimulation of the type-1 receptor (Stewart and Rotwein, 1996), the induction of apoptosis by IGFBP-3 could be explained by an IGF-dependent (sequestering IGF-I) or IGF-independent mechanism. While some studies suggest the IGF-dependent pathway (Rajah *et al.*, 1997), others show the latter (Gill *et al.*, 1997).

## Evidence of an IGFBP-3 binding protein (BP-3BP)

MEBo microsomes (Pocius *et al.*, 1984) were solubilized with SDS non-reducing loading buffer and loaded onto a SDS-PAGE gel. The gels were electroblotted to nitrocellulose and iodinated rhIGFBP-3, with and without excess unlabelled rhIGFBP-3, applied to the blots. The specific binding of $^{125}$I rhIGFBP-3 to membrane-bound microsomal proteins is shown in Fig. 19.6. Solubilized membrane proteins from cow mammary microsomes showed a number of BP-3BP that range from approximately 65 to <18 kDa. A comparison of cow microsomes with those of a mouse immortalized mammary cell line (COMMA-D1) (Danielson *et al.*, 1984) shows that there are differences in the occurrence of BP-3-R proteins between the membrane preparations. COMMA-D1 microsomes showed only lower molecular weight BP-3BP (Fig. 19.6) which agrees with data from previous publications using human breast cancer mammary epithelial cell lines (Oh *et al.*, 1993). Of interest to us was that cow microsomes show BP-3BP of higher molecular weights.

Because bovine milk fat globule membranes (MFGM) are products of merocrine secretion from lactating mammary epithelial cells, they represent an abundant and non-invasive source of epithelial apical plasma membrane. We prepared MFGM from bulk tank milk and confirmed the presence of the ~65 kDa BP-3BP on MFGM as well as other BP-3BP. Cohen (1998) indicated that he had shown a number of proteins that bound to IGFBP-3. One of these was identified as transferrin. We have now shown that the ~65 kDa BP-3BP that appears in bovine mammary membranes is not transferrin, but lactoferrin. The ~65 kDa BP-3BP (shown in Fig. 19.6) is immunoprecipitatable with polyclonal antibody against bovine lactoferrin (b-Lf; kindly donated by F. Schanbacher, Ohio State University). We have subsequently shown that [$^{125}$I]rhIGFBP-3 binds to both apo- and holo-Lf from a number of species, but not to any species transferrins. Because Lf is a positively charged protein that is inherently sticky, we showed that lysozyme, also a positively charged protein, did not show IGFBP-3

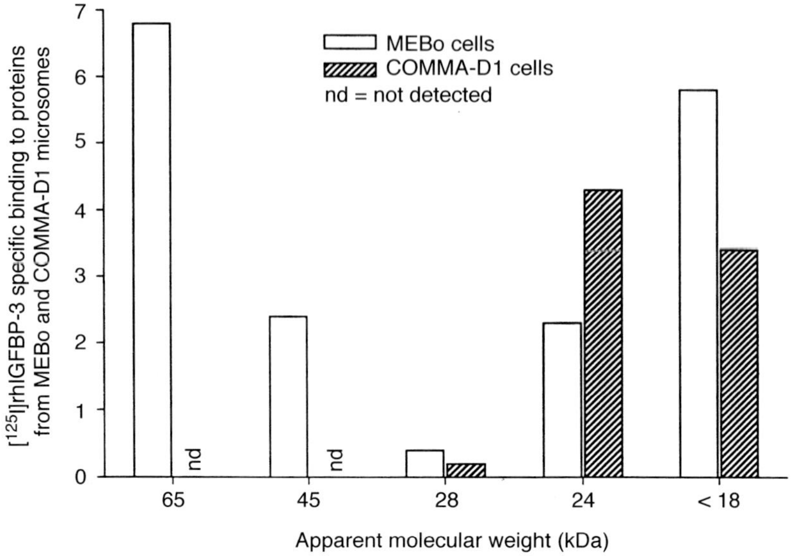


**Fig. 19.6.** Specific binding of [$^{125}$I] recombinant human insulin-like growth factor binding protein 3 (rhIGFBP-3) to proteins solubilized from cellular microsomes. Data is relative specific binding that is corrected for non-specific binding with excess unlabelled rhIGFBP-3 to Western blots from microsomes isolated from mammary epithelial cells from bovine (MEBo) and COMMA-D1 cells.

binding. Reverse experiments with [$^{125}$I]b-Lf (provided by F. Schanbacher) show specific binding to IGFBP-3, but not to any other IGFBP found in bovine milk or serum (IGFBP-2, -4, -5). Additional studies have documented that not only does Lf bind to IGFBP-3, but that if IGF is bound to IGFBP-3, Lf competes for binding with a $K_a$ of 1 μM and will displace bound IGF with an apparent $K_d$ of 10 μM (Baumrucker *et al.*, 1999).

## Lactoferrin

As true for IGFBP-3, nuclear appearance of Lf has also been documented (Garre *et al.*, 1992). Lf is a highly positively-charged ~80 kDa iron-binding glycoprotein that exhibits a high degree of homology with transferrin (Baker *et al.*, 1998). Although the protein has been sequenced and characterized by X-ray crystallography (Lonnerdal and Iyer, 1995), its biological function(s) remain largely elusive. Milks of different species vary in the relative content of Lf and transferrin. Human milk has the highest level of Lf (1–6 mg ml$^{-1}$) with little transferrin present, while bovine milks have low levels of Lf (0.1 mg ml$^{-1}$) and detectable amounts of transferrin.

Lf content of mammary secretions varies with developmental stage. The message is highly expressed and the protein is present in high concentration during the final prepartum stages (human, 10 mg ml$^{-1}$; bovine, 1–2 mg ml$^{-1}$) and at higher concentrations (20–50 mg $ml^{-1}$) in both human and bovine mammary gland during involution. Lf mRNA and protein concentration is low during lactation (Nuijens *et al.*, 1996). It is thought that mammary cell Lf secretion occurs via the normal apical secretory pathway since Lf has been shown to be co-localized with casein micelles in secretory vesicles (Neville *et al.*, 1998), but these studies were conducted during lactation, and pre-partum and involution mechanisms may be different.

### *Correlation between IGFBP-3 and Lf*

Some interesting physiological connections have been reported between IGFBP-3 and Lf. For example, fluctuation in the levels of Lf in the bovine reproductive cycle corresponds well to that of IGFBP-3 (Campbell *et al.*, 1991). In certain pathological conditions such as inflammation, an increased secretion of IGFBP-3 (Schuster *et al.*, 1995) accompanies the elevated secretion of Lf. Our recent studies examining the interaction between Lf and IGFBP-3 (Baumrucker *et al.*, 1999) showed that IGFBP-3 nuclear localization in MCF-7 breast cancer cells is Lf-dependent and requires treatment of the cells with all trans retinoic acid, an agent that is known to up-regulate IGFBP-3 expression while inhibiting growth of these cells.

Thus, the synthesis and secretion of mammary cell Lf will probably have a role in the IGF system and, ultimately, cell growth and survival. We believe that the enhanced production of Lf during the immediate pre-partum period and during involution (Nuijens *et al.*, 1996) will free IGFs (both IGF-I and IGF-II) from their association with IGFBP-3. This mechanism could provide an increased stimulation from type 1 IGF receptor and lead to increased cell growth and decreased apoptosis.

## IGFBP-3 nuclear localization sequence (NLS)

In addition to the discovery of BP-3BP, new evidence has shown that IGFBP-3 protein has an NLS (Radulescu, 1994) that is critical for the biological activity of this protein (Zacksenhaus *et al.*, 1993). The amino acid residues in human IGFBP-3 that comprise the NLS are also found in the bovine IGFBP-3 (Fig. 19.7). Over 50% of known nuclear proteins have the NLS sequence; however, only 4% of the proteins deposited in the SWISS-PROT database have this motif (Radulescu, 1994).

## Nuclear appearance of IGFBP-3:Lf

To demonstrate that IGFBP-3 is found in the mid-lactation MEBo cell nucleus, MEBo cells were incubated with exogenous biotinylated rhIGFBP-3 for 24 h. The cells were fixed in 4% formalin, permeabilized with 0.1% tritonX100, and stained with avidin–biotin conjugate and BCIP/NBT substrate. Figure 19.8 shows that the nuclei of the MEBo cells show the stain indicating the presence of the labelled nuclear IGFBP-3.

**Human IGFBP-3 Accession X64875**
MQRARPTLWAAALTLLVLLRGPPVARAGASSGGLGPVVRCEPCDARALAQCAPPP
AVCAELVREPGCGCCLTCALSEGQPCGIYTERCGSGLRCQPSPDEARPLQALLDG
RGLCVNASAVSRLRAYLLPAPPAPGNASESEEDRSAGSVESPSVSSTHRVSDPKF
HPLHSKIIIIKKGHAKDSQRYKVDYESQSTDTQNFSSESKRETEYGPCRREMEDTLN
HLKFLNVLSPRGVHIPNCD***KK**GFYKKKQCRPS**KGRK***RGFCWCVDKYGQPLPGYTT
KGKEDVHCYSMQSK

**Bovine IGFBP-3 Accession M76478**
MLRAPPRLWAAALTALTLLRGPPAARAGAGTMGAGPVVRCEPCDARAVAQCAPPP
PSPPCAELVRDAGCGCCLTCALREGQPCGVYTERCGSGLRCQPPPGDPRPLQALL
DGRGLCANASAVGRLRPYLLPSASGNGSESEEDHSMGSTENQAGPSTHRVPVSK
FHPIHTKMDVIKKGHAKDSQRYKVDYESQSTDTQNFSSESKRETEYGPCRREMED
TLNHLKFLNMLSPRGIHIPNCD***KK**GFYKKKQCRPS**KGRK***RGFCWCVDKYGQPLPG
FDVKGKGDVHCYSMESK

**Fig. 19.7.** Presence of a nuclear localization sequence (NLS) in bovine insulin-like growth factor binding protein 3 (IGFBP-3) amino acid sequence. Italic underlined sequences are the NLS. Bold residues are the specific residues that comprise the NLS recognition signal: (i) two adjacent basic amino acids; (ii) a spacer of ten residues; (iii) at least three basic amino acids within the next five residues following the space. Data from GenBank.

## Conclusions

When many researchers entered the somatomedin C arena in the early 1980s, they perhaps thought that this was yet another hormonal ligand that would clearly answer the questions posed by the bST application to lactating dairy cows. The new discoveries since that time that have expanded the IGF system components and discoveries of biological actions have expanded our thinking about the regulation of bovine mammary tissue in growth, differentiation, and involution. The IGF system, as perhaps a model for many endocrine, paracrine and autocrine factors, demonstrates that complexity and the potential for levels of regulation that exist in the mammary gland. Each level of complexity provides more opportunities for control and perhaps selection as we seek to understand efficient milk production.

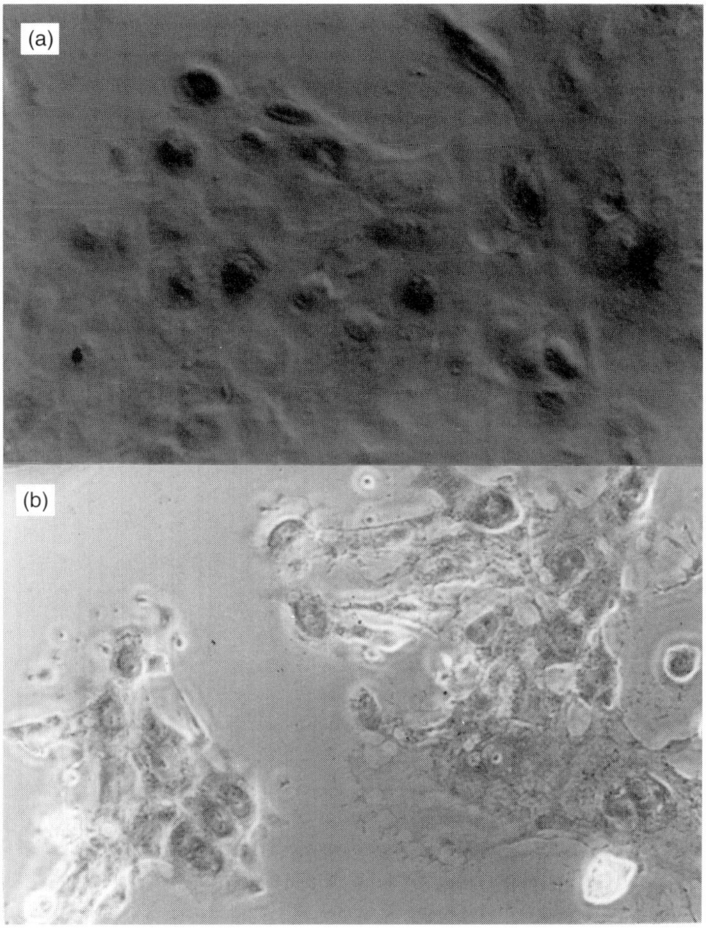


**Fig. 19.8.** Demonstration of the appearance of biotinylated-recombinant human insulin-like growth factor binding protein 3 (rhIGFBP-3) in the nucleus of mammary epithelial cells from bovine cells. Cells were cultured for 8 days in serum-free media (SFM) and biotinylated-rhIGFBP-3 (100 μg) was added to the cells in fresh SFM for 4 h. (a) Cells stained with avidin–biotin–alkaline phosphatase conjugate BCIP/NBT substrate; (b) unstained cells.

## References

Anonymous (1992) Report on the nomenclature of the IGF binding proteins. *Journal of Clinical Endocrinology and Metabolism* 74, 1215–1216.

Baker, E.N., Anderson, B.F., Baker, H.M., Ross, T.A., MacGillivray, S.A., Moore, S.A., Peterson, N.A., Shewry, S.C. and Tweedie, J.W. (1998) Three-dimensional structure of lactoferrin. In: Spik, G., Dominique, L., Maxurier, J. and Pierce, A. (eds) *Advances in Lactoferrin Research*. Plenum Press, New York, pp. 1–14.

Bauman, D.E. and Vernon, R.G. (1993) Effects of exogenous bovine somatotropin on lactation. *Annual Review of Nutrition* 13, 437–461.

Baumrucker, C.R., Deemer, K.P., Walsh, R., Riss, T.L. and Akers, R.M. (1988) Primary culture of bovine mammary acini on a collagen matrix. *Tissue and Cell* 20, 541–554.

Baumrucker, C.R., Campana, W.M., Gibson, C.A. and Kerr, D.E. (1993) Insulin-like growth factors (IGFs) and IGF binding proteins in milk: Sources and functions. *Endocrine Regulations* 27, 157–172.

Baumrucker, C.R., Gibson, C.A., Shang, Y., Schanbacher, F. and Green, M.H. (1999) Lactoferrin specifically binds to IGFBP-3 to cause competitive displacement of IGF from IGF:IGFBP-3 complexes and induces internalization and nuclear localization of lactoferrin:IGFBP-3 complexes in mammary epithelial cells. *The Endocrine Society Annual Meetings* San Diego, California, no. p2–584, p.404.

Baxter, R.C., Binoux, M.A., Clemmons, D.R., Conover, C.A., Drop, S.L.S., Holly, J.M.P., Mohan, S., Oh, Y. and Rosenfeld, R.G. (1998) Recommendations for nomenclature of the insulin-like growth factor binding protein superfamily. *Endocrinology* 139, 4036.

Benito, M., Valverde, A.M. and Lorenzo, M. (1996) IGF-I: A mitogen also involved in differentiation processes in mammalian cells. *International Journal of Biochemistry and Cell Biology* 28, 499–510.

Bork, P. (1998) The modular architecture of a new family of growth regulators related to connective tissue growth factor. *FEBS Journal* 327, 125–130.

Brem, G., Hartl, P., Besenfelder, U., Wolf, E., Zinovieva, N. and Pfaller, R. (1994) Expression of synthetic cDNA sequences encoding human insulin-like growth factor-1 (IGF-1) in the mammary gland of transgenic rabbits. *Gene* 149, 351.

Buckbinder, L., Talbott, R., Velasco-Miguel, S., Takenaka, I., Faha, B., Seizinger, B.R. and Kley, N. (1995) Induction of the growth inhibitor IGF-binding protein 3 by p53. *Nature* 377, 646–649.

Campana, W.M. and Baumrucker, C.R. (1994) Autocrine insulin-like growth factor II stimulates growth and IGF binding proteins through the IGF-I receptor in mammary epithelial cells. *The Endocrine Society Annual Meeting* Annual Meeting, Anaheim, no. 130, p. 233 (Abstract).

Campana, W.M., Skaar, T.C., Gibson, C.A. and Baumrucker, C.R. (1994) Insulin stimulated growth through autocrine IGF-II secretion and the IGF-I receptor in a mouse mammary epithelial cell line. *Endocrine Journal* 2, 883–889.

Campbell, P.G., Skaar, T.C., Vega, J.R. and Baumrucker, C.R. (1991) Secretion of insulin-like growth factor-I (IGF-I) and IGF-binding proteins from bovine mammary tissue *in vitro*. *Journal of Endocrinology* 128, 219–228.

Clemmons, D.R. (1997) Insulin-like growth factor binding proteins and their role in controlling IGF actions. *Cytokine and Growth Factor Reviews* 8, 45–62.

Clemmons, D.R., Camacho-Hübner, C., Coronado, E. and Osborne, C.K. (1990) Insulin-like growth factor binding protein secretion by breast carcinoma cell lines: Correlation with estrogen receptor status. *Endocrinology* 127, 2679–2686.

Cohen, P. (1998) Cell regulatory functions of IGFBP-3. *80th Annual Endocrine Meetings* Program & Abstracts, no. S20–3, p. 20 (Abstract).

Cohen, P., Peehl, D.M., Graves, H.C.B. and Rosenfeld, R.G. (1994) Biological effects of prostate specific antigen as an insulin-like growth factor binding protein-3 protease. *Journal of Endocrinology* 142, 407–415.

Cohick, W.S. (1998) Symposium: Growth hormone and insulin-like growth factors – Role of the insulin-like growth factors and their binding proteins in lactation. *Journal of Dairy Science* 81, 1769–1777.

Cohick, W.S. and Turner, J.D. (1998) Regulation of IGF binding protein synthesis by a bovine mammary epithelial cell line. *Journal of Endocrinology* 157, 327–336.

Collett-Solberg, P.F. and Cohen, P. (1996) The role of the insulin-like growth factor binding proteins and the IGFBP proteases in modulating IGF action. *Growth and Growth Disorders* 25, 591–614.

Conover, C.A. and De Leon, D.D. (1994) Acid-activated insulin-like growth factor-binding protein- 3 proteolysis in normal and transformed cells. Role of cathepsin D. *Journal of Biological Chemistry* 269, 7076–7080.

Danielson, K.G., Oborn, C.J., Durban, E.M., Butel, J.S. and Medina, D. (1984) Epithelial mouse mammary cell line exhibiting normal morphogenesis in vivo and functional differentiation in vitro. *Proceedings of the National Academy USA* 81, 3756–3760.

Daughaday, W.H. and Rotwein, P. (1989) Insulin-like growth factors I and II. Peptide, messenger ribonucleic acid and gene structures, serum, and tissue concentrations. *Endocrine Reviews* 10, 68–91.

Dehoff, M.H., Elgin, R.G., Collier, R.J. and Clemmons, D.R. (1988) Both type I and II insulin-like growth factor receptor binding increase during lactogenesis in bovine mammary tissue. *Endocrinology* 122, 2412–2417.

Dennis, P.A. and Rifkin, D.B. (1991) Cellular activation of latent transforming growth factor β requires binding to the cation-independent mannose 6-phosphate/insulin- like growth factor type II receptor. *Proceedings of the National Academy of Sciences USA* 88, 580–584.

Donovan, S.M., Hintz, R.L. and Rosenfeld, R.G. (1995) Investigation into the potential physiological sources of rat milk IGF-I and IGF-binding proteins. *Journal of Endocrinology* 145, 569–578.

Fielder, P.J., Thordarson, G., English, A., Rosenfeld, R.G. and Talamantes, F. (1992) Expression of a lactogen-dependent insulin-like growth factor-binding protein in cultured mouse mammary epithelial cells. *Endocrinology* 131, 261–267.

Figueroa, J.A. and Yee, D. (1992) The insulin-like growth factor binding proteins (IGFBPs) in human breast cancer. *Breast Cancer Research and Treatment* 22, 81–90.

Fowlkes, J.L. (1997) Insulin-like growth factor-binding protein protelysis, an emerging paradim in insulin-like growth factor physiology. *Trends in Endocrinology and Metabolism* 8, 299–306.

Funston, R.N., Moss, G.E. and Roberts, A.J. (1995) Insulin-like growth factor-I (IGF-I) and IGF-binding proteins in bovine sera and pituitaries at different stages of the estrous cycle. *Endocrinology* 136, 62–68.

Garre, C., Bianchi-Scarra, G., Sirito, M., Musso, M. and Ravazzolo, R. (1992) Lactoferrin binding sites and nuclear localization in K562(S) cells. *Journal of Cell Physiology* 153, 477–482.

Gibson, C.A., Staley, M.D. and Baumrucker, C.R. (1999) Identification of IGF binding proteins in bovine milk and the demonstration of IGFBP-3 synthesis and release by bovine mammary epithelial cells. *Journal of Animal Science* 77, 1547–1557.

Gill, Z.P., Perks, C.M., Newcomb, P.V. and Holly, J.M. (1997) Insulin-like growth factor-binding protein (IGFBP-3) predisposes breast cancer cells to programmed cell death in a non-IGF-dependent manner. *Journal of Biological Chemistry* 272, 25602–25607.

Giudice, L.C. (1995) IGF binding protein-3 protease regulation: How sweet it is! *Journal of Clinical Endocrinology and Metabolism* 80, 2279–2281.

Glimm, D.R., Baracos, V.E. and Kennelly, J.J. (1988) Effect of bovine somatotropin on the distribution of immunoreactive insulin-like growth factor-1 in lactating bovine mammary tissue. *Journal of Dairy Science* 71, 2923–2935.

Glimm, D.R., Baracos, V.E. and Kennelly, J.J. (1992) Northern and in situ hybridization analyses of the effects of somatotropin on bovine mammary gene expression. *Journal of Dairy Science* 75, 2687–2705.

Gucev, Z.S., Oh, Y., Kelley, K.M. and Rosenfeld, R.G. (1996) Insulin-like growth factor binding protein 3 mediates retinoic acid- and transforming growth factor 2-induced growth inhibition in human breast cancer cells. *Cancer Research* 56, 1545–1550.

Hadsell, D.L., Campbell, P.G. and Baumrucker, C.R. (1990) Characterization of the change in type I and II insulin-like growth factor receptors of bovine mammary tissue during the pre- and postpartum periods. *Endocrinology* 126, 637–643.

Hadsell, D.L., Greenberg, N.M., Fligger, J.M., Baumrucker, C.R. and Rosen, J.M. (1996)

Targeted expression of des(1–3) human insulin-like growth factor I in transgenic mice influences mammary gland development and IGF-binding protein expression. *Endocrinology* 137, 321–330.

Hodgkinson, S., Fowke, P., Al Somai, N. and McQuoid, M. (1995) Proteins in tissue extracts which bind insulin-like growth factor binding protein-3 (IGFBP-3). *Journal of Endocrinology* 145, R1–R6.

Huynh, H.T., Robitaille, G. and Turner, J.D. (1991) Establishment of bovine mammary epithelial cells (MAC-T): An *in vitro* model for bovine lactation. *Experimental Cell Research* 197, 191–199.

Huynh, H., Yang, X.F. and Pollak, M. (1996) Estradiol and antiestrogens regulate a growth inhibitory insulin-like growth factor binding protein 3 autocrine loop in human breast cancer cells. *Journal of Biological Chemistry* 271, 1016–1021.

Hwa, V., Oh, Y. and Rosenfeld, R.G. (1999) Insulin-like growth factor binding protein (IGFBP) superfamily: Structure, function and evolution. *Endocrine Society 81st Meeting* (Abstract).

Imai, Y., Busby, W.H., Jr, Smith, C.E.C.J.B., Garmong, A.J., Horwitz, G.D., Rees, C. and Clemmons, D.R. (1997) Protease-resistant form of insulin-like growth factor-binding protein 5 is an inhibitor of insulin-like growth factor-I actions on porcine smooth muscle cells in culture. *Journal of Clinical Investigation* 100, 2596–2605.

Jones, J.I. and Clemmons, D.R. (1995) Insulin-like growth factors and their binding proteins: Biological actions. *Endocrine Reviews* 16, 3–34.

Jones, J.I., D'Ercole, A.J., Camacho-Hübner, C. and Clemmons, D.R. (1991) Phosphorylation of insulin-like growth factor (IGF)-binding protein 1 in cell culture and *in vivo*: Effects on affinity for IGF-I. *Proceedings of the National Academy of Sciences USA* 88, 7481–7485.

Kang, J.X., Li, Y. and Leaf, A. (1997) Mannose-6-phosphate/insulin-like growth factor-II receptor is a receptor for retinoic acid. *Proceedings of the National Academy of Sciences USA* 94, 13671–13676.

Kelley, K.M., Oh, Y., Gargosky, S.E., Gucev, Z., Matsumoto, T., Hwa, V., Ng, L., Simpson, D.M. and Rosenfeld, R.G. (1996) Insulin-like growth factor-binding proteins (IGFBPs) and their regulatory dynamics. *International Journal of Biochemistry and Cell Biology* 28, 619–637.

Lamson, G., Giudice, L.C. and Rosenfeld, R.G. (1991) A simple assay for proteolysis of IGFBP-3. *Journal of Clinical Endocrinology and Metabolism* 72, 1391–1393.

Lamson, C.P., Okjima, T. and Rosenfeld, R.G. (1993) Transfection of the human insulin-like growth factor (IGF) binding protein-3 gene into BALB/c fibrosis inhibits cell growth. *Molecular Endocrinology* 137, 3206–3212.

Leal, S.M., Liu, Q.J., Huang, S.S. and Huang, J.S. (1997) The type V transforming growth factor β receptor is the putative insulin-like growth factor-binding protein 3 receptor. *Journal of Biological Chemistry* 272, 20572–20576.

LeRoith, D. (1991) *Insulin-like Growth Factors. Molecular and Cellular Aspects.* CRC, Boca Raton, Florida.

LeRoith, D., Adamo, M.L., Kato, H. and Roberts, C.T. (1993) Divergence of insulin and insulin-like growth factor-1 signalling pathways. *Pediatric and Adolescent Endocrinology* 24, 346–358.

LeRoith, D., Werner, H., Neuenschwander, S., Kalebic, T. and Helman, L.J. (1995) The role of the insulin-like growth factor-I receptor in cancer. *Annals of the New York Academy of Science* 766, 402–408.

Li, W., Fawcett, J., Widmer, H.-R., Fielder, P.J., Rabkin, R. and Keller, G.-A. (1998) Nuclear transport of insulin-like growth factor-I and insulin-like growth factor binding protein-3 in opossum kidney cells. *Endocrinology* 138, 1763–1766.

Lonnerdal, B. and Iyer, S. (1995) Lactoferrin: molecular structure and biological function. *Annual Review of Nutrition* 15, 93–110.

Mccusker, R.H., Camacho-Hübner, C., Bayne, M.L., Cascieri, M.A. and Clemmons, D.R. (1990) Insulin-like growth factor (IGF) binding to human fibroblast and glioblastoma cells: The modulating effect of cell released IGF binding proteins (IGFBPs). *Journal of Cell Physiology* 144, 244–253.

McGrath, M.F., Collier, R.J., Clemmons, D.R., Busby, W.H., Jr, Sweeny, C.A. and Krivi, G.G. (1991) The direct *in vitro* effect of insulin-like growth factors (IGFs) on normal bovine mammary cell proliferation and production of IGF binding proteins. *Endocrinology* 129, 671–678.

McGuire, M.A., Bauman, D.E., Dwyer, D.A. and Cohick, W.S. (1995) Nutritional modulation of the somatotropin/insulin-like growth factor system: Response to feed deprivation in lactating cows. *Journal of Nutrition* 125, 493–502.

Murphy, L.J. (1998) Insulin-like growth factor-binding proteins: functional diversity or redundancy? *Journal of Molecular Endocrinology* 21, 97–107.

National Research Council (1988) *Nutrient Requirements of Dairy Cattle*, 6th edn. National Academy Press, Washington, DC.

Neuenschwander, S., Roberts, C.T., Jr and LeRoith, D. (1995) Growth inhibition of MCF-7 breast cancer cells by stable expression of an insulin-like growth factor I receptor antisense ribonucleic acid. *Endocrinology* 136, 4298–4303.

Neville, M.C., Chatfield, K., Hausen, L., Lewis, A., Monks, J., Nuijens, J., Ollivier-Bousquet, M., Schanbacher, F., Sawicki, V. and Zhang, P. (1998) Lactoferrin secretion into mouse milk. In: Spik, G.L.D., Mazurier, J. and Pierce, A. (eds) *Advances in Lactoferrin Research*. Plenum Press, New York, pp. 141–153.

Nishimoto, I. (1993) The IGF-II receptor system: A G protein-linked mechanism. *Molecular Reproduction and Development* 35, 398–407.

Nissley, P., Kiess, W. and Sklar, M. (1993) Developmental expression of the IGF-II/mannose 6-phosphate receptor. *Molecular Reproduction and Development* 35, 408–413.

Nuijens, J.H., van Berkel, P.H.C. and Schanbacher, F.L. (1996) Structure and biological actions of lactoferrin. *Journal of Mammary Gland Biology and Neoplasia* 1, 285–295.

Oh, Y. and Rosenfeld, R.G. (1998) IGF-independent actions of IGF binding proteins. *80th Annual Endocrine Society Meetings* no. 520–2, p.3, 30(Abstract)

Oh, Y., Muller, H.L., Pham, H. and Rosenfeld, R.G. (1993a) Demonstration of receptors for insulin-like growth factor binding protein-3 on Hs578T human breast cancer cells. *Journal of Biological Chemistry* 268, 26045–26048.

Oh, Y., Müller, H.L., Lamson, G. and Rosenfeld, R.G. (1993b) Insulin-like growth factor (IGF)-independent action of IGF- binding protein-3 in Hs578T human breast cancer cells. Cell surface binding and growth inhibition. *Journal of Biological Chemistry* 268, 14964–14971.

Oka, Y., Rozek, L.M. and Czech, M.P. (1985) Direct demonstration of rapid insulin-like growth factor II receptor internalization and recycling in rat adipocytes. *Journal of Biological Chemistry* 260, 9435–9442.

Oscar, T.P., Baumrucker, C.R. and Etherton, T.D. (1986) Insulin binding to bovine mammary membranes: comparison of microsomes versus smooth membranes. *Journal of Animal Science* 62, 179–186.

Peralta Soler, A., Alemany, J., Smith, R.M., de Pablo, F. and Jarett, L. (1990) The state of differentiation of embryonic chicken lens cells determines insulin-like growth factor 1 internalization. *Endocrinology* 127, 595–603.

Pocius, P.A., Dreets, J.M., Devery-Pocius, J.E. and Baumrucker, C.F. (1984) Techniques for preparation of bovine mammary smooth membranes. *Journal of Dairy Science* 67, 2055–2061.

Politis, I., Lachance, E., Block, E. and Turner, J.D. (1988) Plasmin and plasminogen in bovine milk: a relationship with involution? *Journal of Dairy Science* 72, 900–912.

Politis, I., Zavizion, B., White, J.H., Goldberg, J.J., Baldi, A. and Akers, R.M. (1995) Hormonal and extracellular matrix regulation of plasminogen activator in a bovine mammary epithelial cell line. *Endocrine Journal* 3, 345–350.

Prosser, C.G. (1996) Insulin-like growth factors in milk and mammary gland. Anonymous

Prosser, C.G. and Fleet, I.R. (1992) Secretion of insulin-like growth factor II into milk. *Biochemistry Biophysics Research Communications* 183, 1230–1237.

Radulescu, R.T. (1994) Nuclear localization signal in insulin-like growth factor- binding protein type 3. *Trends in Biochemical Science* 19, 278.

Radulescu, R.T. and Wendtner, C.M. (1993) Hormone and growth factor subunits: A novel perception of cell growth regulation. *Journal of Endocrinology* 139, 1–7.

Rajah, R., Katz, L., Nunn, S., Solberg, P., Beers, T. and Cohen, P. (1995) Insulin-like growth factor binding protein (IGFBP) proteases: functional regulators of cell growth. *Progress in Growth Factor Research* 6, 273–284.

Rajah, R., Valentinis, B. and Cohen, P. (1997) Insulin like growth factor (IGF)-binding protein-3 induces apoptosis and mediates the effects of transforming growth factor-β1 on programmed cell death through a p53- and IGF-independent mechanism. *Journal of Biological Chemistry* 272, 12181–12188.

Rechler, M.M. (1993) Insulin-like growth factor binding proteins. *Vitamins and Hormones* 47, 1–114.

Roberts, A.J., Nugent, R.A., III, Klindt, J. and Jenkins, T.G. (1997) Circulating insulin-like growth factor I, insulin-like growth factor binding proteins, growth hormone, and resumption of estrus in postpartum cows subjected to dietary energy restriction. *Journal of Animal Science* 75, 1909–1917.

Ronge, H., Blum, J.W., Clement, C., Jans, F., Leuenberger, H. and Binder, H. (1988) Somatomedin C in dairy cows related to energy and protein supply and to milk production. *Animal Production* 47, 165–183.

Rosenfeld, R.G. (1998) The blind men and the elephant – a parable for the study of insulin-like growth factor binding proteins. *Endocrinology* 139, 5–6.

Ruoslahti, E. and Pierschbacher, M.D. (1987) New perspectives in cell adhesion: RGD and integrins. *Science* 238, 491–497.

Schedlich, L.J., Young, T.F., Firth, S.M. and Baxter, R.C. (1998) Insulin-like growth factor-binding protein (IGFBP)-3 and IGFBP-5 share a common nuclear transport pathway in T47D human breast carcinoma cells. *Journal of Biological Chemistry* 273, 18347–18352.

Schuster, D.E., Kehrli, M.E., Jr and Baumrucker, C.R. (1995) Relationship of inflammatory cytokines, growth hormone, and insulin-like growth factor-1 to reduced performance during infectious disease. *Proceedings of the Society of Experimental Biology and Medicine* 210, 140–149.

Seewaldt, V.L., Dietze, E.C., Johnson, B.S., Collins, S.J. and Parker, M.B. (1999) Retinoic acid-mediated G1-S-phase arrest of normal human mammary epithelial cells is independent of the level of p53 protein expression. *Cell Growth and Differentiation* 10, 49–59.

Shang, Y., Baumrucker, C.R. and Green, M.H. (1998) c-Myc is a major mediator of the synergistic growth inhibitory effects of retinoic acid and interferon in breast cancer cells. *Journal of Biological Chemistry* 273, 30608–30613.

Shang, Y., Baumrucker, C.R. and Green, M.H. (1999) The induction and activation of STAT1 by all-trans-retinolic acid are mediated by RARβ signaling pathways in breast cancer cells. *Oncogene* 18, 6725–6732.

Sharma, B.K., Vandehaar, M.J. and Ames, N.K. (1994) Expression of insulin-like growth factor-1 in cows at different stages of lactation and in late lactation cows treated with somatotropin. *Journal of Dairy Science* 77, 2232–2241.

Skaar, T.C. and Baumrucker, C.R. (1993) Regulation of insulin-like growth factor binding protein secretion by a murine mammary epithelial cell line. *Experimental Cell Research* 209, 183–188.

Staley, M.D. (1998) IGFBP-3 iodination and proteolysis in bovine serum and milk and mammary cell cultures. MS Thesis, Penn State University, pp. 1–229.

Stewart, C.E.H. and Rotwein, P. (1996) Insulin-like growth factor-II is an autocrine survival factor for differentiating myoblasts. *Journal of Cell Biology* 271, 11330–11338.

Taylor, S.I., Najjar, S., Cama, A. and Accili, D. (1996) Structure and function of the insulin receptor. In: Anonymous. *Insulin-like Growth Factors: Molecular and Cellular Aspects*, pp. 221–243.

Thissen, J.-P., Ketelslegers, J.-M. and Underwood, L.E. (1994) Nutritional regulation of the insulin-like growth factors. *Endocrine Reviews* 15, 80–101.

Valentinis, B., Bhala, A., DeAngelis, T., Baserga, R. and Cohen, P. (1996) The human insulin-like growth factor (IGF) binding protein- 3 inhibits the growth of fibroblasts with a targeted disruption of the IGF-I receptor gene. *Molecular Endocrinology* 9, 361–367.

Vega, J.R., Gibson, C.A., Skaar, T.C., Hadsell, D.L. and Baumrucker, C.R. (1991) Insulin-like growth factor (IGF)-I and -II and IGF binding proteins in serum and mammary secretions during the dry period and early lactation in dairy cows. *Journal of Animal Science* 69, 2538–2547.

Vicini, J.L., Buonomo, F.C., Veenhuizen, J.J., Miller, M.A., Clemmons, D.R. and Collier, R.J. (1991) Nutrient balance and stage of lactation affect responses of insulin, insulin-like growth factors I and II, and insulin-like growth factor-binding protein 2 to somatotropin administration in dairy cows. *Journal of Nutrition* 121, 1656–1664.

Werner, H. and LeRoith, D. (1996) The role of the insulin-like growth factor system in human cancer. *Advances in Cancer Research* 68, 183–223.

White, M.F. and Kahn C.R. (1994) The insulin signaling system. *Journal of Biological Chemistry* 269, 1–4.

Williams, G.T., Abler, A.S. and Lau, L.F. (1992) Regulation of gene expression by serum growth factors. In: Stein, G.S. and Lian, J.B. (eds) *Molecular and Cellular Approaches to the Control of Proliferation and Differentiation*. Academic Press, San Diego, pp. 115–162.

Yamanaka, Y., Wilson, E.M., Rosenfeld, R.G. and Oh, Y. (1997) Inhibition of insulin receptor activiation by insulin-like growth factor binding proteins. *Journal of Biological Chemistry* 272, 30729–30734.

Yateman, M.E., Claffey, D.C., Cwyfan Hughes, S.C., Frost, V.J., Wass, J.A.H. and Holly, J.M.P. (1993) Cytokines modulate the sensitivity of human fibroblasts to stimulation with insulin-like growth factor-I (IGF-I) by altering endogenous IGF-binding protein production. *Journal of Endocrinology* 137, 151–159.

Yee, D., Paik, S., Lebovic, G.S., Marcus, R.R., Faconi, R.E., Cullen, K.J., Lippman, M.E. and Rosen, N. (1989) Analysis of insulin-like growth factor I gene expression in malignancy: Evidence for a paracrine role in human breast cancer. *Molecular Endocrinology* 3, 509–517.

Zacksenhaus, E., Bremmer, R., Phillips, R.A. and Gallie, B.L. (1993) A bipartite nuclear localization signal in the reinoblastoma gene product and its importance in biological action. *Molecular Cell Biology* 13, 4588–4597.

Zapf, J. (1995) Physiological role of the insulin-like growth factor binding proteins. *European Journal of Endocrinology* 132, 645–654.

# 20 Integrating the Effects of Genotype and Nutrition on Utilization of Body Reserves During Lactation of Dairy Cattle

J.P. McNamara

*Department of Animal Sciences, Washington State University, Pullman, Washington, USA*

## Introduction

Mammals have evolved to store several basic nutrients in order to minimize the effects of variation in environmental supply (Pond, 1984). Body fat is the largest storage organ, often containing several months worth of maintenance energy requirements. Body protein is also stored, in the form of blood and organ proteins, but the largest mass is muscle protein. Reserves of body protein can often be more important to the function of the animal than body fat. During periods of deficit of even a few amino acids, body proteins must be broken down to supply the proper balance of amino acids for synthesis of critical regulatory, enzymatic or structural proteins. In addition, because of the strict requirement for glucose by several organs, especially the central nervous system, body protein serves as a reservoir of glucose. Deficiencies of glucose supply longer than roughly 1 day can only be supplied by gluconeogenesis from amino acids released from visceral or muscle protein, and muscle is by far the largest source.

In animal agriculture, management of body reserves has become most important in the feeding and care of dairy and beef cattle. Dairy animals of even average milk producing ability undergo a period of deficit of both energy and amino acids in early lactation. Research has discovered basic principles and some parameters describing storage and use of body fat and protein during lactation. However, a quantitative description of genetic, endocrine and neural regulation of the ability to store and utilize body reserves is lacking. From a practical feeding management standpoint, we routinely use subjective measures of body fat and protein storage (usually termed 'body condition') in our feeding programmes. It would be advantageous to have a more precise description of body fat and protein use.

To do this we need a more complete mechanistic description of storage and use of body fat and protein in lactating animals. Various empirical equations have been developed to describe these interactions in lactating cows. However, the most efficient utilization of natural resources requires a better understanding of the genetic and physiological regulation of body reserves. Use of a mechanistic, bio-mathematical model of nutrient use may assist us in our discussion of this topic. This chapter is a summary of recent work with the purpose of defining quantitative regulatory mechanisms and descriptions of body reserve storage and use in lactation. A brief overview of practical application will be given to set the frame of reference for mechanistic studies. Discussion will be in reference to challenge and improvement of a deterministic, dynamic, mechanistic model of metabolism in dairy cattle (Baldwin *et al.*, 1987a,b; Baldwin, 1995; McNamara and Baldwin, 1995, 2000). The context will be at the biochemical pathway level, integrating characteristics such as maximal velocities and substrate sensitivities to genetic and nutritional control of body reserve use. The goal is to help define specific research hypotheses to identify mechanisms of body fat and protein storage and use; and help define application strategies that will improve the overall efficiency of natural resource use in milk production.

## Body fat reserves and metabolism in dairy cattle

From a practical point of view, a wide range of fat stored prior to first calving is related to optimal milk production: either too little or too much reduces milk production. Producers know that improper management of adipose tissue leads to immediate problems such as reduced feed intake, dystocia and related post-partum reproductive problems, metabolic acidosis, displaced abomasum, reduced fertility and reduced production (Garnsworthy, 1988; Jaquette *et al.*, 1988; Jones and Garnsworthy, 1989; Gearhart *et al.*, 1990; Waltner *et al.*, 1993; Heuer *et al.*, 1999).

One of the first major efforts to codify knowledge into a practical application was the development of the 'body condition score' (BCS) system. Developed independently in several locations, the principle was to provide an easily conducted, somewhat objective measure by which an observer may codify the amount of fat and muscle storage on an animal. A common system is a 5-point scale, where 1 is assigned to an extremely emaciated animal showing prominent skeletal structures with only skin covering, and a 5 is assigned to an animal so covered with fat that bones are barely visible and lumps of fat may be seen (Wildman *et al.*, 1982). This system has been used in many survey studies to devise equations relating BCS to various nutritional management strategies (Garnsworthy, 1988; Gearhart *et al.*, 1990; Waltner *et al.*, 1993; Heuer *et al.*, 1999). This system was initially (and with some truth) criticized as being too subjective to use in scientific investigations. However, it was the use of this system that helped uncover and define some of the relationships between pre-partum nutrition and milk production in the subsequent lactation. It also had utility in defining the effect of a too rapid rate of growth of dairy heifers on subsequent diminution of milk production potential.

There is now fair agreement between a 'unit of BCS' and an amount of body weight or body fat; estimates range from approximately 25 to 30 kg of body fat per unit BCS if body weight is used in the equation and 40–50 kg per unit of BCS if it is not (Garnsworthy, 1988; Waltner *et al.*, 1994; Komagiri and Erdman; 1997). Thus this rather aggregate measure, when applied to several animals over a sufficiently long period of time (more than at least 1 month), has been quite helpful in defining use of body reserves in various experimental situations.

One focus in a study of BCS was to investigate the relationship between BCS (fat storage) at calving and feed intake and milk production during lactation. In dairy animals, there has been a negative effect of too much fat at calving, known as 'fat cow syndrome', such that overly fat animals display an increased incidence of a variety of post-partum metabolic and reproductive disorders (Garnsworthy, 1988; Jones and Garnsworthy, 1989; McNamara, 1994); this phenomenon is also present in lactating swine (McNamara and Boyd, 1999). The hypothesis was that the extra body fat is rapidly mobilized due to the initialization of the lactation hormone complex and enhanced by the negative energy balance in early lactation. Due to the high amounts of circulating free fatty acids, neural pathways regulating feed intake depress feed intake from a point that it may have otherwise reached.

I have no argument with this theory or phenomenon, and it has proved its utility. As producers have paid more attention to proper pre-partum nutrition, the incidence of fat-cow related health problems has diminished. However, through the years the concept was applied to all dairy animals as gospel, that is 'a BCS of 3.5 or more should be avoided as it will lead to fat-cow syndrome and increased problems'. However, many observations in university and field situations in the US belied the generality of this. In fact, for many animals it was more of a challenge to achieve sufficient body fatness prior to the subsequent lactation, and milk production was compromised. This was thought by some, including me, to be due to the fact the genetic ability of these Holstein dairy cows to produce milk had 'outstripped' their ability to recover body fat in late lactation. Also, several large herd survey studies demonstrated that, although there was a positive association with fatness at calving and disease, the actual incidence of 'fat cows' was quite small (less than 10%; Jaquette *et al.*, 1988; Gearhart *et al.*, 1990; Waltner *et al.*, 1993; Heuer *et al.*, 1999). This is likely because of the increased milk production ability and the improved management compared to the industry 10–20 years earlier. Thus, the biological principle of 'too much fat is bad' was not to be discarded, but rather we sought a more specific definition of what 'too much' really was for various populations.

We measured the changes in BCS of cows in one herd (over 200 lactation records) to develop equations relating BCS to milk production and days in milk. In addition, we asked the question mathematically of 'What is the relationship of BCS at calving to subsequent milk production?' For animals in this herd, which averaged 9541 kg of 3.5% fat corrected milk in 305 days (range 8826–10818) during this study, in order to maximize milk production, BCS needed to be between approximately 3.5 and 4.0 (Fig. 20.1). Below that and milk production was lower. However, for older cows more body fat did not decrease milk production. The BCS range of these cows only went up to 4.5, and there were very few cows at this level. It is also interesting to note that a BCS at calving of 4.5 was associated with the same milk production as that at a BCS of 3.0. This herd-level study simply suggests that we should move beyond any one 'target' BCS and rather look at body fat storage and use as a continuum. The genetic potential, body size and available feed all need to be considered when interpreting BCS or making recommendations about desired BCS.

Another endeavour in the last 20 years of studying body reserves has been to develop useful and reasonable techniques to measure reserves in live animals. Several studies used dilution of body water with deuterium oxide to derive equations describing body fat and protein. There was a tremendous amount of work on this technique to

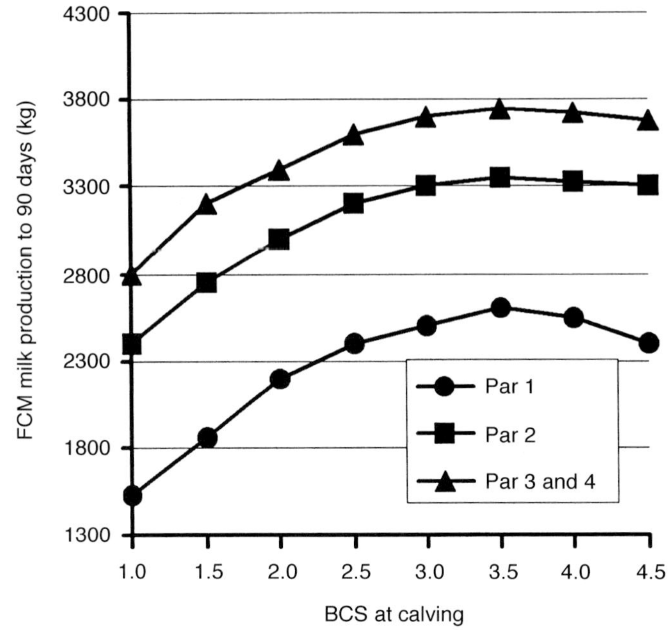


**Fig. 20.1.** Relationship of body condition score (BCS) at calving with production of fat-corrected milk (FCM) (from Waltner *et. al.*, 1993). Note that milk is maximized at a score of 3.5–4 for first lactation animals, while for second and higher lactation, increasing body fat does not have a negative effect on milk production.

measure body water as a way to estimate body fat (see Andrews *et al.*, 1995; Komagiri and Erdman, 1997; Komagiri *et al.*, 1998 and references therein). The literature is too large to expand upon here, but in summary, this technique has been useful, but has as drawbacks the need for more sophisticated equipment and intense labour.

Another technique now in use is the measurement of fat cell size. A small biopsy of subcutaneous adipose tissue is taken and cells are isolated and the average diameter is measured. Because body lipid mass in sub-adult and adult ruminants is primarily a function of adipose hypertrophy, the relationship of fat cell size to body lipid mass is quite strong in lactating dairy cattle (Robelin *et al.*, 1989; Waltner *et al.*, 1994). This technique has the benefits of much reduced cost, and a characteristic of the adipose tissue is being directly measured. Thus it is more likely to have a greater utility than the body water technique, especially in early lactation. In addition to being useful to estimate total body fat, it teaches us something about the cellular changes going on as well. An important point from the discussion above is that regardless of the metabolic or molecular techniques which may be used to define phenotypic or genotypic attributes, we have to estimate total body fat at some point. Otherwise, we do not have a complete picture of what the mechanistic elements mean in relation to total body energy use.

My colleagues and I were able to conduct several investigations covering from 1983 to the present on the rates of biochemical pathways in adipose tissue of dairy cattle varying in genetic background, parity and nutritional environment. We defined a time-course of adaptations in lipogenesis, lipolysis and esterification throughout lactation (McNamara and Hillers, 1989; McNamara, 1994). Lipogenesis decreased 90% in all cows at day 15 of lactation, but even by day 30 was starting to recover (Fig. 20.2). Cows of high milk-producing ability had lower rates than did cows of average ability. Lipogenesis in all cows stayed markedly elevated throughout lactation. Low energy intake diminished lipogenesis in cows of average milking ability, but not in superior ones. Lipolysis increases during lactation (Fig. 20.3), and as lactation progresses, rates of fatty acid release in adipose tissue stay elevated (McNamara and Hillers, 1989; McNamara, 1994). There was a faster rate of lipolysis in high merit versus low merit cows, and little effect of dietary energy intake on lipolysis rates. These genetic differences in lipogenesis have been described elsewhere as changes in substrate sensitivities and maximal velocities (McNamara *et al.*, 1991a; McNamara and Baldwin, 2000). This was the first demonstration that genetic selection for milk production altered enzymatic expression in adipose tissue of dairy cattle.

This continued maintenance of high rates of lipolysis, even in positive energy balance, demonstrates the adaptations of endocrine and kinetic systems to maintain a high rate of milk-fat output. Because these processes are so essential to survival, there is a very strict and redundant series of controls exerted on these reactions. It is likely that the sympathetic nervous system helps maintain a higher rate of lipolysis, driven by the need for fat by the mammary gland (McNamara and Murray, 1994). Such control of

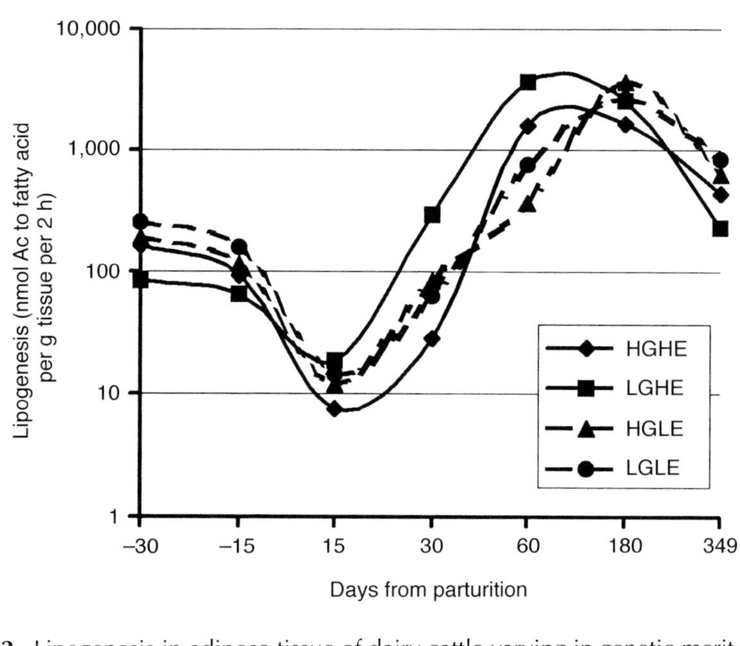


**Fig. 20.2.** Lipogenesis in adipose tissue of dairy cattle varying in genetic merit and dietary intake. Note logarithmic scale. Animals of high merit (HGHE) had lower rates of lipogenesis than lower merit animals (LGHE) when fed normally. Lower energy intake decreased lipogenesis in LG animals (LGLE) but not in HG cows. Note the classic metabolic overshoot of lipogenesis by day 60 and sustained high rates during all of lactation.

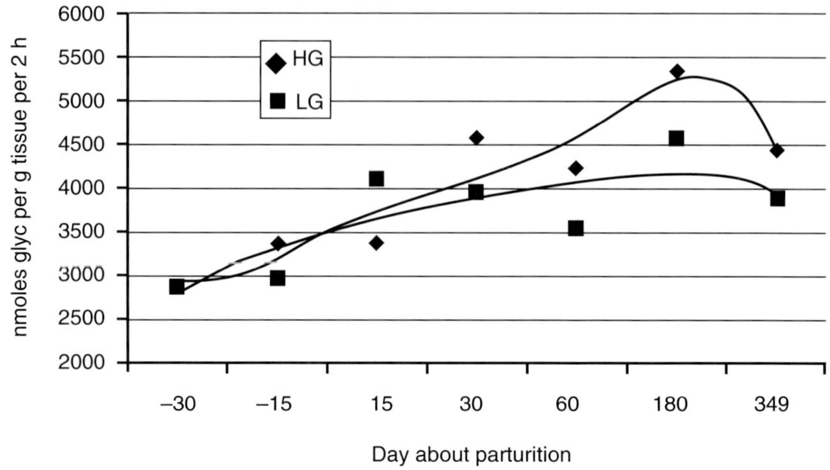


**Fig. 20.3.** Catecholamine stimulated lipolysis in adipose tissue of lactating dairy cattle varying in genetic merit. Note that high genetic merit animals (HG) had higher rates than low merit (LG) animals, even through late lactation. Also note that lipolysis remained elevated during late lactation, even in positive energy balance. This is likely to supply the continued demand for milk fat precursors, and results in a sustained increase in lipid turnover.

adipose tissue accumulation by the sympathetic nervous system has been demonstrated during growth and obesity (Knehans and Romsos, 1983; Dulloo and Miller, 1985). Differential control of tissue metabolism by the sympathetic nervous system during pregnancy and lactation also occurs; for example, there is less activity by the brown fat in rodents, probably to conserve energy (Trayhurn and Richard, 1985). Although this regulation is obviously genetically dictated, the heritability of nervous regulation is not known, and would be difficult to determine for lactating cattle. Rodent models of growth and lactation may have utility for study of this question, as may various potential genetic 'knockout' constructs.

In order to understand this regulation of body reserve use, we have run studies to define how the integration of lipogenesis, esterification and lipolysis occurs under various situations of dietary energy intake, genetic ability of the animal, environmental stress, hormonal management, and stage of lactation. This was done with the purpose of challenging present understanding as explicit in the research model of metabolism in dairy cattle introduced above. The following equations from that model provide a quantitative framework for our discussion of genetic and environmental regulation of body reserve use.

## *Adipose tissue equations*

The reactions are summed into the balance of body fat stores: Ts = integration (ITs, DTs); where ITs is the initial amount of body triglycerides storage (at the beginning of lactation), and DTs = FaTs + AcTs – TsFa. The notation DTs is the integration of the reactions within a given time period (for example, 1 day of a 305-day lactation). The notation FaTs indicates the esterification of preformed fatty acids into triglycerides, AcTs is the integrated total of acetate formation into fatty acids and then triglycerides, and TsFa is the rate of triglyceride hydrolysis to free fatty acids.

The amount of ITs can represent a function of genetic makeup of the animal, measured in such ways as composition of gain to a given day of age for first-lactation animals. The amount of ITs can be measured either directly by slaughter analysis, or more usefully by indirect means, such as determining body water or fat cell size (Brown *et al.*, 1989; Robelin *et al.*, 1989; Waltner *et al.*, 1994; Andrews *et al.*, 1995). Using equations we derived from slaughter trials, the amount of fat can be related to the body weight and condition score or vice versa (Waltner *et al.*, 1994). Incorporation of body condition score into various lactational models is already underway, but accurate descriptions will require estimates of body fat. In most models, body condition score is calculated from body fat predicted from basic energetic use, and not the other way around.

Most equations in the model are either substrate saturation (Michaelis–Menten) or mass action. Thus, the maximal velocity of a reaction for a given animal may be seen as a phenotypic trait, for example the $V_{max}$ for lipogenesis in an aggregate biochemical model can represent such mechanisms as the amount of enzyme complex, total number of fat cells, and activity of the enzyme. The substrate sensitivity constant (*K*s) can initially be a phenotypic trait, representing again the sensitivity of the enzyme complexes for a given reaction. In addition, the *K*s can be variable, and affected by nutritional environment. Biologically, for example, we know that a decreased amount of glucose (even in ruminants) will decrease the carbon flux through lipogenic pathways. Due to decreased insulin or increased glucagon or catecholamines, the enzyme will have decreased activity (increase in *K*s) and eventually enzyme synthesis will decrease (decreased $V_{max}$). Thus, substrate concentration is a direct effect of nutritional programme, and over time, may alter parameter values describing phenotypic variables as well.

The equation for lipogenesis from acetate is:

$$\text{AcTs} = \text{VAcTs} / (1.0 + \text{KAcTs} / \text{cAc} + \text{K1AcTs} / (\text{Ahor2} \times \text{cGl} )).$$

In this equation, VAcTs is the maximal velocity of acetate conversion to triglyceride, KAcTs is the substrate sensitivity constant for Ac, K1AcTs is the sensitivity constant for glucose (cGl), Ahor2 (anabolic hormone) represents the effect of insulin on acetate conversion to lipid, and cGl is the concentration of glucose. Additionally, Ahor2 is a function of concentration of glucose compared with the reference concentration ([cgl]/[rcgl]), reflective of glucose supply and/or energy balance. Thus, most of our basic principles concerning lipogenesis are explicit in this equation. Readers interested in a more detailed treatment of adipose biochemistry should see Chapters 12 and 16 in Baldwin (1995).

The equation describing esterification of fatty acids to triacylglycerols is :

$$\text{FaTs} = \text{VFaTs} \times (\text{EBW**0.75}) / (1.0 + \text{KFaTs} / \text{cFa} + \text{K1FaTs} / (\text{Ahor} \times \text{cGl})).$$

This represents esterification from fatty acids (direct from diet or from re-circulation) as a function of maximal velocity (VFaTs), metabolic body size (EBW**0.75; ** indicates raising to an exponent), sensitivity to circulating fatty acids (KFaTs), and sensitivity to glucose (K1FaTs), which can be altered by anabolic hormone. Note that the 'Ahor' in this equation is not the same 'Ahor2' in the lipogenesis equations, recognizing the potential differential responsiveness of lipogenesis and esterification to insulin or circulating glucose. This initial hypothesis was based on work done in other species or in preliminary trials. This differential control of lipogenesis and esterification was confirmed with studies of adipose tissue metabolism (McNamara and Hillers, 1989; McNamara *et al.*, 1991a). Thus the model explicitly describes hormonal control of major pathways and has the 'mathematical' as well as biochemical attributes to test hypotheses concerning actions of various hormones. When more data become available, other genetically controlled hormonal actions can be included.

The reaction describing hydrolysis of triacylglycerols to fatty acids and glycerol is:

TsFa = VTsFa × (EBW**0.75) × CHOR1 × T3 / (1.0 + (cFa / K1TsFa ) ** EXP10 + (KTsFa /cTs )** THETA1) ;

in which lipolysis (TsFa) is a function of metabolic body size, maximal velocity, sensitivity to catabolic hormone (CHOR1, norepinephrine), injected thyroid hormone (T3), and feedback from circulating fatty acids (K1TsFa; note this is set to be an inhibition on lipolysis as Fa concentration increases). The exponents EXP10 and THETA1 alter the sensitivity of the reaction to the substrate under the exponent. It is likely that this parameter also represents genetically controlled differences in total enzyme present or pool size of substrate. Note that there is a 'protected', very low amount of body fat (KTsFa/cTs) which, if approached, will rapidly diminish rates of lipolysis. The amount of protected body fat is set very low, representing basic membrane lipid, so the animal does not remove all lipid from the body. Note also that lipolysis is controlled by the catabolic hormone and not the anabolic hormone directly.

Thus, the current model recognizes and allows the testing of hypotheses on the differential nature of metabolic control. It is not explicit that the effect of circulating glucose, for example, must have an equally opposite effect on lipogenesis and lipolysis; this captures the concept of homeorhetic control of metabolism quite well (see Chapter 18). The effect can have differential sensitivity based on stage of lactation, rate of milk production, or genetic differences in adipose phenotype. Also note that with use of various exponents (the value of which can be changed from simulation to simulation), the control of sub-components of this equation can alter the sensitivity of response to various substrates as necessary. Studies may be designed to measure these parameters in breeds of cattle widely varying in milk production or growth ability.

## Body protein reserves in lactating dairy cattle

Although the primary function of skeletal muscle is mobility, it also provides a nutritional function during periods of negative nitrogen or glucose balance. In lactating dairy cows, a quantitative appraisal of nitrogen and glucose balance demonstrates that significant amounts of body protein must be used for a period of time in early lactation. There was some research interest in this question over 20 years ago, which led to the term 'protein reserves' (Botts *et al.*, 1979). This was a misnomer in that mammals do not store protein separately like triglyceride in adipose tissue. Nevertheless, protein, in the form of either visceral or muscle protein, is available for mobilization to supply essential amino acids and glucose during periods of deficit.

There were hurdles that diminished research interest in this area. One was the inconsistent response in milk protein output in experiments designed to alter 'protein reserves' pre-partum. This was partly due to the lower production level of the animals at that time, such that the amount of body protein at calving was probably not a limiting factor in the 'average' cow (Botts *et al.*, 1979). Experiments were often done with the idea that a greater amount of 'protein reserves' would increase milk protein output regardless of the potential production rates of the cow (i.e. it was assumed that 'protein reserves' were always limiting). It is probably more exact to test the hypothesis that maximal lactational protein output may be limited in situations in early lactation if protein intake is not adequate, and in those situations a greater amount of body protein at calving may help allow the maximal milk protein output. In addition, the glucose balance of the animals in these early studies was not often controlled for, and the role of body protein for gluconeogenic precursor supply was ignored, adding to variation in experimental results. Another hurdle was technological, in that measurement of body protein content and rates of body protein synthesis and degradation were quite variable, difficult and costly to make. Progress has been made on this front, but technical precision and cost is still a major block to a better quantitative understanding in this area.

Muscle protein accretion is the sum of the integration of protein synthesis and degradation (Waterlow, 1995). Amino acid storage and release from the muscle is also important in metabolic health, as acidosis results from excess oxidation of ketogenic amino acid carbon as well as from oxidation of long-chain fatty acids. The high-producing lactating dairy cow in early lactation can lose significant amounts of body protein (Andrews *et al.*, 1995; Meijer *et al.*, 1995; Komagiri and Erdman, 1997), and not all of these amino acids appear in milk protein. Most present models treat these pathways at a very aggregate level, with admittedly inadequate equation forms and parameter values. Again, this is due to a lack of good data on these processes, not to a lack of appreciation of their importance. Equations below are critical ones for which we need information.

## Muscle tissue and amino acid metabolism equations

Accretion of muscle tissue is summarized as: DPb = AaPb − PbAa; in which amino acids are either synthesized into muscle (AaPb) or removed from muscle in proteolysis (PbAa). Muscle protein synthesis is described as : AaPb = VAaPb / (1.0 + KAaPb / (Ahor×cAa)); where the maximal velocity of muscle growth is VAaPb, and the sensitivity (KAaPb) to amino acid (cAa) supply can be changed by anabolic hormone, Ahor. Note this is a different Ahor from that used for lipogenesis but the same one used for esterification. Thus is recognized the differential sensitivity of control by insulin on a synthetic reaction (such as that making fatty acids from carbohydrate breakdown intermediates), versus the two esterification reactions (peptide bond and triacylglycerol synthesis). This also allows the same hormone to have responses in the same direction, but with differential sensitivity in different tissues. These parameters all describe genetic traits and it may be possible to define phenotypes in such a manner.

Muscle protein breakdown (PbAa) in the model is explicitly a mass action phenomenon with a mass action constant k value (kPbAa), which does not change within a simulation, but can be changed to simulate various potential rates: proteolysis (PbAa) = kPbAa × Pb. Information may be found in scientific literature suggesting that the process is more complex than this; however, there are no data in lactating dairy cattle which allow direct observation of any further complexity, thus it can not be explicit in the model (Baldwin, 1995). Again, for a research model, it directs scientists to an area in dire need of better information.

It would be relatively straightforward to estimate the muscle protein pool size at the beginning of lactation and at key points during lactation on various nutritional regimens. If we used animals of different breeds and/or animals of widely varying genetic ability within a breed, we could begin to define the genetically or phenotypically controlled aspects of these parameters. It will be more difficult to sort out genetic effects on muscle protein synthesis and breakdown due to limitations in measuring these pathways. However, studies done in the context of comparing genetically different groups of animals could easily be designed into research on muscle protein turnover. Presently data are very limiting, and although regulatory mechanisms involved in protein synthesis and degradation are qualitatively understood, a quantitative description is lacking. The unexplained variation between protein input and output in dairy cattle could easily be explained within the standard deviation of estimates of amino acids coming from muscle protein turnover.

The most limiting amino acids are methionine, lysine, histidine, phenylalanine, glutamine and glutamate, defined by the amount needed for milk protein versus that absorbed by the gut. They decrease from 16 to 25% in the blood of cows in early lactation (Meijer *et al.*, 1995), and the amino acid profile in muscle during this stage resembles that in starvation. The free pool of glutamine (which together with its direct derivatives, glutamate and proline make up 25–30% of milk casein) in the muscle decreases by 25% during this time. Muscles make up 60–70% of the empty body weight of dairy cattle. The limited number of estimates of the rate of muscle protein turnover, most extrapolated from the rat or ewe, would suggest the muscle protein turnover rate is approximately 3% day$^{-1}$ (Baldwin, 1995; Waterlow, 1995). At this rate, for 80 kg of protein (~360 kg muscle in a 600 kg empty body weight cow) this is 2.4 kg of muscle protein synthesized and broken down every day. Yet even a cow producing 50 kg of milk at 3.2% protein makes only 1.6 kg of milk protein day$^{-1}$! Thus, the use of limiting amino acids by the muscle every day exceeds that of the mammary gland by at least 50%. During early lactation, net muscle breakdown is positive, so as to supply amino acids from protein synthesis and gluconeogenesis. Recent estimates have shown that this can be up to 20 kg of protein in high-producing animals in the first 5 weeks of lactation (Komagiri and Erdman, 1997). However, there is wide variation in error in these measurements, and more work is clearly needed. Quite recent work, published as yet only in abstract, used 3-methylhistidine to creatinine ratio for an estimate of muscle protein turnover. They demonstrated a large increase during lactation compared with the dry period, with the rough estimate of muscle protein turnover potentially at 5% day$^{-1}$ (Overton *et al.*, 1998).

It is important to note that the amino acids released from the muscle do not always match the pattern of those incorporated, as approximately one in seven histidines are methylated, and branched-chain amino acids can be oxidized at varying rates (Danfaer, 1990; Baldwin, 1995; Waterlow, 1995; Overton *et al.*, 1999). There are wide ranges in the estimates of gluconeogenesis coming from amino acids in lactating ruminants, from 2 to 40% (see Overton *et al.*, 1999, and several references therein). It is likely that a significant part of this variation is due to differences in muscle protein turnover and availability of gluconeogenic amino acids. Causes of these differences could include parity, stage of lactation, genetic makeup, as well as the amino acid and glucose status of the animals.

This muscle is regenerated during middle to late lactation (even more so in first and second parity animals). At 20 kg of protein to be regained and assuming 150 days to accumulate it, then to the 1.6 kg day$^{-1}$ of turnover we add 133 g (20,000 g per 150 days) of accumulation. To put this in perspective, this amount is more than the average observed increase in milk protein due to increasing dietary protein by 2.8 percentage points (Santos *et al.*, 1998; NRC 1989). Thus we propose that variation in muscle protein turnover during lactation, among parities and across diets, can significantly affect our ability to predict milk protein output. Also, simulation analyses performed with several model systems, including the National Research Council (1989), the Cornell Net Carbohydrate and Protein System (Kohn *et al.*, 1998) and the mechanistic model 'Molly' (Baldwin, 1995), demonstrate inadequacies in describing body muscle and fat use during lactation.

In a recent comparison, it was shown that in order for the Cornell system to predict outputs from inputs, the 'average daily gain' function was severely underestimated (5 kg day$^{-1}$; Kohn *et al.*, 1998). This system has been well designed and evaluated, and is useful for describing nutrient use for milk production in dairy cattle. Daily gain (or loss) in lactating animals can be both fat and protein. This lack of knowledge of muscle metabolism limits our ability to accurately predict animal nutrient use.

## Integration of nutrient interconversions in a model of dairy cattle metabolism-utility for study and application of body reserve management

We have used the mechanistic, dynamic 'Molly' model of Dr Baldwin and colleagues in studies designed to evaluate and challenge behaviour of this model (Baldwin *et al.*, 1987a,b; Baldwin, 1995; McNamara and Baldwin, 1995, 2000). Two studies were performed; the first was to describe differences in adipose tissue metabolism due to genetic selection (key results are presented in Figs 20.3 and 20.4). In the second trial, dairy cattle were fed diets that varied in fat from whole cottonseeds and from ruminally protected fats (McNamara *et al.*, 1991a, 1995; Harrison *et al.*, 1995). Intake and milk output were measured over the course of the lactation, and biopsies of adipose tissue were taken. *In vitro* substrate saturation studies were performed on the adipose tissue and parameters describing lipogenesis, esterification and lipolysis were determined (McNamara *et al.*, 1995). Several simulations were then conducted; much of this has been reported (McNamara *et al.*, 1991a; McNamara and Baldwin, 1995, 2000). The 'bottom line' was that the model could predict milk component output from dietary input within 5% of the observed means (Table 20.1) across a wide range of dietary inputs and milk outputs. On normal-energy diets, the body fat changes were also simulated

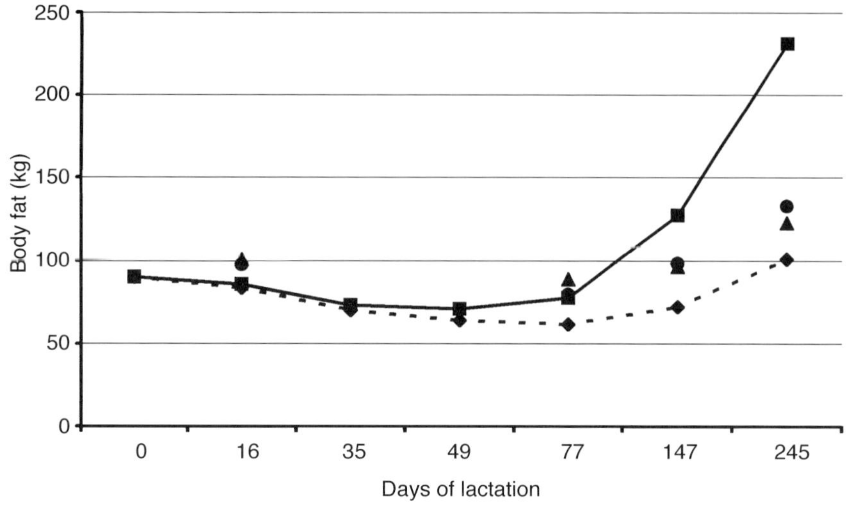


**Fig. 20.4.** Simulated and observed body fat in lactating dairy cattle fed control (2.5% lipid) or high-fat (6.0% lipid) diets. Dietary carbohydrate was approximately 3 percentage units less on high-fat diet and intakes were approximately 1 kg $day^{-1}$less on high-fat diet from day 17 to day 100, then they were similar. Simulated control diets are dotted lines and diamonds, observed control data are triangles; simulated high-fat diets are solid line and boxes, observed high-fat diet data are circles.

quite well (Table 20.2). This demonstrates the utility of mechanistic models for describing genetic differences in body reserve use.

However, the second trial was more extensive and allowed a more severe challenge of our knowledge as described in the model. When feeding cows greater amounts of fat, body fat accumulation was over-simulated at a fast rate (Table 20.2 and Fig. 20.4; McNamara and Baldwin, 2000). Body fat was simulated adequately on control diets but over-simulated on high-energy diets, compared with observed data (open squares and solid circles, Fig. 20.4). Thus the model can balance for input–output, but a severe challenge shows it is inadequate in describing protein and lipid metabolism over longer periods of time. It is telling that the Cornell model system, which was constructed independently, with a different objective than the 'research' Molly model, also has a bias in the 'average daily gain' function in order to balance inputs and outputs in milk (Kohn *et al.*, 1998).

The lack of precision in simulating the fat reserve led us to examine the use of body protein as one of the sources of inadequacy. This is primarily because protein turnover accounts for a large (~20%), variable and little-defined cost of basal metabolism (Baldwin, 1995; Waterlow, 1995). Figure 20.5 shows the use of body protein in the simulations described above. The 15–24 kg of body protein simulated to be lost by day 77 (Fig. 20.5) is remarkably close to the 20 kg reported in one of the few direct observations of protein loss in lactating dairy cattle (Komagiri and Erdman, 1998). It is also telling to note that the model described a greater loss of body protein on the animals fed the high-fat diet. In the experiment, dietary carbohydrate percentages were 3–4%

**Table 20.1.** Observed and simulated data from dairy cattle fed rations varying in fat content.

| Variable | Control[a] | | WCS | | WCSFA | |
|---|---|---|---|---|---|---|
| | Obs[b] | Sim | Obs | Sim | Obs | Sim |
| *Weeks 3 to 17* | | | | | | |
| DM intake (kg $day^{-1}$) | 23.0 | 23.0 | 23.9 | 23.9 | 21.3 | 21.3 |
| Milk production (kg $day^{-1}$) | 43.3 | 43.9 | 43.1 | 44.1 | 41.1 | 41.6 |
| Milk fat (kg $day^{-1}$) | 1.39 | 1.48 | 1.44 | 1.53 | 1.53 | 1.48 |
| Milk protein (kg $day^{-1}$) | 1.32 | 1.37 | 1.29 | 1.39 | 1.21 | 1.33 |
| *Weeks 18 to 35* | | | | | | |
| DM intake (kg $day^{-1}$) | 21.6 | 21.6 | 22.8 | 22.8 | 22.7 | 22.7 |
| Milk production (kg $day^{-1}$) | 33.4 | 33.6 | 33.8 | 33.5 | 35.4 | 32.8 |
| Milk fat (kg $day^{-1}$) | 1.16 | 1.09 | 1.17 | 1.12 | 1.22 | 1.15 |
| Milk protein (kg $day^{-1}$) | 1.11 | 1.04 | 1.14 | 1.04 | 1.12 | 1.01 |

[a] From McNamara and Baldwin (2000). Dairy cattle were fed lucerne hay (23% of dry matter (DM)), grass silage (23% of DM) and 54% concentrate diets (control); same diet with 12.7% of whole cottonseeds in the concentrate, replacing maize (WCS); and WCS plus an additional 1.89% of ruminally protected long-chain fatty acids (WCSFA) from 17 to 305 days in milk (DIM). Simulations were run on the mechanistic model of Baldwin (1995) from 1 to 245 DIM. Intakes of nutrients were simulated explicitly. Initial body weight and fat, and pre-treatment milk yields were used to set initial parameter values.
[b] Obs = observed values; Sim = simulated values.

**Table 20.2.** Observed and simulated data on body fat in dairy cattle.

| Body lipids[c] (kg) | Control[a] | | WCS | | WCSFA | |
|---|---|---|---|---|---|---|
| | Obs[b] | Sim | Obs | Sim | Obs | Sim |
| *Time* | | | | | | |
| Week 2 | 101 | 69 | 109 | 84 | 98 | 71 |
| Week 9 | 90 | 55 | 104 | 85 | 81 | 64 |
| Week 17 | 98 | 65 | 105 | 110 | 100 | 93 |
| Week 35 | 125 | 101 | 134 | 186 | 135 | 205 |

[a] From McNamara and Baldwin (2000). Dairy cattle were fed lucerne hay (23% of dry matter (DM)), grass silage (23% of DM) and 54% concentrate diets (control); same diet with 12.7% of whole cottonseeds in the concentrate, replacing maize (WCS); and WCS plus an additional 1.89% of ruminally protected long-chain fatty acids (WCSFA) from 17 to 305 days in milk (DIM). Simulations were run on mechanistic model of Baldwin (1995) from 1 to 245 DIM. Intakes of nutrients were simulated explicitly. Initial body weight and fat, and pre-treatment milk yields were used to set initial parameter values.
[b] Obs = observed values; Sim = simulated values.
[c] Body fat was biopsied at times indicated. Fat cell size was determined microscopically after cell fixation and dispersion. Equations to predict body fat from body weight and fat cells size, validated using cows from this herd, were used (Waltner *et al.*, 1994). The standard deviation of these measurements is 15 kg of body fat.

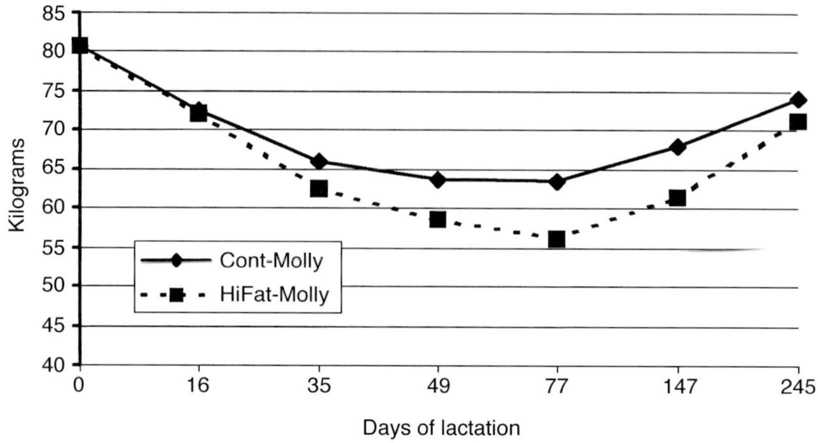


**Fig. 20.5.** Simulated body protein in lactating dairy cattle fed control (2.5% lipid) or high-fat (6.0% lipid) diets. Dietary carbohydrate was approximately 3 percentage units less on high-fat diet and intakes were approximately 1 kg day$^{-1}$ less on high-fat diet from day 17 to day 100, then they were similar. Note model describes a greater use of body protein on the high-fat diet, probably in response to the smaller intake of glucose and amino acids.

less on the high-fat diet, and voluntary feed intake was depressed for several weeks. The model would suggest, even when total energy was not limiting, that the apparent glucose deficit (and potential nitrogen deficit) dictated a larger mobilization of body protein. So there is confidence that the model is behaving adequately, but the precision is not yet adequate. Experiments designed to measure protein use will be essential if we are to describe genetic differences in body reserves.

This combination of animal experiments and model testing allows us to define the parameters and fluxes we need to know if we are to describe metabolism in the cow. These experiments and model simulations lead to the hypotheses proposed herein on muscle metabolism in lactation. A more precise description of the regulation of muscle metabolism will improve our ability to predict responses to nutritional, genetic and pharmaceutical management of milk production.

## Future research focus

Future research is likely to concentrate on two general areas, by nature of the objective of the research. One will be a continued search for molecular or cellular markers of important traits relating to body fat and protein use. Another will be refined precision in model descriptions of animal level use of fat and protein stores in practical situations. Several suggested areas that will help us improve efficiency of natural resource use by dairy animals are included below.

Questions on which we should focus would include: which genes or phenotypes provide the most efficient integration of mammary demand, muscle and adipose responses, feed intake ability? We need identification of genotypes, by DNA fingerprinting or other methods, which relate to most efficient use of nutrients, including potential storage of body fat and protein. How do we identify the critical pathways on which to focus our research efforts? We know most of the biochemical pathways involved. We need to continue to focus on the control systems of lipogenesis, lipolysis, protein synthesis and proteolysis. We need better methods to assess total body protein synthesis and breakdown. Indirect methods described above will be helpful but naturally limited in precision. A few well-designed experiments using stable isotopes of amino acids in dairy cattle to define equations describing these pathways will be very helpful. Less costly and invasive techniques can then be used in more practical situations to define the effects of breeding and nutritional strategies.

We must do research in the context of improving integrative models which explicitly include the complexity of biochemical and endocrine interactions among tissues and dynamically over time. Demand for glucose by the mammary gland affects several pathways in adipose and muscle tissue, from kinetic thermodynamics to altering endocrine regulation. These controlling functions are, in turn affected by genotype, stage of lactation and nutritional environment, often in subtle ways; various systems may become limiting to milk production as lactation progresses. Using examples given above, research on how insulin or catecholamines alter rates of muscle proteolysis and adipose lipolysis will be very useful. Research needs to identify how critical variables change during lactation, and estimate the parameters which define the biochemical and endocrine systems. A description of experimental designs and data needed have been presented previously (McNamara *et al.*, 1991b).

## References

Andrews, S.M., Erdman, R.A. and Waldo, D.R. (1995) Prediction of body composition from deuterium oxide and urea dilution in dairy cows at three physiological stages. *Journal of Dairy Science* 78; 1083–1095.

Baldwin, R.L. (1995) *Modelling Ruminant Digestion and Metabolism.* Chapman & Hall, New York.

Baldwin, R.L., France, J. and Gill, M. (1987a) Metabolism of the lactating cow. I. Animal elements of a mechanistic model. *Journal of Dairy Research* 54, 77–105.

Baldwin, R.L., France, J., Beever, D.E., Gill, M. and Thornley, J.H.M. (1987b) Metabolism of the lactating cow. III. Properties of mechanistic models suitable for evaluation of energetic relationships and factors involved in the partition of nutrients. *Journal of Dairy Research* 54, 133–145.

Botts, R.L., Hemken, R.W. and Bull, L.S. (1979) Protein reserves in the lactating dairy cow. *Journal of Dairy Science* 62, 433–450.

Brown, D.L., Taylor, S.J., De Peters, E.J. and Baldwin, R.L. (1989) Influence of sometribove, USAN (recombinant methionyl bovine somatotropin) on the body composition of lactating cattle. *Journal of Nutrition* 119, 633–638.

Danfaer, A. (1990) *A Dynamic Model of Nutrient Digestion and Metabolism in Lactating Dairy Cows.* National Institute of Animal Sciences Report, Foulum, Denmark.

Dulloo, A.G. and Miller, D.S. (1985) Increased body fat due to elevated energetic efficiency following chronic administration of inhibitors of sympathetic nervous system activity. *Metabolism* 34, 1061–1065.

Garnsworthy, P.C. (1988) The effect of energy reserves at calving on performance of dairy cows. In: Garnsworthy, P.C. (ed.) *Nutrition and Lactation in the Dairy Cow.* Butterworths, London, UK, pp.151–170.

Gearhart, M.A., Curtis, C.R., Erb, H.N., Smith, R.D., Snifffen, C.J., Chase, L.E. and Cooper, M.D. (1990) Relationship of changes in condition score to cow health in holsteins. *Journal of Dairy Science* 73, 3132–3139.

Harrison, J.H., McNamara, J.P., Waltner, R.L.S.S., Loney, K.A., Riley, R.E. and Cronrath, J.D. (1995) Effect of whole cottonseeds and calcium salts of long-chain fatty acids on performance of lactating dairy cows. *Journal of Dairy Science* 78, 181–183.

Heuer, C., Schukken, Y.H. and Dobbelaar, P. (1999) Postpartum body condition score and results from the first test day milk as predictors of disease, fertility, yield and culling in commercial dairy herds. *Journal of Dairy Science* 82, 295–304.

Jaquette, R.D., Rakes, A.H. and Croom, W.J., Jr (1988) Effects of body condition and protein on milk fat depression in early lactation dairy cows. *Journal of Dairy Science* 71, 2123–2131.

Jones, G.P. and Garnsworthy, P.C. (1989) The effects of dietary energy content on the response of dairy cows to body condition at calving. *Animal Production* 49, 183–190.

Knehans, A.W. and Romsos, D.R. (1983) Norepinephrine turnover in obese (ob/ob) mice: effects of age, fasting and cold. *American Journal of Physiology* 242, E253–E261.

Kohn, R.A., Kalscheur, K.F. and Hanigan, M. (1998) Evaluation of models for balancing the protein requirements of dairy cows. *Journal of Dairy Science* 81, 3402–3414.

Komagiri, M.V.S. and Erdman, R.A. (1997) Factors affecting body tissue mobilization in early lactation dairy cows 1. Effect of dietary protein on mobilization of fat and protein. *Journal of Dairy Science* 80, 929–937.

Komagiri, M.V.S., Casper, D.P. and Erdman, R.A. (1998) Factors affecting body tissue mobilization in early lactation dairy cows 2. Effect of dietary fat on mobilization of body fat and protein. *Journal of Dairy Science* 81, 169–175.

McNamara, J.P. (1994) Lipid metabolism in adipose tissue during lactation: a model of a metabolic control system. *Journal of Nutrition* 124, 1383S–1391S .

McNamara, J.P. and Baldwin, R.L. (1995) Modeling metabolism in lactation: estimation of critical parameters in lipid metabolism. In: Danfaer A. and Lescoat, P. (eds) *Proceedings of the IVth International Workshop on Modelling Nutrient Utilisation in Farm Animals.* Denmark Institute of Animal Science, Foulum, Denmark, pp. 63–84.

McNamara, J.P. and Baldwin, R.L. (2000) Estimation of parameters describing lipid metabolism in lactation: challenge of existing knowledge described in a model of metabolism. *Journal of Dairy Science* 83, 1–16.

McNamara, J.P. and Boyd, R.D. (1999) Quantitative regulation by endocrine systems. In: Kyriazakis, I. (ed.) *A Quantitative Biology of the Pig.* CAB International, Wallingford, pp.199–225.

McNamara, J.P. and Hillers, J.K. (1989) Regulation of bovine adipose tissue metabolism during lactation 5. Relationships of lipid synthesis and lipolysis with energy intake and utilization. *Journal of Dairy Science* 72, 407–418.

McNamara, J.P. and Murray, C.E. (1994) Adaptations in sympathetic nervous system activity in white adipose tissue during pregnancy and lactation: role in regulating adipose tissue metabolism. *Journal of Dairy Science* 77(Suppl. 1), 226.

McNamara, J.P., Kincaid, R.L., Harrison, J.H., Waltner, S.S. and Becker-Khaleel, B. (1991a) Challenging and improving models of energy metabolism: Parameters of lipid metabolism during pregnancy and lactation in dairy cattle. In: Wenk, C. and Boessinger, M. (eds) *Energy Metabolism in Farm Animals.* EAAP Pub. No. 58, ETH-Zentrum, Zurich, Switzerland, pp. 32–35.

McNamara, J.P., Pettigrew, J.E., Baldwin, R.L., Close, W.H., Walker, J. and Oltjen, J.W. (1991b) Information needed for mathematical modeling of energy metabolism. In: Wenk, C. and

Boessinger, M. (eds) *Energy Metabolism in Farm Animals*. EAAP Pub. No. 58, ETH-Zentrum, Zurich, Switzerland, pp. 468–472.

McNamara, J.P., Harrison, J.H. and Kincaid, R.L. (1995) Lipid metabolism in adipose tissue of cows fed high fat rations high in lactation. *Journal of Dairy Science*. 78, 2782–2796.

Meijer, G.A.L., Van der Meulen, J., Bakker, J.G.M., Van Der Koelen, C.M. and Van Vuuren, A.M. (1995) Free amino acids in plasma and muscle of high yielding dairy cows in early lactation. *Journal of Dairy Science* 78, 1131–1141

National Research Council (1989) *Nutrient Requirements of Dairy Cattle*, 6th rev. edn. National Academy Press, Washington, DC.

Overton, T.R., Drackley, J.K., Douglas, G.N., Emmert, L.S. and Clark, J.H. (1998) Hepatic gluconeogenesis and whole-body protein metabolism of periparturient dairy cows as affected by source of energy and intake of the prepartum diet. *Journal of Animal Science* 76(Suppl. 1), 295.

Overton, T.R., Drackley, J.K., Otteman-Abbamonte, C.J., Denise Beaulieu, A., Emmert, L.S. and Clark, J.H. (1999) Substrate utilization for hepatic gluconeogenesis is altered by increased glucose demand in ruminants. *Journal of Animal Science* 77, 1940–1951.

Pond, C.M. (1984) Physiological and ecological importance of energy storage in the evolution of lactation: evidence for a common pattern of anatomical organization of adipose tissue in mammals. *Symposium Zoology London* 51, 1–32.

Robelin, J., Chilliard, Y. and Agabriel, J. (1989) Estimation of body lipids and proteins of holstein, charolais and limousine cows by dilution technique and adipose cell size. In: Wenk, C. and Boessinger, M. (eds) *Energy Metabolism in Farm Animals*. EAAP Pub. No. 43, Pudoc, Wageningen, The Netherlands, pp. 370–373.

Santos, F.A.P., Santos, J.E.P., Theurer, C.B. and Huber, J.T. (1998) Effects of rumen-undegradable protein on dairy cow performance: a 12-year literature review. *Journal of Dairy Science* 81, 3182–3213.

Trayhurn, P. and Richard, D. (1985) Brown adipose tissue thermogenesis and the energetics of pregnancy and lactation in rodents. *Biochemical Society Transactions* 13, 826–829.

Waltner, S.S., McNamara, J.P. and Hillers, J.K. (1993) Relationships of body condition score to production variables in high producing holstein dairy cattle. *Journal of Dairy Science* 76, 3410–3419.

Waltner, S.S., McNamara, J.P., Hillers, J.K. and Brown, D.L. (1994) Validation of indirect measures of body fat in lactating dairy cattle. *Journal of Dairy Science* 77, 2570.

Waterlow, J.C. (1995) Whole-body protein turnover in Humans – past, present, and future. *Annual Review of Nutrition* 15, 57–92.

Wildman, E.E., Jones, G.M., Wagner, P.E., Boman, R.L., Trout, H.F., Jr and Lesch, T.N. (1982) A dairy cow body condition scoring system and its relationship to selected production characteristics. *Journal of Dairy Science* 65, 495–503.

# VII Ruminant Physiology and Genetics

# 21 Genetic Manipulation of Ruminant Biochemistry and Physiology for Improved Productivity: Current Status and Future Potential

K.A. WARD

*CSIRO Animal Production, LB1, Delivery Centre, Blacktown, Australia*

## Introduction

The provision of an adequate supply of food during the next 50 years is one of the major challenges that face mankind. We have two related problems, namely, maintaining the current food supply to ensure an adequate diet for our present population of about 6 billion people and then increasing production levels to provide for the increased population to AD 2050, which may reach about 8–9 billion people. While improvements in production, storage and distribution procedures could have a major impact on current food availability, it is highly likely that increased productivity will also be required over the next 5 decades. Over the past 100 years, total world food productivity has increased at just under 3% per annum, due to the implementation of farm mechanization, irrigation, the widespread use of fertilizers, treatments for plant and animal parasites and infectious diseases, and greatly improved methods of food storage and distribution. We might expect these past innovations to help maintain current levels of production and to provide further increases in productivity when applied to farms that are not yet using advanced farming technology. However, there are growing signs that these approaches are reaching their current limits because of environmental factors such as water availability, soil degradation, increasing disease resistance and more subtle deleterious climatic changes. They may therefore not be capable of fully supporting the increases in productivity that will be required by AD 2050.

A powerful method for the long-term improvement in plant and animal productivity stems from the use of genetic selection, in which a desirable genotype is chosen as the source of parental stock for the next generation. This approach has been shown to be capable of increasing the productivity of specific traits in plants and animals by 2–4% per annum, with the rate of increase sustainable in most cases for many years

Italics have been used when referring to natural genes, but not when referring to recombinant molecules.

(Lindsay, 1998; Tribe, 1998) and it is clearly one of the most important tools currently available for addressing the requirement for more food over the next 50 years. Recently, its potential has been greatly enhanced by the new techniques of genetic engineering which allow the direct manipulation of the genotype. This is achieved by enabling purified genes, isolated and characterized by recombinant DNA technology, to be added as naked DNA directly to the genomes of most domestic animals and plants. Using these techniques, it is now possible to consider introducing to a plant or animal a single gene of major effect on the phenotype. In animals, the techniques stem from the pioneering work of several laboratories (for review, see Palmiter and Brinster, 1986) which demonstrated that it was possible to introduce to the mouse genome small pieces of recombinant DNA. This was followed shortly thereafter by the pioneering experiments which demonstrated that it was possible to alter the phenotype of an animal by this technique (Palmiter *et al.*, 1982, 1983). When the technology was subsequently shown to be applicable to domestic animals (Hammer *et al.*, 1985), the potential for practical manipulation of livestock genomes became a reality.

This chapter will attempt to address the current status of genetic engineering research directed towards the modification of the biochemistry and physiology of ruminant livestock for improved farm productivity. The potential targets for such modifications are the endocrine system, intermediary metabolism and the animal's defences against disease. The genes for transfer are currently selected on the basis of known function and predicted effect on phenotype, although the identification and isolation of genes by genome mapping is now becoming a genuine practical possibility. It is worth noting that no transgenic animal with altered physiology or modified biochemistry has yet been introduced to the farmyard on a commercial basis, despite the fact that research in these areas has been very active for more than 10 years. This highlights the difficulties inherent in introducing to animals novel genetic properties that disrupt the homeostasis of a carefully balanced physiology derived from existing gene combinations that have been optimized by many years of selective breeding.

## Transgenic technology

The application of genetic engineering to ruminants has largely been achieved by the use of pronuclear injection of single-cell embryos, although it appears likely that the rapidly emerging technique of nuclear transfer may soon replace it. An excellent review of microinjection technology and its history of development is provided in Palmiter and Brinster (1986) and hence its details will only be briefly summarized here. The technique, pioneered in laboratory mice, involves the introduction of a small quantity of recombinant DNA into one of the pronuclei of single-cell embryos by microinjection, achieved by the insertion of a fine needle into the embryo pronucleus and the injection into this organelle of about 2 picolitres of solution. The embryos are then briefly cultured and survivors transferred to recipient mothers with reproductive cycles suitably synchronized to accept the embryos. In a small number of these embryos, the injected DNA becomes integrated into the embryo genome and an animal born from such an embryo contains the new piece of DNA in all its cells. Such an animal is called 'transgenic'. The process is not very efficient and in the case of ruminants, about 0.9% of injected embryos produce transgenic animals (Wall, 1996). This is due in part to the presence of dense cytoplasmic granules in ruminant embryos which tend to obscure the pronuclei during the microinjection procedure. While the problem is less apparent in sheep and goats than in cattle and can be partially overcome by centrifugation of the embryos prior to microinjection, it presents difficulties in all species of larger animals.

Recently, a novel technique known as nuclear transfer has been developed and may soon prove to be the method of choice for the introduction of transgenes to ruminants. It involves the introduction of a transgene to cells maintained in cell culture and the use of one of these cells as a source of genetic information to programme the development of an enucleated oocyte. Based on the pioneering work of Willadsen (1986), the recent technology has been developed by Campbell *et al.* (1996) and Wilmut *et al.* (1997) who have discovered that the donor cells must be in the G0 stage of the cell cycle for maximum efficiency of the process. The nuclei of these cells are apparently reprogrammed by the cytoplasm of metaphase II enucleated oocytes, allowing them to undergo the full developmental programme normally reserved for zygotes derived by sperm-mediated fertilization. A crucial aspect of the recent discoveries is the finding that somatic cell lines derived from adult tissue can serve as nuclear donors in this process (Wilmut *et al.*, 1997); a remarkable finding recently confirmed in laboratory mice (Wakayama *et al.*, 1998). The technique has also now been extended to the production of transgenic cattle (Cibelli *et al.*, 1998), indicating its probable wide applicability to many domestic animal species. This latter paper also questions the need for donor cells to be in the G0 phase of the cell cycle; a crucial finding if confirmed since it has significant implications for the intellectual property rights associated with the widespread commercial use of the technology.

There are many clear advantages to the use of nuclear transfer compared with microinjection as a method of introducing transgenes to domestic animals. Since the transgene is first introduced to cells in culture, this overcomes the highly inefficient and time-consuming step of using fertilized embryos as the target for gene integration. Cultured cells can be obtained in large numbers and recombinant DNA is readily introduced to these cells by a variety of well-proven techniques. Selected cells can be made into clonal lines and evaluated for site and stability of integration, and in some cases can even be tested for expression, all under conditions of cell culture prior to using them for the production of transgenic animals. In addition, for cells in culture, techniques are available to carry out homologous recombination between the recombinant DNA and the equivalent homologous gene in the target cell, thus allowing specific genes to be inactivated or replaced by variant alleles. This adds a powerful new capability to gene manipulation of domestic animals, since the potential is for genes to be added, inactivated or replaced in specific animal lines. At its current stage of development, nuclear transfer is still in an experimental phase but it appears well-placed eventually to replace microinjection as the method of choice in introducing transgenes to domestic animals.

The techniques of microinjection and nuclear transfer are at present the two preferred methods for the production of transgenic ruminants. In the future, it is possible that transposons, which are a modification of retroviral technology, may provide a third capability, but other methods such as cytoplasmic injection, ballistic guns and sperm-mediated transfer do not seem relevant to ruminants.

## The modification of the endocrine system

The endocrine system of ruminants is an obvious target for direct manipulation by genetic engineering because small changes in the concentration of a single hormone can result in large physiological changes to the whole animal. An obvious target is the suite of peptide hormones comprising growth hormone (GH), growth hormone releasing factor (GRF), insulin-like growth factor-I (IGF-I), insulin and thyrotrophic hormone, because these hormones are widely believed to control body growth and tissue development and hence are likely to influence body growth, carcass composition and feed utilization efficiency. Most research in this area has been carried out on GH itself, for two reasons. Firstly, the administration of exogenous growth hormone to animals has been shown to increase feed utilization efficiency, increase growth and produce a leaner carcass composition (for reviews see Spencer, 1985; Sejrsen, 1986; Florini, 1998). Secondly, the pioneering research of Palmiter and his colleagues (Palmiter *et al.*, 1982, 1983) in transgenic mice demonstrated that novel genes encoding GH and inserted into the genotype could alter the phenotype of transgenic animals.

In domestic animals, the bulk of the early data have been generated in swine and sheep, although there is some recent information pertaining to fish. It is now clear from these results that when the level of circulating GH is elevated, there are a number of physiological changes that occur in all the animals studied, but that in addition there are changes to phenotype that appear to be species specific. The results also demonstrate that it is important to avoid consistently high levels of the hormone, identifying the need for precise regulation of the expression of the transgene. The major observations from these studies are summarized below.

The GH-related transgenes that have been used in ruminant animals are shown in Table 21.1, together with the number of transgenic animals produced and expressing these genes. Three of these genes have been designed directly on the basis of the work of Palmiter and colleagues (Palmiter *et al.*, 1982, 1983) and consist of a GH-coding sequence joined to a regulatory sequence derived from a metallothionein gene. However, in later experiments, modification to this basic design was carried out in attempts to provide more precise control of the expression of the gene. In total, 35 founder transgenic sheep have been produced, with 12 of these animals expressing the inserted gene.

The transgenic sheep showed a large depletion in body fat, an increase in muscle mass and an increase in the size of the visceral organs. The growth rates for all the transgenic sheep except those containing the gene oMT-GH10 (see below), were either not increased or in some cases even decreased (Rexroad *et al.*, 1988, 1990; Murray *et al.*, 1989; Nancarrow *et al.*, 1991). This is consistent with two studies of exogenous GH administered to lambs, both of which reported no significant increase in average daily gain (Muir *et al.*, 1983; Wagner and Veenhuizen, 1988). Interestingly, a third study found a significant weight gain over controls but this did not result in an increased carcass weight (Johnson *et al.*, 1985).

In sheep, severe health problems were noted, including the development of diabetes (Ward *et al.*, 1989; Rexroad *et al.*, 1990; Nancarrow *et al.*, 1991). The results obtained with metallothionein-regulated GH-encoding genes suggested that the elevated GH caused a shift in body metabolism towards protein synthesis at the expense of fat and carbohydrates, consistent with the results obtained by exogenous administra-

**Table 21.1.** Transgenic ruminants containing various GH-related transgenes.

| Gene | Transgenics produced | Transgenics expressing | Reference |
|---|---|---|---|
| mMT-hGH | 1 | 0 | Hammer *et al.* (1985)<br>Rexroad *et al.* (1990) |
| mMT-bGH | 2 | 2 | Rexroad *et.al.* (1990) |
| mMT-hGRF | 9 | 1 | Rexroad *et al.* (1990) |
| mTF-bGH | 11 | 3 | Rexroad *et al.* (1988) |
| mAL-hGRF | 4 | 2 | Rexroad *et al.* (1990) |
| oMT-oGH9 | 7 | 3 | Murray *et al.* (1989) |
| oMT-oGH10 | 1 | 1 | Ward and Brown (1999) |
| Totals | 35 | 12 | |

Promoter sequences: MT = metallothionein; TF = transferrin; AL = albumin.
Gene sequences: GH = growth hormone; GRF = growth hormone releasing factor.

tion of GH. However, sustained elevation of GH concentrations was deleterious to the health of the animals. Accordingly, attempts have been made to regulate the expression of the transgene. Thus, attempts have been made to improve gene regulation by changing the promoter used to drive expression of the exogenous GH or to change the hormone from GH to one of the other growth-regulating peptides. The range of transgenes that have been constructed and introduced to transgenic sheep is shown in Table 21.1. None of these gene modifications appear to be any more successful than the original metallothionein-regulated GH-encoding genes. However, a novel modification to the ovine metallothionein-GH construction, giving rise to the gene oMTSGH10, appears to be promising. This gene, oMTSGH10, is based on the gene oMTSGH9, which has been described by Shanahan *et al.* (1989). OMTSGH10 has been constructed in a way that was designed to reduce the level of constitutive and inducible transcription both constitutively and when subjected to zinc induction (Ward and Brown, 1999). In contrast to its progenitor, oMTSGH10 has very low expression in transgenic mice and shows no evidence of zinc inducibility. It has been introduced into one transgenic sheep, from which a line of animals has been made and used to examine the effects of the gene on growth rate and body composition. While some of the data are still being collected, it is already clear that the animals are in good health, grow from the age of about 4 months onwards at a rate between 15% and 20% faster than controls and have a much leaner carcass (Ward and Brown, 1999). Feed utilization efficiency has been measured for the period from 3.5 months to 7 months of age and was the same as that of controls during this period. This will need to be repeated at later stages of growth. To date, the gene has only been successfully inserted into a single founder animal, from which a line of sheep has been bred through three generations. More founder animals will clearly need to be generated before it can be unequivocally concluded that the advantageous growth characteristics shown by the current line of animals is likely to be a universal finding in all animals containing the oMTSGH10 gene, but when considered together with the results obtained in transgenic mice, the research looks very encouraging.

A recent approach that has been carried out in swine may also prove to be useful for ruminants. In this case (Pursel *et al.*, 1996, 1998), the IGF-I gene is controlled by the avian skeletal α-actin promoter. Using this transgene, expression of IGF-I is primarily confined to the muscle cells and has its effect directly on this tissue. In consequence, average daily weight gain and feed efficiency do not differ from control animals, but the transgenic animals had 11.4% and 3.3% less fat and 5.1% and 2.2% more protein in the eviscerated carcasses of males and females, respectively (Pursel *et al.*, 1998). The IGF-I concentrations of plasma were between 9% and 10% higher than in control pigs and the general health of the animals was good. The avian skeletal α-actin promoter has also been used in another approach recently described in mice. In this case, the promoter is joined to the gene sequence encoding GRH and the fusion gene inserted into mice by injecting the DNA directly into regenerating quadriceps muscle tissue. This results in the uptake of some of the DNA into the muscle cells, which then produce and secrete GRH in levels sufficient to elevate GH and produce increased body growth (Draghia-Akli *et al.*, 1997).

The above results together indicate that the concept of increasing body growth and feed utilization efficiency while simultaneously improving the carcass composition of domestic animals to make it more suitable for human consumption remains a worthwhile and achievable goal. The difficulty that must still be overcome is the adverse effect on health that results in animals that produce GH at high levels. However, optimization of GH-encoding genes is progressing to the stage where transgenic animals with genuine commercial application are close to reality and might be expected to be in field trials within the next few years.

## The modification of intermediary metabolism

The evolution of domestic animals has been a complex process involving both natural selection and the deliberate selection of specific traits relevant to increased productivity. During this process there has been a loss of biochemical capacity, presumably because this was advantageous for the survival of animals in their natural environment, and as a result, some of the important productivity traits of modern domestic animals are now rate-limited by dietary nutrient supply. Once lost, the genetic information for such biochemistry is essentially impossible to restore by conventional breeding procedures. However, these pathways are still fully functional in prototrophic bacteria and hence can be transferred back to animal species by transgenic technology.

One of the areas where this approach might prove useful in increasing productivity is the introduction of biochemical pathways to allow the biosynthesis of rate-limiting amino acids. The more obvious targets are the supply of cysteine to sheep for increased wool growth and the supply of threonine and lysine to swine to supplement diets aimed at optimal body growth. Most progress has been made in the introduction of a functional cysteine biosynthetic pathway to ruminants and this will be described in some detail as an example of the possibilities that are opened by transgenic technology.

Under some pasture conditions, the amino acid cysteine is rate-limiting for wool growth and it has been shown that when the amino acid is infused abomasally or intravenously into sheep, wool growth could be increased substantially (Reis, 1979). The simple procedure of feeding additional cysteine to deficient animals is not effective because most ingested nutrients are rapidly metabolized by the ruminal microflora in sheep and the sheep is lacking the genes necessary for the enzymes of cysteine biosynthesis. The relevant biosynthetic pathway in *Escherichia coli* is shown in Fig. 21.1. It can be broadly divided into a pathway for the reduction of sulphur to an active form of reduced sulphide and a carbon pathway in which, firstly, the amino acid serine is converted to O-acetylserine in the presence of acetyl-CoA and the enzyme serine transacetylase (SAT) and then the O-acetylserine is converted to cysteine in the presence of sulphide and the enzyme O-acetylserine sulphydrylase (OAS). The sulphide concentration in the sheep rumen has been reported to be in the range of 0.6 $\mu$g ml$^{-1}$ to 288 $\mu$g ml$^{-1}$ (Bray and Till, 1975), suggesting that a functional biosynthetic pathway might be possible in sheep if the enzymes SAT and OAS could be provided in the ruminal epithelial cells.

In *E. coli*, the *cysE* gene encodes the enzyme SAT and the *cysK* or *cysM* genes encode variants of the enzyme OAS. Accordingly, the *cysE* and *cysK* (Denk and Bock, 1987; Byrne *et al.*, 1988) genes were isolated and modified for expression in eukaryotic organisms (Leish *et al.*, 1993). The structure of the gene, named MTCEK1, is shown in Fig. 21.2 and consists of the *cysE* and *cysK* coding sequences, each regulated independently by a sheep metallothionein-Ia (MT-Ia) promoter sequence and each containing exon 5 of the sheep growth hormone gene spliced downstream of the bacterial coding sequences (Leish *et al.*, 1993).

When tested in eukaryotic cells in tissue culture, this gene could be transcribed and translated into the relevant bacterial enzymes and these could be synthesized at high levels (Ward and Nancarrow, 1991; Leish *et al.*, 1993). The gene was then introduced into transgenic mice where it was shown to be expressed at high levels in several tissues including the small intestine (Ward *et al.*, 1994). When this intestinal tissue was incubated with sulphide, the synthesis of cysteine was clearly demonstrated. The most convincing demonstration of the functionality of the cysteine pathway in transgenic mice, however, was obtained by a dietary study in which transgenic mice and appropriate control mice were placed on a synthetic diet which was supplemented with $Na_2S$ but in which the sulphur amino acids cysteine and methionine were reduced in concentration to low levels. After 7 days on this diet, substantial hair loss and weight loss was experienced by the control animals but the transgenic animals continued to grow normally and did not lose any hair (Ward *et al.*, 1994).

**Fig. 21.1.** The carbon pathway portion of the cysteine biosynthetic pathway in *Escherichia coli.*

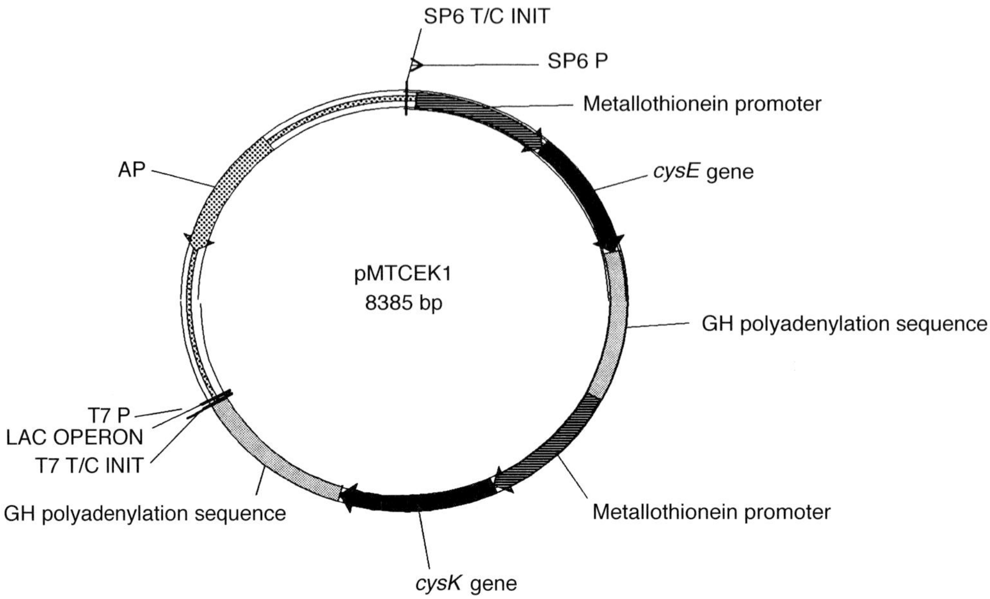


**Fig. 21.2.** The gene MTCEK1 which encodes the cysteine biosynthetic pathway shown in Fig. 21.1.

While the pathway for cysteine biosynthesis appears to operate effectively in transgenic mice, it has not yet been shown to function in ruminants, despite being introduced into 28 different primary transgenic sheep. A summary of the results obtained in our own laboratory and that of Rogers *et al.* (Sivaprasad *et al.*, 1992; Bawden *et al.*, 1995) is provided in Ward *et al.* (1998). Briefly, these show that 28 transgenic sheep with the genes encoding cysteine biosynthesis have been produced. These animals contain the bacterial coding sequences for the enzymes SAT and OAS, isolated either from *E. coli* or from *Salmonella typhimurium* (Sivaprasad *et al.*, 1992) and regulated by three different eukaryotic promoters. While some low-level expression has been detected (Bawden *et al.*, 1995; Ward *et al.*, 1998), no useful expression of the genes has been obtained in any of these animals. In addition, the number of transgenic animals produced as a percentage of embryos microinjected is low compared with the efficiency obtained for other genes (Ward *et al.*, 1998). A possible explanation for these results is that high levels of expression of a cysteine biosynthesis pathway in sheep embryos are lethal so that the only transgenic animals obtained are those in which the genes have been inserted in a region of the genome that prevents their expression or at best allows only low levels of expression. Two suggestions for the way this might occur are that the cellular level of acetyl-CoA becomes unacceptably depleted during early embryogenesis (Ward *et al.*, 1998) or that the level of cysteine rises to toxic levels (Bawden *et al.*, 1995). At present there is no convincing evidence to support either hypothesis and resolution of the alternatives will require further research. However, regardless of which explanation is correct for the difficulties experienced in duplicating in sheep the functionality of the pathway that was obtained in transgenic mice, a promoter needs to be found that regulates expression of the genes only to the rumen epithelium and which also prevents such expression until after the birth of the animal. In this way, the pathway should be able to function effectively in sheep, assuming that it is not totally incompatible with the intermediary metabolism of adult ruminants.

The research directed towards the introduction of a cysteine biosynthetic pathway is a well-advanced example of the use of genetic engineering to modify the biochemical capacity of an animal, but several other projects are in earlier stages of development. It has been proposed to introduce to non-ruminants the pathways for the biosynthesis of the amino acids threonine and lysine (Rees *et al.*, 1990), since these need to be added to cereal protein for maximal utilization. An interesting and valuable aspect of this paper was the use of computer simulation to predict the flux of biosynthetic products produced by the introduction of a novel pathway to an animal. The general concepts of the cysteine biosynthesis project described above are equally applicable to threonine and lysine, but the pathways are more complex and hence the number of genes that need to be isolated, modified and transferred to the target animal increases substantially. However, it is now possible to construct genes encoding proteins with multiple active sites, thus reducing the number of separate genes that need to be introduced to the target animal (Robinson and Sauer, 1998). It is also now possible to utilize from the encephalomyocarditis virus a DNA sequence called an internal ribosomal entry site (IRES). This sequence, when interspersed between two coding sequences of a single mRNA, allows the initiation of translation to occur both at the start of the mRNA and also internally within the mRNA, resulting in the production of two separate polypeptides (Kim *et al.*, 1992).

Another project which is in its early stages of development involves introducing to ruminants a functional glyoxylate cycle (Byrne, 1990; Ward and Nancarrow, 1991) which would allow such animals to synthesize glucose from the abundant supplies of acetate they receive from the rumen. Ruminants are substantially less efficient in their utilization of feedstuff compared with monogastric animals and one of the major reasons for this is thought to be the lack of glucose available to them for direct absorption from the gut (Bergman, 1975; Van Soest, 1982). There are several tissues in ruminants that could conceivably benefit from the ability to synthesize glucose directly from acetate. These include the mammary epithelium, because of the high carbohydrate content of milk, and the sheep wool follicle, which has an extremely active pentose phosphate pathway thought to be involved in the maintenance of the cellular redox potential during keratin protein biosynthesis (Chapman and Ward, 1979).

A functional glyoxylate cycle requires the presence of two enzymes, isocitrate lyase which cleaves isocitrate to succinate and glyoxylate, and malate synthase, which catalyses the fusion of glyoxylate and acetyl-CoA to form malate (Ward and Nancarrow, 1991). The *aceA* and *aceB* genes of *E. coli* encode these two enzymes and both genes have been isolated, characterized and assembled into a fusion gene construct MTAceAB1 (Byrne, 1990) which is similar in structure to that of the gene MTCEK1 encoding cysteine biosynthesis (Leish *et al.*, 1993) (Fig. 21.2), except that the *cysE* and *cysK* coding sequences have been replaced by the *aceA* and *aceB* coding sequences. This DNA has been introduced into mammalian cells in culture and shown to be actively transcribed and translated into the appropriate enzymes (Byrne, 1990; Ward and Nancarrow, 1991), a valuable piece of information that demonstrates that the glyoxylate cycle can apparently co-exist with existing biochemistry in a mammalian cell without sequestering the new enzymes in specific organelles. The same gene has also been inserted into transgenic mice and has been shown to be expressed in a variety of tissues including the liver and small intestine (Saini *et al.*, 1996). The level of expression in these animals was not as high as that found for the gene MTCEK1 and the reason for this is not yet known. It may be that more animals have to be produced in order to generate one with a high level of expression. However, it is interesting that attempts so far in our laboratory to insert the gene into sheep have been unsuccessful. It is conceivable that high expression of a glyoxylate cycle in an animal may not be well tolerated, in which case the gene will need to be constructed with promoters that can be regulated and which can direct expression to specific tissues such as the mammary epithelium, the ruminal epithelium and the skin, where the increased production of glucose might be advantageous. This research will benefit from the nuclear transfer technology as it matures into a useful method for transgenic animal production, because the pathway can be introduced and examined for functional integration within its new genome prior to its transfer to animals.

## The modification of digestion

Another innovative approach to the modification of the biochemical potential of domestic animals is the attempt to alter the process of digestion. This work is primarily directed towards monogastric animals, but the overall concept may also prove to be of value in ruminants. The current work involves the secretion of cellulase enzymes into the digestive system of monogastric animals to provide a mechanism for the utilization of cellulose and xylan, which is inefficiently degraded by microbial fermentation in the hindgut (Hazlewood *et al.*, 1989; Forsberg *et al.*, 1993). The enzyme chosen for introduction to the digestive tract was the bacterial endoglucanase E from *Clostridium thermocellum* (Hall *et al.*, 1988, 1990). The DNA encoding this enzyme has been isolated, characterized, fused to a pancreas-specific enhancer region of the elastase I gene and inserted into the genomes of transgenic mice. These animals produced the cellulase mRNA only in the pancreas and secreted enzymically active cellulase protein into the small intestine (Hall *et al.*, 1993). This is a finding of some significance to the concept of digestion modification, because it demonstrates not only that the bacterial enzyme is stable in the environment of the small intestine but that a mechanism has been discovered for the secretion of heterologous proteins from the exocrine pancreas into the lumen of the small intestine. While the level of expression in these animals is currently too low for effective practical applications, current research in this project is aimed at increasing this expression to that approximating the expression of the normal elastase I gene (Ali *et al.*, 1995).

## The modification of disease resistance in animals

Transgenic technology has the potential to alter the disease resistance of domestic animals in novel ways, although progress so far has been limited. The goal can hypothetically be achieved either by modification of the immune system itself or by the introduction to domestic animals of gene sequences encoding unique proteins which can provide resistance to various diseases. Some progress has been made in both these areas, but no commercially useful animals have yet been produced. Thus, it has been shown that it is possible to isolate portions of functionally-rearranged immunoglobulin genes encoding the antibody recognition to a specific antigen and to introduce this DNA to transgenic mice (Brinster *et al.*, 1983; Ritchie *et al.*, 1984; Storb *et al.*, 1986). These mice then produce antibodies to the bacterial antigen without ever being exposed to it. The results to date have been encouraging but have yet to be extended to domestic animals. One of the problems encountered in mice was that the expression of the transgene had a suppressive effect on the production of other immunoglobulins, thus potentially reducing the overall disease resistance of the transgenic animals. Nevertheless, it is clearly an approach which needs further examination as a practical method for altering disease resistance.

A quite different approach currently in progress within CSIRO in Australia is the introduction to animals of a gene encoding a protein which has inherent disease-resistance properties. This approach is being pursued in an attempt to increase the resistance of sheep to attack by blowfly larvae. In this project, the protein under investigation is the tobacco chitinase, which has been shown to have significant larvicidal activity against blowfly larvae. It is the aim of this research to provide sheep with the ability to produce the chitinase protein in their sweat gland secretions in quantities sufficient to inhibit the larval attack when blowfly eggs hatch on the skin surface. At present, various forms of the chitinase are being evaluated for larvicidal activity to identify the most active form needed to insert into transgenic animals. However, to date the results obtained are not encouraging (Dr A.G. Brownlee, CSIRO Animal Production, personal communication), since they suggest that animals will need to produce very high levels of the chitinase in order to be protected from blowfly larvae. Nevertheless, the overall concept underlying this research is sound and should be pursued further, particularly since similar approaches have proven to be highly effective in the protection of plants from insect predators.

## Conclusions

The modification of ruminant physiology and biochemistry by the use of transgenic techniques offers a wide range of possibilities. While the research is proving difficult to bring from laboratory to the practical farming environment, nevertheless, several projects appear to be making significant progress towards this goal. The most advanced are those that involve the modification of growth and carcass and it is possible that genetically-modified pigs and sheep could be in field trial within the next few years. The introduction to the farm environment of animals with modified biochemical properties is still in the research phase and likely to remain at this point for some time yet. While it is apparent that new pathways interacting with existing intermediary metabolism can be introduced to animals, species-specific difficulties are clearly a factor in slowing the progression of the work from the laboratory to the field environment. At present, the introduction of the cysteine biosynthetic pathway to sheep appears to be the project closest to application, but there remain several crucial areas of research yet to be completed before this can take place. The introduction of the glyoxylate cycle to ruminants is much further from completion but holds the promise of a greater long-term impact on animal agriculture.

Nuclear transfer and animal cloning are going to provide a major stimulus to the application of transgenic techniques to domestic animals, because these techniques allow the introduction and preliminary testing of DNA constructions in cells in culture and the subsequent use of such cells to produce viable animals containing the recombinant DNA. It has been apparent for some years now that the ability to produce whole organisms from cultured cells has been one of the major advantages held by those scientists working with plants compared with those working with animals, allowing the genetic modification of plants to proceed very quickly in recent years.

## Acknowledgements

I am grateful to my colleagues Drs Jim Murray, Colin Nancarrow, Alan Brownlee, Zdenka Leish and Mr Bruce Brown for their help in the original research described in this paper. I am also indebted to the expert technical assistance of Mrs Nola Rigby, Mrs Cathy Pruss, Miss Tania Radziewic, Mr Alan Fawcett, Mr John O'Grady, Mrs Jenny Gordon, Mr Peter Mitchell and Mr Jim Marshall.

Some of the research described in this paper was supported by the International Wool Secretariat.

## References

Ali, S., Hall, J., Soole, K.L., Fontes, C.M.G.A., Hazlewood, G.P., Hirst, B.H. and Gilbert, H.J. (1995) In: Petersen, S.B., Svensson, B. and Pedersen, S. (eds) *Targeted expression of microbial cellulases in transgenic animals. Carbohydrate bioengineering. Proceedings of an International Conference Elsinore, Denmark, 23–26 April, 1995.* Elsevier Science Publishers, Amsterdam, The Netherlands, pp.279–293.

Bawden, C.S., Sivaprasad, A.V., Verma, S.K., Walker, S.K. and Rogers, G.E. (1995) Expression of bacterial cysteine biosynthesis genes in transgenic mice and sheep: toward a new in vivo amino acid biosynthesis pathway and improved wool growth. *Transgenic Research* 4, 87–104.

Bergman, E.N. (1975) Production and utilization of metabolites by the alimentary tract as measured in portal and hepatic blood. In: McDonald, I.W. and Warner, A.C.I. (eds) *Digestion and Metabolism in the Ruminant.* The University of New England Publishing Unit, Armidale, Australia, pp. 292–305.

Bray, A.C. and Till, A.R. (1975) Metabolism of sulphur in the gastro-intestinal tract. In: McDonald, I.W. and Warner, A.C.I. (eds) *Digestion and Metabolism in the Ruminant.* The University of New England Publishing Unit, Armidale, Australia, pp. 243–260.

Brinster, R.L., Ritchie, K.A., Hammer, R.E., O'Brien, R.L., Arp, B. and Storb, U. (1983) Expression of a microinjected immunoglobulin gene in the spleen of transgenic mice. *Nature* 306, 332–336.

Byrne, C.R. (1990) The isolation of bacterial acetate metabolism genes and their expression in a eukaryote. PhD Thesis Macquarie University, Sydney, Australia.

Byrne, C.R., Monroe, R.S., Ward, K.A. and Kredich, N.M. (1988) DNA sequences of the *csyK* regions of *S. typhimurium* and *E. coli* and linkage of the *cysK* regions to *ptsH. Journal of Bacteriology* 170, 3150–3157.

Campbell, K.H.S., McWhir, J., Ritchie, K.A. and Wilmut, I. (1996) Sheep cloned by nuclear transfer from a cultured cell line. *Nature* 380, 64–66.

Chapman, R.E. and Ward, K.A. (1979) Histological and biochemical features of the wool fibre and follicle. In: Black, J.L. and Reis, P.J. (eds) *Physiological and Environmental Limitations to Wool Growth.* University of New England Publishing Unit, Armidale, Australia, pp. 193–208.

Cibelli, J.B., Stice, S.L., Golueke, P.J., Kane, J.J., Jerry, J., Blackwell, C., Ponce de Leon, F.A. and Robl, J.M. (1998) Cloned transgenic calves produced from nonquiescent fetal fibroblasts. *Science* 280 1256–1258.

Denk, D. and Bock, A. (1987) L-cysteine biosynthesis in *Escherichia coli*: nucleotide sequence and expression of the serine acetyltransferase (*cysE*) gene from the wild type and a cysteine-excreting mutant. *Journal of General Microbiology* 133, 515–525.

Draghia-Akli, R., Li, X. and Schwartz, R.J. (1997) Enhanced growth by ectopic expression of growth hormone releasing hormone using an injectable myogenic vector. *Nature Biotechnology* 15, 1285–1294.

Florini, J.R. (1998). Hormonal control of muscle cell growth. *Journal of Animal Science* 61 (suppl. 2), 21–38.

Forsberg, C.W., Cheng, K.J., Krell, P.J. and Phillips, J.P. (1993) Establishment of rumen microbial gene pools and their manipulation to benefit fibre digestion by domestic animals. *Proceedings 7th World Conference on Animal Production, Edmonton, Alberta, Canada,* 1, 281–316.

Hall, J., Barker, P., Hazlewood, G.P. and Gilbert, H.J. (1988). Conserved reiterated domains in *Clostridium thermocellum* endoglucanases are not essential for catalytic activity. *Gene* 69, 29–38.

Hall, J., Hazlewood, G.P., Azim Surani, M., Hirst, B.H. and Gilbert, H.J. (1990) Eukaryotic and prokaryotic signal peptides direct secretion of a bacterial endoglucanase by mammalian cells. *The Journal of Biological Chemistry* 265, 19996–19999.

Hall, J., Ali, S., Azim Surani, M., Hazlewood, G.P., Clark, A.J., Simons, J.P., Hirst, B.H. and Gilbert, H.J. (1993) Manipulation of the repertoire of digestive enzymes secreted into the gastrointestinal tract of transgenic mice. *Biotechnology* 11, 376–379.

Hammer, R.E., Pursel, V.G., Rexroad, C.E., Jr, Wall, R.J., Bolt, D.J., Ebert, K.M., Palmiter, R.D. and Brinster, R.L. (1985) Production of transgenic rabbits, sheep and pigs by microinjection. *Nature* 315, 680–683.

Hazlewood, G., Mann, S. and Gilbert, H. (1989). Improving forage digestion by genetic engineering. *AFRC News*, July, 12–13.

Johnson, I.D., Hart, I.C. and Butler-Hogg, B.W. (1985) The effects of exogenous bovine growth hormone and bromocryptine on growth, body development, fleece weight and plasma concentrations of growth hormone, insulin and prolactin in female lambs. *Animal Production* 41, 207–217.

Kim, D.G., Kang, H.M., Jang, S.K. and Shin, H.S. (1992) Construction of a bifunctional mRNA in the mouse by using the internal ribosomal entry site of the Encephalomyocarditis virus. *Molecular and Cellular Biology* 12, 3636–3643.

Leish, Z., Byrne, C.R., Hunt, C.L. and Ward, K.A. (1993) Introduction and expression of the bacterial genes cysE and cysK in eukaryotic cells. *Applied Environmental Microbiology* 59, 892–898.

Lindsay, D. (1998) The Australian livestock industries: a case study in non-genetic factors that control genetic improvement. *Proceedings of 6th World Congress on Genetics Applied to Livestock Production, Armidale, Australia,* 23, 3–9.

Muir, L.A., Wien, S., Duquette, P.F., Rickes, E.L. and Cordes, E.H. (1983) Effects of exogenous growth hormone and diethylstilbestrol on growth and carcass composition of growing lambs. *Journal of Animal Science* 56, 1315–1323.

Murray, J.D., Nancarrow, C.D., Marshall, J.T., Hazelton, I.G. and Ward, K.A. (1989) Production of transgenic sheep by microinjection of ovine metallothionein-ovine growth hormone fusion genes. *Reproduction, Fertility and Development* 1, 147–155.

Nancarrow, C.D., Marshall, J.T.A., Clarkson, J.L., Murray, J.D., Millard, R.M., Shanahan, C.M., Wynn, P.C. and Ward, K.A. (1991) Expression and physiology of performance regulating genes in transgenic sheep. *Journal of Reproduction and Fertility* 43(suppl), 277–291.

Palmiter, R.D. and Brinster, R.L. (1986) Germline transformation of mice. *Annual Review of Genetics* 20, 465–500.

Palmiter, R.D., Brinster, R.L., Hammer, R.E., Trumbauer, M.E., Rosenfeld, M.G., Birnberg, N.C. and Evans, R.M. (1982) Dramatic growth of mice that develop from eggs microinjected with metallothionein-growth hormone GH fusion genes stimulate growth in mice. *Nature* 300, 611–615.

Palmiter, R.D., Norstedt, R.E., Gelinas, R.E., Hammer, R.E. and Brinster, R.L. (1983) Metallothionein-human GH fusion genes stimulate growth in mice. *Science* 222, 809–814.

Pursel, V.G., Coleman, M.E., Wall, R.J., Elsasser, T.H. and Haden, M. (1996) Regulatory avian skeletal alpha-actin directs expression of insulin-like growth factor-1 to skeletal muscle of transgenic pigs. *Theriogenology* 45, 348.

Pursel, V.G., Bee, G., Mitchell, A.D., Wall, R.J., Wells, K.D., Elsasser, T.H., Solomon, M.B., Coleman, M.E., DeMayo, F. and Schwartz, R.J. (1998) Expression of insulin-like growth factor-1 in skeletal muscle of transgenic swine. In: Murray, J.D., Anderson, G.B., McGloughlin, M.M. and Oberbauer, A.M. (eds) *Transgenic Animals in Agriculture.* CAB International, Wallingford, UK, pp. 131–144.

Rees, W.D., Flint, H.J. and Fuller, M.F. (1990) A molecular biological approach to reducing dietary amino acid needs. *Biotechnology* 8, 629–633.

Reis, P.J. (1979). Effect of amino acids on the growth and properties of wool. In: Black, J.L. and Reis, P.J. (eds) *Physiological and Environmental Limitations to Wool Growth.* University of New England Publishing Unit, Armidale, Australia, pp. 223–242.

Rexroad, C.E., Jr, Behringer, R.R., Bolt, D.J., Miller, K.F., Palmiter, R.D. and Brinster, R.L. (1988) Insertion and expression of a growth hormone fusion gene in sheep. *Journal of Animal Science* 6, 267.

Rexroad, C.E., Jr, Hammer, R.E., Behringer, R.R., Palmiter, R.D. and Brinster, R.L. (1990) Insertion, expression and physiology of growth-regulating genes in ruminants. *Journal of Reproduction and Fertility* Suppl. 41, 119–124.

Ritchie, K.A., Brinster, R.L. and Storb, U. (1984) Allelic exclusion and control of endogenous immunoglobulin gene rearrangement in *K* transgenic mice. *Nature* 312, 517–520.

Robinson, C.R. and Sauer, R.T. (1998). Optimizing the stability of single-chain proteins by linker length and composition mutagenesis. *Proceedings of the National Academy of Sciences USA,* 95, 5929–5934.

Saini, K.S., Byrne, C.R., Leish, Z., Pruss, C.A., Rigby, N.W., Brownlee, A.G., Nancarrow, C.D. and Ward, K.A. (1996) Introduction and expression of the bacterial glyoxylate cycle genes in transgenic mice. *Transgenic Research* 5, 467–473.

Sejrsen, K. (1986) Endocrine mechanisms underlying genetic variation in growth in ruminants. *Proceedings of the 3rd World Congress on Animal Genetics Applied to Animal Production.* University of Nebraska, Lincoln, Nebraska, 11, 261–270.

Shanahan, C.M., Rigby, N.W., Murray, J.D., Marshall, J., Townrow, C., Nancarrow, C.D. and Ward, K.A. (1989) Regulation of expression of a sheep growth hormone fusion gene in transgenic mice. *Molecular and Cellular Biology* 9, 5473–5479.

Sivaprasad, A.V., Kuczek, E.S., Bawden, C.S. and Rogers, G.E. (1992) Coexpression of the cysE and cysM genes of Salmonella typhimurium in mammalian cells: a step towards establishing cysteine biosynthesis in sheep by transgenesis. *Transgenic Research* 1, 79–92.

Spencer, G.S.G. (1985) Hormonal systems regulating growth: A review. *Livestock Production Science* 12, 31–46.

Storb, U., Pinkert, C., Arp, B., Engler, P., Gollahon, K., Manz, J., Brady, W. and Brinster, R.L. (1986) Transgenic mice with lambda and kappa genes encoding antiphosphorylcholine antibodies. *Journal of Experimental Medicine* 164, 627–641.

Tribe, D.E. (1998) How Sapiens is Homo? *Proceedings of 6th World Congress on Genetics Applied to Livestock Production, Armidale, Australia* 23, 10–15.

Van Soest, P.J. (1982) *Nutritional Ecology of the Ruminant.* Durham and Downey, Portland.

Wagner, J.F. and Veenhuizen, E.L. (1988) Growth performance, carcass deposition and plasma hormone levels in wether lambs when treated with growth hormone and thryoprotein. *Journal of Animal Science* 47 (suppl.1), 397.

Wakayama, T., Perry, A.C.F., Zuccotti, M., Johnson, K.R. and Yanagimachi, R. (1998) Full-term development of mice from enucleated oocytes injected with cumulus cell nuclei. *Nature* 394 369–374.

Wall, R.J. (1996) Transgenic livestock: progress and prospects for the future. *Theriogenology* 45, 57–68.

Ward, K.A. and Brown, B.W. (1999) The production of transgenic domestic livestock: successes, failures and the need for nuclear transfer. *Reproduction, Fertility and Development* 10, 659–665.

Ward, K.A. and Nancarrow, C.D. (1991) The genetic engineering of production traits in domestic animals. *Experientia* 47, 913–922.

Ward, K.A., Nancarrow, C.D., Murray, J.D., Wynn, P.C., Speck, P. and Hales, J.R.S. (1989) The physiological consequences of growth hormone fusion gene expression in transgenic sheep. *Journal of Cellular Biochemistry* 13, 164.

Ward, K.A., Leish, Z., Bonsing, J., Nishimura, N., Cam, G.R., Brownlee, A.G. and Nancarrow, C.D. (1994) Preventing hairloss in mice. *Nature* 371, 563–564.

Ward, K.A., Leish, Z., Brownlee, A.G., Nancarrow, C.D. and Brown, B.W. (1998) The utilization of bacterial genes to modify domestic animal biochemistry. In: Murray, J.D., Anderson, G.B., McGloughlin, M.M. and Oberbauer, A.M. (eds) *Transgenic Animals in Agriculture* CAB International, Wallingford, UK, pp. 157–176.

Willadsen, S.M. (1986) Nuclear transplantation in sheep embryos. *Nature* 320, 63–65.

Wilmut, I., Schnieke, A.E., McWhir, J., Kind, A.J. and Campbell, K.H.S. (1997) Viable offspring derived from fetal and adult mammalian cells. *Nature* 385, 810–813.

# 22 Genetics of Rumen Microorganisms: Gene Transfer, Genetic Analysis and Strain Manipulation

H.J. FLINT AND K.P. SCOTT

*Rowett Research Institute, Bucksburn, Aberdeen, UK*

## Introduction

Interest in the genetics of rumen microorganisms was first sparked by the prospect of creating manipulated strains that might be used to improve rumen function (Smith and Hespell, 1983). While some progress has been achieved towards this end, it is now more widely recognized that molecular genetics has a vital role to play in understanding the dynamics and diversity of rumen microbial communities, in understanding the functioning of enzyme systems and in unravelling the evolution of rumen microorganisms. In addition natural horizontal gene transfer is a potentially important, but little studied, factor in the adaptation and evolution of the rumen community and might also be involved in disseminating antibiotic resistance genes, or possibly even transgenes derived from modified feed plants or microbial additives, to different gut microorganisms. Furthermore, since the ruminant harbours human pathogens such as *Escherichia coli*, the rumen and hindgut are potential sites for exchange of pathogenicity determinants including toxin genes.

Because this chapter concentrates on transfer genetics it deals almost exclusively with the bacterial flora of the rumen. Despite the absence of work on gene transfer in rumen eukaryotes, there has been considerable effort in the isolation of genes from anaerobic rumen fungi and more recently from anaerobic protozoa which is leading to information on gene expression signals and codon usage in these organisms as well as casting light upon their evolutionary origins. These areas are beyond the scope of the present review, but have been discussed in several recent reviews and articles (Trinci *et al.*, 1994; Hespell *et al.*, 1996; Orpin and Joblin, 1997).

## Natural gene transfer in the rumen microbial community

### *Horizontal transfer of antibiotic resistance and other genes*

Gene transfer under ruminal conditions has been studied in only a few cases. Scott and Flint (1995) showed that plasmid transfer could occur between ruminal strains of *E. coli* added to whole rumen contents, while an earlier study (Smith, 1975) showed transfer of antibiotic resistances between *E. coli* strains *in vivo* in the rumen of starved cattle. Studies with obligately anaerobic rumen species have so far been confined to pairwise matings *in vitro*. In *Prevotella ruminicola* 223/M2/7, a transferable *tetQ* tetracycline resistance gene is located on a 19.5 kb plasmid, pRRI4 (Flint *et al.*, 1988), but in other ruminal *Prevotella* strains *tetQ* may be chromosomally located (Nikolich *et al.*, 1994) (Table 22.1). The pRRI4 *tetQ* determinant shares 97% sequence identity with chromosomally located *tetQ* determinants found in human colonic *Bacteroides* (Nikolich *et al.*, 1994). pRRI4 could also be transferred to human colonic *Bacteroides* in laboratory matings (Shoemaker *et al.*, 1992). Conjugative R plasmids carrying tetracycline resistance are also reported in rumen strains of *Enterococcus faecium* (Laukova *et al.*, 1990) and in *Lactobacillus* spp. (Kmet *et al.*, 1989).

In *Butyrivibrio fibrisolvens*, tetracycline resistance ($Tc^R$) could be transferred at remarkably high frequencies (up to $10^{-1}$ per recipient) in laboratory matings between a donor strain, 1.230, and the recipient 2221 (Scott *et al.*, 1997). This transfer was chromosomally mediated, and was ascribed to a novel self-transmissible element of 40–50 kb. The donor was found to carry two $Tc^R$ genes – a non-transmissible gene whose sequence is 100% identical to that of *tet*(O) from *Streptococcus pneumoniae*, and a transmissible gene that represents a new class of ribosome protection type $Tc^R$ gene, designated *tet*(W) (Barbosa *et al.*, 1999). Tet W is most closely related to Tet O and Tet M, but shares only 68% amino acid sequence identity with them (Fig. 22.1). Genes hybridizing with *tet*(W) were also found in ruminal $Tc^R$ strains of *Selenomonas ruminantium* and *Mitsuokella multiacidus*, and show remarkably close sequence identity (99–100%) with *B. fibrisolvens tet*(W), which presumably reflects recent rapid transfer between these different genera. Interestingly, sequences flanking *tet*(W) in *Selenomonas* and *Mitsuokella* did not hybridize with corresponding regions from *Butyrivibrio*, suggesting that different genetic elements are involved in transfer of the determinant in the different species (Barbosa *et al.*, 1999).

**Table 22.1.** Distribution and location of tetracycline resistance determinants among $Tc^R$ strains of ruminal obligate anaerobes.

| | TetQ | TetO | TetW | Reference |
|---|---|---|---|---|
| *Prevotella ruminicola* | Pl, Chr[a] | | | Flint *et al.* (1988); Nikolich *et al.* (1994) |
| *Butyrivibrio fibrisolvens* | | Chr | Chr | Barbosa *et al.* (1999) |
| *Selenomonas ruminantium* | | | Pl, Chr | Barbosa *et al.* (1999) |
| *Mitsuokella multiacidus* | | | Chr | Barbosa *et al.* (1999) |

[a] Pl, plasmid encoded; Chr, chromosomally encoded.

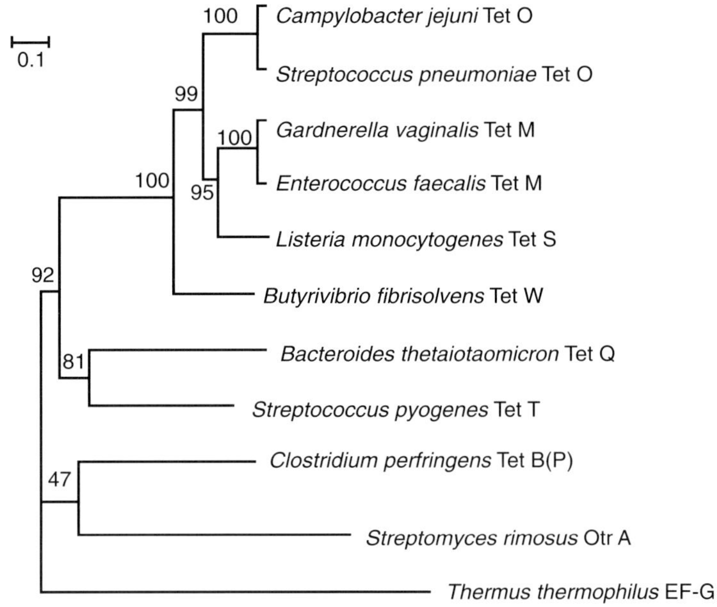


**Fig. 22.1.** Unrooted phylogenetic tree showing the evolutionary relationships between ribosome protection-type Tc$^R$ proteins. The scale bar refers to amino acid substitutions per position.

There are few reports of gene transfer events between rumen bacteria that involve traits other than antibiotic resistances. Lactate utilization was apparently transferred between *S. ruminantium* strains, although the mechanism of transfer has not been established (Gilmour *et al.*, 1996). It has also been suggested that amylolytic activity is transferred between strains of rumen *Lactobacillus* spp. (Kmet *et al.*, 1989).

## Agents of gene transfer

### Bacteriophages

The rumen contains a large and diverse bacteriophage population, and temperate bacteriophage have been identified in *Prevotella* spp., *B. fibrisolvens*, *Ruminococcus* spp. and *Streptococcus bovis* (Klieve *et al.*, 1989). Initial electron micrograph studies showed that most were tailed bacteriophage. Pulsed field gel electrophoresis (PFGE) analysis illustrates the diversity of the phage present (sizes range from 10 kb to 850 kb) and phage numbers were estimated to range from $3 \times 10^9$ to $1.6 \times 10^{10}$ phage ml$^{-1}$ rumen fluid (Klieve and Swain, 1993). The phage population varies greatly between animals and over time, suggesting that phage lysis may be a significant factor affecting the turnover of different bacterial populations in the rumen. PFGE has also been used to identify specific lytic bacteriophage which are active at certain times, perhaps in response to dietary changes which may affect the availability of phage receptor sites on the bacterial surfaces (Swain *et al.*, 1996).

Phage can mediate gene transfer events through a variety of mechanisms including specialized and generalized transduction and transfection. Potentially the most significant transfer events are those involving toxin genes, and the verocytotoxin genes found in *E. coli* strains including serotype O157 are known to have been acquired through bacteriophage-mediated transfer (Saunders *et al.*, 1999). To what extent the diversity of verotoxigenic coliform bacteria is enhanced by transfers occurring in the ruminant gut is an unanswered, but important, question.

### Plasmids

The incidence of plasmids varies greatly between different rumen species (Ogata *et al.*, 1996). Few strains of rumen *Prevotella/Bacteroides* and of *Ruminococcus* spp. appear to harbour plasmids, and there are no reports of plasmids from *Fibrobacter succinogenes* strains. However, a high proportion of *Selenomonas* and *Butyrivibrio* strains carry plasmids (e.g. Zhang *et al.*, 1991). *B. fibrisolvens* 2221 carries a very large plasmid of 300 kb (Teather, 1982; Scott *et al.*, 1997) and as many as four plasmids ranging up to 40 kb in size are present in some *Selenomonas* strains (Fig. 22.2, Fliegorova *et al.*, 1998). Traits encoded on these larger plasmids might have a significant role in the ability of the rumen population to adapt to environmental and dietary change. Among the smaller plasmids, which are cryptic and of interest mainly as the basis for vectors for genetic work, some appear likely to replicate via a single-stranded intermediate (e.g. pOM1;

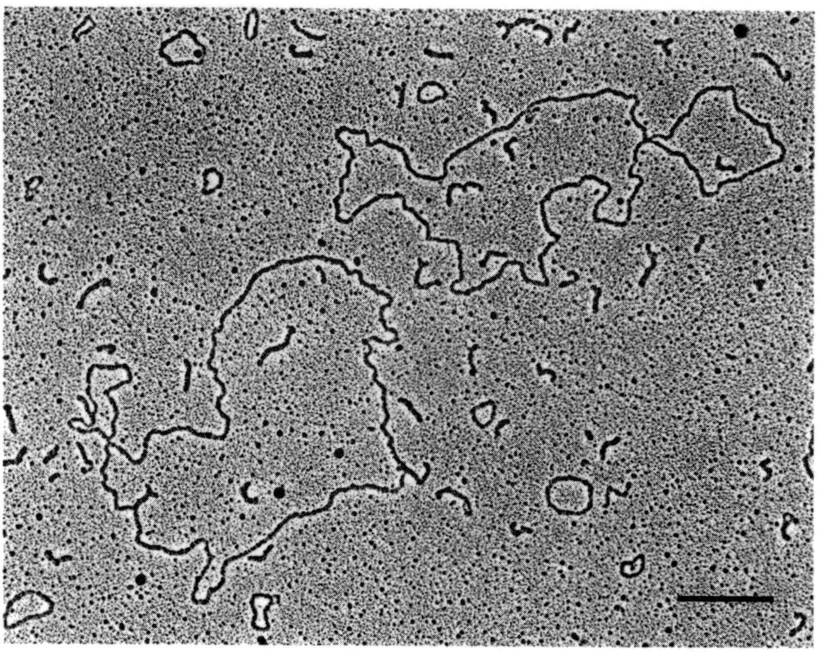

**Fig. 22.2.** Electron micrograph of plasmids purified from the sheep isolate *Selenomonas ruminantium* 5521cl. Three plasmids of 30 kb, 2.4 kb and 1.4 kb are visible. (Courtesy of K. Fliegerova.)

Hefford *et al.*, 1997) while others show theta-mode replication (e.g. pRJF2; Kobayashi *et al.*, 1995).

### *Transposable elements*

Chromosomal elements are important agents of antibiotic resistance transfer in Gram-positive bacteria and in *Bacteroides* from gut environments other than the rumen (for review see Salyers and Shoemaker, 1995). The mobile $Tc^R$ chromosomal element Tn*But1230* identified recently in *B. fibrisolvens* 1.230 (Scott *et al.*, 1997) is inferred to be 40–50 kb in size, with a preferred insertion site in the chromosome of *B. fibrisolvens* 2221. Since no evidence was found of hybridization with Tn*916* or Tn*5253* this appears to be a novel chromosomal element. Tn*916* can also transfer to *B. fibrisolvens*, but appears unable to transfer out of this species (Hespell and Whitehead, 1991a,b) while Tn*1545* can transfer to *Eubacterium cellulosolvens* (Anderson *et al.*, 1998).

### *Free DNA and natural transformation*

Turnover of free DNA in the rumen is assumed to be very rapid, but it is not ruled out that some fraction of released DNA is protected from degradation, e.g. by feed or soil components, and brief survival may be sufficient to yield transformants (Mercer *et al.*, 1999). There is currently very little information on the natural transformability of rumen bacteria, but the ability to take up and incorporate foreign DNA (competence) is common in other groups of bacteria (Lorenz and Wackernagel, 1994). Recent work shows that *S. bovis* JB1 is transformable using protocols developed for natural transformation of *Streptococcus gordonii* (Mercer *et al,* 1999). Transformation of rumen microorganisms by DNA of non-microbial (e.g. plant) origin is expected to require either the generation of a plasmid capable of replication in the bacterial host, or some DNA homology with the bacterial chromosome, and would be predicted to occur only under very special circumstances.

## Barriers to gene transfer

### *Nucleases and restriction modification systems*

Elucidation of restriction/modification systems can help to facilitate laboratory manipulation through protection of donor DNA or identification and isolation of nuclease deficient strains. Morrison *et al.* (1992, 1994) isolated two restriction endonucleases from *Ruminococcus flavefaciens* FD-1 and an *Mbo*I isoschizomer was found in a *Ruminococcus albus* strain (Miyagi *et al.*, 1998). Restriction systems have also been identified in *S. bovis* (Vanat *et al.*, 1993), *F. succinogenes* (Lee *et al.*, 1992) and in *S. ruminantium* (Pristas *et al.*, 1995). Non-specific nuclease activity can be detected in rumen fluid and in the supernatants of many rumen bacterial cultures, and was particularly evident in strains of *F. succinogenes* and some *Prevotella* strains (Flint and Thomson, 1990; Morrison *et al.*, 1994). In addition to its specific type II endonuclease, *Fsu*1, a highly active DNaseA was found in *F. succinogenes* which rapidly degraded DNA, even from the same strain of *Fibrobacter,* into small fragments of 8–10 bp (Lee *et al.*, 1992).

Although the function of restriction enzymes is presumably to protect the cell from foreign DNA, they do not provide a complete barrier against bacteriophage, or against single-stranded DNA entering via conjugation (Matic *et al.*, 1996).

### *Gene transfer and the evolution of rumen microorganisms*

It seems likely that horizontal gene transfer has played a role in the long-term evolution of rumen microorganisms in addition to its role in short-term adaptation to environmental change. Sequences of many polysaccharidase enzymes, for example family 11 xylanases, show very significant similarities between rumen bacteria (*Ruminococcus* spp.), rumen fungi (*Neocallimastix*) and even rumen protozoa such as *Polyplastron multivesiculatum* (Devillard *et al.*, 1999). It is an attractive hypothesis to suppose that anaerobic protozoa acquired fibre-degrading genes from ingested cellulolytic bacteria, but whether such acquisition occurred in the rumen, in the digestive tracts of other herbivores or elsewhere is not possible to decide, since related family 11 xylanases are also present in non-rumen bacteria and fungi.

Recent analyses of 16SrDNA sequences have revealed enormous genetic diversity among cultured isolates of species of rumen bacteria (Avgustin *et al.*, 1994; Willems *et al.*, 1996), and an even greater diversity among sequences amplified directly from rumen samples without cultivation (Whitford *et al.*, 1998). The dominant populations appear to be *Prevotella*/*Bacteroides* and low G+C content Gram-positive bacteria, but it is clear that the full diversity of these groups is not reflected by existing cultured strains. This has important consequences both for genetic work *in vitro*, and for attempts to reintroduce strains into the rumen.

## *In vitro* manipulation and genetic analysis in ruminal bacteria

### *Vector systems and gene expression*

Although initial attempts at gene transfer in rumen bacteria were mostly aimed at strain manipulation, there is increasing recognition that important information on gene function in rumen organisms can only be gained from genetic analysis, e.g. via insertional mutagenesis. Furthermore, the design of modified strains that would be acceptable for release would require detailed knowledge of transfer genetics and the stability of constructs. Most attempts to introduce genetic material into rumen bacteria have begun with the construction of vectors based on native plasmids or the use of plasmid vector systems designed for other organisms (Table 22.2). Largely because of the relative inefficiency of the available transfer systems, there are very few cases in which genes have been targeted to the chromosome of ruminal bacteria either via homologous recombination, using suicide vectors, or by means of transposons or bacteriophage vectors.

#### S. ruminantium

There have been no reports of the successful introduction of plasmid vectors into this species to date. Several small native plasmids have been analysed by sequencing (e.g. Nakamura *et al.*, 1999), including pJDB23 which was shown to be capable of replication in *E. coli* (Attwood and Brooker, 1992).

#### B. fibrisolvens

Introduction of plasmid vectors into *B. fibrisolvens* is complicated by strain diversity (Manarelli, 1988; Willems *et al.*, 1996). Electroporation protocols fail to work with all

**Table 22.2.** Plasmid vectors for ruminal bacteria.

| Vector | Plasmid replicon/s | Selectable marker [a] | Hosts | Size (kb) | Other characteristics | Reference |
|---|---|---|---|---|---|---|
| pBHerm | pRJF1 (*Butyrivibrio fibrisolvens*) OB156)/pUC118 | Em$^R$ (Ap$^R$) | *B. fibrisolvens, Escherichia coli* | 9.3 | | Beard *et al.* (1995) |
| pSMerm1 | pOM1 (*B. fibrisolvens* Bu49)/pUC19 | Em$^R$ (Ap$^R$) | *B. fibrisolvens, E. coli* | 7.8 | | Hefford *et al.* (1997) |
| pYK4 | pRJF2 (*B. fibrisolvens*)/ pUC18 | Em$^R$ (Ap$^R$) | *B. fibrisolvens, E. coli* | 7.9 | | Kobayashi *et al.* (1995) |
| pRRI207 | pRRI12 (*Prevotella ruminicola*)/ pHG165 | Em$^R$ (Ap$^R$) | *Bacteroides, E. coli* | 11 | Mobilizable by pRK2013 | Thomson *et al.* (1992) |
| pRH3 | pRRI2 (*P. ruminicola*)/ pBluescript | Tc$^R$ (Ap$^R$) | *Bacteroides, E. coli* | 8.8 | Non-mobilizable. Multiple cloning site | Daniel *et al.* (1995) |
| pTC-COW | pB8–51 (*Bacteroides*)/ pBR 328 | Tc$^R$ Cm$^R$(Ap$^R$) | *Bacteroides, Prevotella bryantii B$_1$4, E. coli* | 13.3 | Mobilizable in *Bacteroides* | Gardner *et al.* (1996) |
| pVA838 | | Em$^R$ (Ap$^R$,Cm$^R$) | G + bacteria, *E. coli* | 9.2 | Mobilizable by pVA797 | Macrina *et al.* (1982) |
| pTRW10 | pVA838 | Em$^R$ (Ap$^R$) | G + bacteria, *E. coli* | 7.1 | Multiple cloning site, Mobilizable by pVA797 | Wykoff and Whitehead (1997) |

[a] Resistances shown in parenthesis allow selection in *E. coli* but not in the alternative host.

strains (Beard *et al.*, 1995) perhaps in part because of differences in the production of extracellular polysaccharides. *B. fibrisolvens* shuttle vectors have been created by fusing the native *B. fibrisolvens* plasmids, pRJF1 and pRJF2, with *E. coli* plasmids (Ware *et al.*, 1992; Hefford *et al.*, 1993; Beard *et al.*, 1995; Kobayashi *et al.*, 1995). A fluoroacetate dehalogenase gene was introduced into such a shuttle vector (pBH*erm*; Fig. 22.3) downstream of an erythromycin-resistance promoter and was electroporated into *B. fibrisolvens* OB156. The resulting cultures were able to degrade the toxin fluoroacetate (Gregg *et al.*, 1994a). The modified strains were found to persist for at least 5 months following introduction into the rumen of a sheep. When trial sheep were inoculated with four modified *B. fibrisolvens* strains the sheep did not exhibit strong symptoms of fluoroacetate poisoning provided the modified bacteria were allowed to establish for 5 weeks before challenging with fluorocetate (Gregg *et al.*, 1998). It is significant that these modified bacteria were able to establish and be maintained in the rumen in competition with the resident flora.

The same shuttle vector was also used to introduce a xylanase gene from *Neocallimastix patriciarum* into *B. fibrisolvens* OB156 (Xue *et al.*, 1997). In this case the

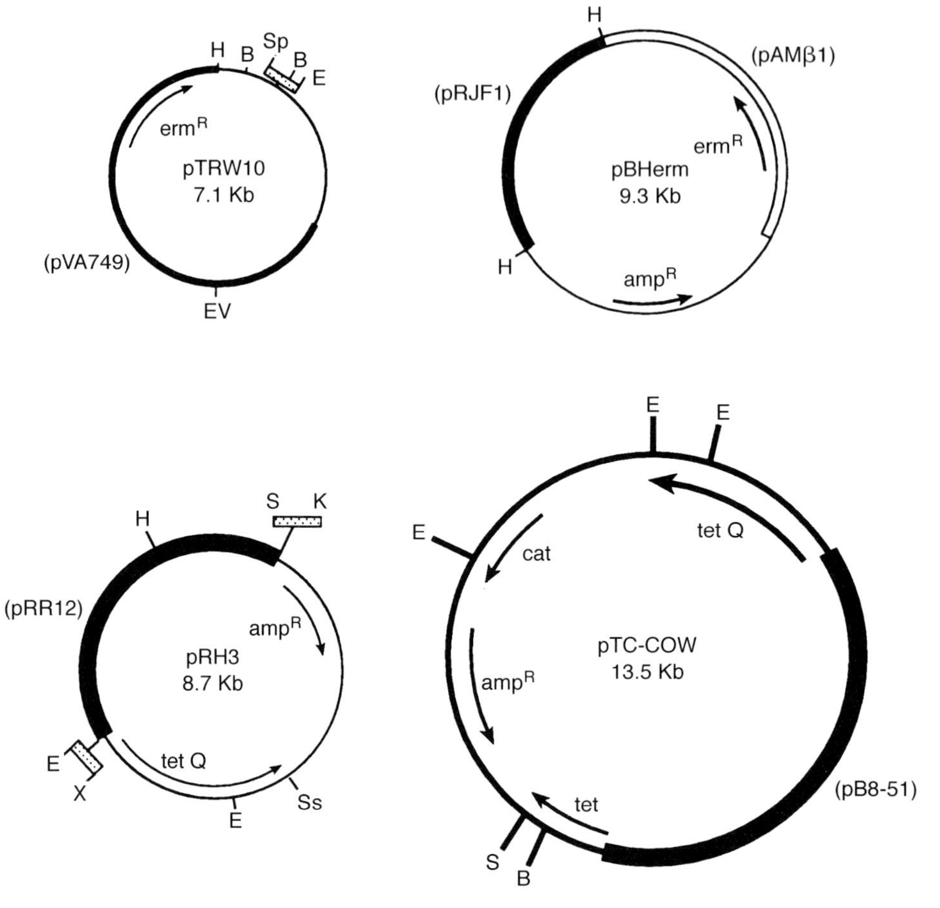


**Fig.22. 3.** Selected shuttle vectors available for the manipulation of rumen bacteria. pBHerm – *Butyrivibrio fibrisolvens* (Beard *et al.*, 1995); pTC-COW – *Bacteroides–Prevotella* (Gardner *et al.*, 1996); pTRW10 – *Streptococcus bovis* (Wykoff and Whitehead, 1997); pRH3 – *Bacteroides–Prevotella* (Daniel *et al.*, 1995).

gene was inserted downstream of a *B. fibrisolvens* xylanase promoter and transformants apparently exhibited 100-fold greater xylanase activity than the parent culture. The cryptic *B. fibrisolvens* plasmid pOM1 (Hefford *et al.*, 1997) differs from the pRJF series in that it replicates by a rolling circle mechanism, and thus is compatible with pRJF-based vectors (Hefford *et al.*, 1997).

Transposon Tn*916*, which can be transferred by conjugation from *Enterococcus faecalis* to *B. fibrisolvens* (Hespell and Whitehead, 1991b), has been used to mobilize plasmid vectors into non-transformable strains of *B. fibrisolvens*. Plasmids based on a *Staphylococcus aureus* plasmid pUB110 were thus mobilized between *B. fibrisolvens* strains at frequencies of $10^{-6}$ to $10^{-7}$ (Clark *et al.*, 1994) but in the absence of antibiotic selection the introduced plasmids were only maintained in under 50% of the population.

The potential of the native *B. fibrisolvens* transposable elements (Scott *et al.*, 1997), which are capable of transfer between distantly related strains, is clearly worth exploring.

### Ruminococcus *spp.*

Two plasmid vectors developed for use in other Gram-positive bacteria have been introduced into four different *R. albus* strains by electroporation (Cocconcelli *et al.*, 1992) and an efficiency of $3 \times 10^5$ transformants $\mu g^{-1}$ was achieved with one of the plasmids, pSC22. A low frequency of transfer of the broad host range plasmid pAMβ1 was also achieved into *R. albus* by conjugation from *Bacillus thuringiensis* BT351 (Aminov *et al.*, 1994). Despite these encouraging developments there have been no reports of introduction of vectors into *R. flavefaciens* strains and no studies on gene inactivation or the expression of genes introduced into *Ruminococcus* spp.

### Prevotella/Bacteroides

No discussion of this group can ignore the extraordinary genetic diversity among rumen strains revealed by 16S rDNA sequence analysis (Avgustin *et al.*, 1997; Whitford *et al.*, 1998; Wood *et al.*, 1998). Cultured rumen *Prevotella* isolates have been reclassified into four species (Avgustin *et al.*, 1997) and further subdivision may well be justified. Thus it may not be easy to obtain vector contructs that are successful in all strains. Electrotransformation of *Prevotella bryantii* $B_14$ was demonstrated with the native plasmid pRRI4, which carries a *tet*(Q) marker, using plasmid DNA extracted from the same strain background (Thomson and Flint, 1989). Transformation of the same strain by vectors that carry the *tet*(Q) marker, using DNA derived from *E. coli* has, however, been unsuccessful (Shoemaker *et al.*, 1991; Thomson *et al.*, 1992) probably due to restriction barriers. Introduction of vector constructs into *P. bryantii* $B_14$ has been achieved by a conjugation procedure that relies on the ability of certain conjugative transposons present in *Bacteroides* spp. to mobilize *Bacteroides* plasmids. By this means the vector pRDB5, which carries the *Bacteroides* plasmid replicon pB8–51, has been transferred into *P. bryantii* from *Bacteroides uniformis* (Shoemaker *et al.*, 1991), and another vector pTC-COW is now also available that carries a second selectable marker, $Cm^R$, in addition to *tet*(Q) (Fig. 22.3; Gardner *et al.*, 1996). pTC-COW has been used to transfer a hybrid *P. ruminicola* CMCase gene into *P. bryantii* $B_14$, but expression was not observed, possibly because the *P. ruminicola* promoter used failed to express in *P. bryantii* $B_14$ (Gardner *et al.*, 1996).

Naturally occurring plasmids have been reported in a few *Prevotella* strains (Flint and Stewart, 1987; Ogata *et al.*, 1996). pRRI2, a 3.4 kb plasmid from *P. ruminicola* 223/M2/7, has been used as the basis for the vectors pRRI207 ($Em^R$ marker) and pRH3 (*tet*(Q) marker; Fig. 22.3) which replicate in a range of *Bacteroides* strains and in the *Prevotella/Bacteroides* strain 2202 (Thomson *et al.*, 1992; Daniel *et al.*, 1995). pRH3 has been used to express a cellulase/xylanase gene from *P. ruminicola* 23 and two xylanase genes (*xynA* and *xynB*) from *P. bryantii* $B_14$ in *Bacteroides vulgatus* and in *Bacteroides/Prevotella* strain 2202. Interestingly, the *P. bryantii* xylanase genes were not expressed in *Bacteroides* hosts when introduced on the plasmid pTC-COW (T.R. Whitehead and H.J. Flint, unpublished observations). This may indicate that a fortuitous plasmid promoter is driving their expression in the pRH3 construct.

Some work has also been done on the exploitation of *Prevotella* bacteriophage genes in construction of vectors for chromosomal integration. Following identification of the excisionase and integrase genes from the *P. ruminicola* bacteriophage ØAR29 (Gregg *et al.*, 1994b), these genes were cloned into a *P. ruminicola* shuttle vector which was electroporated into a different *P. ruminicola* host strain, AR20, possibly resulting in integration into the chromosome (reviewed in Vercoe and White, 1996).

### F. succinogenes

There are no reports of indigenous plasmids or of successful attempts at gene transfer into this important cellulolytic species.

### S. bovis

*S. bovis* offers the advantages of being aerotolerant and of being able to support replication of many plasmids used in related Gram-positive bacteria. Hespell and Whitehead (1991b) were able to transfer the transposon Tn*916* and the plasmid pAMβ1 into *S. bovis* JB1 by conjugation from *E. faecalis.* The vector plasmid pVA838 (9.2 kb) has been used to express a cellulase (*endA*) and a bifunctional xylanase/ β(1,3–1,4) glucanase (*xynD*) from *R. flavefaciens* following electroporation into *S. bovis* (Whitehead, 1992; Whitehead and Flint, 1995; Ekinci, 1997). An improved version of pVA838 (pTRW10; Fig. 22.3) (Wykoff and Whitehead, 1997) was also used to express a *S. bovis* β(1,3–1,4)-glucanase (Ekinci *et al.*, 1997) and the green fluorescent protein from the jellyfish *Aequoria victoria* (Scott *et al.*, 1998) in Gram-positive bacteria including *S. bovis*. Another strategy is to fuse pIL253, a high copy number vector derived from pAMβ1, with pUC18 constructs to create shuttle constructs capable of replication in *E. coli* and *S. bovis* (Ekinci *et al.*, 1997). *S. bovis* JB1 produces a native secreted β(1,3–1,4)-glucanase whose gene has been isolated and sequenced (Ekinci *et al.*, 1997) allowing the construction of translational fusions in which the N terminal regions of the secreted enzyme drive expression of foreign gene products in *S. bovis* (M.S. Ekinci *et al.*, unpublished results).

Suicide constructs have also been used to inactivate an intracellular α-amylase in *S. bovis* helping to elucidate the role of this enzyme (Brooker *et al.*, 1995; Brooker and McCarthy, 1997).

## *Gene expression and regulation*

Although there has been little functional analysis of gene expression signals, putative promoters, transcriptional terminators, ribosome binding sites and signal peptides have generally been identified in rumen organisms through their homology with similar sequences in non-rumen relatives (reviewed by Vercoe and White, 1996; Teather *et al.*, 1997). Genes from *Prevotella* spp. resemble those from *Bacteroides* spp. in lacking obvious *E. coli*-like consensus ribosome binding sites, but have nevertheless been cloned by activity screening in *E. coli*. While genes from many different rumen species appear to be expressed from their own promoters in *E. coli*, the lack of expression of a *P. ruminicola* 23 promoter in *P. bryantii* $B_1$4 (Gardner *et al.*, 1996) suggests that promoter recognition may be highly specific in this group. Codon usage follows characteristic patterns for several rumen species, e.g. *R. flavefaciens* polysaccharidase genes use CAG almost exclusively rather than CAA to code for glutamine, whereas *S. bovis* favours CAA (Ekinci *et al.*, 1997; Kirby *et al.*, 1997). Instability of sequences cloned in *E. coli* appears to be a problem in some species, notably *Ruminococcus* spp. where the recovery of full-length cellulase genes may have been limited (Kirby *et al.*, 1997). Among rumen eukaryotes, both the protozoa and the fungi show exceptionally low G+C contents and there are early indications that genes from ruminal protozoa show highly biased codon usage (Eschenlauer *et al.*, 1998). Such differences in codon usage and promoter recognition can complicate heterologous expression of cloned genes.

### Genomic analysis

The uncertain recovery of genes by activity screening, and the relative inefficiency of piecemeal sequencing, argue that genome sequencing of some of the major ruminal bacteria will be worthwhile. The first to be considered are probably cellulolytic bacteria, where it is likely that only a small proportion of the significant genes required for fibre breakdown have so far been identified. *F. succinogenes* belongs to a little-studied group of bacteria and has a relatively small genome size (3.5 mega base pairs) making it an attractive proposition (Aminov, 1998). Genome sequencing would ultimately provide the most definitive information on gene flow between rumen organisms.

## Biotechnology

### Exploitation of genes isolated from ruminal microorganisms

The most obvious category of exploitable genes from rumen organisms are those responsible for the rapid breakdown of plant cell wall material, which have potential in animal feed pretreatment, paper pulp treatment, food processing and textile manufacture. The rumen fungi in particular have yielded cloned xylanases whose specific activities are higher than for any other xylanase available (reviewed by Selinger *et al.*, 1996). Enzymes of comparable activity have not so far been reported from rumen bacteria, but relatively few gene products have been purified to date. Although their catalytic domains generally show close similarities with those of non-rumen microorganisms, some rumen polysaccharidases show unique organization (e.g. Flint *et al.*, 1997). In the case of cellulases, recent evidence points to the importance of enzyme complexes both in rumen fungi and in *Ruminococcus* spp. (Fanutti *et al.*, 1995; Flint, 1997; Kirby *et al.*, 1997) and the organization of multiple enzyme subunits is likely to be critical for maximum activity. Nevertheless it has been shown that the activity of an endoglucanase from *P. bryantii* against crystalline cellulose can be enhanced by fusion to a cellulose-binding domain from the non-rumen bacterium *Thermomonospora fusca* (Maglione *et al.*, 1992). Exploitation of ruminal polysaccharidase genes has resulted in their expression in an increasingly wide range of bacterial, fungal and plant hosts, and in cultured mammalian cells as illustrated in Table 22.3.

There are clearly potential applications for many other genes from rumen organisms ranging from proteinases and lipases to restriction endonucleases and methylases (Selinger *et al.*, 1996). Bacteriocins have also attracted recent interest. These small proteinaceous antibiotics may play an important role in competition between bacteria for particular niches, and may have potential for exploitation as inhibitors of undesirable

**Table 22.3.** Exploitation of polysaccharidase genes from rumen microorganisms – heterologous expression systems.

| Expression host | Gene product | Species of origin | Reference |
|---|---|---|---|
| *Escherichia coli* | Products of all expressed genes cloned to date | | |
| *Streptococcus bovis* | Endoglucanase | *Ruminococcus flavefaciens* | Whitehead and Flint (1995) |
| *Bacteroides vulgatus* | Xylanase | *R. flavefaciens* | Ekinci (1997) |
| | Endoglucanase/xylanase | *Prevotella ruminicola* | Daniel *et al.* (1995) |
| *Butyrivibrio fibrisolvens* | Xylanase | *Neocallimastix patriciarum* | Xue *et al.* (1997) |
| | Xylanase | *Eubacterium ruminantium* | Kobayashi *et al.* (1998) |
| *Lactococcus lactis* | β(1,3–1,4)glucanase | *Streptococcus bovis* | Ekinci *et al.* (1997) |
| *Enterococcus faecalis* | β(1,3–1,4)glucanase | *S. bovis* | Ekinci *et al.* (1997) |
| *Saccharomyces cerevisiae* | Endoglucanase | *B. fibrisolvens* | van Rensburg and van Zyl (1994) |
| | Cellodextrinase | *R. flavefaciens* | van Rensburg *et al.* (1995) |
| Tobacco | Xylanase | *R. flavefaciens* | Herbers *et al.* (1996) |
| | β(1–3,1–4)glucanase | *R. flavefaciens* | Herbers *et al.* (1996) |
| | Xylanase | *Ruminococcus albus* | Ohmiya *et al.* (1997) |
| | Cellulase | *R. albus* | Kawazu *et al.* (1996) |
| Brassica | Xylanase | *N. patriciarum* | Liu *et al.* (1997) |
| Chinese hamster ovary cells | Xylanase | *Fibrobacter succinogenes S85* | Zhang *et al.* (1998) |

bacteria. Bacteriocins have been identified in facultative rumen inhabitants such as *S. bovis* (Iverson and Mills, 1976) and *Staphylococcus* spp. (Laukova and Marekova, 1993). In addition Odenyo *et al.* (1994) found that *R. flavefaciens* FD-1 was completely eliminated when cultured with *R. albus* 8, probably as a result of bacteriocin production. Bacteriocin-like activities were also found in 25/50 *Butyrivibrio* strains tested (Kalmokoff *et al.*, 1996); the inhibitory activities ranged from narrow to broad spectrum, and some inhibitors were shown to be proteinaceous, oxygen sensitive and resistant to gastric proteases. The inhibitory activity from *B. fibrisolvens* AR10 was due to a small protein, consisting of 20 predominantly hydrophobic amino acids (Kalmokoff and Teather, 1997).

Ruminal isolates of *Pseudomonas aeruginosa,* which were recently found to inhibit *E. coli* O157, produced pyocins which were active against different strains of *P. aeruginosa*. The inhibitory activity against *E. coli* is, however, due to the production of the inhibitory pigments pyocyanin and fluorescein (Duncan *et al.*, 1997).

### *Use of genetically modified or selected strains as rumen inoculants*

The use of isolated strains of rumen microorganisms as rumen inoculants has received some attention, mainly as a means of overcoming antinutritional factors. Thus *Megasphaera* might be used to prevent acidosis (Kung and Hession, 1995) while strains capable of detoxifying plant components (e.g. *Synergistes jonesii*) also have potential (Odenyo *et al.*, 1997). Inoculation with active fibre-degrading bacteria has also been considered (Saluzzi *et al.*, 1995). In addition rumen organisms may find application outside the rumen, e.g. *S. bovis* as a silage inoculant or obligate anaerobes for anaerobic digestion, biomass conversion or chemical production.

Genetic modification offers the potential for designing strains for particular purposes, e.g. *B. fibrisolvens* strains capable of degrading fluoroacetate (Gregg *et al.*, 1994a), or low pH tolerant rumen bacteria that deliver fibrolytic enzymes. Practical difficulties to be overcome include protecting inocula against oxygen and the establishment of sufficient populations within the rumen, which may require introduction of multiple strains for reasons discussed earlier. In some cases long-term survival of manipulated strains may be hindered by their inability to compete with the native flora and the inclusion of genes encoding bacteriocins has been suggested (Teather and Forster, 1998). Live preparations of non-rumen organisms are used extensively as feed additives (reviewed in Nagaraja *et al.*, 1997) and genetic modification might be used to introduce desired characteristics into such additives. Expression of a cloned xylanase in *Bacteroides thetaiotaomicron* led to a strain (BTX) with possible application as a rumen additive, although survival in rumen contents *in vitro* required provision of chondroitin sulphate as an energy source (Cotta *et al.*, 1997). The main impediment to the use of modified strains in ruminant nutrition, however, concerns safety and regulatory issues.

### *Regulatory and safety issues*

Apart from any issues of efficacy, it is still unclear whether it will become acceptable to release genetically modified rumen microorganisms as silage or rumen additives. Given the readiness with which certain resistance markers have been found to transfer within the rumen community (Barbosa *et al.*, 1999) it seems desirable to find alternative selection markers for strains intended for release. Chromosomal insertion of novel sequences is probably the most desirable way forward in the environmental release of genetically modified microorganisms, and also increases the stability of the new trait. The potential for colonization of other hosts, and any possible deleterious consequences of such colonization, is also a major issue that has prevented release of *B. fibrisolvens* strains engineered to degrade fluoroacetate (Gregg *et al.*, 1994a).

Ruminants are already being fed a wide variety of transgenic plant material. In the future ruminant diets may contain plants that have been modified deliberately to improve animal nutrition and rumen fermentation, e.g. to deliver feed enzymes, or essential amino acids. On the other hand a whole array of genetic modifications currently being made to crop and forage plants have non-nutritional aims, from protection against insect pests to drought resistance, and it is important to ensure that these have no unpredicted deleterious effects on the consuming animal (Flint and Chesson, 1999). A more general concern is that organisms in the rumen microflora might somehow acquire transgenes from the diet, perhaps leading to further transfer of these genes to transient members of the rumen flora that can inhabit the digestive tracts of other animals and of man. More basic information on ruminal gene transfer will help to estimate the probability (or improbability) of such events.

## Conclusions

The past decade has seen the isolation of more than 100 genes, mainly encoding polysaccharidases, from ruminal bacteria and fungi. By contrast, progress in studies on gene transfer has been relatively slow, although significant progress has been made with the development of vectors and the identification of plasmids, transposable elements, and bacteriophage. Gene transfer and insertional inactivation remain the key to analysing gene function and regulation in important ruminal microorganisms, without which the rewards of developments in rapid sequencing, including genome sequencing, will not be realized.

## Acknowledgement

The authors wish to acknowledge the support of the Scottish Executive Rural Affairs Department and the Ministry of Agriculture, Fisheries and Food (UK).

## References

Aminov, R.I. (1998) Complete sequence analysis of rumen cellulolytic bacterial genomes. *MIE Bioforum* 98 (abstract), p. 178.

Aminov, R.I., Kaneichi, K., Miyagi, T., Sakka, K. and Ohmiya, K. (1994) Construction of genetically marked *Ruminococcus albus* strains and conjugal transfer of plasmid pAMβ1 into them. *Journal of Fermentation and Bioengineering* 78, 1–5.

Anderson, K.L., Megehee, J.A. and Varel, V.H. (1998) Conjugal transfer of transposon Tn1545 into the cellulolytic bacterium *Eubacterium cellulosolvens. Letters in Applied Microbiology* 26, 35–37.

Attwood, G.T. and Brooker, J.D. (1992) Complete nucleotide sequence of a *Selenomonas ruminantium* plasmid and definition of a region necessary for its replication in *Escherichia coli. Plasmid* 28, 123–129.

Avgustin, G., Wright, F.W. and Flint, H.J. (1994) Genetic diversity and phylogenetic relationships among strains of *Prevotella (Bacteroides) ruminicola* from the rumen. *International Journal of Systematic Bacteriology* 44, 246–255.

Avgustin, G., Wallace, R.J. and Flint, H.J. (1997) Phenotypic diversity among ruminal isolates of *Prevotella ruminicola*: proposal of *Prevotella brevis* sp. nov., *Prevotella bryantii* sp. nov., and *Prevotella albensis* sp. nov. and redefinition of *Prevotella ruminicola. International Journal of Systematic Bacteriology* 47, 284–288.

Barbosa, T.M., Scott, K.P. and Flint, H.J. (1999) Evidence for recent intergeneric transfer of a new tetracycline resistance gene, *tet* (W), isolated from *Butyrivibrio fibrisolvens*, and the occurrence of *tet*(O) in ruminal bacteria. *Environmental Microbiology* 1, 53–64.

Beard, C.E., Hefford, M.A., Forster, R.J., Sontakke, S., Teather, R.M. and Gregg, K. (1995) A stable and efficient transformation system for *Butyrivibrio fibrisolvens* OB156. *Current Microbiology* 30, 105–109.

Brooker, J.D. and McCarthy, J.M. (1997) Gene knockout of the intracellular amylase gene by homologous recombination in *Streptococcus bovis*. *Current Microbiology* 35, 133–138.

Brooker, J.D., Lum, D.K., Thomson, A.M. and Ward, H.M. (1995) A gene-targeting suicide vector for *Streptococcus bovis*. *Letters in Applied Microbiology* 21, 292–297.

Clark, R.G., Cheng, K.-J., Selinger, L.B. and Hynes, M.F. (1994) A conjugative transfer system for the rumen bacterium, *Butyrivibrio fibrisolvens*, based on Tn*916*-mediated transfer of the *Staphylococcus aureus* plasmid pUB110. *Plasmid* 32, 295–305.

Cocconcelli, P.S., Ferrari, E., Rossi, F. and Bottazzi, V. (1992) Plasmid transformation of *Ruminococcus albus* by means of high-voltage electroporation. *FEMS Microbiology Letters* 94, 203–208.

Cotta, M.A., Whitehead, T.R. and Rasmussen, M.A. (1997) Survival of the recombinant *Bacteroides thetaiotaomicron* strain BTX in *in vitro* rumen incubations. *Journal of Applied Microbiology* 82, 743–750.

Daniel, A.S., Martin, J., Vanat, I., Whitehead, T.R. and Flint, H.J. (1995) Expression of a cloned cellulase/xylanase gene from *Prevotella ruminicola* in *Bacteroides vulgatus*, *Bacteroides uniformis* and *Prevotella ruminicola*. *Journal of Applied Bacteriology* 79, 417–424.

Devillard, E., Newbold, C.J., Scott, K.P., Forano, E., Wallace, R.J., Jouany, J.-P. and Flint, H.J. (1999) A family 11 xylanase from the ruminal protozoan *Polyplastron multivesiculatum*. *FEMS Microbiology Letters* 181, 145–152.

Duncan, S.H., Doherty, C.J., Govan, J.R.W. and Stewart, C.S. (1997) Rumen isolates of *Pseudomonas aeruginosa* antagonistic to an *Escherichia coli* O157 strain. Abstract 97th Annual Meeting, ASM, Miami, p. 419.

Ekinci, M.S. (1997) Heterologous expression of genes in the anaerobic bacterium *Streptococcus bovis*. PhD thesis, University of Aberdeen.

Ekinci, M.S., McCrae, S.I. and Flint, H.J. (1997) Isolation and overexpression of a gene encoding an extracellular β-(1,3–1,4)-glucanase from *Streptococcus bovis* JB1. *Applied and Environmental Microbiology* 63, 3752–3756.

Eschenlauer, S.C.P., McEwan, N.R., Calza, R.E., Wallace, R.J., Onodera, R. and Newbold, C.J. (1998) Phylogenetic position and codon usage of two centrin genes from the rumen ciliate protozoan, *Entodinium caudatum*. *FEMS Microbiology Letters* 166, 147–154.

Fanutti, C., Ponyi, T., Black, G.W., Hazlewood, G.P. and Gilbert, H.J. (1995) The conserved noncatalytic 40-residue sequence in cellulases and hemicellulases from anaerobic fungi functions as a protein docking domain. *Journal of Biological Chemistry* 270, 29314–29322.

Fliegorova, K., Benada, O. and Flint, H.J. (1998) Large plasmids in ruminal strains of *Selenomonas ruminantium*. *Letters in Applied Microbiology* 26, 243–247.

Flint, H.J. (1997) The rumen microbial ecosystem – some recent developments. *Trends in Microbiology* 5, 483–488.

Flint, H.J. and Chesson, A. (1999) The impact of gene technology used on raw materials for animal feedstuffs. *Proceedings from the 25th International Dairy Congress 1998 vol. III Future Milk Farming*. International Dairy Books, Denmark, pp. 98–109.

Flint, H.J. and Stewart, C.S. (1987) Antibiotic resistance patterns and plasmids of ruminal strains of *Bacteroides ruminicola* and *Bacteroides multiacidus*. *Applied Microbiology and Biotechnology* 26, 450–455.

Flint, H.J. and Thomson, A.M. (1990) Deoxyribonuclease activity in rumen bacteria. *Letters in Applied Microbiology* 11, 18–21.

Flint, H.J., Thomson, A.M. and Bisset, J. (1988) Plasmid-associated transfer of tetracycline resistance in *Bacteroides ruminicola*. *Applied and Environmental Microbiology* 54, 855–860.

Flint, H.J., Whitehead, T.R., Martin, J.C. and Gasparic, A. (1997) Interrupted catalytic domain structures in xylanases from two distantly related strains of *Prevotella ruminicola*. *Biochimica et Biophysica Acta* 1337, 161–165.

Gardner, R.G., Russell, J.B., Wilson, D.B., Wang, G-R. and Shoemaker, N.B. (1996) Use of a modified *Bacteroides-Prevotella* shuttle vector to transfer a reconstructed β-1,4-D-endoglucanase gene into *Bacteroides uniformis* and *Prevotella ruminicola* $B_14$. *Applied and Environmental Microbiology* 62, 196–202.

Gilmour, M., Mitchell, W.J. and Flint, H.J. (1996) Genetic transfer of lactate-utilizing ability in the rumen bacterium *Selenomonas ruminantium. Letters in Applied Microbiology* 22, 52–56.

Gregg, K., Cooper, C.L., Schafer, D.J., Sharpe, H., Beard, C.E., Allen, G. and Xu, J. (1994a) Detoxification of the plant toxin fluoroacetate by a genetically modified rumen bacterium. *Biotechnology* 12, 1361–1365.

Gregg, K., Kennedy, B.G. and Klieve, A.V. (1994b) Cloning and DNA sequence analysis of the region containing *attP* of the temperate phage ØAR29 of *Prevotella ruminicola* AR29. *Microbiology* 140, 2109–2114.

Gregg, K., Hamdorf, B., Henderson, K., Kopecny, J. and Wong, C. (1998) Genetically modified ruminal bacteria protect sheep from fluoroacetate poisoning. *Applied and Environmental Microbiology* 64, 3496–3498.

Hefford, M.A., Teather, R.M. and Forster, R.J. (1993) The complete nucleotide sequence of a small cryptic plasmid from a rumen bacterium of the genus *Butyrivibrio. Plasmid* 29, 63–69.

Hefford, M.A., Kobayashi, Y., Allard, S.E., Forster, R.J. and Teather, R.M. (1997) Sequence analysis and characterization of pOM1, a small cryptic plasmid from *Butyrivibrio fibrisolvens,* and its use in construction of a new family of cloning vectors for Butyrivibrios. *Applied and Environmental Microbiology* 63, 1701–1711.

Herbers, K., Flint, H.J. and Sonnewald, V. (1996) Apoplastic expression of the xylanase and β(1,3–1,4) glucanase domains of the *xynD* gene from *Ruminococcus flavefaciens* leads to functional polypeptides in transgenic tobacco plants. *Molecular Breeding* 2, 81–87.

Hespell, R.B. and Whitehead, T.R. (1991a) Conjugal transfer of Tn*916*, Tn*916*ΔE and pAMβ1 from *Enterococcus faecalis* to *Butyrivibrio fibrisolvens* strains. *Applied and Environmental Microbiology* 57, 2703–2709.

Hespell, R.B. and Whitehead, T.R. (1991b) Introduction of Tn*916* and pAMβ1 into *Streptococcus bovis* JB1 by conjugation. *Applied and Environmental Microbiology* 57, 2710–2713.

Hespell, R.B., Akin, D.E. and Dehority, B.A. (1996) Bacteria, fungi and protozoa of the rumen. In: Mackie, R.I., White, B.A. and Isaacson, R.E. (eds) *Gastrointestinal Microbiology*, Vol. 2, Chapman & Hall Microbiology Series, New York pp. 59–141.

Iverson, W.J. and Mills, N.F. (1976) Characterization of *Streptococcus bovis* bacteriophages. *Canadian Journal of Microbiology* 22, 847–852.

Kalmokoff, M.L. and Teather, R.M. (1997) Isolation and characterization of a bacteriocin (Butyrivibriocin AR10) from the ruminal anaerobe *Butyrivibrio fibrisolvens* AR10: Evidence in support of the widespread occurrence of bacteriocin-like activity among ruminal isolates of *B. fibrisolvens. Applied and Environmental Microbiology* 63, 394–402.

Kalmokoff, M.L., Bartlett, F. and Teather, R.M. (1996) Are ruminal bacteria armed with bacteriocins? *Journal of Dairy Science* 79, 2297–2306.

Kawazu, T., Ohta, T., Ito, K., Shibata, M., Kimura, T., Sakka, K. and Ohmiya, K. (1996) Expression of a *Ruminococcus albus* cellulase gene in tobacco suspension cells. *Journal of Fermentation and Bioengineering* 82, 205–209.

Kirby, J., Martin, J.C., Daniel, A.S. and Flint, H.J. (1997) Dockerin-like sequences in cellulases and xylanases from the rumen cellulolytic bacterium *Ruminococcus flavefaciens. FEMS Microbiology Letters* 149, 213–219.

Klieve, A.V. and Swain, R.A. (1993) Estimation of ruminal bacteriophage numbers by pulsed-field gel electrophoresis and laser densitometry. *Applied and Environmental Microbiology* 59, 2299–2303.

Klieve, A.V., Hudman, J.F. and Bauchop, T. (1989) Inducible bacteriophages from ruminal bacteria. *Applied and Environmental Microbiology* 55, 1630–1634.

Kmet, V., Javorsky, P., Nemcova, R., Kopecny, J. and Boda, K. (1989) Occurrence of conjugative amylolytic activity in rumen Lactobacilli. *Zentralblat fur Microbiology* 144, 53–57.

Kobayashi, Y., Forster, R.J., Hefford, M.A., Teather, R.M., Wakita, M., Ohmiya, K. and Hoshino, S. (1995) Analysis of the sequence of a new cryptic plasmid, pRJF2, from a rumen bacterium of the genus *Butyrivibrio*: Comparison with other *Butyrivibrio* plasmids and application in the development of a cloning vector. *FEMS Microbiology Letters* 130, 137–144.

Kobayashi, Y., Okuda, N., Matsumoto, M., Inoue, K., Wakita, M. and Hoshino, S. (1998) Constitutive expression of a heterologous *Eubacterium ruminantium* xylanase gene (*xynA*) in *Butyrivibrio fibrisolvens*. *FEMS Microbiology Letters* 163, 11–17.

Kung, L. Jr, and Hession, A.O. (1995) Preventing *in vitro* lactate accumulation in ruminal fermentations by inoculation with *Megasphaera elsdenii*. *Journal of Animal Science* 73, 250–256.

Laukova, A. and Marekova, M. (1993) Antimicrobial spectrum of bacterocin-like substances produced by ruminal staphylococci. *Folia Microbiologia* 38, 74.

Laukova, A., Kuncova, M. and Kmet, V. (1990) Isolation of several conjugative plasmids of the rumen bacteria *Enterococcus faecium*. *Biologia* (Bratislava) 45, 533–538.

Lee, S.F., Forsberg, C.W. and Gibbins, A.M. (1992) Type II DNA restriction-modification system and an endonuclease from the ruminal bacterium *Fibrobacter succinogenes* S85. *Journal of Bacteriology* 174, 5275–5283.

Liu, J., Selinger, L.B., Cheng, K.J., Beauchamin, K.A. and Moloney, M.M. (1997) Plant seed oil bodies as an immobilisation matrix for a recombinant xylanase from the rumen fungus *Neocallimastix patriciarum*. *Molecular Breeding* 3, 463–470.

Lorenz, M.G. and Wackernagel, W. (1994) Bacterial gene transfer by natural genetic transformation in the environment. *Microbiological Reviews* 58, 563–602.

Macrina, F.L., Tobian, J.A., Jones, K.R., Evans, R.P. and Clewell, D.B. (1982) A cloning vector able to replicate in *Escherichia coli* and *Streptococcus sanguis*. *Gene* 19, 345–353.

Maglione, G., Matsushita, O., Russell, J.B. and Wilson, D.B. (1992) Properties of a genetically reconstructed *Prevotella ruminicola* endoglucanase. *Applied and Environmental Microbiology* 58, 3593–3597.

Mannarelli, B.M. (1988) Deoxyribonucleic acid relatedness among strains of the species *Butyrivibrio fibrisolvens*. *International Journal of Systematic Bacteriology* 38, 340–347.

Matic, I., Taddei, F. and Radman, M. (1996) Genetic barriers among bacteria. *Trends in Microbiology* 4, 69–73.

Mercer, D.K., Scott, K.P., Bruce-Johnson, W.A., Glover, L.A. and Flint, H.J. (1999) Fate of free DNA and transformation of the oral bacterium *Streptococcus gordonii* DL1 by plasmid DNA in human saliva. *Applied and Environmental Microbiology* 64, 6–10.

Mercer, D.K., Melville, C.M., Scott, K.P. and Flint, H.J. (1999) Natural genetic transformation in the rumen bacterium *Streptococcus bovis* JB1. FEMS *Microbiology Letters* 179, 485–490.

Miyagi, T., Javorsky, P., Pristas, P., Karita, S., Sakka, K. and Ohmiya, K. (1998) Partial purification and characterization of *Ral*F40I, a class II restriction endonuclease from *Ruminococcus albus* F-40, which recognizes and cleaves 5'-/GATC-3'. *FEMS Microbiology Letters* 164, 215–218.

Morrison, M., Mackie, R.I. and White, B.A. (1992) Partial characterization of a deoxyribonucleic acid restriction endonuclease from *Ruminococcus flavefaciens* FD-1 and its inhibition by site specific adenine methylation. *Applied and Environmental Microbiology* 58, 66–69.

Morrison, M., Mackie, R.I. and White, B.A. (1994) The restriction endonuclease *Rfl*FII, isolated from *Ruminococcus flavefaciens* FD-1, recognizes the sequence 5'-AGTACT-3', and is inhibited by site-specific adenine methylation. *FEMS Microbiology Letters* 122, 181–186.

Nagaraja, T.G., Newbold, C.J., Von Neval, C.J. and Demeyer, D.I. (1997) Manipulation of ruminal fermentation. In: Hobson, P.N. and Stewart, C.S. (eds) *The Rumen Microbial Ecosystem*, Vol. 2. Blackie Academic and Professional, London, pp. 523–632.

Nakamura, M., Nagamine, T., Ogata, K., Tajima, K., Aminov, R.I. and Benno, Y. (1999) Sequence analysis of small cryptic plasmids isolated from *Selenomonas ruminantium* S20. *Current Microbiology* 38, 107–112.

Nikolich, M.P., Shoemaker, N.B., Wang, G.R. and Salyers, A.A. (1994) Evidence for natural horizontal transfer of tetQ between bacteria that normally colonise humans and bacteria that normally colonise livestock. *Applied and Environmental Microbiology* 60, 3255–3260.

Odenyo, A.A., Mackie, R.I., Stahl, D.A. and White, B.A. (1994) The use of 16S rRNA-targeted oligonucleotide probes to study competition between ruminal fibrolytic bacteria: development of probes for *Ruminococcus* species and evidence for bacteriocin production. *Applied and Environmental Microbiology* 60, 3688–3696.

Odenyo, A.A., Osuji, P.O., Karanfil, O. and Aidinew, K. (1997) Microbiological evaluation of *Acacia angustissima* as a protein supplement for sheep. *Animal Feed Science and Technology* 65, 99–112.

Ogata, K., Aminov, R.I., Nagamine, T., Benno, Y., Sekizaki, T., Mitsumori, M., Minato, H. and Itabashi, H. (1996) Structural organization of pRAM4, a cryptic plasmid from *Prevotella ruminicola. Plasmid* 35, 91–97.

Ohmiya, K., Suu, J.-L., Karita, S., Kawazu, T., Kimura, T. and Sakka, K. (1997) Gene expression of cellulase and xylanase in tobacco and cell wall digestion by domain shuffled enzymes. *Reproduction Nutrition and Development* Suppl., 53–54.

Orpin, C.G. and Joblin, K.N. (1997) The rumen anaerobic fungi. In: Hobson P.N. and Stewart C.S. (eds) *The Rumen Microbial Ecosystem*, 2nd edn. Blackie Academic and Professional, London, pp. 140–195.

Pristas, P., Vanat, I. and Javorský, P. (1995) Isolation and characterization of a new restriction endonuclease, *Sru30*DI, from *Selenomonas ruminantium. Gene* 158, 139–140.

Saluzzi, L., Stewart, C.S., Flint, H.J. and Smith, A. (1995) Plasmalogens of microbial communities associated with barley straw and clover in the rumen. *FEMS Microbiology Ecology* 17, 47–56.

Salyers, A.A. and Shoemaker, N.B. (1995) Conjugative transposons: the force behind the spread of antibiotic resistance genes among *Bacteroides* clinical isolates. *Anaerobe* 1, 143–150.

Saunders, J.R., Sergeant, M.J., McCarthy, A.J., Mobbs, K.J., Hart, C.A., Marks, T.S. and Sharp, R.J. (1999) Genetics and molecular ecology of *Escherichia coli* O157. In: Stewart, C.S. and Flint, H.J. (eds) Escherichia coli *O157 in Farm Animals*. CAB International, Wallingford, UK.

Scott, K.P. and Flint, H.J. (1995) Transfer of plasmids between strains of *Escherichia coli* under rumen conditions. *Journal of Applied Bacteriology* 78, 189–193.

Scott, K.P., Barbosa, T.M., Forbes, K.J. and Flint, H.J. (1997) High-frequency transfer of a naturally occurring chromosomal tetracycline resistance element in the ruminal anaerobe *Butyrivibrio fibrisolvens. Applied and Environmental Microbiology* 63, 3405–3411.

Scott, K.P., Mercer, D.K., Glover, L.A. and Flint, H.J. (1998) The green fluorescent protein as a visible marker for lactic acid bacteria in complex ecosystems. *FEMS Microbiology Ecology* 26, 219–230.

Selinger, L.B., Forsberg, C.W. and Cheng, K.-J. (1996) The rumen: a unique source of enzymes for enhancing livestock production. *Anaerobe* 2, 263–284.

Shoemaker, N.B., Anderson, K.L., Smithson, S.L., Wang, G.-R. and Salyers, A.A. (1991) Conjugal transfer of a shuttle vector from the human colonic anaerobe *Bacteroides uniformis* to the ruminal anaerobe *Prevotella (Bacteroides) ruminicola* $B_14$. *Applied and Environmental Microbiology* 57, 2114–2120.

Shoemaker, N.B., Wang, G. and Salyers, A.A. (1992) Evidence for natural transfer of a tetracycline resistance gene between bacteria from the human colon and bacteria from the bovine rumen. *Applied and Environmental Microbiology* 58, 1313–1320.

Smith, M.G. (1975) *In vivo* transfer of R factors between *Escherichia coli* inoculated into the rumen of sheep. *Journal of Hygiene Cambridge* 75, 363–370.

Smith, C.J. and Hespell, R.B. (1983) Prospects for development and use of recombinant deoxyribonucleic acid techniques with ruminal microorganisms. *Journal of Dairy Science* 66, 1536–1546.

Swain, R.A., Nolan, J.V. and Klieve, A.V. (1996) Natural variability and diurnal fluctuations within the bacteriophage population of the rumen. *Applied and Environmental Microbiology* 62, 994–997.

Teather, R.M. (1982) Isolation of plasmid DNA from *Butyrivibrio fibrisolvens. Applied and Environmental Microbiology* 43, 298–302.

Teather, R.M., Hefford, M.A. and Forster, R.J. (1997) Genetics of rumen bacteria. In: Hobson, P.N. and Stewart, C.S. (eds) *The Rumen Microbial Ecosystem*, 2nd edn. Blackie Academic and Professional, London, pp. 427–466.

Teather, R.M. and Forster, R.J. (1998) Manipulating the rumen microflora with bacteriocins to improve ruminant production. *Canadian Journal of Animal Science* 78 (suppl), 57–69.

Thomson, A.M. and Flint, H.J. (1989) Electroporation induced transformation of *Bacteroides ruminicola* and *Bacteroides uniformis* by plasmid DNA. *FEMS Microbiology Letters* 61, 101–104.

Thomson, A.M., Flint, H.J., Béchet, M., Martin, J. and Dubourguier, H.C. (1992) A new *Escherichia coli*: *Bacteroides* shuttle vector, pRRI207, based on the *Bacteroides ruminicola* plasmid replicon pRRI2. *Current Microbiology* 24, 49–54.

Trinci, A.P.J., Davies, D.R., Gull, K., Lawrence, M.I., Nielsen, B.B., Rickers, A. and Theodorou, M.K. (1994) Anaerobic fungi in herbivorous animals. *Mycological Research* 98, 129–152.

Van Rensburg, P. and van Zyl, W.H. (1994) Expression of the *Butyrivibrio fibrisolvens* encloded-β-1,4-glucanase gene together with the *Erwinia* pectate lyase and polygalacturonase genes in *Saccharomyces cerevisiae. Current Genetics* 27, 17–22.

Van Rensburg, P., van Zyl, W.H. and Pretorius, I.S. (1995) Expression of the *Ruminococcus flavefaciens* cellodextrinase gene in *Saccharomyces cerevisiae. Biotechnology Letters* 17, 481–486.

Vanat, I., Pristas, P., Kutejova, E., Judova, J., Godany, A. and Javorsky, P. (1993) *Sbv*I restriction endonuclease from *Streptococcus bovis. Letters in Applied Microbiology* 17, 297–299.

Vercoe, P.E. and White, B.A. (1996) Genetics of ruminal anaerobic bacteria. In: Mackie, R.I., White, B.A. and Isaacson, R.E. *Gastrointestinal Microbiology*, Vol. 2. Chapman & Hall Microbiology Series, New York, pp. 321–372.

Ware, C.E., Bauchop, T., Hudman, J.F. and Gregg, K. (1992) Cryptic plasmid pBf1 from *Butyrivibrio fibrisolvens* AR10: Its use as a replicon for recombinant plasmids. *Current Microbiology* 24, 193–197.

Whitehead, T.R. (1992) Genetic transformation of the ruminal bacteria *Butyrivibrio fibrisolvens* and *Streptococcus bovis* by electroporation. *Letters in Applied Microbiology* 15, 186–189.

Whitehead, T.R. and Flint, H.J. (1995) Heterologous expression of an endoglucanase gene from the ruminal anaerobe *Ruminococcus flavefaciens* 17 in *Streptococcus bovis* and *Streptococcus sanguis. FEMS Microbiology Letters* 126, 165–169.

Whitford, M.F., Forster, R.J., Beard, C.E., Gong, J.H. and Teather, R.M. (1998) Phylogenetic analysis of rumen bacteria by comparative sequence and analysis of cloned 16S rRNA genes. *Anaerobe* 4,153–163.

Willems, A., Amat-Marco, M. and Collins, M.D. (1996) Phylogenetic analysis of *Butyrivibrio* strains within the *Clostridium* subphylum of the gram-positive bacteria. *International Journal of Systematic Bacteriology* 46, 195–199.

Wood, J., Scott, K.P., Avgustin, G., Newbold, C.J. and Flint, H.J. (1998) Estimation of the relative abundance of different *Bacteroides* and *Prevotella* ribotypes in gut samples by restriction enzyme profiling of PCR-amplified 16S rRNA gene sequence. *Applied and Environmental Microbiology* 64, 3683–3689.

Wykoff, H.A. and Whitehead, T.R. (1997) Improved electroporation protocol and vectors for *Streptococcus bovis. World Journal of Microbiology and Biotechnology* 13, 269–272.

Xue, G.-P., Johnson, J.S., Bransgrove, K.L., Gregg, K., Beard, C.E., Dalrymple, B.P., Gobius, K.S. and Aylward, J.H. (1997) Improvement of expression and secretion of a fungal xylanase in the rumen bacterium *Butyrivibrio fibrisolvens* OB156 by manipulation of promoter and signal sequences. *Journal of Biotechnology* 54, 139–148.

Zhang, N., Attwood, G.T., Lockington, R.A. and Brooker, J.D. (1991) Genetic diversity in ruminal isolates of *Selenomonas ruminantium*. *Current Microbiology* 22, 279–284.

Zhang, J.X., Krell, P.J., Phillips, J.P. and Forsberg, C.W. (1998) A truncated bacterial xylanase expressed in mammalian cells exhibits increased sensitivity to proteases. *Biotechnology Letters* 20, 1001–1005.

# 23 Nutrient–Gene Interactions: Future Potential and Applications

P.B. CRONJÉ

*Department of Animal and Wildlife Sciences, University of Pretoria, Pretoria, South Africa*

## Introduction

It is commonly accepted that the aim of the discipline of animal science is to increase the efficiency of human edible protein production. As a result, past research has been directed at either improving input efficiencies (i.e. feed characterization) or output efficiencies (i.e. breed characterization and genetic selection). Despite these efforts, the accuracy with which current systems can predict individual and breed-related responses to differing nutritional inputs is still not adequate for practical purposes. The reason for this inability to predict nutrition–genotype interactions becomes evident when research priorities are considered within the context of the perceived primary limiting factors in different farming systems.

## Consequences of research paradigms for response prediction

In many regions of the sub-tropics, and in many of the developing areas of the world, ruminant nutrition is primarily based on natural pasture resources or crop residues. In many cases, these nutritional resources can be characterized as being of low quality and subject to substantial fluctuations in availability. Because attempts to improve the quality or quantity of nutrient supplies are constrained by variable rainfall and other climatic factors, the emphasis of research strategies in such circumstances has generally been directed at improvement of efficiency by genetic rather than nutritional means. By way of example, research in South Africa has been dominated by attempts to produce animals that are capable of higher production than indigenous animals but are better adapted to the prevalent climate than their high-producing exotic counterparts. Programmes designed to produce a breed adapted to the tropical climate of the northern part of South Africa by cross-breeding the indigenous Afrikander breed with European breeds began in 1930 at Messina experimental station, and were continued from 1936 onwards at Mara experimental station. This breeding scheme culminated in

 *Ruminant Physiology: Digestion, Metabolism, Growth and Reproduction* (ed. P.B. Cronjé)

the identification of an optimum cross-breed consisting of 5/8 Afrikander and 3/8 exotic blood (Shorthorn or Hereford) which was later developed into a breed known as the Bonsmara. These cross-breeding trials were then repeated at a number of other locations that differed with respect to climate and nutritive value of feed resources: projects aimed at characterizing and improving the productive attributes of indigenous Afrikander cattle were conducted at Potchefstroom, Koopmansfontein, Armoedsvlakte, Matoppos, Omatjenne and KwaZulu-Natal (Naudé, 1965). Similarly, cross-breeding projects aimed at producing sheep breeds adapted to different regions resulted in the development of breeds such as the Dormer (Dorset Horn × German Merino), Dorper (Dorset Horn × Blackhead Persian), Afrino (25% Ronderib Afrikander, 25% Merino, 50% South African Mutton Merino) and Dohne Merino (Peppin-style Merino × German Mutton Merino). This research policy, which accorded priority to the empirical evaluation of breeds within numerous different nutritional environments, to a large extent obviated the need for developing methods of predicting the responses of different genotypes to different planes of nutrition. As a consequence, local feeding standards were never developed, and the accuracy of prediction of genotype–nutrition interactions remains poor. In regions where the nutrient supply fluctuates, however, the nature of genotype–nutrition interactions is of primary importance: the multitude of failed aid projects in Africa and elsewhere where exotic 'high-producing' breeds were introduced in misguided attempts to improve animal productivity (see Ørskov, 1993) bears witness to this.

In contrast to the above, research in developed countries of the world and temperate climatic regions tends to have been focused to a greater extent on feed evaluation than breed evaluation. This is perhaps best illustrated by the construction of calorimeters in several countries during the period corresponding to the proliferation of breed evaluation projects in South Africa: Berlin (1885), Pennsylvania State University (1903), Rostock (1905), Copenhagen (1912), California (1935), Cornell University (1948), Rowett Research Institute (1952) and Hannah Research Institute (1954). With the advent of calorimetry came the introduction of stochastic concepts of the energy value of feeds such as Metabolizable Energy and Net Energy that were to form the basis of feeding standards. These feeding standards were designed to predict the amount of nutrients required for a defined level of production, and reflect an underlying paradigm which implies that the availability of ruminant nutrient resources is not a limiting factor. It is important to realize that these feeding systems were never designed to predict the level of performance that would result from a defined level of nutrition or from a defined genotype, and as a consequence, are of inadequate accuracy for this purpose. This is perhaps understandable in the light of the fact that resources available for ruminant nutrition in the past were of relatively high quality and in relatively constant supply in these regions. Recently, however, a series of reforms to the common agricultural policy of the European Union has encouraged lower stocking densities and a greater reliance on pasture-based farming systems. Under these circumstances, the choice of appropriate genotype takes on added importance (Sinclair *et al.*, 1998), and the ability to predict performance responses to defined levels of nutrition becomes critical (BBSRC, 1998). Unfortunately, feeding standards that were designed to predict levels of nutrition required for defined levels of performance cannot be used to predict performance responses with an acceptable degree of accuracy.

## Current accuracy of prediction of production responses to nutrition

Prediction of growth responses using feeding standards has been reported to be associated with errors ranging from 42% (Rohr *et al.*, 1983) to 80% (Oldham, 1988) for cattle, and of 35% for sheep (Rohr *et al.*, 1983). MacRae *et al.* (1988) summarized the results of 14 experiments in which different levels of dietary protein were fed to dairy cattle, and concluded that the observed milk yield responses were less than 50% of those predicted by the British protein-feeding system in 57% of the experiments studied. The accuracy of prediction of the British metabolizable-energy system for dairy cattle is such that 35% of cows fed according to current recommendations are underfed (BBSRC, 1998). Errors of this magnitude apply equally to all feeding systems currently in use.

The need for an accurate system for the prediction of nutrient responses is highlighted by a recent review of the British feeding system for dairy cows (BBSRC, 1998) that recommends that a fundamental change in approach be adopted. In the proposed system, nutrient supply from the diet would be described in terms of individual nutrient fluxes, and the physical and biological characteristic of the individual cow would be explicitly recognized. While some of the error term for prediction associated with current systems undoubtedly lies with variation in the accuracy of quantification of feed nutrient content, it would appear that individual genetic differences in the partitioning of available nutrients between different tissues or products represent a major source of error. It is evident that a concerted effort to quantify and describe the physiological basis of genotype–nutrient interactions for individual animals and different breeds will be necessary before the accuracy of response predictions can be improved to the level necessary for meaningful practical application.

The inadequacies of feeding systems have stimulated substantial research investment in the development of mechanistic computer simulation models in recent years. Most models use a derivative of the following equation to simulate sigmoidal input–output response relationships:

$$\text{Output} = V_{max} / (1 + (K_m / \text{Input})^z)$$

Recent reviews of current models (France *et al.*, 1995; Sauvant, 1995) indicate that although there is general consensus on the use of sigmoidal response functions, there is by no means consensus on which parameters of this equation should be used to modulate changes in nutrient partitioning. For instance, in lactation models, adjustments to either $V_{max}$, $K_m$ or the power function z have been used by various authors to accommodate changes in the nature of relationships between input and output which are known to be induced by homeostatic or homeorhetic mechanisms (see Sauvant, 1994). While identical changes in output for a defined level of input can be induced by appropriate changes to any of these parameters, it is important to note that each parameter will induce different changes in the nature of the response curve. A decrease in the parameter $K_m$ will shift the response curve to the left (Fig. 23.1). This phenomenon results in increased sensitivity, as the point of maximum output response occurs at lower input levels. An increase in $V_{max}$ will increase the maximum response possible (Fig. 23.2). This phenomenon is known as increased responsiveness, and differs from increased sensitivity in that the input required for half-maximal output remains constant. The distinction between these two concepts has far-reaching implications,

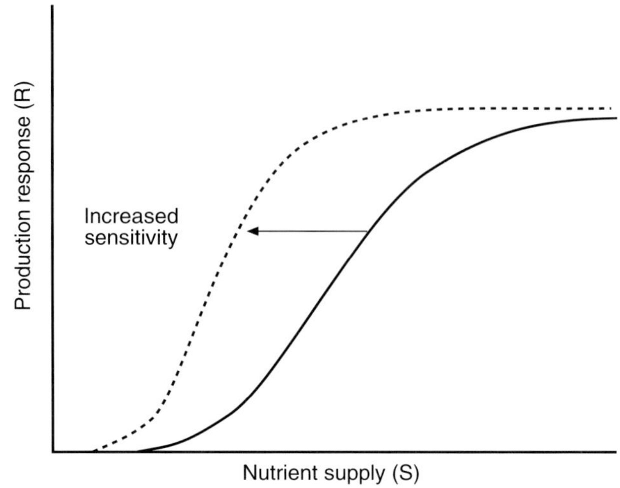


**Fig. 23.1.** Sigmoidal response curve described by the equation: Production response = $V_{max}$ / (1 + ($K_m$ / Nutrient supply)$^z$). The dotted line indicates the effect of a decrease in the value assigned to $K_m$.

particularly in the context of the possible effects of genetic selection on production characteristics. Assume, for example, that genetic increases in milk yield could be mediated via either increased responsiveness or increased sensitivity to plane of nutrition of mechanisms controlling nutrient partitioning to the mammary gland. An increase in either responsiveness or sensitivity would result in greater partitioning of nutrients to milk production at high levels of nutrition (see Figs 23.1 and 23.2), but in the case of increased sensitivity, this would persist at low levels of nutrition (Fig. 23.1). If extended to sufficiently low levels of nutrition, diversion of nutrients to productive functions such as milk yield, fibre production or muscle growth could compromise nutrient supply for other essential functions such as reproduction. It is thus pertinent to discuss the possible effects of genetic selection on the nature of responses to nutrient availability.

## Genotype–nutrition interactions

In the wild, evolutionary genetic changes in animals are profoundly influenced by the constancy of nutritive resources. In environments where there is relatively little seasonal change in nutrient availability, there is little need for expression of mechanisms that facilitate adaptive flexibility (adaptability) to changes in the plane of nutrition. Under these circumstances selection pressure tends to promote the development of specialized mechanisms that result in maximum utilization of a specific resource (adaptation). The fact that many of the world's endangered species are found in environments where nutrient supply is relatively constant (e.g. tropical rain forests) suggests that specialization (adaptation) has occurred at the expense of the capacity to adapt (adaptability). In terms of the nature of response functions as discussed above, this would suggest an increase in responsiveness and in sensitivity of regulatory mechanisms to a specific level

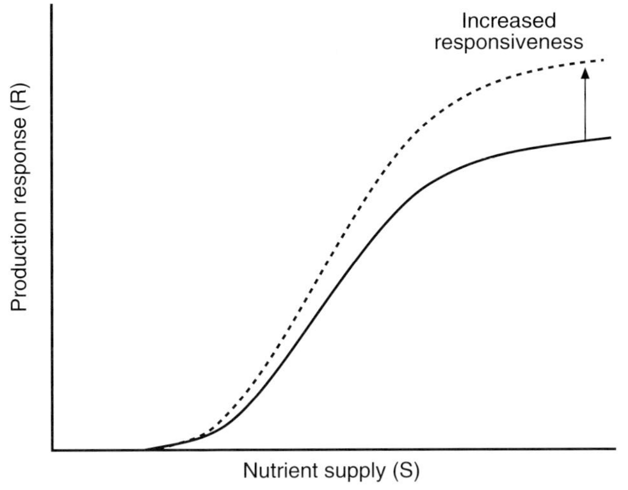


**Fig. 23.2.** Sigmoidal response curve described by the equation: Production response = $V_{max}$ / (1 +($K_m$ / Nutrient supply)$^z$). The dotted line indicates the effect of an increase in the value assigned to $V_{max}$.

of nutrition (Fig. 23.3). On the other hand, animals subject to large seasonal and periodic (drought-induced) fluctuations in the availability of nutrients are far more likely to evolve towards optimum utilization of the long-term mean nutrient supply, and selection pressure would favour adaptability at the expense of specialization. Adaptation to the long-term mean nutrient supply would increase nutrient partitioning to endogenous reserves as a precautionary measure against potential future decreases in exogenous nutrient supply, and result in a decrease in sensitivity and responsiveness to plane of nutrition (Fig. 23.3) for non-essential productive functions. Although these singular and integrated genotype–nutrition interactions are evident throughout biology, it is conspicuous that little information exists as to whether genetic selection for high production in domestic species has changed sensitivity to level of nutrition.

## Physiological consequences of genetic selection for productive attributes

The fact that almost all the indigenous livestock breeds of the sub-tropics and the developing regions of the world exhibit lower productive responses for milk yield or growth rate at high planes of nutrition than their specialized exotic counterparts indicates that genetic selection for single productive attributes has increased responsiveness. Of greater concern is the question of whether genetic selection for single production characteristics also increases sensitivity. It is self-evident that the long-term sustainability of animal production in regions subject to frequent droughts and seasonal changes in the availability of nutrient resources for ruminants will to a large extent be determined by the ability of livestock to adapt to periods of nutrient restriction.

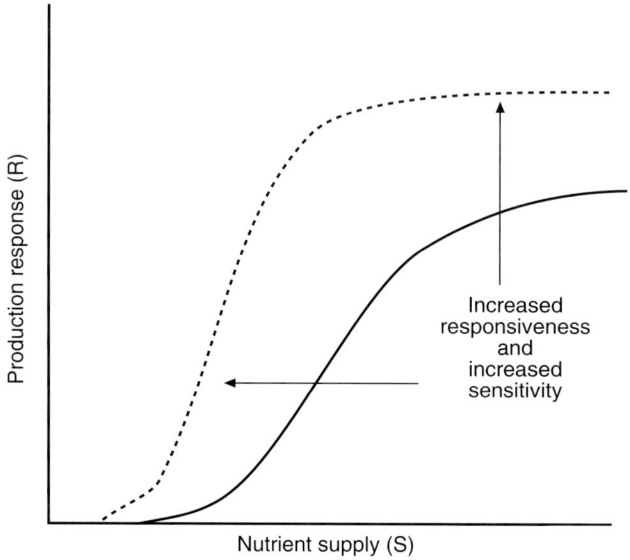


**Fig. 23.3.** Possible effects of genetic selection pressure on production responses to level of nutrition under conditions where the supply of nutrition is relatively constant.

The Angora goat is a good model of a breed that has been subject to intense single-trait genetic selection for fibre production. Although the rate of mohair production in this breed by far surpasses that of sheep selected for high wool-growth rates, it exhibits a high rate of abortions and a low survival rate under conditions of cold-stress. Comparisons between Angora goats and goats selected for meat production (Cronjé, 1992a,b) and between Angora phenotypes that differ in respect of fibre production rates (Cronjé, 1995) have established that the primary cause of the high incidence of abortions and cold-stress fatalities in animals selected for high fibre-growth rates is an inability to maintain blood glucose concentrations when the plane of nutrition is decreased. It would appear that Angora goats selected for high levels of mohair production preferentially deposit a greater proportion of ingested nitrogen as fibre, resulting in insufficient labile protein reserves and decreased substrate availability for gluconeogenesis when the plane of nutrition is decreased (Cronjé, 1998). This evidence suggests that in this instance, the effects of genetic selection have been to decrease the ability of the animal to adapt to changes in plane of nutrition, and indicates that this was probably mediated through increased sensitivity as well as increased responsiveness of fibre production to plane of nutrition (Fig. 23.3). Evidence exists that a similar situation prevails in Merino sheep selected for high wool-production rates (Cronjé and Smuts, 1994; Herselman *et al.*, 1998).

Phenotypic and genetic correlations between milk yield and gross feed efficiency are often high (Veerkamp and Emmans, 1995), suggesting that selection for milk yield automatically increases the partial efficiency of conversion of nutrients to milk. However, several reviews have shown that there is little evidence to indicate that genetic variation exists with respect to the partial efficiency of conversion of absorbed nutrients to milk (Blake and Custodio, 1984; Bauman *et al.*, 1985; Veerkamp and Emmans, 1995). Blake and Custodio (1984) concluded that feed efficiency is an artefact of the selection response to milk yield. Although large genetic differences have been observed for gross efficiency, it would appear that 'high genetic merit animals are more efficient because they partition the available energy differently from low genetic merit cows, and not because the processes used to transform consumed feed into product have become more efficient'(Veerkamp and Emmans, 1995). Because total energy expenditure during early lactation exceeds intake capacity in high-producing dairy cows, these genetic differences in nutrient partitioning are manifest as increased subsidy of nutrient requirements for milk production by tissue reserves (Blake and Custodio, 1984). More specifically, selection for milk yield is considered to increase the cow's ability to mobilize adipose tissue reserves in early lactation and to replace them in late lactation (Bauman *et al.*, 1985). The fact that there is evidence to suggest that continued selection pressure for milk yield has caused a decrease in reproduction rates in dairy breeds (Muller *et al.*, 1999) indicates that genetic selection has had the effect of increasing sensitivity as well as increasing responsiveness of milk production to plane of nutrition. The fact that reproduction rates of 'high-producing' genotypes typically decrease to levels lower than that of existing 'low-producing' indigenous animals when introduced into developing areas (Zarate, 1996) corroborates this.

In summary, there is evidence to indicate that genetic selection for production rates under conditions where nutrition is not a limiting factor has increased the sensitivity of production response mechanisms and resulted in a decreased ability to adapt to variations in the plane of nutrition. As lactation is the single most energetically demanding physiological phase in the female reproductive cycle, misdirection of resources between milk and endogenous reserves will typically impact negatively on reproductive rates. The principal pathway sustaining responses to selection for milk yield appears to be via an 'endocrine-controlled accession of body tissue reserves' (Blake and Custodio, 1984). It is evident that a better understanding of the genetic and nutritional basis of hormonal regulation of nutrient partitioning will be a key factor if the accuracy of prediction of feed efficiency is to be improved.

## Genotype differences in endocrine control of nutrient partitioning during lactation

Insulin and the glucose transporters (GLUT) play a major role in the regulation of adipose tissue metabolism during lactation. Insulin is secreted by the pancreas in response to an increased uptake of glucose or glucose precursors into the blood. Insulin binds with its receptor on adipose and muscle cells and induces the translocation of GLUT4 glucose transporters from the interior of the cell to the cell membrane where they facilitate the uptake of glucose. In addition to this, the binding of insulin to its receptor also induces increased activity of many enzymes necessary for the further metabolism of intracellular glucose. In the lactating animal, glucose is partitioned away from uptake by these tissues and towards uptake by the mammary gland by decreasing the sensitivity of adipose and muscle tissue to insulin. Decreased insulin sensitivity results in less glucose being taken up by tissues in which glucose uptake is dependent on insulin-mediated recruitment of GLUT4 glucose transporters (adipose tissue and muscle). In this way, the availability of circulating glucose for uptake by tissues such as the mammary gland in which glucose uptake is determined by the concentration gradient (and not by insulin) is increased. Because of the effect of insulin on intracellular enzyme activity, decreased insulin sensitivity also decreases the rate of triglyceride synthesis from glucose metabolism and accelerates the rate of fat breakdown in adipose cells.

Recent work in our laboratory (P.B. Cronjé, E. Vlok and M. de Jager, unpublished results) suggests that the insulin sensitivity may play a key role in regulating genotype-related differences in nutrient partitioning. Responses to plane of nutrition in lactating Indigenous goats were compared with those in Saanen–Indigenous cross-breed goats. The Saanen cross-breed produced more milk than the Indigenous genotype at the same plane of nutrition. The consequence of this difference in nutrient partitioning is reflected in the fact that the Saanen cross-breed lost 19% of initial body mass by week 10 of lactation, while the Indigenous goat regained initial body mass by week 6 of lactation (Fig. 23.4). Plasma glucose concentrations were depressed to a lesser extent by insulin in the Saanen cross-breed than in the Indigenous goat (Fig. 23.5). This indicates that skeletal muscle and adipose tissues in the Saanen cross-breed genotype were less sensitive to insulin than in the Indigenous genotype and represents a possible mechanism whereby genotype-related differences in nutrient partitioning are mediated. Differences in insulin receptor concentrations have recently been observed in different breeds of sheep (Wylie *et al.*, 1998), and it has been suggested that variations in GLUT4 gene expression may be the major determinant of insulin sensitivity in humans (Charron *et al.*, 1999). This evidence suggests that genetic increases in feed efficiency or milk yield may, in part, be mediated by decreased insulin sensitivity.

Growth hormone or bovine somatotropin (ST) has been shown to change nutrient partitioning in dairy cattle (Bauman and Vernon, 1993) and thus also represents a likely candidate for regulation of genotype-related differences in nutrient partitioning.

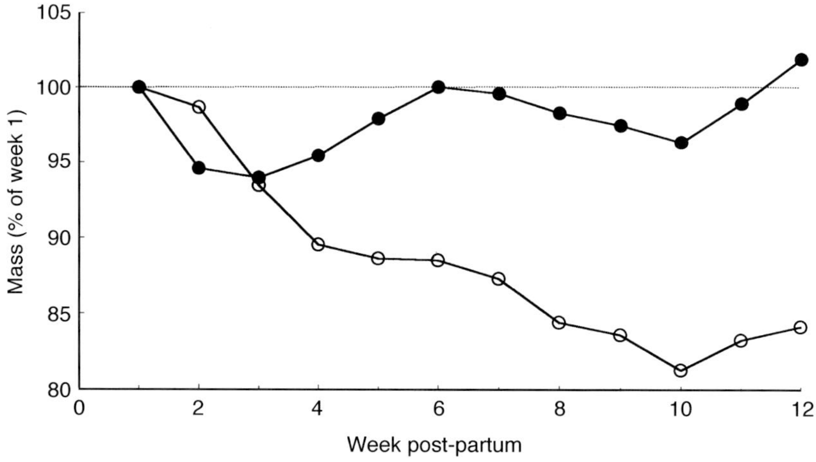


**Fig. 23.4.** Mean liveweight of Indigenous (●) and Saanen cross-breed (○) goats during lactation.

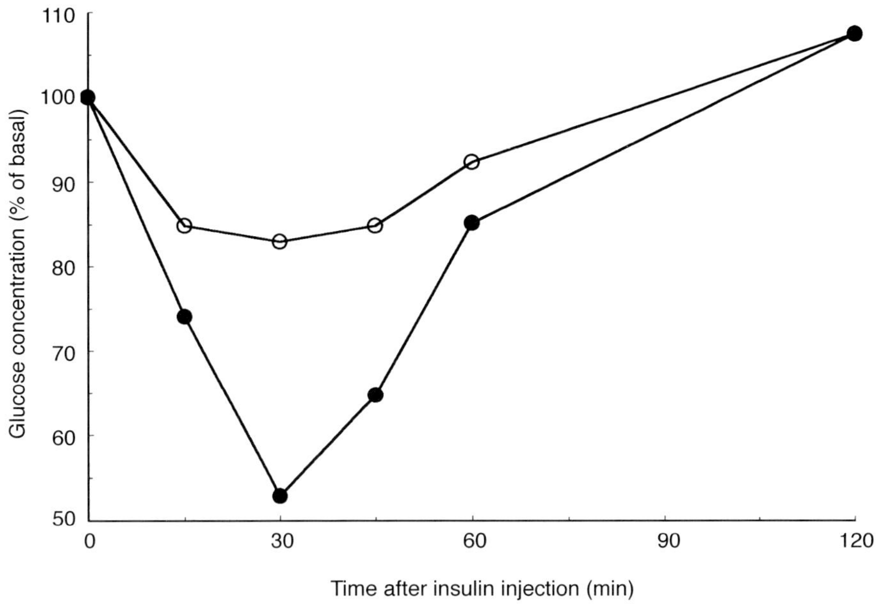


**Fig. 23.5.** Plasma glucose concentrations following an intravenous insulin challenge in Indigenous (●) and Saanen cross-breed (○) goats during week 7 of lactation.

The metabolic effects of ST administration in dairy cattle include reduced insulin sensitivity (Rose *et al.*, 1996), reduced expression of GLUT4 mRNA in adipose and muscle tissue (Zhao *et al.*, 1996) and increased mobilization of adipose tissue reserves (Etherton and Bauman, 1998). It also probable that many of the metabolic effects of ST are as a result of ST-induced secretion of insulin-like growth factor-I (IGF-I) from the liver. IGF-I is mitogenic, and has been implicated in preventing programmed cell death in mammary cells (apoptosis) and may therefore play an important role in lactation persistency (Cohick, 1998). Genetic increases in milk yield may be mediated by differences in the ST–IGF hormonal axis, as Gallo *et al.* (1997) have shown that Holstein cows of low estimated breeding value respond differently to ST than those of high breeding value. There is also evidence that ST concentration and ST responses to challenges are associated with genetic merit for milk yield in male and juvenile cattle (Woolliams and Lovendahl, 1991).

Leptin is a hormone that is secreted by adipocytes, and is thought to function as a sensor and regulator of body energy stores. Leptin is known to regulate insulin secretion and affect insulin sensitivity, and it has been suggested that leptin functions to orchestrate the complex array of signals which regulate nutrient partitioning (Houseknecht and Portocarrero, 1998). Furthermore, leptin is known to influence appetite and also modulate ST secretion, and could represent a link between adipose tissue reserves, voluntary intake and milk yield. This may explain why selection for milk yield and milk components has been reported to influence voluntary intake at various stages of lactation in dairy cows (Veerkamp *et al.*, 1994; Akerlind *et al.*, 1999). In humans, leptin concentrations have been found to have a heritability ($h^2$) of 0.39

(Rotimi *et al.*, 1997). Although data on leptin in domestic animals are only now starting to emerge, the fact that plasma leptin concentrations in sheep are highly variable and repeatable (Chilliard *et al.*, 1998) suggests that leptin expression may be related to genetic differences in nutrient partitioning.

The above discussion has emphasized some of the most important mechanisms regulating nutrient partitioning during lactation. Although differences in concentrations of hormones and their receptors may explain why differences in responsiveness exist between different genotypes at the same plane of nutrition, the postulated existence of differences in sensitivity implies that differences exist with respect to the way in which endocrine factors respond to changes in nutritional stimuli. Genetic differences in milk yield may be a function not only of differences in genes coding for these endocrine factors but also differences in factors which regulate gene expression.

## Nutrient–gene interactions

Genes determine the maximum possible rates of formation of gene products but, in most if not all situations, these maximum rates will not be attained. Observed responses to genetic selection are therefore most probably not a function of the animal's genes *per se*, but of genetic variation in the extent to which genes are expressed. There is increasing evidence to show that post-transcriptional regulation of RNA expression by nutrients in the cytoplasm of the cell exerts a major influence over the expression of many genes (Hesketh *et al.*, 1998). The total DNA in the genome of cattle consists of approximately 3,000,000,000 base pairs. Of this, less than 5% consists of functional genes. In early studies, it was thought that the nucleotide sequences in these and the untranslated regions (UTR) which flank functional RNA sequences were redundant and of no importance, but it is now becoming evident that UTR are highly important for nutrient-mediated regulation of gene expression. Hesketh *et al.* (1998) have proposed that an understanding of the interaction between nutrients and gene expression will provide a basis for determining nutritional requirements of humans on an individual basis. In the present context, this may represent the key to understanding why nutrient partitioning varies between individual animals and genotypes, and so enable the accuracy of prediction of input–output responses to be improved.

One of the best known physiological responses to selection for milk yield is known to be a complex genetic mechanism for maximizing the amount and availability of adipose tissue during early lactation (Bauman *et al.*, 1985). Several studies have also established that the activity of lipogenic enzymes varies between genotypes (see Chapter 20). Differences in the sensitivity of adipose tissue enzyme responses to nutritional supply could be mediated via differences in endocrine responses to nutrients or by differences in the direct effects of nutrients on enzyme gene-expression.

Polymorphisms (variations in DNA sequences) exist for most of the genes coding for endocrine factors involved with lactation. These differences are functionally significant, as polymorphisms for the genes coding for growth hormone releasing hormone, growth hormone, the growth hormone receptor and Pit-1, a pituitary specific transcription factor controlling the expression of ST, have been associated with milk traits in dairy cattle (for review see Parmentier *et al.*, 1999). Although it is not yet known exactly how these small differences in the DNA sequence of genes coding for the same hormone cause differences in milk yield between individuals and breeds, it is possible that they may differ with respect to sensitivity to transcription factors such as nutrients. Differential sensitivities of polymorphisms to nutrition have been demonstrated for IGF-I and ST (Duan, 1998; Gootwine *et al.*, 1998), and may represent a possible explanation of why genotypes differ in their sensitivity to variations in nutritional supply.

Differences in sensitivity to nutrition could also be mediated by nutrient–gene interactions at a more direct level. Recent studies (Clarke and Abraham, 1992; Towle, 1995; Girard *et al.*, 1997) have shown that a metabolite of glucose, probably glucose-6-phosphate, directly regulates the expression of the rate-limiting enzymes for lipogenesis, acetyl-CoA carboxylase and fatty acid synthase, as well as other key enzymes such as phosphofructokinase and pyruvate kinase. Expression of fatty acid synthase is also regulated by poly-unsaturated fatty acids (Clarke and Abraham, 1992), and long-chain fatty acids have been shown to regulate gene expression of carnitine palmitoyl transferase and hydroxy-methyl-glutaryl-CoA synthase, both of which are critical for lipid metabolism. In addition to absorbed nutrients, other dietary components such as the fat-soluble vitamins may play a role in the direct regulation of adipose tissue metabolism, as retinoic acid has been shown to regulate the expression of glycerol-P-dehydrogenase (Clarke and Abraham, 1992). Furthermore, the predominant control of GLUT-4, which is critical for the intracellular supply of glucose for metabolism by these enzymes, is now thought to be linked to the intracellular metabolism of glucose (Charron *et al.*, 1999).

Whereas past research dealing with homeostasis has been concentrated on endocrine regulation of metabolism, the significance of these recently discovered nutrient–gene interactions is such that Girard *et al.* (1997) have suggested that insulin only has a potentiating role in adipose tissue, and Charron *et al.* (1999) have suggested that GLUT-4 gene expression is consistent with metabolic rather then hormonal regulation. Clearly, this invokes the question of whether a paradigm shift is not called for in the field of domestic animal physiology research: a shift in emphasis towards individual nutrient concentrations and metabolism would certainly be complementary to the individual nutrient approach which has been recommended for future ruminant nutrition research by the BBSRC (1998). In the context of nutrition–genotype interactions, it is not unlikely that the differences in input–output response sensitivity which appear to be induced by genetic selection could be directly related to the sensitivity of different polymorphisms to regulation of gene-expression by individual nutrients. Although the exact location and associations of UTR with functional genes is still unclear, the finding that polymorphisms in the UTR of mRNA influence susceptibility to hyperlipidaemia in humans (see Hesketh *et al.*, 1998) indicates that markers for sensitivity, or variations in nutrient–gene interaction, can be developed.

## Conclusions

In the past, research programmes aimed at increasing the efficiency of animal production systems have tended to concentrate on either breed improvement or feed evaluation. Unfortunately, neither of these approaches has proved to be of much value for predicting animal responses to changes in nutrition. It is evident that more information on the nature and extent of interactions between genotype and nutrition is required in order to enable accurate prediction of animal responses to be made. There is a substantial body of evidence indicating that different genotypes respond differently to changes in nutrition. One of the most promising recent developments in this regard is increasing evidence of regulation of gene expression by individual nutrients, and it is proposed that differences in the sensitivity of gene-regulatory mechanisms to nutrients may underlie variation of responses between individuals and genotypes. As gene expression is, in many instances, a function of nutrient concentrations, an integrated approach encompassing the disciplines of physiology, molecular genetics and nutrition is called for if we are to increase the accuracy of response prediction in dairy cattle.

## References


Akerlind, M., Holtenius, K., Bertilsson, J. and Emanuelson, M. (1999) Milk composition and feed intake in dairy cows selected for high or low milk fat percentage. *Livestock Production Science* 59, 1–11.

Bauman, D.E. and Vernon, R.G. (1993) Effects of exogenous bovine somatotropin on lactation. *Annual Review of Nutrition* 13, 437–461.

Bauman, DE., McCutcheon, S.N., Steinhour, W.D., Eppard, P.J. and Sechen, S.J. (1985) Sources of variation and prospects for improvement of productive efficiency in the dairy cow: a review. *Journal of Animal Science* 60, 583–592.

BBSRC (1998) Biotechnology and biological sciences research council technical committee on responses to nutrients report no 11. *Responses in the Yield of Milk Constituents to Intake of Nutrients by Dairy Cows.* CAB International, Wallingford.

Blake, R.W. and Custodio, A.A. (1984) Feed efficiency: a composite trait of dairy cattle. *Journal of Dairy Science* 67, 2075–2083.

Charron, M.J., Katz, E.B. and Olson, A.L. (1999) GLUT-4 gene regulation and manipulation. *Journal of Biological Chemistry* 274, 3253–3256

Chilliard, Y., Bocquier, F., Delavaud, C., Guerre-Millo, M., Bonnet, M., Martin, P., Faulconnier, Y. and Ferlay, A. (1998) Leptin in ruminants: effects of species, breed, adiposity, photoperiod, beta-agonists and nutritional status. In: *Proceedings of the Cornell Conference for Feed Manufacturers.* Rochester, USA.

Clarke, S.D. and Abraham, S. (l992) Gene expression: nutrient control of pre- and posttranscriptional events. *Federation of American Societies for Experimental Biology Journal* 6, 3146–3152.

Cohick, W.S. (1998) Role of the insulin-like growth factors and their binding proteins in lactation. *Journal of Dairy Science* 81, 1769–1777.

Cronjé, P.B. (1992a) Differences in nitrogen and urea metabolism between goats bred for fibre production (Angora goat) or meat production (Boer goat). *South African Journal of Animal Science* 22, 143–148.

Cronjé, P.B. (1992b) Glucose metabolism and adrenal function in goats bred for fibre production (Angora goat) or meat production (Boer goat). *South African Journal of Animal Science* 22, 149–153.

Cronjé, P.B. (1995) Differences in glucose response to insulin stimulation and fasting between Angora goats of two different phenotypes. *Small Ruminant Research* 16, 121–127.

Cronjé, P.B. (l998) Possible causes of differences in glucose metabolism between Angora goats of two different phenotypes. *Animal Science* 66, 705–711.

Cronjé, P.B. and Smuts, M. (1994) Nutrient partitioning in Merino rams with different wool growth rates. *Animal Production* 59, 55–60.

Duan, C. (1998) Nutritional and developmental regulation of insulin-like growth factors in fish. *Journal of Nutrition* 128, 306S–314S.

Etherton, T.D. and Bauman, D.E. (1998) Biology of somatotropin in growth and lactation of domestic animals. *Physiological Reviews* 78, 745–761.

France, J., Dijkstra, J. and Dhanoa, M.S. (1995) Growth functions and their application in animal science. In: *Proceedings of a Symposium on Methods in Modelling Herbivore Nutrition.* Institut National Agronomique Paris-Grignon, Paris.

Gallo, L., Carnier, P., Cassandro, M., Mantovani, R. and Bittante, G. (1997) Genetic and non-genetic effects on the response of Holstein cows to slow-release somatotropin injection. *Livestock Production Science* 49, 269–275.

Girard, J., Chatelain, F., Boillot, J., Prip-Buus, C., Thumelin, S., Pégorier, J.P., Foufelle, F. and Ferré, P. (1997) Nutrient regulation of gene expression. *Journal of Animal Science* 75 (suppl. 2), 46–57.

Gootwine, E., Suttie, J.M., McEwan, J.C., Veenvliet, B.A., Littlejohn, R.P., Fennessy, P.F. and Montgomery, G.W. (1998) The physiological effects of natural variation in growth hormone gene copy in ram lambs. *Domestic Animal Endocrinology* 14, 381–390.

Herselman, M.J., Olivier, J.J. and Snyman, M.A. (1998) Studies on small ruminant breeds with inherent differences in fibre production and ewe productivity 1. Relationship between ewe productivity and wool production potential. *South African Journal of Animal Science* 28, 1–8.

Hesketh, J.E., Vasconcelos, M.H. and Bermano, G. (1998) Regulatory signals in messenger RNA: determinants of nutrient-gene interaction and metabolic compartmentation. *British Journal of Nutrition* 80, 307–321.

Houseknecht, K.L. and Portocarrero, C.P. (1998) Leptin and its receptors: regulators of whole-body energy homeostasis. *Domestic Animal Endocrinology* 15, 457–475.

MacRae, J.C., Buttery, P.J. and Beever, D.E. (1988) Nutrient interactions in the dairy cow. In: Garnsworthy, P.C. (ed.) *Nutrition and Lactation in the Dairy Cow.* Butterworths, London, pp. 55–75.

Müller, U., Leucht, W., Reinecke, P. and Dalle, T. (1999) Breeding and biological evaluation of yield selection in dairy cattle. *Animal Research and Development* 49, 51–62.

Naudé, R.T. (1965) A review of research work on the growth of farm animals in Southern Africa – 1890 to 1965. *Proceedings of the South African Society of Animal Production* 4, 18–44.

Oldham, J.D. (1988) Nutrient allowances for ruminants. In: Haresign, W. and Cole, D.J.A. (eds) *Recent Advances in Animal Nutrition.* Butterworths, London, p. 147.

Ørskov, E.R. (1993) *Reality in Development Aid with Emphasis on Livestock.* Rowett Research Services Ltd, Aberdeen, 88 pp.

Parmentier, I., Portelle, D., Gengler, N., Prandi, A., Bertozzi, C., Vleurick, L., Glison, R. and Renaville, R. (1999) Candidate gene markers associated with somatotropic axis and milk selection. *Domestic Animal Endocrinology* 17, 139–148.

Rohr, K., Schafft, H. and Lebzien, P. (1983) Critical analysis of present protein allowances for growing ruminants. In: Arnal, M., Pion, R. and Bonin, D. (eds) *Proceedings of the Fourth International Symposium on Protein Metabolism and Nutrition.* INRA, Versailles, p. 449.

Rose, M.T., Obara, Y., Fuse, H., Itoh, F., Ozawa, A., Takahashi, Y., Hodate, K. and Ohashi, S. (1996) Effect of growth hormone releasing factor on the response to insulin of cows during early and late lactation. *Journal of Dairy Science* 79, 1734–1745.

Rotimi, C., Luke, A., Li, Z.H., Compton, J., Bowsher, R. and Cooper, R. (1997) Heritability of plasma leptin in a population sample of African-American families. *Genetic Epidemiology* 14, 255–263.

Sauvant, D. (1994) Modelling homeostatic and homeorhetic regulations in lactating animals. *Livestock Production Science* 39, 105–113.

Sauvant, D. (1995) A comparative evaluation of models of lactating ruminants. In: *Proceedings of a Symposium on Methods in Modelling Herbivore Nutrition.* Institut National Agronomique Paris-Grignon, Paris.

Sinclair, K.D., Yildiz, S., Quintans, G. and Broadbent, P.J. (1998) Annual energy intake and the performance of beef cows differing in body size and milk potential. *Animal Science* 66, 643–655.

Towle, H.C. (1995) Metabolic regulation of gene transcription in mammals. *Journal of Biological Chemistry* 270, 23235–23238.

Veerkamp, R.F. and Emmans, G.C. (1995) Sources of genetic variation in energetic efficiency of dairy cows. *Livestock Production Science* 44, 87–97.

Veerkamp, R.F., Simm, G. and Oldham, J.D. (1994) Effects of interaction between genotype and feeding system on milk production, feed intake, efficiency and body tissue mobilization in dairy cows. *Livestock Production Science* 39, 229–241.

Woolliams, J.A. and Lovendahl, P. (1991) Physiological attributes of male and juvenile cattle differing in genetic merit for milk yield: a review. *Livestock Production Science* 29, 1–16.

Wylie, A.R.G., McGrattan, P.D. and Nelson, J. (1998) Insulin receptor characteristics in ruminant tissues: effects of breed, sex and stage of development. In: McCracken, K.J., Unsworth, E.F. and Wylie, A.R.G. (eds) *Energy Metabolism of Farm Animals*. CAB International, Wallingford, pp. 111–114.

Zarate, A.V. (1996) Breeding strategies for marginal regions in the tropics and subtropics. *Animal Research and Development* 43/44, 99–118.

Zhao, F., Moseley, W.M., Tucker, H.A. and Kennelly, J.J. (1996) Regulation of glucose transporter gene expression in mammary gland, muscle, and fat of lactating cows by administration of bovine growth hormone and bovine growth hormone-releasing factor. *Journal of Animal Science* 74, 183–189.

# VIII Host Resistance to Parasites and Pathogens

# 24 Host Resistance to Gastrointestinal Parasites of Sheep

S.J. McClure, D.L. Emery and J.W. Steel

*CSIRO Animal Production, McMaster Laboratory, Prospect, New South Wales, Australia*

## Introduction

Sustainable parasite control requires an understanding of the mechanisms involved in the immunological resistance, and of how such mechanisms can be induced.

We will concentrate here on those nematodes whose parasitic stage is solely restricted to the gastrointestinal tract. Relatively isolated from the systemic immune system, this presents a particular set of problems to the host immune system and to livestock managers. Some of the aspects of the acquisition of immunity and mechanisms of resistance to these species may also operate in infections with parasites such as lungworm (*Dictyocaulus*) and tapeworms (*Taenia*) which have a gastrointestinal stage to their life cycle.

The most common pathogenic nematode species of the ruminant gastrointestinal tract are *Teladorsagia* (*Ostertagia*), *Haemonchus* (*H*) (abomasum), *Trichostrongylus* (*T*) (abomasum and small intestine), *Nematodirus*, *Cooperia* (small intestine) and *Oesophagostomum* (large intestine). Some occur and are equally pathogenic in more than one ruminant species, while others are host specific. These parasites have a direct life cycle consisting of free-living stages on pasture (egg to infective larvae, L3) and, after ingestion, parasitic stages (L4 to adult) in the host gastrointestinal tract. They do not have a tissue migratory phase. There are, however, some differences in their relationship with the host gut tissue, and therefore with the host immune system. *Haemonchus* attaches to the mucosa of the abomasum and sucks blood. *Ostertagia* and *Oesophagostomum* larvae penetrate into the abomasal glands or colonic mucosa before emerging and residing on the mucosal surface. *Trichostrongylus* live in mucus-covered tunnels eroded on the surface of intestinal villi or abomasal folds.

 *Ruminant Physiology: Digestion, Metabolism, Growth and Reproduction* (ed. P.B. Cronjé)

## Gastrointestinal defence mechanisms to parasitic nematodes of sheep

### *Inductive and effector phenomena*

Most adult ruminants exhibit naturally-acquired protective immunity to the endemic-species of gastrointestinal nematodes. However, immunologically naive animals are vulnerable to infection and disease for a period, as protective immunity to most worms takes several months to develop. The speed with which immunity develops is influenced by the dose of larvae ingested (Dobson *et al.*, 1990). As protection develops, animals sequentially acquire the ability to reject incoming larvae (after 5–7 weeks of continuous exposure), depress fecundity (after 10–12 weeks), and finally to expel adult worms (16–20 weeks later). The development of immunity against worms may require the reversal of an active mechanism (termed oral tolerance) which suppresses immune reactivity to ingested proteins in food. Reversal of oral tolerance allows the induction of hypersensitivity responses associated with worm rejection. It is most likely that the mucosal inflammation induced by parasites and the presence of infectious microorganisms (e.g. bacteria) are central to the induction of acquired mucosal immunity. However, the mucosal immune system is slower to invoke hypersensitivity responses than antimicrobial immunity and restricts these to the site of worm establishment. While our knowledge may be deficient concerning precisely how mucosal-dwelling parasites induce protective immunity, there is no reason to think that there are significant differences between species of worms in the type of host immune response they induce, although the relative importance of various components of that response may differ. Despite considerable research and understanding about the responses occurring at the time of immune rejection of worms, most of these responses are associations or correlations only at this stage – we do not know whether they are causal or casual. A few have been examined in *in vivo* depletion studies in an attempt to address this question of which responses are protective. It is known that immunity to re-infection is triggered by antigens on incoming worms to which the animal is sensitized. This induces the release of inflammatory mediators which can cause rejection of the target species or unrelated worms attempting to establish in the same location or downstream of it (Dineen *et al.*, 1977).

Sheep that are fully immune to a roundworm species and which have been exposed within the previous 7 weeks ('hyper-immune') can mount a 'rapid rejection' response, preventing establishment of *Haemonchus* third-stage larvae within 30 min (Jackson *et al.*, 1988) and eliminating incoming *Trichostrongylus* larvae from the entire 15 m of the small intestine within 2 h (Wagland *et al.*, 1996). This is thought to be a mucosal mast-cell-mediated hypersensitivity response, and is associated with large numbers of mucosal mast cells and globule leucocytes, and with local production of mast cell protease. If this hyperimmune state has lapsed, or previous exposure was insufficient to generate this degree of immunity, rejection takes somewhat longer – approximately 5–14 days (McClure *et al.*, 1992). It is then associated with increases in local lymphocytic and humoral immune responses, in addition to the mast cell involvement.

### *Cellular responses*

Observations of infections with various worm species suggest that in immune sheep which have not mounted a 'rapid rejection' of incoming larvae, worm challenge induces a rapid increase (3–5 days after challenge) in lymphoid cell availability to the gut (especially those lymphocytes expressing the gamma delta T-cell receptor, granulocytes and antigen-presenting dendritic cells) and rapid sequestration and activation of memory cells in lamina propria (Dawkins *et al.*, 1989; Buddle *et al.*, 1992; McClure *et al.*, 1992; Bendixsen *et al.*, 1995; Pfeffer *et al.*, 1996; S.J. McClure, unpublished observations). This appears to be followed by increased activation of lymphocytes in the draining node and return of memory cells, initially activated and later (> 7 days) resting, to the blood (Adams and Cripps, 1977; Haig *et al.*, 1989; Emery *et al.*, 1991; S.J. McClure, unpublished observations).

There are also changes in the local nerves. The autonomic innervation of the gastrointestinal tract is complex and extensive, consisting of both extrinsic and intrinsic nerves. The density of intrinsic nerve cell bodies within the gut is very high, with numbers comparable to the total number of neurons within the spinal cord. Sheep immune to *Trichostrongylus colubriformis* showed increases in the number and metabolic activity of enteric nerve fibres following challenge infection, suggesting that the nervous system can be primed by exposure to nematodes in a similar manner to the priming of the immune system (Stewart *et al.*, 1995b). In addition, immune and nervous systems can synthesize and respond to shared chemical mediators. Thus it is probable that the local and central nervous systems have a role in integrating the anamnestic immunological, muscular and physiological changes that follow worm challenge into a coordinated and flexible protective response.

It is not yet clear which of the many associated cellular responses are protective, and some have been depleted *in vivo* in an attempt to further define the protective mechanism. Depletion of CD4+ helper T-cells during challenge of sheep or goats immunized by viable infection or non-viable vaccines impaired the rejection of *Haemonchus contortus*, suggesting a role for CD4+ cells in protection (Gill *et al.*, 1993; Karanu *et al.*, 1997). Depletion of CD8+ or WC1/Tcrγδ+ T-cells during induction of immunity to *T. colubriformis* both resulted in enhanced rejection of worms, suggesting that these cells may be involved in the slowness of normal induction (McClure *et al.*, 1995). The conventional method for reducing resistance to worms is administration of glucocorticoids, but a range of cellular responses is affected, and a specific mechanism has not been identified.

### *Humoral responses*

#### *Systemic*

Protective immunity in sheep against abomasal and intestinal worm species is associated with early increases in worm-specific antibody and IgA concentrations in local efferent lymph (Smith *et al.*, 1984, 1985; S.J. McClure, unpublished observations), and with elevated levels of all isotypes of antibody in serum (McClure *et al.*, 1992; Pfeffer *et al.*, 1996; Shaw *et al.*, 1998). However, with the possible exception of IgA antibody in *Ostertagia* infection (Stear *et al.*, 1996), serum antibody level in sheep exposed to viable infection is not consistently predictive of protective immunity. It may be that serum antibody is particularly relevant for protection against the blood-sucking worms, as immunity to *H. contortus* can be transferred passsively with serum in sheep vaccinated with novel antigens (Smith, 1993).

Rapid rejection is associated with a temporary decrease in serum antibody level coinciding with a rise in mucus levels of IgG antibody (McClure *et al.*, 1992), presumably reflecting a protease-induced exit of blood proteins into the gut (Miller, 1996).

#### *Local*

Immune sheep after challenge have increased concentrations of worm-specific antibody in gut tissue and mucus, with all isotypes represented in the increase, and an increase in the number of cells with surface-bound IgE (McClure *et al.*, 1992; Pfeffer *et al.*, 1996). There are suggestions that IgG1 and IgE correlate best with protection against *Trichostrongylus* species, but more studies are required to confirm this.

### *Neuropeptides*

The role of neuropeptides in immunity to worms has not been investigated *in vivo.* However, the neuropeptides employed by the enteric nervous system (Substance P, somatostatin, vaso-intestinal peptide and β-endorphin) rendered mucosal mast cells more sensitive to limiting concentrations of worm antigens *in vitro*. These peptides also enhanced the *in vitro* proliferation to worm antigen of lymphocytes from mesenteric lymph node and prefemoral efferent lymph of immune sheep but did not affect specific antibody production by these cells (Stewart *et al.*, 1995a, 1996).

### *Inflammatory mediators*

Local tissue concentrations and secretion into the intestine of mediators such as leucotrienes, 5-hydroxytryptamine and histamine increase during immune rejection of *T. colubriformis* (Steel *et al.*, 1990; Jones *et al.*, 1990). The administration of glucocorticoids immediately before challenge abrogates immunity to *H. contortus* and *T. colubriformis*; however, injection of more specific antagonists of leucotrienes, histamine, platelet-activating factor or phosphodiesterases failed to affect worm rejection (Adams, 1988; Jackson *et al.*, 1988; Emery and McClure, 1995).

### *'Non-specific' mechanisms*

A number of local mechanisms important in the rejection of gut parasites are non-specific in effect but immunologically specific in induction. They are thus adaptive responses, developing after exposure to the antigens, and possibly only after exposure to the viable parasite. They may partly explain the observation that, to date, non-viable vaccines are less protective against browsing worms than are viable infection or irradiated larval vaccines.

### *Mucus*

Mucus *per se* has been proposed to have a protective role in trapping incoming worms and preventing their establishment (reviewed by Rothwell, 1989). Protective immunity to *T. colubriformis* in Merinos is associated with increased numbers of goblet cells and increased quantities of mucus in the jejunum (R.G. Windon, personal communication). The quality of the mucus also changes in sheep immune to gastrointestinal nematodes, with altered composition of the muco-polysaccharides and -proteins, the leucotriene content (Douch *et al.*, 1983; Jones *et al.*, 1990, 1994), and the content of various as yet unidentified molecules with inhibitory effects on nematodes. Mucus also stabilizes and prolongs the biological activity of inflammatory mediators (W.O. Jones, unpublished observations). The induction of these mechanisms is not well understood.

### *Peristalsis*

Immunity in sheep to gut worms appears to be associated with increased enteric nerve fibre number and metabolic activity (Stewart *et al.*, 1995b), and with hypertrophy and hyper-contractility of local smooth muscle (Tremain and Emery, 1994; S.J. McClure, unpublished observations). In rodents, gut smooth muscle function is subject to modulation by the immune system, with T lymphocytes, particularly those expressing CD4, implicated in the alteration of smooth muscle contractility seen by 6 days after infection with *Trichinella spiralis* (Vallance and Collins, 1998). These adaptive changes in gastrointestinal motility persist for some time after exposure to worms, and in addition to hustling incoming larvae, may have wider and more long-term repercussions in the physiology and function of the gastrointestinal tract.

### *Epithelial sloughing and proliferation*

If immediate rejection of incoming larvae does not occur, expulsion of *T. colubriformis* from immune sheep is associated with transient loss of the jejunal epithelium at 4–5 days after intra-duodenal infection, effectively dislodging the larvae, which do not embed in the sub-epithelial mucosa (McClure *et al.*, 1992). The epithelium is repaired within approximately 24 h.

### *Fluid and electrolyte movement into lumen*

Mediators such as mast cell protease disrupt cell junctions, increasing mucosal permeability (Miller, 1996). This both increases the fluid content of the lumen and allows access of potentially protective host molecules to the worm.

Thus the host mounts a wide range of immune and inflammatory responses to a challenge infection with worms, and probably no one response is essential or sufficient on its own for protection.

## Influences of parasites on gastrointestinal function of host

Effects of gastrointestinal parasitism on the digestive physiology of ruminants have been reviewed most recently by van Houtert and Sykes (1996) and Coop and Holmes (1996) with particular emphasis on quantitative aspects of digestion and absorption. Depression in food consumption is frequently, but not invariably, observed, and is dependent on the level of intake of infective larvae of small-intestinal (Steel *et al.*, 1980) and abomasal (Symons *et al.*, 1981) parasites. The mechanisms responsible are poorly understood although abdominal pain, gut inflammation, pH changes, changes in digesta flow rate, changes in protein to energy ratios of absorbed nutrients together with changes in cholecystokinin secretion have all been implicated (Symons, 1985), as well as neuropeptides and possibly some cytokines (van Houtert and Sykes, 1996). During continuous infections, appetite begins to return to normal after 10–12 weeks of larval intake. Although anorexia clearly has a profound effect on overall nutrient absorption, numerous pair-feeding studies have demonstrated that it does not entirely account for the impaired productivity of parasitized animals.

Impaired uptake of nutrients at the site of small-intestinal parasite infections is attributable to deficiencies in digestive enzymes, proliferation of undifferentiated non-absorbing cells on the villi, and reduced surface area due to villous atrophy (Steel and Symons, 1982). However, the overall capacity of the ruminant small intestine to digest and absorb protein appears to be generally unaffected by parasitism at proximal sites (Poppi *et al.*, 1986; Kimambo *et al.*, 1988a). The increased flow of nitrogen observed at the ileum during the first 12 weeks of *T. colubriformis* infection (Steel, 1974; Poppi *et al.*, 1986; Kimambo *et al.*, 1988a) and at the abomasum and terminal ileum during *Teladorsagia circumcincta* (Steel, 1978) and *H. contortus* (Rowe *et al.*, 1988) infection have therefore been attributed to increased endogenous nitrogen secretion into the gastrointestinal tract. The principal sources of this endogenous nitrogen are increases at the site of infection in gastroenteric plasma and blood protein loss, turnover and loss into the gut of mucosal epithelia, and mucoprotein secretion by goblet cells. These changes appear to be associated in turn, not surprisingly, with an increased rate of protein synthesis by gastrointestinal tissues (Steel and Symons, 1982) such that there is a diversion of amino acids away from muscle, bone and wool growth into protein synthesis for processes essential for survival. Although often termed 'pathophysiological' these responses may indeed be an unavoidable component of the development of immune competence, and it is significant to note that they appear to peak at a time (around 12–14 weeks) when manifestations of resistance mechanisms in weaner lambs, such as declining faecal egg count and the progressive expulsion of resident worms, are clearly evident (Steel *et al.*, 1980; Poppi *et al.*, 1986; Kimambo *et al.*, 1988a).

## Cost to the host of protective immunity

The sequelae of physiological and immunological events responsible for the expulsion of resident adult worm populations in primary infections of young sheep correlates closely in time with maximal effects on liveweight gain and wool growth, particularly for small-intestinal infections (Steel *et al.*, 1980; Kimambo *et al.*, 1988a). Continued infection beyond this time is accompanied by a progressive recovery in productive performance and gastrointestinal function. Whether animals which have acquired a protective immunity to worm establishment experience a physiological and production cost during persistent exposure to infective larvae, because of hypersensitivity responses, has been of considerable interest since reduced wool growth and elevated plasma pepsinogen concentrations have been recorded in mature 'resistant' sheep under challenge (Barger 1982). Daily challenge of immune sheep for 10 weeks with *T.*

*colubriformis* larvae, following a 24-week parasite-free period, elicited a small increase in plasma nitrogen loss into the gut during the first 9 days, but no other change in nitrogen flow in digesta, digestibility or retention, and no excretion of worm eggs in faeces (Kimambo *et al.*, 1988b). Other studies have recorded immediate but transient increases in enteric plasma loss 4–6 days after a single challenge of *T. colubriformis* given to lambs made solidly immune by vaccination with irradiated larvae (Steel *et al.*, 1990), and in 'immune' ewes taken off contaminated pasture and given daily challenge with mixed trichostrongyle larvae, predominantly of *Ostertagia circumcincta* (Yakoob *et al.*, 1983). However, in young sheep made highly resistant to infection by vaccination with irradiated larvae, continuous challenge with *T. colubriformis* larvae for 20 weeks had no effect on rates of liveweight gain or wool growth, despite evidence of immunological reactivity in terms of elevated serum antibody levels and total absence of faecal egg count (Wagland *et al.*, 1982). Protective immunity in terms of larval exclusion and prevention of worm establishment therefore seems to be effected through changes in the composition of intestinal mucus, including the content of 'larval migration inhibitory factor', without any detectable metabolic or production cost (Kimambo and MacRae, 1988).

## The non-specific enhancement of immunity

Two features of immunity to worms, namely the role of inflammation in induction, and the expulsion of parasites via a pharmacologically-mediated non-specific effector response, raise the possibility of augmenting the levels of resistance to parasites. Feasible means to increase mucosal inflammation include concurrent (controlled) infections or infestations, immunomodulators or the judicious use of nutritional supplements. It is already appreciated that *Trichostrongylus* and particularly *Ostertagia* reduce the establishment of *Haemonchus* (Dobson and Barnes, 1995) and that genetically-selected resistant sheep mount more rapid and pronounced inflammatory responses to gastrointestinal nematodes (GIN) parasites (Windon, 1996). Irradiated *Haemonchus* or *Trichostrongylus* larval vaccines also induced about 30% protection against the heterologous parasite.

Several other manipulations have been found to reduce parasitism:

**1.** Ground navy beans, given as 10% of pelleted diets and constituting 50% of digestible dietary protein, significantly ($P < 0.05$) reduced parasitism in sheep given *T. colubriformis* L3 or harbouring a mature infection (Ad) (Fig. 24.1). Sheep were fed the diet for up to 22 days from the day of infection (L3) or from 3 weeks into a primary infection (Ad), and in both situations, parasitism was reduced by > 60%. It is hypothesized that lectins within the navy beans increased mucosal inflammation in the same way that soybean diets can induce diarrhoea in feedlot cattle.
**2.** Variable levels of resistance, ranging from 35 to 50%, can be achieved against challenge with *T. colubriformis* by oral delivery of attenuated *Salmonella typhimurium* aro A (H.N. Brahmbatt, unpublished observations) or intraperitoneal inoculation of liposomes (D.L. Emery, unpublished observations).
**3.** Supplementaton of sheep diets with Mo also decreased levels of parasitism following challenge (Suttle *et al.*, 1992a, b; McClure *et al.*, 1999).

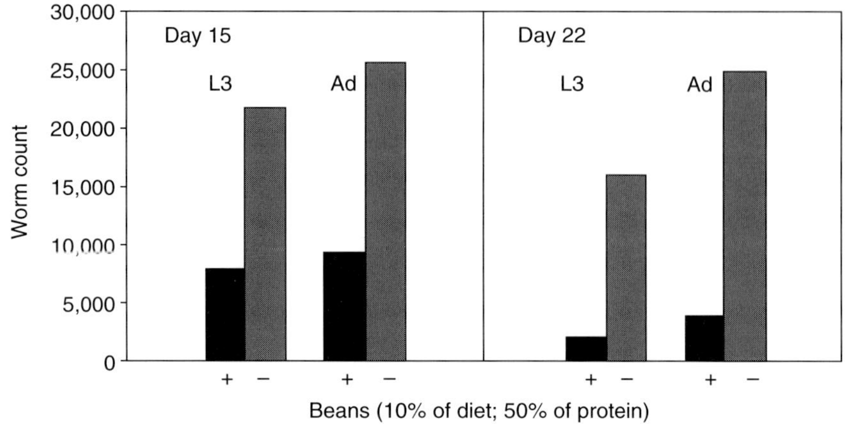


**Fig. 24.1.** A total of 32 sheep were fed a normal ration (−, 16 sheep) or a ration in which navy beans (10% wt/wt) provided 50% of dietary protein (+, 16 sheep). Eight sheep from each group were infected with 20,000 *Trichostrongylus colubriformis* L3 21 days prior to initiation of feeding, to examine effects on established adult worms (Ad), and the remaining sheep were infected with 20,000 *T. colubriformis* L3 coincidentally with the diet. Four sheep from each of the four treatment groups were killed for worm counts after 15 and 22 days of the two rations.

Although the 'protective' effects of these relatively diverse treatments could act directly or indirectly on the gut immune responses of the host, the exact mechanisms are not resolved at present.

## Conclusions

Unlike some internal parasites, GIN of ruminants have a simple life-cycle, and during their parasitic phase reside only in the gastrointestinal tract of the host. They therefore present a difficult immunological problem for the host, requiring protective immunity from an organ normally restrained from reacting to locally-presented molecules, and one with a good barrier between local immune system and pathogen. The gut immune responses associated with rejection of these parasites are physiologically complex and appear to contain much redundancy in that there is no evidence that any one mechanism is essential for protection. Protective immune responses against GIN parasites which are allergic in nature are also highly regulated within the local mucosal environment to avoid systemic sequelae. Mucosal immune responses come at considerable cost to the host in terms of cell and protein loss, reduced nutrient absorption and increased metabolic demand for nutrients; a cost exacerbated by the damage to gastrointestinal function due directly to the worm.

It should be remembered that these mechanisms are subject to a number of influences. These include physiological factors such as age, liveweight and sex, and external factors such as nutrition and weaning stress. The nutritional requirements for optimal mucosal immunity appear to be greater than currently recognized, and additional to those required to meet the demands of growth and production. The efficacy of the mucosal immune responses can also be enhanced by altering the local immunological environment, for example the degree of inflammation. Thus gut immunity is not an isolated component of ruminant physiology, but is highly integrated with the nervous, digestive and endocrine systems, and underlaid by a common cell biology regulating signalling, activation, metabolism, replication and differentiation. In such a situation, intervention which is directed towards addressing any single factor contributing to susceptibility will predictably have limited prospects of success.

## References

Adams, D.B. (1988) The effect of dexamethasone on a single and a superimposed infection with *Haemonchus contortus* in sheep. *International Journal for Parasitology* 18, 575–579.

Adams, D.B. and Cripps, A.W. (1977) Cellular changes in the intestinal lymph of sheep infected with the enteric nematode, *Trichostrongylus colubriformis. Australian Journal of Biology and Medical Science* 55, 509–522.

Barger, I.A. (1982) Helminth parasites and animal production. In: Symons, L.E.A., Donald, A.D. and Dineen, J.K. (eds) *Biology and Control of Endoparasites.* Academic Press, Sydney, pp. 133–155.

Bendixsen, T., Emery, D.L. and Jones, W.O. (1995) The sensitization of mucosal mast cells during infections with *Trichostrongylus colubriformis* or *Haemonchus contortus* in sheep. *International Journal for Parasitology* 25, 741–748.

Buddle, B.M., Jowett, G., Green, R.S., Douch, P.G.C. and Risdon, P.L. (1992) Association of blood eosinophilia with the expression of resistance in Romney lambs to nematodes. *International Journal for Parasitology* 22, 955–960.

Coop, R.L. and Holmes, P.H. (1996) Nutrition and parasite interaction. *International Journal for Parasitology* 26, 951–962.

Dawkins, H.J.S., Windon, R.G. and Eagleson, G.K. (1989) Eosinophil responses in sheep selected for high and low responsiveness to *Trichostrongylus colubriformis. International Journal for Parasitology* 19,199–205.

Dineen, J.K., Gregg, P., Windon, R.G., Donald, A.D. and Kelly, J.D. (1977) The role of immunologically specific and non-specific components of resistance in cross-protection of intestinal nematodes. *International Journal for Parasitology* 7, 211–215.

Dobson, R.J. and Barnes, E.H. (1995) Interaction between *Ostertagia circumcincta* and *Haemonchus contortus* infection in young lambs. *International Journal for Parasitology* 25, 495–501.

Dobson, R.J., Waller, P.J. and Donald, A.D. (1990) Population dynamics of *Trichostrongylus colubriformis* in sheep: The effect of infection rate on the establishment of infective larvae and parasite fecundity. *International Journal for Parasitology* 20, 347–352.

Douch, P.G.C., Harrison, G.B.L., Buchanan, L.L. and Greer, K.S. (1983) In vitro bioassay of sheep gastrointestinal mucus for nematode paralysing activity mediated by substances with some properties characteristic of SRS-A. *International Journal for Parasitology* 13, 207–212.

Emery, D.L. and McClure, S.J. (1995) Studies on the inhibition of rejection of *Trichostrongylus colubriformis* larvae from immune sheep. *International Journal for Parasitology* 25, 761–764.

Emery, D.L., Bendixsen, T. and McClure, S.J. (1991) The use of electroblotted antigens of *Trichostrongylus colubriformis* to induce proliferative responses in sensitized lymphocytes from sheep. *International Journal for Parasitology* 21, 179–185.

Gill, H.S., Watson, D.L. and Brandon, M.R. (1993) Monoclonal antibody to $CD4^+$ T cells abrogates genetic resistance to *Hamonchus contortus* in sheep. *Immunology* 78, 43–49.

Haig, D.M., Windon, R., Blackie, W., Brown, D. and Smith, W.D. (1989) Parasite-specific T cell responses of sheep following live infection with the gastric nematode *Haemonchus contortus*. *Parasite Immunology* 11, 463–477.

Jackson, F., Miller, H.R.P., Newlands, G.J.F., Wright, S.E. and Hay, L.A. (1988) Immune exclusion of *Haemonchus contortus* larvae in sheep: dose dependency, steroid sensitivity and persistence of the response. *Research in Veterinary Science* 44, 320–323.

Jones, W.O., Windon, R.G., Steel, J.W. and Outteridge, P.M. (1990) Histamine and leukotriene concentrations in duodenal tissue and mucus of lambs selected for high and low responsiveness to vaccination and challenge with *Trichostrongylus colubriformis*. *International Journal for Parasitology* 20, 1075–1079.

Jones, W.O., Emery, D.L., McClure, S.J. and Wagland, B.M. (1994) Changes in inflammatory mediators and larval inhibitory activity in intestinal contents and mucus during primary and challenge infections of sheep with *Trichostrongylus colubriformis*. *International Journal for Parasitology* 24, 519–525.

Karanu, F.N., McGuire, T., Davis, W., Besser, T.E. and Jasmer, D.P. (1997) CD4+ lymphocytes contribute to protective immunity induced in sheep and goats by *Haemonchus contortus* gut antigens. *Parasite Immunology* 19, 435–445.

Kimambo, A.E. and MacRae, J.C. (1988) Measurement *in vitro* of a larval migration inhibitory factor in gastrointestinal mucus of sheep made resistant to the roundworm *Trichostrongylus colubriformis*. *Veterinary Parasitology* 28, 213–222.

Kimambo, A.E., MacRae, J.C., Walker, A., Watt, C.F. and Coop, R.L. (1988a) Effect of prolonged infection with *Trichostrongylus colubriformis* on the performance and nitrogen metabolism of growing lambs. *Veterinary Parasitology* 28, 191–203.

Kimambo, A.E., MacRae, J.C. and Dewey, P.J.S. (1988b) The effect of daily challenge with *Trichostrongylus colubriformis* larvae on the nutrition and performance of immunologically resistant sheep. *Veterinary Parasitology* 28, 205–212.

McClure, S.J., Emery, D.L., Wagland, B.M. and Jones, W.O. (1992) A serial study of rejection of *Trichostrongylus colubriformis* by immune sheep. *International Journal for Parasitology* 22, 227–234.

McClure, S.J., Davey, R.J., Lloyd, J.B. and Emery, D.L. (1995) Depletion of IFN-γ, CD8+ or Tcr γδ+ cells *in vivo* during primary infection with an enteric parasite (*Trichostrongylus colubriformis*) enhances protective immunity. *Immunology and Cell Biology* 73, 552–555.

McClure, S.J., McClure, T.J. and Emery, D.L. (1999) Effects of molybdenum intake on primary infection and subsequent challenge by the nematode parasite *Trichostrongylus colubriformis* in Merino lambs. *Research in Veterinary Science* 67, 17–22.

Miller, H.R.P. (1996) Mucosal mast cells and the allergic response against nematode parasites. *Veterinary Immunology and Immunopathology* 54, 331–336.

Pfeffer, A., Douch, P.G.C., Shaw, R.J., Gatehouse, T.K., Rabel, B., Green, R.S., Shirer, C.L., Jonas, W.E. and Bisset, S. (1996) Sequential cellular and humoral responses in the abomasal mucosa and blood of Romney sheep dosed with *Trichostrongylus axei*. *International Journal for Parasitology* 26, 765–773.

Poppi, D.P., MacRae, J.C., Brewer, A. and Coop, R.L. (1986) Nitrogen transactions in the digestive tract of lambs exposed to the intestinal parasite *Trichostrongylus colubriformis*. *British Journal of Nutrition* 55, 593–602.

Rothwell, T.L.W. (1989) Immune expulsion of parasitic nematodes from the alimentary tract. *International Journal for Parasitology* 19, 139–168.

Rowe, J.B., Nolan, J.V., Chaneet, G. de, Teleni, E. and Holmes, P.H. (1988) The effect of haemonchosis and blood loss into the abomasum on digestion in sheep. *British Journal of Nutrition* 59, 125–139.

Shaw, R.J., Gatehouse, T.K. and McNeill, M.M. (1998) Serum IgE responses during primary and challenge infections of sheep with *Trichostrongylus colubriformis*. *International Journal for Parasitology* 28, 293–302.

Smith, W.D. (1993) Protection in lambs immunized with *Haemonchus contortus* gut membrane proteins. *Research in Veterinary Science* 54, 94–101.

Smith, W.D., Jackson, F., Jackson, E., Williams, J. and Miller, H.R.P. (1984) Manifestations of resistance to ovine ostertagiasis associated with immunological responses in the gastric lymph. *Journal of Comparative Pathology* 94, 591–602.

Smith, W.D., Jackson, F., Jackson, E. and Williams, J. (1985) Age immunity to *Ostertagia circumcincta*: comparison of the local immune responses of $4\frac{1}{2}$ and 10-month-old lambs. *Journal of Comparative Pathology* 95, 235–245.

Stear, M.J., Park, M. and Bishop, S.C. (1996) The key components of resistance to *Ostertagia circumcincta* in lambs. *Parasitology Today* 12, 438–441.

Steel, J.W. (1974) Pathophysiology of gastrointestinal nematode infections in the ruminant. *Proceedings of the Australian Society of Animal Production* 10, 139–147.

Steel, J.W. (1978) Inter-relationships between gastrointestinal helminth infection, nutrition and impaired productivity in the ruminant. In: Farrell, D.J. (ed.) *Recent Advances in Animal Nutrition in Australia.* The University of New England, Armidale, Australia, pp. 98–109.

Steel, J.W. and Symons, L.E.A. (1982) Nitrogen metabolism in nematodosis of sheep in relation to productivity. In: Symons, L.E.A., Donald, A.D. and Dineen, J.K. (eds) *Biology and Control of Endoparasites.* Academic Press, Sydney, pp. 235–256.

Steel, J.W., Symons, L.E.A. and Jones, W.O. (1980) Effects of level of larval intake on the productivity and physiological and metabolic responses of lambs infected with *Trichostrongylus colubriformis. Australian Journal of Agricultural Research* 31, 821–838.

Steel, J.W., Jones, W.O. and Wagland, B.M. (1990) The response of immune sheep to challenge with *Trichostrongylus colubriformis*: enteric plasma loss and secretion of biogenic amines. *International Journal for Parasitology* 20, 1067–1073.

Stewart, M.J., Emery, D.L. and McClure, S.J. (1995a) Effects of four neuropeptides in lymphocytes from sheep. *Regional Immunology* 6, 264–269.

Stewart, M.J., McClure, S.J., Rothwell, T.L.W. and Emery, D.L. (1995b) Changes in the enteric innervation of the sheep during primary or secondary immune responses to the nematode *Trichostrongylus colubriformis. Regional Immunology* 6, 460–470.

Stewart, M.J., Emery, D.L., McClure, S.J. and Bendixsen, T. (1996) The effects of four neuropeptides on the degranulation of mucosal mast cells from sheep. *Immunology and Cell Biology* 74, 255–257.

Symons, L.E.A. (1985) Anorexia; occurrence, pathophysiology and possible causes in parasitic infections. In: Baker, J.R. and Muller, R. (eds) *Advances in Parasitology*, Vol. 24. Academic Press, London, pp. 103–133.

Symons, L.E.A., Steel, J.W. and Jones, W.O. (1981) Effects of level of larval intake on the productivity and physiological and metabolic responses of lambs infected with *Ostertagia circumcincta. Australian Journal of Agricultural Research* 32, 139–148.

Suttle, N.F., Knox, K.W., Angus, K.W., Jackson, F. and Coop, R.L. (1992a) Effects of dietary molybdenum on nematode and host during *Haemonchus contortus* infection in lambs. *Research in Veterinary Science* 52, 230–235.

Suttle, N.F., Knox, K.W., Jackson, F., Coop, R.L. and Angus, K.W. (1992b) Effects of dietary molybdenum on nematode and host during *Trichostrongylus vitrinus* infection in lambs. *Research in Veterinary Science* 52, 224–229.

Tremain, S.A. and Emery, D.L. (1994) Myoelectric activity of the small intestine during parasite rejection. *30th Annual Scientific Meeting of the Australian Society for Parisitology.* Australian Society for Parasitology, Sydney, p. 43.

Vallance, B.A. and Collins, S.M. (1998) The effect of nematode infection upon intestinal smooth muscle function. *Parasite Immunology* 20, 249–253.

van Houtert, M.F.J. and Sykes, A.R. (1996) Implications of nutrition for the ability of ruminants to withstand gastrointestinal nematode infections. *International Journal for Parasitology* 11, 1151–1168.

Wagland, B.M., Steel, J.W. and Dineen, J.K. (1982) Effect of continuous challenge with *Trichostrongylus colubriformis* on liveweight gain and wool growth of resistant sheep. *Proceedings of the Australian Society of Animal Production* 14, 642.

Wagland, B.M., Emery, D.L. and McClure, S.J. (1996) Studies on the host–parasite relationship between *Trichostrongylus colubriformis* and susceptible and resistant sheep. *International Journal for Parasitology* 26, 1279–1286.

Windon, R.G. (1996) Genetic control of resistance to helminths in sheep. *Veterinary Immunology and Immunopathology* 54, 245–254.

Yakoob, A., Holmes, P.H. and Armour, J. (1983) Pathophysiology of gastrointestinal trichostrongyles in sheep: plasma losses and changes in plasma pepsinogen levels associated with parasite challenge of immune animals. *Research in Veterinary Science* 34, 305–309.

# 25 Host Resistance to Fleece Rot and Blowfly Strike

I.G. Colditz[1] and R.L. Tellam[2]

[1]*CSIRO Animal Production, Pastoral Research Laboratory, Armidale, New South Wales, Australia;* [2]*CSIRO Tropical Agriculture, Longpocket Laboratory, Indooroopilly, Queensland, Australia*

## Introduction

Interest in resistance of sheep and cattle to external parasites arose soon after emergence of blowfly strike and cattle tick as diseases of major economic importance in Australia. It is most likely that blowflies were introduced to Australia on several occasions: *Lucilia sericata* from Britain during the early days of European settlement; and *Lucilia cuprina* from South Africa or India during the latter stages of the 19th century (Gilruth *et al.*, 1933). Blowfly strike was sporadic until 1903 when it became widespread in New South Wales and Victoria, extending throughout the range of sheep over the next few years (Gilruth *et al.*, 1933). Emergence of the disease coincided with introductions of American Merinos from Vermont, which had a high grease content in the fleece and pronounced skin wrinkle (Belschner, 1966). With the realization that body conformation was a predisposing factor for susceptibility to blowfly strike (Seddon *et al.*, 1931a) and that these predisposing traits were heritable (Seddon *et al.*, 1931b), interest turned to breeding sheep resistant to the disease and to surgical removal of skin folds in the breech (mulesing). Wetting of fleece was found to induce bacterial growth (Stuart, 1894), which in turn induced green or red discoloration of wool (Seddon and McGrath, 1929) and dermatitis (Bull, 1931). These studies led to the conclusion that predisposing conditions for blowfly strike are wool characteristics and body conformations that favour, principally through susceptibility to wetting and resultant dermatitis, attraction of flies, oviposition and subsequent nutrition of larvae (Gilruth *et al.*, 1933). Thus fleece rot and blowfly strike occur as a disease complex, with the strongest interdependence when blowfly strike occurs over the shoulders, back and flanks (body strike).

This early work provided the basis for the next 70 years of research into resistance to blowfly strike. Major areas of research have been genetic studies on direct and indirect selection for resistance to blowfly strike, mechanisms of resistance in selected flocks, innate immunity, acquired immunity and vaccines. Progress in these areas will be considered in turn.

## Direct and indirect selection for resistance to blowfly strike

Hayman (1953) confirmed Seddon's observations on heritability of resistance to fleece rot. Quantitatively, McGuirk and Atkins (1984) estimated a heritability of susceptibility to fleece rot of 0.40 in unselected medium-wool Peppin Merinos from 15 years of field data at Trangie Agricultural Research Centre in New South Wales. Heritability of susceptibility to blowfly strike in the same environment was between 0.26 and 0.37 (Gilmour and Raadsma, 1986; Raadsma, 1991). Experimental conditions for inducing fleece rot and fly strike were developed at Trangie using artificial wetting by overhead sprinklers of sheep temporarily housed indoors (McGuirk *et al.*, 1978). Fleece rot and subsequent body strike resulting from artificial wetting provided the basis for establishing two selection lines with resistance and susceptibility to the disease complex (McGuirk *et al.*, 1978). These selection lines have been the subjects of intensive study of both wool characteristics and physiological responses associated with resistance. Heritability of susceptibility to fleece rot under artificial wetting was comparable (Raadsma *et al.*, 1989) to that seen under field conditions (Raadsma and Rogan, 1987). Following 17 years of selection, the lines had diverged at an annual rate of 2.8% for natural fleece rot and 0.4% for natural body strike (Mortimer *et al.*, 1998). With the absence of a random-bred control line in this study it is unclear whether there has been a greater impact of selection on resistance or susceptibility to the disease complex. The sporadic occurrence in many of the grazing zones of Australia of weather conditions that induce fleece rot and that accompany fly waves stimulated research on correlated wool traits that could provide the basis for indirect selection for resistance. Wettability of the fleece (Raadsma, 1989), objectively measured wool colour (Raadsma and Wilkinson, 1990) and fibre characteristics (Raadsma, 1993) are genetically correlated with resistance to the disease complex. Despite these findings there has been limited selection for these correlated traits in Merino breeding programmes other than through the long-held husbandry practice of culling susceptible sheep on the basis of conformational faults, subjectively assessed wool characteristics and wool discoloration associated with fleece rot.

## Mechanisms of resistance in the Trangie selection flock

Freshly hatched larvae feed on proteins in exudates associated with bacterial dermatitis (Gilruth *et al.*, 1933; Sandeman *et al.*, 1987). Skin inflammatory responses were therefore examined to determine whether underlying differences in plasma leakage could account for differences between the selection lines in susceptibility to blowfly strike. Leakage of radiolabelled albumin from the vascular compartment into skin was measured following intradermal injection of histamine, bradykinin, activated complement, platelet-activating factor or serotonin. Plasma leakage induced by activated complement was greater in susceptible sheep, and for all mediators there was a tendency for plasma leakage to be greater in susceptible sheep (Colditz *et al.*, 1992a). This finding suggested that a smaller plasma leakage response following release of endogenous permeability mediators might contribute to resistance to blowfly strike and is in accord with the nutritive role of blood proteins for larvae. Serum complement C3 levels, however, did not differ between lines (O'Meara and Raadsma, 1995), suggesting that differences might lie at the level of receptor number or function controlling vascular permeability rather than in supply of mediator. Contrary to this finding is the observation that when excretory and secretory products from larvae were injected intradermally, there was a greater wheal response in resistant animals (O'Meara *et al.*, 1992). The wheal response to excretory and secretory products has also been correlated with resistance to fleece rot and blowfly strike in unselected sheep (Broadmeadow, 1988) and has been proposed as a selection marker for resistance to the disease complex (Raadsma *et al.*, 1992).

Cellular components of the inflammatory response in skin and resident leucocytes in skin have also been examined in the resistant and susceptible lines. Intense neutrophil accumulation is a feature of the dermatitis that accompanies fleece rot and fly strike (Bull, 1931; Burrell *et al.*, 1982; Bowles *et al.*, 1992) and there is extensive production of inflammatory cytokines interleukin (IL)-1α, IL-1β, IL-6 and IL-8 (Elhay *et al.*, 1994) at the wound site. Neutrophil accumulation in response to intradermal injection of ovine IL-1β, human IL-8, human IL-1α, human tumour necrosis factor (TNF)-α activated complement, leukotriene $B_4$ and endotoxin from *Pseudomonas aeruginosa* did not differ between lines (Colditz *et al.*, 1994). Leucocyte populations in untreated skin and in skin sites collected 6 h following intradermal injection of TNF-α were examined by immunohistology. WC1+ lymphocytes and eosinophils were more prevalent in skin of susceptible sheep and IgE+ cells (probably mast cells) were more prevalent in skin of resistant animals (Colditz *et al.*, 1994). In an independent study (Nesa, 1994) mast cells were also found to be more prevalent in skin from animals of the resistant genotype. This difference in the prevalence of mast cells may underlie the greater wheal response to intradermal injection of larval antigens in resistant sheep noted above. Furthermore, in an examination of the IgE gene, Engwerda *et al.* (1998) found a restriction fragment length polymorphism between resistant and susceptible lines. Together these findings suggest that the role of mast cells in resistance to blowfly strike deserves further attention.

The response of resistant and susceptible sheep to artificial infestation with larvae has been examined in two experiments. O'Meara *et al.* (1995) found greater exudation of serum proteins onto the skin surface during the first 12 h of infestation in resistant animals, which is in accord with greater skin wheal responses seen in these animals. In the second experiment, the effects of lymphocyte and interferon (IFN)-γ depletion on larval establishment and growth have been studied by treatment of sheep with monoclonal antibodies. In contrast to results seen with the internal parasite, *Trichostrongylus colubriformis,* where depletion of CD8+, WC1+ lymphocytes or IFN-γ enhanced resistance to infection (McClure *et al.*, 1996), depletion of CD4+, CD8+, WC1+ lymphocytes or IFN-γ had no effect on establishment or growth of blowfly larvae (Colditz *et al.*, 1996). Importantly in both artificial infestations, there were no differences between the resistant and susceptible lines in the establishment or growth of larvae.

Lymphocyte subsets in blood do not differ between resistant and susceptible lines (Colditz *et al.*, 1996; McColl *et al.*, 1997). Following intravenous challenge with endotoxin from *P. aeruginosa* there are higher neutrophil counts and monocyte counts in the resistant line (I. Colditz, unpublished findings). Antibody responses to intradermal injection of *P. aeruginosa* antigens are generally greater in resistant sheep (Chin and Watts, 1991; Gogolewski *et al.*, 1996); however, no significant differences between the lines were found in the antibody response to *L. cuprina* antigens (O'Meara *et al.*,

1997). Sheep from the resistant line develop higher titres of antibody to *P. aeruginosa* antigens during simulated fleece rot conditions when live cultures of *P. aeruginosa* are applied epicutaneously to wetted skin (Chin and Watts, 1991). This may result from differences in skin characteristics between the lines resulting in greater uptake of antigens through skin in the resistant animals, or alternatively to differences between lines in immunological recognition and response to *P. aeruginosa* antigens. Systemically administered *P. aeruginosa* vaccines can confer resistance to fleece rot (Burrell, 1985) so the differences in antibody titres noted by Chin and Watts (1991) and Gogolewski *et al.* (1996) may well contribute to the differences in prevalence and severity of fleece rot seen between the selection lines. Taken together, these extensive studies on the Trangie resistant and susceptible lines suggest that reactivity of skin may contribute to resistance but may play a subsidiary role to wool characteristics that predispose to fleece rot.

## Innate immunity in unselected sheep

In a review of fleece rot and body strike, McGuirk and Watts (1983) concluded that there appear to be three barriers providing resistance to blowfly strike: wool, skin and the immune system. Wool characteristics have been mentioned briefly above in relation to the Trangie selection flocks. Physical skin characteristics affecting resistance such as composition of fats and wettability have received less attention than inflammatory reactions in skin. The host response to fleece rot and fly strike results in production of high levels of inflammatory cytokines in skin (Elhay *et al.*, 1994). We therefore examined the direct toxic effects of these endogenous mediators on larvae in an *in vitro* feeding assay (Colditz and Eisemann, 1994). At concentrations that exhibit inflammatory activity in skin (Colditz, 1991; Colditz and Watson, 1992; Mulder and Colditz, 1993), histamine, bradykinin, platelet-activating factor, serotonin, activated complement, leukotriene $B_4$, IL-8, IL-1$\beta$, TNF-$\alpha$ and IFN-$\gamma$ all failed to affect larval growth. Antibody to peritrophic membrane antigens was an exception, inducing dose-dependent inhibition of larval growth. Protease inhibitors in plasma inhibit growth of larvae *in vitro* (Bowles *et al.*, 1990) and anti-inflammatory drugs retard larval growth (O'Sullivan *et al.*, 1984). These findings suggest that, with the probable exception of antibody, the promotion of larval growth by nutrients is greater than the concomitant inhibition induced by protease inhibitors present in the exudate (O'Meara *et al.*, 1995). In accord with the finding in the Trangie selection lines noted above, IgE+ cells are more prevalent in skin of sheep with no history of fleece rot than in sheep with a record of fleece rot (Colditz *et al.*, 1994).

We recently examined factors contributing to the break in wool fibres that accompanies strike. Blowfly strike was induced in a group of five, 18-month-old Merino wethers by application of approximately 500 freshly hatched larvae per day for 8 days to an abraded skin site on the flank of each sheep. Daily feed intakes were recorded and pair-fed sheep were offered the quantity of feed consumed by their pair on the preceding day. A third group of uninfested controls was fed *ad libitum* for the duration of the experiment. Severe strike was induced in four of the infested sheep with one sheep exhibiting only mild strike. Average daily gain was significantly lower in infested sheep than control sheep during week one of infestation. Rectal temperature was elevated in infested sheep from the first day of infestation until 3 days after the last application of infective larvae, which coincided with the treatment of all sheep with the organophosphate Diazinon to terminate infestations. Plasma cortisol levels were elevated in infested sheep from day 2 to day 6 of infestation. The cytokines IL-1β, IL-6, IL-8 and TNF-α were assayed in plasma. IL-6 was elevated from day 2 to day 6 of infestation, whereas no significant differences were observed between treatment groups for concentrations of the other cytokines. Longitudinal growth of wool fibres did not differ between groups, though there was a trend towards less growth in struck and pair-fed sheep than in controls. Staple strength was significantly lower in struck sheep than in control and pair-fed sheep, but did not differ between sites near to and distant from the strike lesion. Taken together these findings suggest that the host response to infestation has systemic consequences that lead to reduced fibre strength throughout the fleece, and that the reduced feed intake accompanying blowfly strike is not primarily responsible for reduced fibre strength. Cortisol and IL-6 are implicated as systemic mediators that contribute to reduced fibre strength; however, the interdependence, pleiotropism and redundancy of mediators associated with stress and inflammatory responses make it unlikely that a single mediator will be identified as accountable for the effect of blowfly strike on fibre strength.

## Acquired immunity in unselected sheep

There have been several studies on the acquisition of immunity following repeated experimental infestation of sheep. Sheep rapidly produce antibodies to larval antigens (O'Donnell *et al.*, 1980; Sandeman *et al.*, 1985; Eisemann *et al.*, 1990; Seaton *et al.*, 1992) and there are dramatic changes in leucocyte populations at the site of infestation (Bowles *et al.*, 1992) and in draining lymph (Bowles *et al.*, 1994). There is, however, very limited effect of such sensitization from experimental infestations on survival and growth of larvae during subsequent challenge. It is noteworthy that larvae exert an immunosuppressive effect on the host, which has been attributed, in part, to excretion of ammonia (Guerrini, 1998). While lymphocyte depletion in the Trangie selection lines failed to affect larval establishment and growth, the effect of immunosuppressive treatments on growth of larvae in unselected sheep has not been examined. An infiltrate of CD4+, WC1+ and CD1+ leucocytes is seen at sites of infestation following immunization with larval antigens (Bowles *et al.*, 1996). However, leucocyte depletion studies to determine the contribution of this infiltrate to reduced larval survival in immunized sheep have not been performed. Thus the role of cellular responses in resistance to blowfly strike remains an open question.

## Development of blowfly vaccines

The failure of repeated larval infestations to promote strong immunity and the success of the hidden antigen approach in the development of the cattle tick vaccine (Willadsen *et al.*, 1989; Tellam *et al.*, 1992) led to studies on gut antigens as candidates for a blowfly vaccine. It was argued that it may be possible to induce a humoral immune response in sheep that produced specific antibodies that, after ingestion by the larvae, bound to the larval gut. These antibodies might have the potential of disrupting the normal functions of the gut in a manner deleterious to the larvae. The strategy relies on the ingestion of sufficient quantities of biologically active antibodies. It was found that larvae feeding on sheep ingest substantial quantities of functional antibodies (Eisemann *et al.*, 1993). High concentrations of ingested antibody persist throughout the larval foregut and the anterior midgut. The quantity of antibody decreases sharply posterior to this region as a consequence of proteolysis. In contrast, only very small quantities of ingested antibody penetrate the wall of the midgut to reach the haemolymph and thence internal tissues. The most promising 'concealed' targets for immunological attack therefore would appear to lie in the anterior midgut. This region is exposed to ingested antibodies and is not protected by an impermeable layer of cuticle as is present in the foregut and hindgut. However, the midgut region is protected by a semi-permeable matrix, the peritrophic membrane, which is thought to have a central role in the facilitation of the digestive process in the gut and protection of the insect from invasion by bacteria.

East *et al.* (1993) used this knowledge in testing crude extracts of peritrophic membrane in sheep vaccination trials. Sera from vaccinated sheep strongly inhibited larval growth in an *in vitro* feeding bioassay (East *et al.*, 1993). Moreover, significant although weaker effects on larval growth were observed using bioassays directly on the backs of sheep. Vaccination trials in sheep were then used as an assay to guide the purification of specific peritrophic membrane antigens. This process, which involved many steps, led to the identification of several potential vaccine antigens but particularly the glycoprotein, peritrophin-95 (Casu *et al.*, 1997). Vaccination of sheep with this antigen led to greater than 50% reduction in larval weights as measured by an *in vitro* larval feeding bioassay. The cDNA coding for this protein has been sequenced and various recombinant proteins produced. These were tested in vaccination trials which showed significant, albeit weak, anti-larval effects. Further studies of both the structure of the recombinant antigens and the nature of the antibody response to the native protein demonstrated that the nature of the oligosaccharides attached to this glycoprotein is an important determinant of the efficacy of the antilarval immune response induced by this antigen. Further studies are under way to ensure the appropriateness of the oligosaccharides attached to the recombinant proteins.

Isolation of specific anti-peritrophin-95 antibodies and their re-constitution into control sera transferred the antilarval growth activity when measured in *in vitro* feeding bioassays. Moreover, higher concentrations of antibody resulted in more severe effects on larval growth. There was also a direct correlation between antibody titre from a number of sheep and the degree of larval growth inhibition measured from each serum in an *in vitro* feeding bioassay (R.L. Tellam and C.H. Eisemann, unpublished results). These experiments and others proved that the antilarval effect was mediated by antibody and that complement was not required. Examination of the peritrophic membrane from larvae feeding on sera from sheep vaccinated with peritrophin-95 showed that the luminal side of the peritrophic membrane was lined with a new layer of material of undefined composition. This layer was impermeable to 6 nm colloidal gold particles that normally freely diffused across the peritrophic membrane. Presumably, the layer inhibited the movement of nutrients into the digestive epithelia and resulted in the starvation of the larvae (R.L. Tellam and C.H. Eisemann, unpublished results). A vaccine based on the peritrophin-95 antigen is being developed but will require greater efficacy to be effective in the field. Combinations of this antigen with antigens possibly involved in larval establishment on the host epidermis (Bowles *et al.*, 1996) may result in a more efficacious vaccine.

## Vaccine delivery

With the realization that antibody held promise as a defence mechanism for protection against blowfly larvae, we turned our attention to investigating immunization protocols for elevating antibody concentrations in skin. Colleagues in our laboratory developed an apparatus for inducing transudation of interstitial fluid by applying a vacuum of −25 kPa for 90 min to the skin surface (Watson *et al.*, 1992). Up to 40 μl of interstitial fluid collects on the skin surface during this procedure (Colditz *et al.*, 1992b). The ratio of $^{125}$I-Ig$G_1$ to $^{111}$In-transferrin in blood was compared with the ratio in normal skin and at skin sites receiving the vascular permeability mediators histamine, bradykinin, activated complement, platelet-activating factor or serotonin to determine whether there is preferential transfer of Ig$G_1$ to these extravascular sites. No evidence was obtained for selective transport of Ig$G_1$ into the dermis or onto the skin surface (Colditz *et al.*, 1992b). Immunoglobulins can be detected on the surface of sheep skin in low concentrations (Lloyd *et al.*, 1979) and our results suggest that for Ig$G_1$ this occurs by filtration rather than by selective transport. It is noteworthy, however, that IgA is selectively transported across sebaceous glands and sweat glands in man (Gebhart and Metze, 1990) and can be detected in skin washings from sheep following infection with *Dermatophilus congolensis* (Sutherland *et al.*, 1987).

The potential to induce local antibody responses in sheep by topical application of antigens was examined in a number of experiments (Colditz and Watson, 1993). Antigen delivery via a Panjet vaccination gun was more effective than topical application of antigen onto skin. Adjuvation of the experimental antigen ovalbumin with immunostimulating complexes (ISCOMs) resulted in further elevation of the antibody response. Antibody titres were higher at locally immunized skin sites than at non-immunized skin sites of the same animals. Isotype analysis of antibody present in vacuum transudates indicated that immunoglobulin (Ig)M and Ig$G_1$ were elevated at the locally immunized skin sites, Ig$G_2$ antibody did not differ between sites and IgA antibody was not detected. Numbers of CD4+, CD8+ and WC1+ lymphocytes were elevated in the dermis at sites of local immunization (I.G. Colditz, D.L. Watson and S.J. McClure 1999, unpublished observations). Similar results have been obtained with recombinant peritrophic membrane antigens PM44, PM48 and PM95 from *L. cuprina* (I.G. Colditz, C.H. Eisemann and R.L. Tellam, unpublished observations). Together these findings demonstrate the potential for elevating the concentration of antibody in skin by local application of antigen.

It has recently been shown that antigens combined with cholera toxin can induce high titres of systemic antibody when applied to unbroken skin in mice (Glenn *et al.*, 1998a,b). No local gross or histological reactions accompanied this transcutaneous immunization, which protected mice from a lethal intranasal challenge with cholera toxin. We have recently observed a systemic antibody response in sheep following topical application of 100 μg of cholera toxin to skin (R.B. Cope and I.G. Colditz, unpublished observations). Interestingly, during fleece rot and dermatophilosis, sheep develop systemic antibody responses to these non-invasive infectious organisms, although sensitization in this natural disease setting is generally thought to be not protective (Sutherland *et al.*, 1987). The adjuvation effect of cholera toxin applied topically to skin may provide a new strategy for elevating antibody responses to vaccines against dermal pathogens and external parasites of sheep and cattle. Furthermore, the potential of cholera toxin to induce local antibody responses in skin also deserves investigation. Percutaneous delivery of vaccines should be very attractive to livestock industries due to ease of application, lower requirement for animal restraint, avoidance of carcass and hide damage and possibly no need for asepsis.

## Conclusions

A century's research on blowfly strike has yielded a great deal of information on the pathogenesis of infestation and the nature of the host–parasite interaction. Despite these advances, the two major goals for research in this field, development (or selection) of resistant animals and the induction of resistance by vaccination, remain unattained. In view of the sporadic occurrence of conditions that permit direct selection for resistance, a major shortcoming of research to date is the absence of a reliable trait for indirect selection for resistance. As discussed by K.A. Ward in Chapter 21, transgenesis may yet yield novel solutions to the quest for resistant animals. For vaccines, solutions to the problem of invoking protective immunity may come not from harnessing components of naturally acquired resistance but by broaching the adapted host parasite interface by vaccination with evolutionarily naïve, concealed antigens.

## Acknowledgements

The collaboration of S. Walkden-Brown, B. Crook, B. Daley and C. Eisemann in studies of the effect of blowfly strike on fibre strength is gratefully acknowledged. This work was supported in part by Australian woolgrowers through the Australian Wool Research and Promotion Organisation and the L.W. Bett trust.

## References

Belschner, H.G. (1966) *Sheep Management and Diseases,* 8th edn. Angus and Robertson, Sydney, 814 pp.

Bowles, V.M., Feehan, J.P. and Sandeman, R.M. (1990) Sheep plasma protease inhibitors influencing protease activity and growth of Lucilia cuprina larvae *in vitro*. *International Journal for Parasitology* 20, 169–174.

Bowles, V.M., Grey, S.T. and Brandon, M.R. (1992) Cellular immune responses in skin of sheep infected with larvae of *Lucilia cuprina*, the sheep blowfly. *Veterinary Parasitology* 44, 151–162.

Bowles, V.M., Meeusen, E.N.T., Chandler, K., Verhagen, A., Nash, A.D. and Brandon, M.R. (1994) The immune response of sheep infected with larvae of the sheep blowfly *Lucilia cuprina* monitored via efferent lymph. *Veterinary Immunology and Immunopathology* 40, 341–352.

Bowles, V.M., Meeusen, E.N.T., Young, A.R., Andews, A.E., Nash, A.D. and Brandon, M.R.

(1996) Vaccination of sheep against larvae of the sheep blowfly (*Lucilia cuprina*). *Vaccine* 14, 1347–1352.

Broadmeadow, M.E. (1988) Studies on resistance of sheep to blowfly strike. Wool Research and Development Council Project DAQ 21P Final Report, Queensland Department of Primary Industries, Brisbane, pp. 25.

Bull, L.B. (1931) Some observations on dermatitis of the folds in the breech of sheep and its possible relationship to blowfly strike. *Australian Veterinary Journal* 7, 143–148.

Burrell, D.H. (1985) Immunisation of sheep against experimental *Pseudomonas aeruginosa* dermatitis and fleece rot associated body strike. *Australian Veterinary Journal* 62, 55–57.

Burrell, D.H., Merritt, G.C., Watts, J.E. and Walker, K.H. (1982) Experimental production of dermatitis in sheep with *Pseudomonas aeruginosa*. *Australian Veterinary Journal* 59, 140–144.

Casu, R., Eisemann, C., Pearson, R., Riding, G., East, I., Donaldson, A., Cadogan, L. and Tellam, R. (1997) Antibody-mediated inhibition of the growth of larvae from an insect causing cutaneous myiasis in a mammalian host. *Proceedings of the National Academy of Sciences USA* 94, 8939–8944.

Chin, J.E. and Watts, J.E. (1991) Dermal and serological response against *Pseudomonas aeruginosa* in sheep bred for resistance and susceptibility to fleece-rot. *Australian Veterinary Journal* 68, 28–31.

Colditz, I.G. (1991) The induction of plasma leakage in skin by histamine, bradykinin, activated complement, platelet-activating factor and serotonin. *Immunology and Cell Biology* 69, 215–219.

Colditz, I.G. and Eisemann, C.H. (1994) The effect of immune and inflammatory mediators on growth of *Lucilia cuprina* larvae *in vitro*. *International Journal for Parasitology* 24, 401–403.

Colditz, I.G. and Watson, D.L. (1992) The effect of cytokines and chemotactic agonists on the migration of T lymphocytes into skin. *Immunology* 76, 272–278.

Colditz, I.G. and Watson, D.L. (1993) Immunisation to induce antibody production in skin of sheep. *Proceedings of Vaccines 93, Lorne House, Victoria* (abstract J94).

Colditz, I.G., Woolaston, R.R., Lax, J. and Mortimer, S.I. (1992a) Plasma leakage in skin of sheep selected for resistance or susceptibility to fleece rot and fly strike. *Parasite Immunology* 14, 587–594.

Colditz, I.G., Altmann, K.G. and Watson, D.L. (1992b) Intradermal and percutaneous transudation of $IgG_1$ and transferrin in sheep. *Immunology and Cell Biology* 70, 323–327.

Colditz, I.G., Lax, J., Mortimer, S.I., Clarke, R.A. and Beh, K.J. (1994) Cellular inflammatory responses in skin of sheep selected for resistance or susceptibility to fleece rot and fly strike. *Parasite Immunology* 16, 289–296.

Colditz, I.G., Eisemann, C.H., Tellam, R.L., McClure, S.J., Mortimer, S.I. and Husband, A.J. (1996) Growth of Lucilia cuprina larvae following treatment of sheep divergently selected for fleece rot and fly strike with monoclonal antibodies to T lymphocyte subsets and interferon gamma. *International Journal for Parasitology* 26, 775–782.

East, I.J., Fitzgerald, C.G., Pearson, R.D., Donaldson, R.A., Vuocolo, T., Cadogan, L.C., Tellam, R.L. and Eisemann, C.H. (1993) *Lucilia cuprina*: inhibition of larval growth induced by immunization of host sheep with extracts of larval peritrophic membrane. *International Journal for Parasitology* 23, 221.

Eisemann, C.H., Johnston, L.A.Y., Broadmeadow, M., O'Sullivan, B.M., Donaldson, R.A., Pearson, R.D., Vuocolo, T. and Kerr, J.D. (1990) Acquired resistance of sheep to larvae of *Lucilia cuprina*, assessed *in vivo* and *in vitro*. *International Journal for Parasitology* 20, 299–305.

Eisemann, C.H., Pearson, R.D., Donaldson, R.A., Cadogan, L.C. and Vuocolo, T. (1993) Uptake and fate of specific antibody in feeding larvae of the sheep blowfly, *Lucilia cuprina*. *Medical and Veterinary Entomology* 7, 177–185.

Elhay, M.J., Hanrahan, C.F., Bowles, V.M., Seow, H., Andrews, A.E. and Nash, A.D. (1994)

Cytokine mRNA expression in skin in response to ectoparasite infection. *Parasite Immunology* 16, 451–461.

Engwerda, C.R., Dale, C.J. and Sandeman, R.M. (1998) IgE, TNFα, IL1β, IL4 and IFNg gene polymorphisms in sheep selected for resistance to fleece rot and flystrike. *International Journal for Parasitology* 26, 787–791.

Gebhart, W. and Metze, D. (1990) Secretory immunoglobulins. In: Bos, J.D. (ed.) *Skin Immune System*. CRC Press, Boca Raton, pp. 197–206.

Gilmour, A.R. and Raadsma, H.W. (1986) Estimating genetic variation and covariation for flystrike incidence in Australian Merino sheep on an underlying normal scale. *3rd World Congress on Genetics applied to Livestock Production, Lincoln, Nebraska, USA, July 16–22, 1986. XII. Biotechnology, selection experiments, parameter estimation, design of breeding systems, management of genetic resources*. p. 460.

Gilruth, J.A., Tillyard, R.J., Seddon, H.R., Gurney, W.B. and Mackerras, I.M. (1933) The sheep blowfly problem, Report No. 1 of the Joint Blowfly Committee. *NSW Department of Agriculture Science Bulletin* 40, 1–136.

Glenn, G.M., Rao, M., Matyas, G.R. and Alving, C.R. (1998a) Skin immunization made possible by cholera toxin. *Nature* 391, 851.

Glenn, G.M., Scharton-Kersten, T., Vassell, R., Mallett, C.P., Hale, T.L. and Alving, C.R. (1998b) Cutting Edge: Transcutaneous immunisation with cholera toxin protects mice against lethal mucosal toxin challenge. *Journal of Immunology* 161, 3211–3214.

Gogolewski, R.P., Nicholls, P.J., Mortimer, S.I., Mackintosh, J.A., Nesa, M., Ly, W. and Chin, J.C. (1996) Serological responses against *Pseudomonas aeruginosa* in Merino sheep bred for resistance or susceptibility to fleece rot and body strike. *Australian Journal of Agricultural Research* 47, 917–926.

Guerrini, V.H. (1998) Excretion of ammonia by *Lucilia cuprina* larvae suppresses immunity in sheep. *Veterinary Immunology and Immunopathology* 56, 311–317.

Hayman, R.H. (1953) Studies on fleece-rot in sheep. *Australian Journal of Agricultural Research* 4, 430–468.

Lloyd, D.H., McEwan Jenkinson, D. and Mabon, R.M. (1979) Location of immunoglobulins in skin of cattle and sheep. *Research in Veterinary Science* 26, 47–52.

McClure, S.J., Davey, R.L., Emery, D.L., Colditz, I.G. and Lloyd, J.B. (1996) *In vivo* depletion of T-cells and cytokines during primary exposure of sheep to parasites. *Veterinary Immunology and Immunopathology* 54, 83–90.

McColl, K.A., Gogolewski, R.P. and Chin, J.C. (1997) Peripheral blood lymphocyte subsets in fleece rot-resistant and -susceptible sheep. *Australian Veterinary Journal* 75, 421–423.

McGuirk, B.J. and Atkins, K.D. (1984) Fleece rot in Merino sheep. I The heritability of fleece rot in unselected flocks of medium-wool Peppin Merinos. *Australian Journal of Agricultural Research* 35, 423–434.

McGuirk, B.J. and Watts, J.E. (1983) Association between fleece, skin and body characters of sheep and susceptibility to fleece rot and body strike. In: Raadsma, H.W. (ed.) *Sheep Blowfly and Flystrike in Sheep. Second National Symposium*. New South Wales Department of Agriculture, Sydney, pp. 367–386.

McGuirk, B.J., Atkins, D.A., Kowal, E. and Thornberry, K. (1978) Breeding for resistance to fleece rot and body strike – the Trangie programme. *Wool Technology and Sheep Breeding* 26(part iv), pp. 17–24.

Mortimer, S.I., Atkins, K.D. and Raadsma, H.W. (1998) Responses to selection for resistance and susceptibility to fleece rot and body strike in Merino sheep. *Proceedings of the 6th World Congress on Genetics applied to Livestock Production*, p. 283.

Mulder, K. and Colditz, I.G. (1993) Migratory responses of ovine neutrophils to inflammatory mediators *in vitro* and *in vivo*. *Journal of Leukocyte Biology* 53, 273–278.

Nesa, M. (1994) The immune and inflammatory responses of Merino sheep bred for resistance and susceptibility to fleece rot and flystrike: the Trangie R and S flocks. MSc Thesis, La Trobe University, Melbourne, Australia.

O'Donnell, I.J., Green, P.E., Connell, J.A. and Hopkins, P.S. (1980) Immunoglobulin G antibodies to the antigens of *Lucilia cuprina* in the sera of fly-struck sheep. *Australian Journal of Biological Sciences* 33, 27–34.

O'Meara, T.J. and Raadsma, H.W. (1995) Phenotypic and genetic indicators of resistance to ectopathogens. In: Gray, G.D., Woolaston, R.R. and Eaton, B.T. (eds) *Breeding for Resistance to Infectious Diseases in Small Ruminants*. ACIAR monograph no. 34, Canberra, pp. 187–218.

O'Meara, T.J., Nesa, M., Raadsma, H.W., Saville, D.G. and Sandeman, R.M. (1992) Variation in skin inflammatory responses between sheep bred for resistance or susceptibility to fleece rot and blowfly strike. *Research in Veterinary Science* 52, 205–210.

O'Meara, T.J., Nesa, M., Seaton, D.S. and Sandeman, R.M. (1995) A comparison of inflammatory exudates released from myiasis wounds on sheep bred for resistance or susceptibility to Lucilia cuprina. *Veterinary Parasitology* 56, 207–223.

O'Meara, T.J., Nesa, M. and Sandeman, R.M. (1997) Antibody responses to *Lucilia cuprina* in sheep selected for resistance or susceptibility to *L. cuprina*. *Parasite Immunology* 19, 535–543.

O'Sullivan, B.M., Hopkins, P.S. and Connell, J.A. (1984) The pathogenesis of flystrike in sheep. *Animal Production in Australia* 15, 171–181.

Raadsma, H.W. (1989) Fleece rot and body strike in Merino sheep. III. Significance of fleece moisture following experimental induction of fleece rot. *Australian Journal of Agricultural Research* 40, 897–912.

Raadsma, H.W. (1991) Fleece rot and body strike in Merino sheep. V. Heritability of liability to body strike in weaner sheep under flywave conditions. *Australian Journal of Agricultural Research* 42, 279–293.

Raadsma, H.W. (1993) Fleece rot and body strike in Merino sheep. VI. Experimental evaluation of some physical fleece and body characteristics as indirect selection criteria for fleece rot. *Australian Journal of Agricultural Research* 44, 915–931.

Raadsma, H.W. and Rogan, I.M. (1987) Genetic variation in resistance to blowfly strike. In: McGuirk, B.J. (ed.) *Merino Improvement Programs in Australia. Proceedings of a National Symposium, Leura, NSW.* Australian Wool Corporation, Melbourne, pp. 321–340.

Raadsma, H.W. and Wilkinson, B.R. (1990) Fleece rot and body strike in Merino sheep. IV Experimental evaluation of traits related to greasy wool colour for indirect selection against fleece rot. *Australian Journal of Agricultural Research* 41, 139–153.

Raadsma, H.W., Gilmour, A.R. and Paxton, W.J. (1989) Fleece rot and body strike in Merino sheep. II. Phenotypic and genetic variation in liability to fleece rot following experimental induction. *Australian Journal of Agricultural Research* 40, 207–220.

Raadsma, H.W., Sandeman, R.M., Sasiak, A.B., Engwerda, C.R. and O'Meara, T.J. (1992) Genetic improvement in resistance to body strike in Merino sheep: where are we at with indirect selection. *Proceedings of the Australian Association for Animal Breeding and Genetics* 10, 143–146.

Sandeman, R.M., Dowse, C.A. and Carnegie, P.R. (1985) Initial characterisation of the sheep immune response to infections of *Lucilia cuprina. International Journal for Parasitology* 15, 181–185.

Sandeman, R.M., Collins, B.J. and Carnegie, P.R. (1987) A scanning electron microscope study of L. cuprina larvae and the development of blowfly strike in sheep. *International Journal for Parasitology* 17, 759–765.

Seaton, D.S., O'Meara, T.J., Chandler, R.A. and Sandeman, R.M. (1992) The sheep antibody response to repeated infection with *Lucilia cuprina. International Journal for Parasitology* 22, 1169–1174.

Seddon, H.R. and McGrath, T.T. (1929) Green colouration in wool. *Agricultural Gazette of New South Wales* 40, 206.

Seddon, H.R., Belschner, H.G. and Mulhearn, C.R. (1931a) Observations on the susceptibility of sheep to blowfly attack. *NSW Department of Agriculture Science Bulletin* 37, 3–23.

Seddon, H.R., Belschner, H.G. and Mulhearn, C.R. (1931b) The influence of breeding as a factor in the sheep blowfly problem. *NSW Department of Agriculture Science Bulletin* 37, 24–30.

Stuart, T.P.A. (1894) On green-producing chromogenic microorganisms in wool. *Journal and Proceedings of the Royal Society of New South Wales* 28, 320 322.

Sutherland, S.S., Ellis, T.M., Robertson, G.M. and Gregory, A.R. (1987) Serum and skin surface antibody responses in Merino sheep given three successive inoculations with *Dermatophilus congolensis. Veterinary Microbiology* 15, 209–218.

Tellam, R.L., Smith, D., Kemp, D.H. and Willadsen, P. (1992) Vaccination against ticks. In: Yong, W.K. (ed.) *Animal Parasite Control Utilizing Biotechnology.* CRC Press, New York, pp. 303–331.

Watson, D.L., Lea, J. and Burke, G.L. (1992) A method for collecting interstitial fluid from skin of sheep. *Australian Veterinary Journal,* 69, 14–15.

Willadsen, P., Riding, G., McKenna, R.V., Kemp, D.H., Tellam, R.L., Nielsen, J.N., Lahnstein, J., Cobon, G. and Gough, J.M. (1989) Immunological control of a parasitic arthropod. Identification of a protective antigen from *Boophilus microplus. Journal of Immunology* 143, 1346–1351.

# 26 Host Resistance to Mastitis

K. Persson Waller

*Swedish University of Agricultural Sciences, Faculty of Veterinary Medicine, Department of Obstetrics and Gynaecology, Uppsala, Sweden*

## Introduction

Udder diseases, mainly mastitis induced by bacterial infections, are major problems for the dairy industry world wide, and the yearly costs are substantial (Philpot, 1984; Nilsson and Holmberg, 1996). Current programmes for control of bovine mastitis have improved the situation substantially, but the incidence of clinical and subclinical mastitis is still high. This indicates problems in the application of the control measures, or limitations in their effects. A need for new and innovative approaches for mastitis control especially concerning preventive measures is obvious. Lately, emphasis has been put on working towards a better understanding of the host resistance against udder infections, in order to find ways to increase the natural ability of the cow to resist, or defend herself, against udder infections.

This paper will consist of a review of the latest knowledge in defence mechanisms of the mammary gland of ruminants. Special emphasis will be put on the distribution and function of various leucocyte populations, and the influence of different inflammatory mediators, such as cytokines, during the lactation cycle. Genetic influence on immunological functions will also be highlighted. Moreover, the influence of nutrition and stress on immunological functions and the defence against udder infections will be discussed, especially in relation to periods of immunosuppression, e.g. around calving, when the risk for mastitis is increased.

## Specific and non-specific defence mechanisms of the mammary gland

### *Mechanisms*

The defence mechanisms can be divided into non-specific (innate) and specific (adaptive/acquired) mechanisms. Leucocytes in udder secretions and udder tissues are important components of both categories. The non-specific response is essential during the early stages of infection and inflammation. Important considerations are anatomical features of the teat end, but also cell populations like macrophages, neutrophils, natural killer-like cells, and soluble factors like lactoferrin, lysozyme and complement. The specific response is mediated mainly by various antibodies (immunoglobulins) and lymphocytes recognizing specific determinants of the pathogen, but also by the effects of macrophages.

The teat end and the teat canal form the first line of defence against infections. A well-functioning sphincter muscle, in combination with keratin lining the teat canal, ensures a secure mechanical closure of the canal (Craven and Williams, 1985). The keratin can trap bacteria entering the canal, and at each milking some of the keratin, with its possible content of bacteria, is desquamated and flushed out. The integrity of the teat skin of the teat end, especially around the teat orifice, is also important in order to minimize the numbers of bacteria present and thus reduce the risk for infection.

The soluble factors involved in the host defence have direct antibacterial properties of varying efficacy, but also the ability to stimulate antibacterial functions of leucocytes, e.g. by opsonization. Moreover, combinations of certain soluble factors have synergistic effects. The concentrations of these substances are most often low in normal milk, but increase during involution of the udder and mastitis. Their importance during udder infections and mastitis is, to some extent, uncertain.

## *Distribution of different leucocyte populations*

The leucocytes, mainly neutrophils, macrophages and lymphocytes, have a key role in the defence of the udder. The proportions of each cell type in the mammary secretions vary with stage of lactation and health status in the gland (Concha, 1986; Paape *et al.*, 1991).

In normal milk and dry secretion, macrophages and lymphocytes are the predominating cell types while the numbers of neutrophils are low. However, during acute inflammation, the numbers of neutrophils increase dramatically. Chemotactic substances attract these cells from the blood to the site of inflammation. Prompt migration of sufficient numbers of neutrophils is essential to the outcome of an infection (Hill, 1994).

Macrophages in tissues and udder secretions originate in blood monocytes and a selection for these cells, as expressed by CD14+ leucocytes, occurs in dry secretion compared with in milk (Persson Waller and Colditz, 1998). An increase in the number of lymphocytes occurs during mastitis, but to a lesser extent compared with the numbers of neutrophils.

Most lymphocytes present in mammary secretions are either T-cells or B-cells (Craven and Williams, 1985; Concha, 1986). T lymphocytes can be further divided into $\alpha\beta$ T-cells consisting either of T-helper (CD4+) or T-suppressor/cytotoxic cells (CD8+), and $\gamma\delta$ T-cells (WC1+). A selective migration to, or retention of, CD8+ cells in favour of CD4+ cells occurs in the mammary gland (Taylor *et al.*, 1994; Guiguen *et al.*, 1996; Persson Waller and Colditz, 1998). The proportions of the different subpopulations in udder secretion and tissues, as well as expression of leucocyte antigens on mammary cells, differ with stage of lactation. In ewes, greater numbers of IL-2R+ and MHCII+ lymphocytes were observed in dry secretions compared with in milk (Persson

Waller and Colditz, 1998). Memory cells of both B- and T-types are usually produced after an infection.

### Leucocyte functions

Phagocytosis and intracellular killing of bacteria are crucial functions of the neutrophils, which is illustrated by a higher prevalence of clinical mastitis when the cellular capacity for phagocytosis is low (Grommers *et al.*, 1989; Hill, 1994). Their antimicrobial effects can be enhanced by opsonins, i.e. antibodies and complement (Craven and Williams, 1985; Paape *et al.*, 1991). The functions of neutrophils may often be reduced in milk as they are affected by local conditions. However, newly recruited cells from the bloodstream are considered to have more efficient antibacterial properties (Herbelin *et al.*, 1997), thus the importance of migration of sufficient numbers of neutrophils. Changes in adhesion molecules, shedding of L-selectin and up-regulation of CD18, occur in ovine neutrophils due to migration from blood into the udder (Persson Waller and Colditz, 1998).

Macrophages are also phagocytic cells but their bactericidal capacity is lower than for neutrophils. The most important function of macrophages is probably their role as initiators of inflammatory processes (Adams and Hamilton, 1988). They recognize antigen, which they phagocytose, process and present to lymphocytes, which, in turn, become activated. Stimulated macrophages release inflammatory mediators important for inflammation and attraction of neutrophils (Adams and Hamilton, 1988). Also bovine mammary macrophages are capable of inducing proliferation of both blood and mammary lymphocytes (Concha and Holmberg, 1990).

As mentioned earlier, lymphocytes can be divided into T-cells, involved in cell-mediated immune response, and B-cells, which are precursors of antibody-producing plasma cells. Recent research has shown the importance of different lymphocyte subpopulations for the mammary immune response. CD8+ T-cells seem to have varying functions, being either of suppressor or cytotoxic type, depending on the stage of lactation (Shafer-Weaver and Sordillo, 1997). T-helper cells (CD4+) play an important role in the immune response by activation of other cells, like lymphocytes and macrophages (Tizard, 1996). The proportion of CD8+ lymphocytes in udder secretion is often larger than the proportion of CD4+ cells, which might have implications on the immune response of the mammary gland (Taylor *et al.*, 1994; Guiguen *et al.*, 1996; Persson Waller and Colditz, 1998). CD8+ cells have been shown to suppress immune function during udder infections (Park *et al.*, 1993). The functions of $\gamma\delta$ T-cells in the udder are not certain but these cells are thought to be associated with protection of epithelial surfaces (Allison and Havran, 1991). Shafer-Weaver *et al.* (1996) found that the percentage of $\gamma\delta$ T lymphocytes decreased significantly in mammary parenchyma during times of increased susceptibility to infections.

Antibody production is the primary role of B-cells. Antigen activation of cellular receptors results in a proliferation into antibody-secreting plasma cells which produce immunoglobulins directed against invading pathogens (Tizard, 1996). They also present antigen to T-helper lymphocytes, which produce cytokines inducing further proliferation and differentiation of B-cells.

## The inflammatory process and influence of inflammatory mediators (e.g. cytokines)

Mastitis, or udder inflammation, is the host response to tissue damage, and is thereby proof of a functioning mammary defence. The inflammatory response is initiated through the production and/or the release of a complex cascade of inflammatory mediators resulting in a sequence of vascular and cellular events inducing the classical symptoms of inflammation. Among the inflammatory mediators responsible for these events are histamine, prostaglandins, leukotrienes and cytokines. To efficiently eliminate pathogens, the inflammatory reaction must be rapidly elicited and leucocytes, especially neutrophils, must quickly migrate in sufficient numbers to the udder.

Cytokines are hormone-like soluble proteins and/or glucoproteins synthesized naturally by both immune and non-immune cells. They have important regulatory functions in most aspects of inflammation and immunity. Many cytokines have been identified, their roles during mastitis have been studied to some extent, and some cytokines have been used in research to treat or prevent mastitis in cows (Nickerson *et al.*, 1989; Daley *et al.*, 1993). Examples of some important cytokines are interferons (IFN), interleukins (IL) and tumour necrosis factors (TNF).

The recruitment of neutrophils is an important part of the defence and cytokines are important for their migration. IFN-γ, IL-1β and TNF-α are cytokines known to stimulate the expression of adhesion molecules on circulating neutrophils and endothelial cells (Brown Cramer, 1992). Following adherence and the production of IL-8 by the endothelium, neutrophils then migrate into the extravascular space. Bovine mammary epithelial cells of lactating cows can produce IL-1 and IL-6, and these cells also have gene expression for multiple cytokines (Okada *et al.*, 1997). Recent studies have suggested a role for cytokines in ruminant mastitis (Shuster *et al.*, 1993; Shuster and Kehrli, 1995; Persson Waller *et al.*, 1997a,b). During endotoxin-induced mastitis in sheep, TNF-α, IL-8 and GM-CSF, but not IL-1β and IFN-γ, seem to be involved in leucocyte accumulation (Persson Waller *et al.*, 1997b). The cytokine concentrations tend to increase just before, or together with, the increase in leucocyte accumulation. Cytokines were usually released during a short period and reached peak concentrations earlier than peak leucocyte infiltration (Persson Waller *et al.*, 1997b). In contrast, in a similar study of bovine endotoxin-induced mastitis, high concentrations of IL-1 and IL-6 were found in milk but only small amounts of TNF-α and no IL-8 (Shuster *et al.*, 1993; Shuster and Kehrli, 1995). However, IL-1 may not be an important mediator of endotoxin-induced inflammation in the bovine udder as intramammary infusion of an IL-1 receptor antagonist had no effect on the inflammatory response to endotoxin (Shuster and Kehrli, 1995; Persson Waller, 1997).

TNF-α, IL-8 and IL-1β were also detected in milk during ovine mastitis caused by experimental infection with *Staphylococcus aureus* or *Escherichia coli* (Persson Waller *et al.*, 1997a). The kinetics of TNF-α and IL-8 influx were similar to that of leucocytes during both infections. However, IL-1β was detected mainly during *S. aureus* mastitis indicating differences in the inflammatory responses elicited by the organisms. TNF-α was also detected during bovine mastitis induced by *E. coli* (Sordillo and Peel, 1992). TNF-α is produced by mononuclear cells and this production is enhanced during the periparturient period which might explain the severe acute phase response of the mammary gland during this period (Sordillo *et al.*, 1995). The presence of TNF-α and IL-6 in sera appears to relate to the severity of clinical symptoms during coliform mastitis (Nakajima *et al.*, 1997).

## Genetic influences on immunological functions and the risk for udder infections

### General aspects

Research on genetic resistance against udder infections and mastitis is important. However, heritability for clinical mastitis has been estimated to be low, on average 0.04 (e.g. Emanuelson *et al.*, 1988; de Haas, 1998). Somatic cell count (SCC), as an indirect measurement of mastitis, has a higher heritability (on average 0.10), and might be a better selective tool than clinical mastitis (e.g. Emanuelson *et al.*, 1988; de Haas, 1998). Moderate genetic correlation has been estimated between clinical mastitis and SCC. A combination of these parameters should therefore be considered for selection towards reduced susceptibility to mastitis. Kehrli and Shuster (1994) argued that too low SCC should be avoided as cows with very low SCC might be more susceptible to mastitis. However, Philipsson *et al.* (1995) reported a linear relationship between sires' breeding values for clinical mastitis and for SCC.

High milk production and various udder traits, such as udder suspensory ligament, fore udder attachment and udder depth, are associated with increased incidence of mastitis (Emanuelson, 1988; Uribe *et al.*, 1995; Alexandersson, 1998). A similar relationship was observed between mastitis traits and milk protein production (de Haas, 1998).

In the Nordic countries, udder traits, cell counts and clinical mastitis are included in the breeding values of bulls in order to improve disease resistance. In the Swedish selection programme, diseases other than mastitis are also included in the breeding values. A combined index of breeding values for mastitis and SCC was introduced in the beginning of this decade (Koenen *et al.*, 1994). Other countries have recently started to include mastitis selection parameters in their breeding programmes.

### Genetics and the immune response

Recently, the genetic influence on immunological parameters has been studied. These are aimed towards the possibility of selecting cows with improved immune response, without negative effects on milk productivity.

Differences in proportions of leucocyte subpopulations may have an important role in the probability of an individual cow developing mastitis, as the proportions of polymorphonuclear leucocytes, macrophages, and CD4+ or CD8+ T-lymphocytes differ significantly between cows (Leitner *et al.*, 1998). Kelm *et al.* (1997) used different measures of mastitis, and a number of immunological assays and molecular markers. The molecular markers accounted for up to 40% of the variation in estimated breeding value (EBV) for measures of mastitis. In cows with a low EBV for somatic cell score, the neutrophils functioned more effectively at maximal immunosuppression, the serum immunoglobulin concentration was lower, and the numbers of circulating mononuclear cells were higher. Significant differences in sire progeny groups for various neutrophil assays have also been observed (Kehrli *et al.*, 1991), and results of Fitzpatrick *et al.* (1998) indicated that immunological assays may be useful in identifying bulls whose progeny would be associated with a higher resistance to intramammary infections with *S. aureus*.

A selection programme for animals resistant to immunosuppression and disease around parturition would be especially desirable. Genetic variability in certain immune functions, i.e. total numbers of neutrophils, neutrophil chemokinesis, assays of neutrophil respiratory burst associated with phagocytosis, and in serum concentrations of immunoglobulins and hemolytic complement activity was observed in periparturient dairy cows (Detilleux *et al.*, 1994). Moreover, Mallard *et al.* (1998) showed that cows at peripartum may be categorized as high or low responders to immunizations and that the heritability for specific antibody responses was moderate to high. These findings may be used to identify animals with high resistance to disease.

The importance of adhesion molecules, essential for leucocyte migration, has recently been highlighted. Genetic leucocyte adhesion deficiencies leading to chronic, or even fatal, infections have been observed in several animal species including cattle (bovine leucocyte adhesion deficiency) (Kehrli *et al.*, 1992). All species affected show signs of chronic and recurrent infections due to a deficiency in the chemotactic and phagocytic properties of leucocytes, particularly neutrophils.

## Influence of stress on immune functions, and periods of immunosuppression

### *Influence of stress*

Stress has a negative effect on the immune response and can make both adults and young animals less resistant to infectious diseases. This is probably mainly due to increased levels of endogenous corticosteroids which can affect both the functions and numbers of leucocytes, and thus increase the host's susceptibilty to infections (Guidry *et al.*, 1976).

Treatment of cows with synthetic glucocorticoids, like dexamethasone, have a negative influence on immune functions, e.g. affecting the numbers and proportions of leucocytes in blood due to down-regulation of adhesion molecules on blood neutrophils (Burton and Kehrli, 1995; Burton *et al.*, 1995). Such changes were associated with leucocytosis and increased shedding of *S. aureus* in subclinically infected mammary glands, which reinforces the potential risk of treating infected animals with potent synthetic glucocorticoids such as dexamethasone.

### *Periods of immunosuppression*

As a result of depressed immune functions, susceptibility to infectious diseases, such as mastitis, is associated with the peripartum period (Sordillo *et al.*, 1997). High blood levels of glucocorticoids are present during this time. Examples of stress factors during this period are parturition, onset of lactation, and changes in feeding and management regimes. Impaired functions of a number of immunological parameters have been observed in peripartal dairy cows. Amongst these are, for example, changes in the white blood cell count and in blood and mammary neutrophil and lymphocyte functions. The cell functions decrease during the weeks before calving, reach their lowest levels at calving, and return to initial preparturient levels one to three weeks after calving (e.g. Saad *et al.*, 1989; Detilleux *et al.*, 1994; Shafer-Weaver *et al.*, 1996; Lee and Kehrli, 1998). A reduced function of the blood cells diminishes their ability to kill bacteria and to stimulate the immune defence. A significant negative correlation between severity of *E. coli* mastitis and the number of circulating mature neutrophils in blood immediately before infection has been demonstrated (Dosogne *et al.*, 1997).

Preferential trafficking of T-cells of the suppressor type rather than of the cytotoxic type into the mammary gland tissues around calving, compared with in mid-lactation, may be another factor responsible for the lowered local immune responsiveness during this period (Sordillo *et al.*, 1995). Moreover, Sordillo *et al.* (1991) reported that the levels of the cytokines IL-2 and IFN-γ in mammary gland secretions decrease before calving. Both cytokines are essential for several important immune functions, and lowered concentrations may contribute to the increased susceptibility to infections.

## Influence of some aspects of nutrition on immunological functions and the risk for udder infections

### *General aspects*

Adequate nutrition is essential for a good immune response and thus for overall health and the ability of the animal to resist diseases. The balance between protein and energy, the feeding routines and eating time are factors influencing animal health by potentially increasing the risk for metabolic disturbances. Differences in feeding strategy were observed between farms with a high or low incidence of clinical mastitis (Hallén Sandgren, 1998). Also, the hygienic quality of the ration is essential.

At the onset of lactation dairy cows are metabolically stressed due to negative energy balance leading to the mobilization of considerable amounts of tissue reserves. This could also influence the immune functions as increased levels of β-hydroxybutyrate and other ketone bodies can have negative effects on cellular functions (e.g. Klucinski *et al.*, 1988; Hoeben *et al.*, 1997). Balanced feeding during the dry period is a prerequisite for good feed consumption after calving thus avoiding metabolic stress and increased risk for disease during the early lactation period. Dairy cows often suffer from hepatic lipidosis during early lactation, increasing the risk for metabolic diseases. This may further decrease the immune response, as a high fat content of the liver is associated with decreased function of granulocytes (Zerbe *et al.*, 1998).

The amino acid glutamine has unique metabolic functions in dairy cows. Demands for glutamine increase with increased metabolic stress at the onset of lactation (Lacey and Wilmore, 1990; Hall *et al.*, 1996). Substantial amounts of glutamine are metabolized in the udder where it is used for synthesis of milk protein, as a source of energy and as substrate for cell proliferation (Meijer *et al.*, 1993). Glutamine is a key substance also for the immune defence, e.g. for the proliferation of lymphocytes and macrophages (Rohde *et al.*, 1996).

In a recent study, K. Holtenius, K. Persson Waller, B. Essén-Gustavsson and P. Holtenius (unpublished data) investigated the connection between metabolic stress around calving/early lactation and the incidence of clinical mastitis in high-producing Swedish cows. All herds had low bulk milk SCC, but had either a low (LT) or a high (HT) treatment incidence of mastitis. The results indicate that cows in the HT group mobilize their tissue reserves of fat and protein to a greater extent, during the weeks after calving, than cows in the LT group. The concentrations of glutamine decreased during late pregnancy and early lactation, and was lower ($P = 0.09$) in cows from HT herds during early lactation. Moreover, the blood concentrations of tryptophane were lower ($P < 0.01$) and the free fatty acids were higher ($P = 0.04$) in HT herds during early lactation. These differences could be due to increased metabolic stress in these herds. The results also indicate that the metabolic changes started earlier than 3 weeks before calving. Preliminary results indicate that there were differences in feeding strategies between the groups: the HT group was fed more concentrates and less roughage (C. Hallén Sandgren, L. Elander and U. Emanuelson, unpublished data). There were probably other management differences as well.

The results of the blood leucocyte counts did also indicate a higher stress level in the HT group, which possibly affected the immune defence of these animals. The numbers of eosinophils were lower ($P = 0.08$) in the HT group, while the numbers of neutrophils were higher ($P = 0.06$) in this group. Lower numbers of eosinophils and higher numbers of neutrophils can be a result of stress-induced release of cortisol. This is associated with a negative effect on the adhesion and migration ability of the neutrophils. An alternative explanation to the higher numbers of neutrophils is an increased recruitment of cells from the bone marrow due to some inflammatory stimulus in the body.

## Micronutrients

A balanced supply of micronutrients (vitamins and trace elements) is essential during periods of immune suppression. Deficiencies in selenium (Se), vitamin E, vitamin A, copper (Cu) and zinc (Zn) have been associated with a negative influence on the immune response in association with mastitis (Reddy and Frey, 1990; Harmon and Torre, 1994; Smith *et al.*, 1997).

Se deficiency is a problem in areas where Se concentrations in soils and pastures are low. Se is an important component of the enzyme glutathione peroxidase, which is essential for the protection of cells and tissues from auto-oxidative damage from production of oxygen radicals (Reddy and Frey, 1990). Se deficiency results in reduced neutrophil migration into the udder and impaired intracellular killing of bacteria (Erskine, 1993; Hogan *et al.*, 1993). Supplementation of Se may improve udder health by reducing the severity and duration of cases of clinical mastitis (Erskine, 1993).

Vitamin E is important for both cellular and humoral immune functions. This may be elicited through its effect on cell membrane stability and regulatory role in biosynthesis of various inflammatory mediators (Reddy and Frey, 1990; Smith *et al.*, 1997). Dietary supplementation of vitamin E is of value, especially as the concentration of this vitamin in fodder decreases with age and length of storage. Moreover, the serum concentration of vitamin E drops around calving, a period of increased susceptibility to disease (Smith *et al.*, 1997). Supplementation with vitamin E increased intracellular killing of bacteria by neutrophils, and reduced the numbers of new intramammary infections (Smith *et al.*, 1997). Dietary supplementation with vitamin E should be considered as a preventive measure in the control of mastitis during the peripartal period.

Vitamin A and the precursor β-carotene are also important for mucosal integrity and stability. Both substances have stimulatory effects on immune cell populations and have been correlated with an increased resistance to disease (Sordillo *et al.*, 1997). Deficiencies were related with severity of mastitis, and there is a negative correlation between plasma vitamin A and somatic cell counts in milk (Johnston and Chew, 1984). Plasma vitamin A and β-carotene concentrations decrease during the periparturient period contributing towards increased susceptibility to new udder infections (Johnston and Chew, 1984).

Limited information is available on the importance of Cu and Zn in minimizing the risk for mastitis in dairy cows. However, Cu and Zn have important biological functions, such as being parts of enzyme systems with antioxidative properties. These systems protect cells and tissues from detrimental effects of oxidative substances released at phagocytosis and killing of bacteria by white blood cells (Reddy and Frey, 1990). A Cu deficiency has also been reported to decrease the antibacterial effects of the immune defence (Reddy and Frey, 1990; Harmon and Torre, 1994). The addition of Cu in the feed gave a considerable decrease in the numbers of infected udder quarters at calving compared with untreated controls (Harmon and Torre, 1994). Moreover, the cell count in milk tended to be lower in the group receiving Cu supplementation. Zn is important for the integrity of the skin, the first barrier against infections, and has also importance for immune functions. Zn deficiency can cause degeneration of lymphoid tissues and have a negative influence on immune cells (Reddy and Frey, 1990; Harmon and Torre, 1994). Few studies have been made to study the relationship between Zn and mastitis. However, deficiency in Zn can predispose cows to secondary infections, which can be reversed by supplementation of Zn (Reddy and Frey, 1990; Harmon and Torre, 1994). Problems associated with Zn deficiency can be exacerbated by having a feeding regime containing high amounts of calcium during early lactation. Plasma Zn concentrations decrease markedly at parturition which may be connected with the immune suppression observed during this period (Goff and Stabel, 1990).

## Conclusions

Proper immune functions are essential for the defence against udder infections. Detailed knowledge about the immune response and important defence factors are essential in order to find new ways for the prevention and treatment of udder infections leading to mastitis. Work should be concentrated on ways of minimizing the negative influence on immune functions and/or ways of stimulating these functions, especially during periods of immune suppression such as around calving. Possibilities of stimulating the non-specific immune response, for example through the use of cytokines and other immunomodulatory substances, should be comprehensively investigated.

As mentioned above, it is important to identify risk factors, which negatively influence the defence mechanisms of the udder. The importance of management and adequate nutrition are some important factors to consider. Provision of suitable dietary supplies of vitamins and trace elements is one important step to ensure a good mammary defence and prevent mastitis. The possibility to find markers for genetic selection of individuals with a well-developed immune system should also be further evaluated.

## References

Adams, D.O. and Hamilton, T.A. (1988) Phagocytic cells: Cytotoxic activities of macrophages. In: Gallin, J.I., Goldstein, I.M. and Snyderman, R. (eds) *Inflammation: Basic Principles and Clinical Correlates*. Raven Press, New York.

Alexandersson, G. (1998) *Genetiska Samband mellan Exteriöra Juveregenskaper och Klinisk Mastit hos Förstakalvande SRB-kor*. Publication 195, Department of Animal Breeding and Genetics, Swedish University of Agricultural Sciences, Sweden.

Allison, J.P. and Havran, W.L. (1991) The immunobiology of T cells with invariant gamma delta antigen receptors. *Annual Review of Immunology* 9, 679–705.

Brown Cramer, E. (1992) Cell biology of phagocyte migration from the bone marrow, out of the bloodstream, and across organ epithelia. In: Gallin, J.I., Goldstein, I.M. and Snyderman, R. (eds) *Inflammation: Basic Principles and Clinical Correlates*. Raven Press, New York, pp. 341–352.

Burton, J.L. and Kehrli, M.E., Jr (1995) Regulation of neutrophil adhesion molecules and shedding of *Staphylococcus aureus* in milk of cortisol- and dexamethasone-treated cows. *American Journal of Veterinary Research* 56, 997–1006.

Burton, J.L., Kehrli, M.E., Jr, Kapil, S. and Horst, R.L. (1995) Regulation of L-selectin and CD18 on bovine neutrophils by glucocorticoids: effects of cortisol and dexamethasone. *Journal of Leukocyte Biology* 57, 317–325.

Concha, C. (1986) Cell types and their immunological functions in bovine mammary tissues and secretions – A review of the literature. *Nordic Veterinary Medicine* 38, 257–272.

Concha, C. and Holmberg, O. (1990) Ability of bovine mammary macrophages to enhance proliferation of autologous blood and mammary secretion lymphocytes. *Journal of Dairy Research* 57, 7–16.

Craven, N. and Williams, M.R. (1985) Defences of the bovine mammary gland against infection and prospects for their enhancement. *Veterinary Immunology and Immunopathology* 10, 71–127.

Daley, M.J., Williams, T., Coyle, P., Furda, G., Dougherty, R. and Hayes, P.W. (1993) Prevention and treatment of *Staphylococcus aureus* infections with recombinant cytokines. *Cytokine* 5, 276–284.

de Haas, Y. (1998) *Genetic Parameters of Mastitis Traits and Protein Production in the First Two Lactations of Swedish Red and White Cows*. Publication no. 17. Department of Animal Breeding and Genetics, Swedish University of Agricultural Sciences, Uppsala, Sweden.

Detilleux, J.C., Koehler, K.J., Freeman, A.E., Kehrli, M.E., Jr and Kelley, D.H. (1994) Immunological parameters of periparturient Holstein cattle: Genetic variation. *Journal of Dairy Science* 77, 2640–2650.

Dosogne, H., Burvenich, C., van Werven, T., Roets, E., Noordhuizen-Stassen, E.N. and Goddeeris, B. (1997) Increased surface expression of CD11b receptors on polymorphonuclear leukocytes is not sufficient to sustain phagocytosis during *Escherichia coli* mastitis in early postpartum dairy cows. *Veterinary Immunology and Immunopathology* 60, 47–59.

Emanuelson, U., Danell, B. and Philipsson, J. (1988) Genetic parameters for clinical mastitis, somatic cell counts, and milk production estimated by multiple-trait restricted maximum likelihood. *Journal of Dairy Science* 71, 467–476.

Erskine, R.J. (1993) Nutrition and mastitis. *Veterinary Clinics of North America Food Animal Practice* 9, 551–559.

Fitzpatrick, J.L., Logan, K.E., Stear, M.J., McGuirk, B. and Platt, D.J. (1998) Immunological responses of Holstein/Friesian cattle to *Staphylococcus aureus*. In: *Proceedings of the 5th International Veterinary Immunology Symposium, November 8–13, Punjab Agricultural University, Ludhiana, India*, p. 117.

Goff, J.P. and Stabel, J.R. (1990) Decreased plasma retinol, α-tocopherol, and zinc concentration during the periparturient period: Effect of milk fever. *Journal of Dairy Science* 73, 3195–3199.

Grommers, F.J., van de Geer, D., van der Vliet, H., Henricks, P.A. and Nijkamp, F.P. (1989) Polymorphonuclear leucocyte function: Relationship between induced migration into the bovine mammary gland and in vitro cell activity. *Veterinary Immunology and Immunopathology* 23, 75–83.

Guidry, A.J., Paape, M.J. and Pearson, R.E. (1976) Effects of parturition and lactation on blood and milk cell concentrations, corticosteroids, and neutrophil phagocytosis in the cow. *American Journal of Veterinary Research*, 37, 1195–1200.

Guiguen, F., Greenland, T., Pardo, E., Panaye, G. and Mornex, J.F. (1996) Flow cytometric analysis of goat milk lymphocytes: Subpopulations and adhesion molecule expression. *Veterinary Immunology and Immunopathology* 53, 173–178.

Hall, J.C., Heel, K. and McCauley, R. (1996) Glutamine: A review. *British Journal of Surgery* 83, 305–312.

Hallén Sandgren, C. (1998) Field experiences of mastitis risk factors with special emphasis on heifer management and replacement. In: *Proceedings of the Nordic Seminar Regarding Future Aspects of Mastitis Prevention, January 21–22, Aarhus, Denmark.*

Harmon, R.J. and Torre, R.M. (1994) Copper and zinc: do they influence mastitis? In: *Proceedings of the 33rd Annual Meeting of the National Mastitis Council, Orlando, Florida*, National Mastitis Council, Inc., Arlington, Virginia, USA, p. 54.

Herbelin, C., Poutrel, B., Gilbert, F.B. and Rainard, P. (1997) Immune recruitment and bactericidal activity of neutrophils in milk of cows vaccinated with staphylococcal alpha-toxin. *Journal of Dairy Science* 80, 2025–2034.

Hill, A.W. (1994) *Escherichia coli* mastitis. In: Gyles, C.L. (ed.) *Escherichia coli in Domestic Animals and Humans*. CAB International, Wallingford, UK, pp. 117–133.

Hoeben, D., Heyneman, R. and Burvenich, C. (1997) Elevated levels of beta-hydroxybutyric acid in periparturient cows and in vitro effect on respiratory burst activity of blood neutrophils. *Veterinary Immunology and Immunopathology*, 58, 165–170.

Hogan, J.S., Weiss, W.P. and Smith, K.L. (1993) Role of vitamin E and selenium in host defense against mastitis. *Journal of Dairy Science* 76, 2795–2803.

Johnston, L.A. and Chew, B.P. (1984) Peripartum changes of plasma and milk vitamin A and β-carotene among dairy cows with or without mastitis. *Journal of Dairy Science*, 67, 1832–1840.

Kehrli, M.E. and Shuster, D.E. (1994) Factors affecting milk somatic cells and their role in health of the bovine mammary gland. *Journal of Dairy Science* 77, 619–627.

Kehrli, M.E., Jr, Weigel, K.A., Freeman, A.E., Thurston, J.R. and Kelley, D.H. (1991) Bovine sire effects on daughters' in vitro blood neutrophil functions, lymphocyte blastogenesis, serum complement and conglutinin levels. *Veterinary Immunology and Immunopathology* 27, 303–319.

Kehrli, M.E., Jr, Shuster, D.E. and Ackermann, M.R. (1992) Leukocyte adhesion deficiency among Holstein cattle. *Cornell Veterinarian* 82, 103–109.

Kelm, S.C., Detilleux, J.C., Freeman, A.E., Kehrli, M.E., Jr, Dietz, A.B., Fox, L.K., Butler, J.E., Kasckovics, I. and Kelley, D.H. (1997) Genetic association between parameters of innate immunity and measures of mastitis in periparturient Holstein cattle. *Journal of Dairy Science* 80, 1767–1775.

Klucinski, W., Degorski, A., Miernik-Degorska, E., Targowski, S. and Winnicka, A. (1988) Effect of ketone bodies on the phagocytic activity of bovine milk macrophages and polymorphonuclear leucocytes. *Zentralblatt für Veterinarmedizin A*, 35, 632–639.

Koenen, E.P.C., Berglund, B., Philipsson, J. and Groen, A.F. (1994) Genetic parameters of fertility disorders and mastitis in the Swedish Friesian breed. *Acta Agricultura Scandinavica* 44, 202–207.

Lacey, J.M. and Wilmore, D.W. (1990) Is glutamine a conditionally essential amino acid? *Nutrition Reviews* 48, 297–309.

Lee, E.K. and Kehrli, M.E., Jr (1998) Expression of adhesion molecules on neutrophils of periparturient cows and neonatal calves. *American Journal of Veterinary Research* 59, 37–43.

Leitner, G., Chaffer, M., Winkler, M., Glickman, A., Weisblit, L., Krifucks, O., Ezra, E. and Saran, A. (1998) The determination of milk leukocyte population in heifers free of udder infection as a tool for the genetic improvement of dairy cows. In: *Proceedings of the 5th International Veterinary Immunology Symposium, November 8–13, Punjab Agricultural University, Ludhiana, India*, p. 119.

Mallard, B.A., Dekkers, J.C., Ireland, M.J., Leslie, K.E., Sharif, S., Van Kampen, C.L., Wagter, L. and Wilkie, B.N. (1998) Alteration in immune responsiveness during the peripartum period and its ramification on dairy cow and calf health. *Journal of Dairy Science* 81, 585–595.

Meijer, G.A., van der Meulen, J. and van Vuuren, A.M. (1993) Glutamine is a potentially limiting amino acid for milk production in dairy cows: A hypothesis. *Metabolism* 42, 358–364.

Nakajima, Y., Mikami, O., Yoshioka, M., Motoi, Y., Ito, T., Ishikawa, Y., Fuse, M., Nakano, K. and Yasukawa, K. (1997) Elevated levels of tumor necrosis factor-alpha (TNF-alpha) and interleukin-6 (IL-6) activities in the sera and milk of cows with naturally occurring coliform mastitis. *Research in Veterinary Science* 62, 297–298.

Nickerson, S.C., Owens, W.E. and Watts, J.L. (1989) Effects of recombinant granulocyte colony-stimulating factor on *Staphylococcus aureus* mastitis in lactating dairy cows. *Journal of Dairy Science* 72, 3286–3294.

Nilsson, L. and Holmberg, O. (1996) Research projects in mastitis. In: *National Veterinary Institute, Research and Development for Animal Health.* Uppsala, Sweden.

Okada, H., Ito, T., Ohtsuka, H., Kirisawa, R., Iwai, H., Yamashita, K., Yoshino, T. and Rosol, T.J. (1997) Detection of interleukin-1 and interleukin-6 on cryopreserved bovine mammary epithelial cells in vitro. *Journal of Veterinary Medical Science* 59, 503–507.

Paape, M.J., Guidry, A.J., Jain, N.C. and Miller, R.H. (1991) Leukocytic defense mechanisms in the udder. *Flemish Veterinary Journal* 62 (Suppl. 1), 95–109.

Park, Y.H., Fox, L.K., Hamilton, M.J. and Davis, W.C. (1993) Suppression of proliferative response of BoCD4+ T lymphocytes by activated BoCD8+ T lymphocytes in the mammary gland of cows with *Staphylococcus aureus* mastitis. *Veterinary Immunology and Immunopathology* 36, 137–151.

Persson Waller, K. (1997) Modulation of endotoxin-induced inflammation in the bovine teat using antagonists/inhibitors to leukotrienes, platelet activating factor and interleukin 1β. *Veterinary Immunology and Immunopathology* 57, 239–251.

Persson Waller, K. and Colditz, I.G. (1998) Expression of surface antigens on blood and mammary leukocytes in lactating and dry ewes. *Veterinary Immunology and Immunopathology* 62, 237–278.

Persson Waller, K., Colditz, I.G. and Seow, H.-F. (1997a) Accumulation of leucocytes and cytokines in the lactating ovine udder during mastitis due to *Staphylococcus aureus* and *Escherichia coli*. *Research in Veterinary Science* 62, 63–66.

Persson Waller, K., Colditz, I.G., Flapper, P. and Seow, H.-F. (1997b) Leukocyte and cytokine accumulation in the ovine teat and udder during endotoxin-induced inflammation. *Veterinary Research Communications* 21, 101–115.

Philipsson, J., Ral, G. and Berglund, B. (1995) Somatic cell count as a selection criterion for mastitis resistance in dairy cattle. *Acta Agricultura Scandinavica* 41, 195–200.

Philpot, W.N. (1984) Economics of mastitis control. *Veterinary Clinics of North America* 6, 233–245.

Reddy, P.G. and Frey, R.A. (1990) Nutritional modulation of immunity in domestic food animals. *Advances in Veterinary Science and Comparative Medicine* 35, 255–281.

Rohde, T., Mclean, D.A. and Klarlund-Petersen, B. (1996) Glutamine, lymphocyte proliferation and cytokine production. *Scandinavian Journal of Immunology* 44, 648–650.

Saad, A.M., Concha, C. and Astrom, G. (1989) Alterations in neutrophil phagocytosis and lymphocyte blastogenesis in dairy cows around parturition. *Zentralblatt für Veterinarmedizin Reihe B*, 36, 337–345.

Shafer-Weaver, K.A. and Sordillo, L.M. (1997) Bovine CD8+ suppressor lymphocytes alter immune responsiveness during the postpartum period. *Veterinary Immunology and Immunopathology* 56, 53–64.

Shafer-Weaver, K.A., Pighetti, G.M. and Sordillo, L.M. (1996) Diminished mammary gland lymphocyte functions parallel shifts in trafficking patterns during the postpartum period. *Proceedings of the Society for Experimental Biology and Medicine* 212, 271–279.

Shuster, D.E. and Kehrli, M.E., Jr (1995) Administration of recombinant human interleukin 1 receptor antagonist during endotoxin-induced mastitis in cows. *American Journal of Veterinary Research* 56, 313–320.

Shuster, D.E., Kehrli, M.E. and Stevens, M.G. (1993) Cytokine production during endotoxin-induced mastitis in lactating dairy cows. *American Journal of Veterinary Research* 54, 80–85.

Smith, K.L., Hogan, J.S. and Weiss, W.P. (1997) Dietary vitamin E and selenium affect mastitis and milk quality. *Journal of Animal Science* 75, 1659–1665.

Sordillo, L.M. and Peel, J.E. (1992) Effect of interferon-γ on the production of tumor necrosis factor during acute *Escherichia coli* mastitis. *Journal of Dairy Science* 75, 2119–2125.

Sordillo, L.M., Redmond, M.J., Campos, M., Warren, L. and Babiuk, L.A. (1991) Cytokine activity in bovine mammary gland secretions during the periparturient period. *Canadian Journal of Veterinary Research* 55, 298–301.

Sordillo, L.M., Pighetti, G.M. and Davis, M.R. (1995) Enhanced production of bovine tumor necrosis factor-alpha during the periparturient period. *Veterinary Immunology and Immunopathology* 49, 263–270.

Sordillo, L.M., Shafer-Weaver, K. and DeRosa, D. (1997) Immunobiology of the mammary gland. *Journal of Dairy Science* 80, 1851–1865.

Taylor, B.C., Dellinger, J.D., Cullor, J.S. and Stott, J. (1994) Bovine milk lymphocytes display the phenotype of memory T cells and are predominantly CD8+. *Cellular Immunology* 156, 245–253.

Tizard, I.R. (1996) *Veterinary Immunology: An Introduction,* 5th edn. W.B. Saunders Company, Philadelphia, USA, 531 pp.

Uribe, H.A., Kennedy, B.W., Martin, S.W. and Kelton, D.F. (1995) Genetic parameters for common health disorders of Holstein cows. *Journal of Dairy Science* 78, 421–430.

Zerbe, H., Schneider, N., Obadnik, C., Schuberth, H.-J., Wensing, T., Kruip, T., Grunert, E. and Leibold, W. (1998) The negative energy balance reduces the neutrophil defense capacity in cows. In: *Proceedings of the 5th International Veterinary Immunology Symposium, November 8–13. Punjab Agricultural University, Ludhiana, India*, p. 185.

# Index

Figures in **bold** indicate major references.
Figures in *italic* refer to diagrams, photographs and tables.